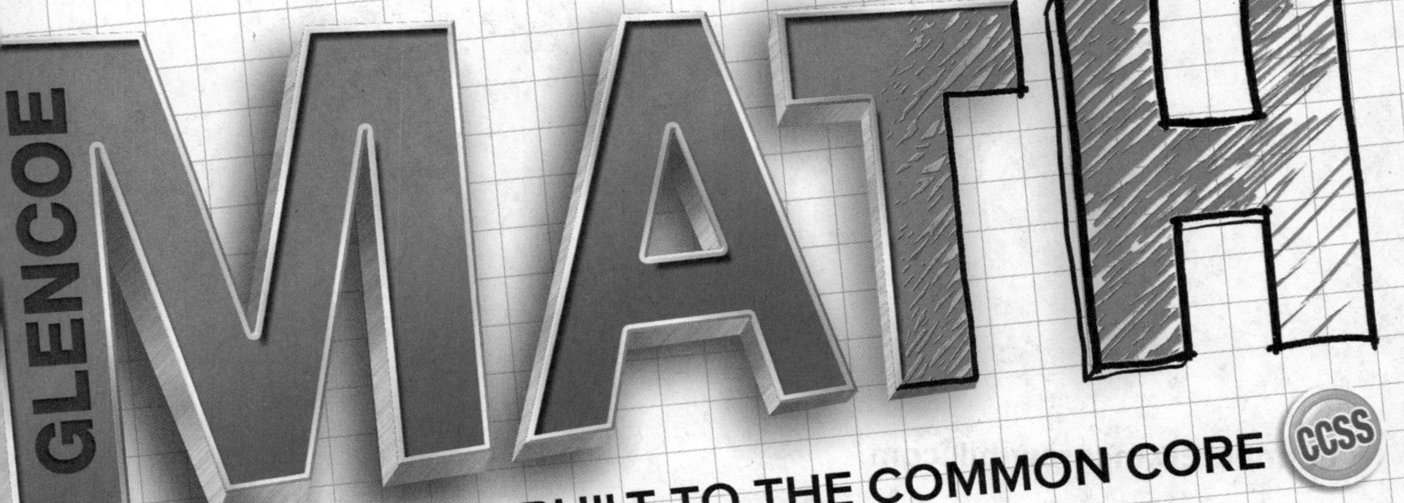

BUILT TO THE COMMON CORE CCSS

AUTHORS
Carter • Cuevas • Day • Malloy
Kersaint • Reynosa • Silbey • Vielhaber

Bothell, WA • Chicago, IL • Columbus, OH • New York, NY

connectED.mcgraw-hill.com

STEM McGraw-Hill is committed to providing instructional materials in Science, Technology, Engineering, and Mathematics (STEM) that give all students a solid foundation, one that prepares them for college and careers in the 21st century.

Send all inquiries to:
McGraw-Hill Education
8787 Orion Place
Columbus, OH 43240

ISBN: 978-0-07-670930-4 (*Volume 2*)
MHID: 0-07-670930-2

Printed in the United States of America.

8 9 10 11 12 LMN 22 21 20 19 18

CONTENTS IN BRIEF

Units organized by CCSS domain

Glencoe Math is organized into units based on groups of related standards called domains. The Standards for MP Mathematical Practices are embedded throughout the course.

Aldo Murillo/Getty Images

MP Mathematical Practices

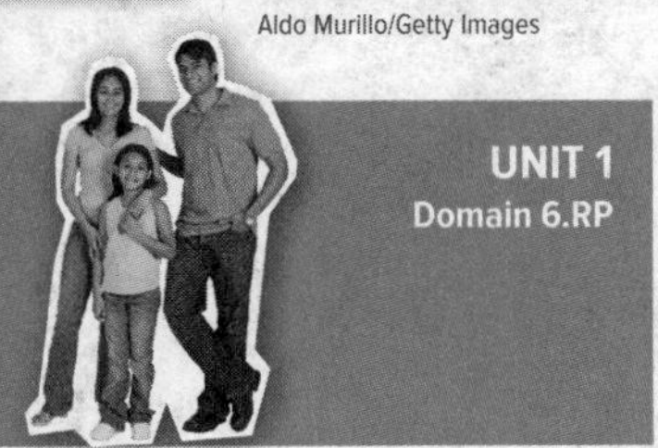

Corbis/PunchStock

Ratios and Proportional Relationships

LatinStock Collection/Alamy

The Number System

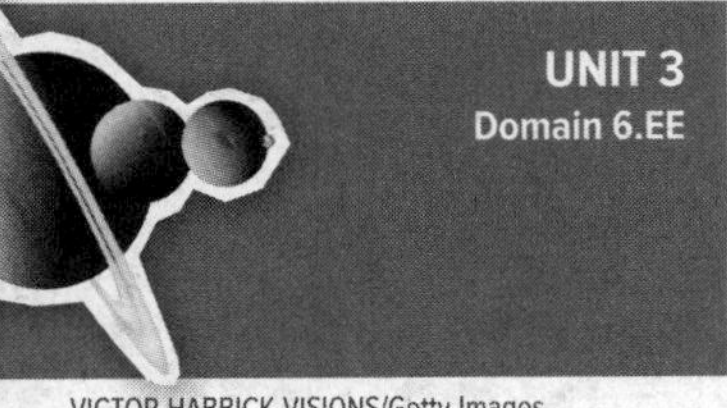

VICTOR HABBICK VISIONS/Getty Images

Expressions and Equations

©Comstock/PunchStock

Geometry

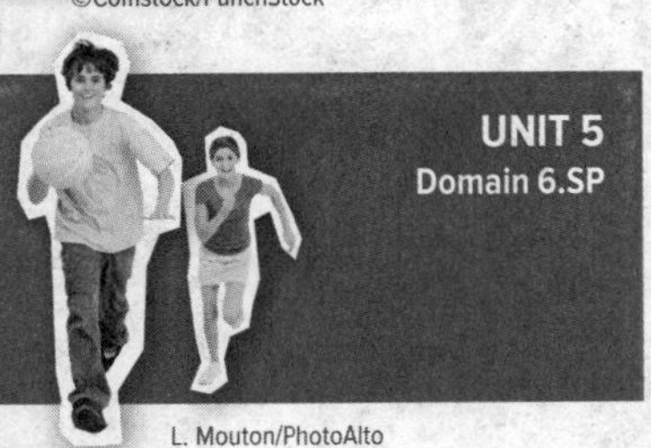

L. Mouton/PhotoAlto

Statistics and Probability

Everything you need,

anytime, anywhere.

With ConnectED, you have instant access to all of your study materials—anytime, anywhere. From homework materials to study guides—it's all in one place and just a click away. ConnectED even allows you to collaborate with your classmates and use mobile apps to make studying easy.

Resources built for you—available 24/7:

- Your eBook available wherever you are
- Personal Tutors and Self-Check Quizzes to help your learning
- An Online Calendar with all of your due dates
- eFlashcard App to make studying easy
- A message center to stay in touch

Go Mobile!

Visit mheonline.com/apps to get entertainment, instruction, and education on the go with ConnectED Mobile and our other apps available for your device.

Go Online!

connectED.mcgraw-hill.com

your Username

your Password

Vocab

Learn about new vocabulary words.

Watch

Watch animations and videos.

Tutor

See and hear a teacher explain how to solve problems.

Tools

Explore concepts with virtual manipulatives.

Sketchpad

Discover concepts using The Geometer's Sketchpad®.

Check

Check your progress.

eHelp

Get targeted homework help.

Worksheets

Access practice worksheets.

UNIT 1 Ratios and Proportional Relationships

UNIT PROJECT PREVIEW
page 2

Chapter 1 Ratios and Rates

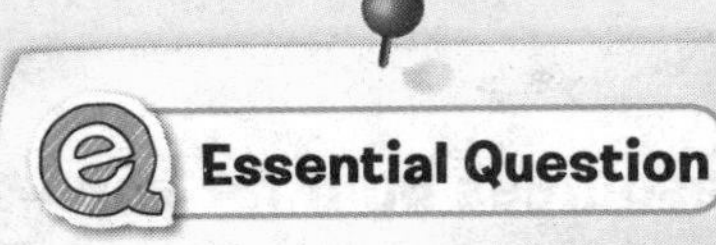

HOW do you use equivalent rates in the real world?

Real World p. 31

Chapter 2
Fractions, Decimals, and Percents

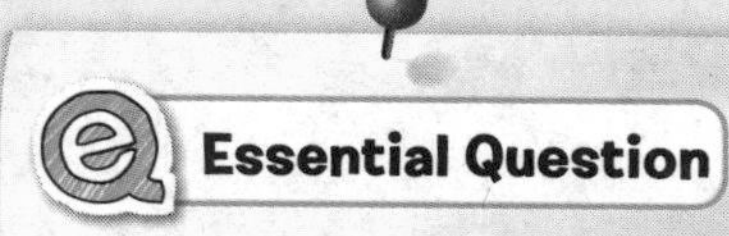

WHEN is it better to use a fraction, a decimal or a percent?

p. 155

People Everywhere

CCSS **UNIT 2** The Number System

UNIT PROJECT PREVIEW page 172

Chapter 3 Compute with Multi-Digit Numbers

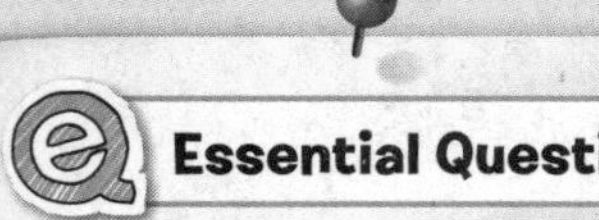

HOW can estimating be helpful?

Chapter 4 Multiply and Divide Fractions

Essential Question

WHAT does it mean to multiply and divide fractions?

Chapter 5
Integers and the Coordinate Plane

Essential Question

HOW are integers and absolute value used in real-world situations?

Real World
p. 387

Get Out the Map!

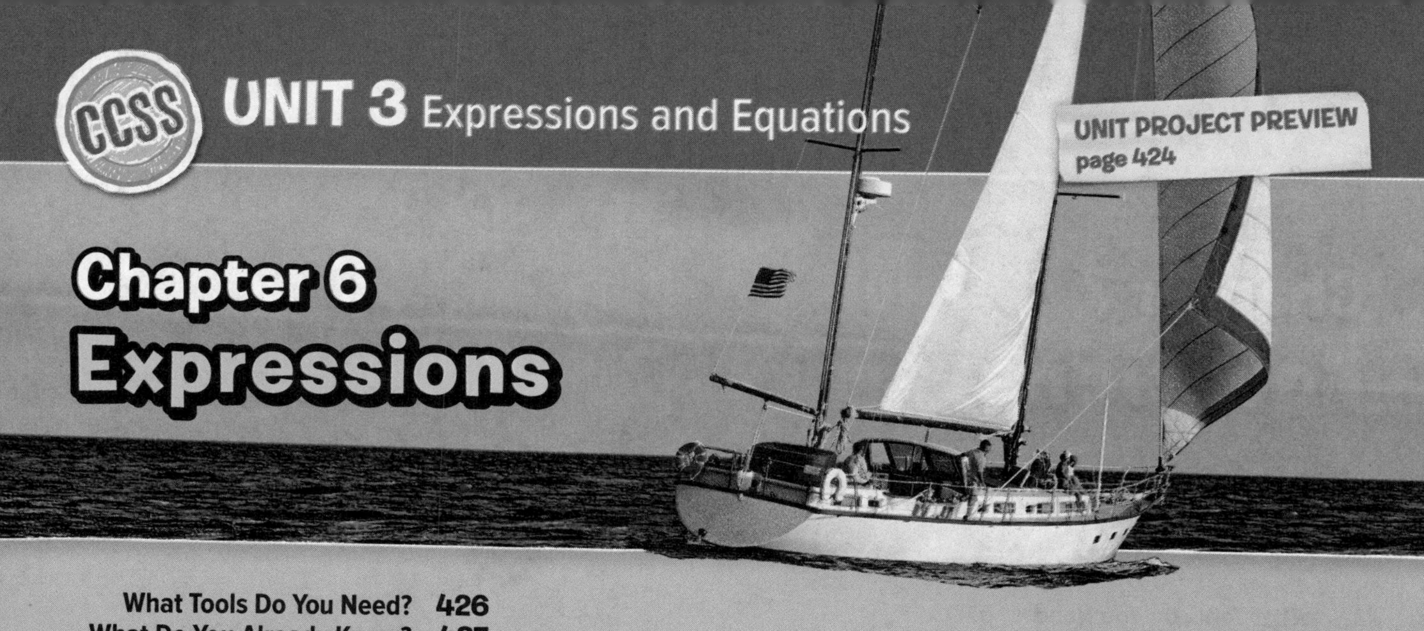

Chapter 6 Expressions

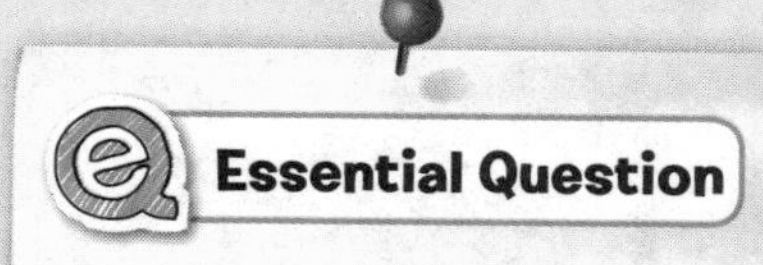

HOW is it helpful to write numbers in different ways?

Chapter 7 Equations

Essential Question

HOW do you determine if two numbers or expressions are equal?

p. 535

Chapter 8 Functions and Inequalities

Essential Question

HOW are symbols, such as <, >, and =, useful?

p. 595

Essential Question

HOW does measurement help you solve problems in everyday life?

Chapter 10 Volume and Surface Area

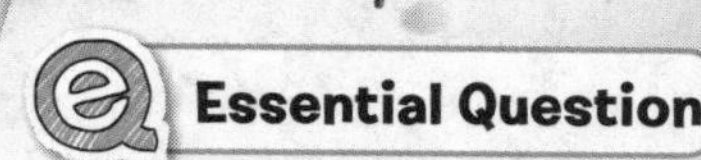

HOW is shape important when measuring a figure?

p. 783

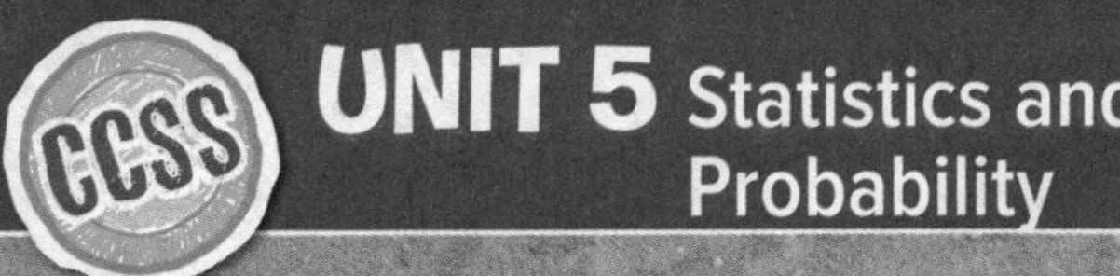

Chapter 11 Statistical Measures

Essential Question

HOW are the mean, median, and mode helpful in describing data?

Chapter 12 Statistical Displays

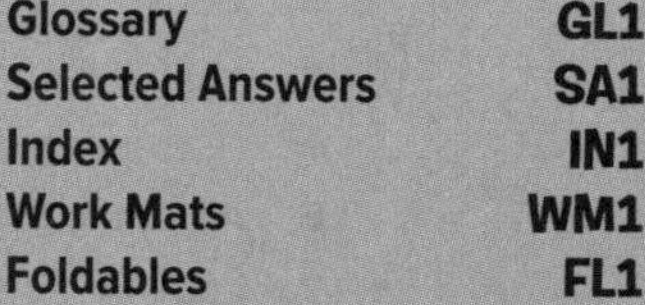

Common Core State Standards for MATHEMATICS, Grade 6

Glencoe Math, Course 1, focuses on four critical areas: (1) using concepts of ratio and rate to solve problems; (2) understanding division of fractions; (3) using expressions and equations; and (4) understanding of statistical reasoning.

Content Standards

Domain 6.RP **Ratios and Proportional Relationships**

- Understand ratio concepts and use ratio reasoning to solve problems.

Domain 6.NS **The Number System**

- Apply and extend previous understandings of multiplication and division to divide fractions by fractions.
- Compute fluently with multi-digit numbers and find common factors and multiples.
- Apply and extend previous understandings of numbers to the system of rational numbers.

Domain 6.EE **Expressions and Equations**

- Apply and extend previous understandings of arithmetic to algebraic expressions.
- Reason about and solve one-variable equations and inequalities.
- Represent and analyze quantitative relationships between dependent and independent variables.

Domain 6.G **Geometry**

- Solve real-world and mathematical problems involving area, surface area, and volume.

Domain 6.SP **Statistics and Probability**

- Develop understanding of statistical variability.
- Summarize and describe distributions.

MP Mathematical Practices

1 Make sense of problems and persevere in solving them.
2 Reason abstractly and quantitatively.
3 Construct viable arguments and critique the reasoning of others.
4 Model with mathematics.
5 Use appropriate tools strategically.
6 Attend to precision.
7 Look for and make use of structure.
8 Look for and express regularity in repeated reasoning.

Track Your Common Core Progress

These pages list the key ideas that you should be able to understand by the end of the year. You will rate how much you know about each one. Don't worry if you have no clue **before** you learn about them. Watch how your knowledge grows as the year progresses!

 I have no clue. — I've heard of it. — I know it!

6.RP Ratios and Proportional Relationships	Before			After		
Understand ratio concepts and use ratio reasoning to solve problems.						
6.RP.1 Understand the concept of a ratio and use ratio language to describe a ratio relationship between two quantities.						
6.RP.2 Understand the concept of a unit rate *a/b* associated with a ratio *a:b* with $b \neq 0$, and use rate language in the context of a ratio relationship.						
6.RP.3 Use ratio and rate reasoning to solve real-world and mathematical problems, e.g., by reasoning about tables of equivalent ratios, tape diagrams, double number line diagrams, or equations. **a.** Make tables of equivalent ratios relating quantities with whole number measurements, find missing values in the tables, and plot the pairs of values on the coordinate plane. Use tables to compare ratios. **b.** Solve unit rate problems including those involving unit pricing and constant speed. **c.** Find a percent of a quantity as a rate per 100 (e.g., 30% of a quantity means 30/100 times the quantity); solve problems involving finding the whole, given a part and the percent. **d.** Use ratio reasoning to convert measurement units; manipulate and transform units appropriately when multiplying or dividing quantities.						

6.NS The Number System	Before			After		
Apply and extend previous understandings of multiplication and division to divide fractions by fractions.						
6.NS.1 Interpret and compute quotients of fractions, and solve word problems involving division of fractions by fractions, e.g., by using visual fraction models and equations to represent the problem.						
Compute fluently with multi-digit numbers and find common factors and multiples.						
6.NS.2 Fluently divide multi-digit numbers using the standard algorithm.						
6.NS.3 Fluently add, subtract, multiply, and divide multi-digit decimals using the standard algorithm for each operation.						
6.NS.4 Find the greatest common factor of two whole numbers less than or equal to 100 and the least common multiple of two whole numbers less than or equal to 12. Use the distributive property to express a sum of two whole numbers 1–100 with a common factor as a multiple of a sum of two whole numbers with no common factor.						

6.NS The Number System *continued*	Before			After		
Apply and extend previous understandings of numbers to the system of rational numbers.						
6.NS.5 Understand that positive and negative numbers are used together to describe quantities having opposite directions or values (e.g., temperature above/below zero, elevation above/below sea level, credits/debits, positive/negative electric charge); use positive and negative numbers to represent quantities in real-world contexts, explaining the meaning of 0 in each situation.						
6.NS.6 Understand a rational number as a point on the number line. Extend number line diagrams and coordinate axes familiar from previous grades to represent points on the line and in the plane with negative number coordinates. **a.** Recognize opposite signs of numbers as indicating locations on opposite sides of 0 on the number line; recognize that the opposite of the opposite of a number is the number itself, e.g., $-(-3) = 3$, and that 0 is its own opposite. **b.** Understand signs of numbers in ordered pairs as indicating locations in quadrants of the coordinate plane; recognize that when two ordered pairs differ only by signs, the locations of the points are related by reflections across one or both axes. **c.** Find and position integers and other rational numbers on a horizontal or vertical number line diagram; find and position pairs of integers and other rational numbers on a coordinate plane.						
6.NS.7 Understand ordering and absolute value of rational numbers. **a.** Interpret statements of inequality as statements about the relative position of two numbers on a number line diagram. **b.** Write, interpret, and explain statements of order for rational numbers in real-world contexts. **c.** Understand the absolute value of a rational number as its distance from 0 on the number line; interpret absolute value as magnitude for a positive or negative quantity in a real-world situation. **d.** Distinguish comparisons of absolute value from statements about order.						
6.NS.8 Solve real-world and mathematical problems by graphing points in all four quadrants of the coordinate plane. Include use of coordinates and absolute value to find distances between points with the same first coordinate or the same second coordinate.						

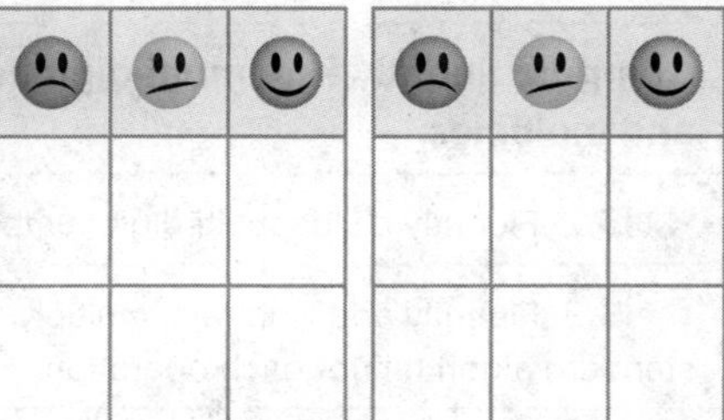

6.EE Expressions and Equations						
Apply and extend previous understandings of arithmetic to algebraic expressions.						
6.EE.1 Write and evaluate numerical expressions involving whole-number exponents.						

6.EE Expressions and Equations ***continued***	Before			After		
6.EE.2 Write, read, and evaluate expressions in which letters stand for numbers. **a.** Write expressions that record operations with numbers and with letters standing for numbers. **b.** Identify parts of an expression using mathematical terms (sum, term, product, factor, quotient, coefficient); view one or more parts of an expression as a single entity. **c.** Evaluate expressions at specific values of their variables. Include expressions that arise from formulas used in real-world problems. Perform arithmetic operations, including those involving whole number exponents, in the conventional order when there are no parentheses to specify a particular order (Order of Operations).						
6.EE.3 Apply the properties of operations to generate equivalent expressions.						
6.EE.4 Identify when two expressions are equivalent (i.e., when the two expressions name the same number regardless of which value is substituted into them).						
Reason about and solve one-variable equations or inequalities.						
6.EE.5 Understand solving an equation or inequality as a process of answering a question: which values from a specified set, if any, make the equation or inequality true? Use substitution to determine whether a given number in a specified set makes an equation or inequality true.						
6.EE.6 Use variables to represent numbers and write expressions when solving a real-world or mathematical problem; understand that a variable can represent an unknown number, or, depending on the purpose at hand, any number in a specified set.						
6.EE.7 Solve real-world and mathematical problems by writing and solving equations of the form $x + p = q$ and $px = q$ for cases in which p, q and x are all nonnegative rational numbers.						
6.EE.8 Write an inequality of the form $x > c$ or $x < c$ to represent a constraint or condition in a real-world or mathematical problem. Recognize that inequalities of the form $x > c$ or $x < c$ have infinitely many solutions; represent solutions of such inequalities on number line diagrams.						
Represent and analyze quantitative relationships between dependent and independent variables.						
6.EE.9 Use variables to represent two quantities in a real-world problem that change in relationship to one another; write an equation to express one quantity, thought of as the dependent variable, in terms of the other quantity, thought of as the independent variable. Analyze the relationship between the dependent and independent variables using graphs and tables, and relate these to the equation.						

6.G Geometry	Before			After		
Solve real-world and mathematical problems involving area, surface area, and volume.						
6.G.1 Find the area of right triangles, other triangles, special quadrilaterals, and polygons by composing into rectangles or decomposing into triangles and other shapes; apply these techniques in the context of solving real-world and mathematical problems.						
6.G.2 Find the volume of a right rectangular prism with fractional edge lengths by packing it with unit cubes of the appropriate unit fraction edge lengths, and show that the volume is the same as would be found by multiplying the edge lengths of the prism. Apply the formulas $V = l\,w\,h$ and $V = b\,h$ to find volumes of right rectangular prisms with fractional edge lengths in the context of solving real-world and mathematical problems.						
6.G.3 Draw polygons in the coordinate plane given coordinates for the vertices; use coordinates to find the length of a side joining points with the same first coordinate or the same second coordinate. Apply these techniques in the context of solving real-world and mathematical problems.						
6.G.4 Represent three-dimensional figures using nets made up of rectangles and triangles, and use the nets to find the surface area of these figures. Apply these techniques in the context of solving real-world and mathematical problems.						

6.SP Statistics and Probability	Before			After		
Develop understanding of statistical variability.						
6.SP.1 Recognize a statistical question as one that anticipates variability in the data related to the question and accounts for it in the answers.						
6.SP.2 Understand that a set of data collected to answer a statistical question has a distribution which can be described by its center, spread, and overall shape.						
6.SP.3 Recognize that a measure of center for a numerical data set summarizes all of its values with a single number, while a measure of variation describes how its values vary with a single number.						
Summarize and describe distributions.						
6.SP.4 Display numerical data in plots on a number line, including dot plots, histograms, and box plots.						
6.SP.5 Summarize numerical data sets in relation to their context, such as by: **a.** Reporting the number of observations. **b.** Describing the nature of the attribute under investigation, including how it was measured and its units of measurement. **c.** Giving quantitative measures of center (median and/or mean) and variability (interquartile range and/or mean absolute deviation), as well as describing any overall pattern and any striking deviations from the overall pattern with reference to the context in which the data were gathered. **d.** Relating the choice of measures of center and variability to the shape of the data distribution and the context in which the data were gathered.						

UNIT 3

Expressions and Equations

Essential Question

HOW can you communicate mathematical ideas effectively?

Chapter 6
Expressions

Numerical and algebraic expressions can be used to represent and solve real-world problems. In this chapter, you will write and evaluate expressions and apply the properties of operations to generate equivalent expressions.

Chapter 7
Equations

Variables are used to represent an unknown number in an expression or equation. In this chapter, you will write and solve one-variable addition, subtraction, multiplication, and division equations.

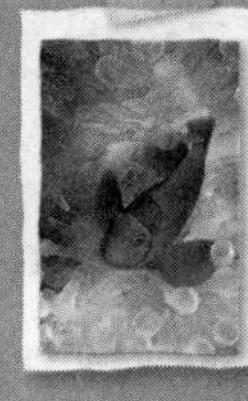

Chapter 8
Functions and Inequalities

Functions can be represented using words, equations, tables, and graphs. In this chapter, you will represent and analyze the relationship between two variables using functions. You will also write, graph, and solve one-variable inequalities.

Unit Project Preview

It's Out of This World The speed at which a planet orbits the sun or a moon orbits a planet is called *orbital velocity* or *orbital speed*. Each planet and moon in our solar system has a different average orbital speed.

Work with a partner. Take turns counting the number of steps you each take while walking in a circle for 10 seconds. Then use the information to find the approximate number of steps you take in 20, 30, and 40 seconds. Write and graph the ordered pairs to represent your walking speed.

At the end of Chapter 8, you'll complete a project to compare two planets' orbits around the sun. Their speed is out of this world!

My Walking Speed

Number of Steps

Time (s)

Chapter 6 Expressions

Essential Question

HOW is it helpful to write numbers in different ways?

Common Core State Standards

Content Standards
6.EE.1, 6.EE.2, 6.EE.2a, 6.EE.2b, 6.EE.2c, 6.EE.3, 6.EE.4, 6.EE.6, 6.NS.3, 6.NS.4

MP Mathematical Practices
1, 2, 3, 4, 5, 6, 7

Math in the Real World

Sailboats can travel at a cruising speed of about 6 knots. In a recent race from the United States to the United Kingdom, a racing sailboat traveled at an average speed of 25.8 knots.

Use the bar diagram below to find the difference between the cruising speed and the racing sailboat's speed.

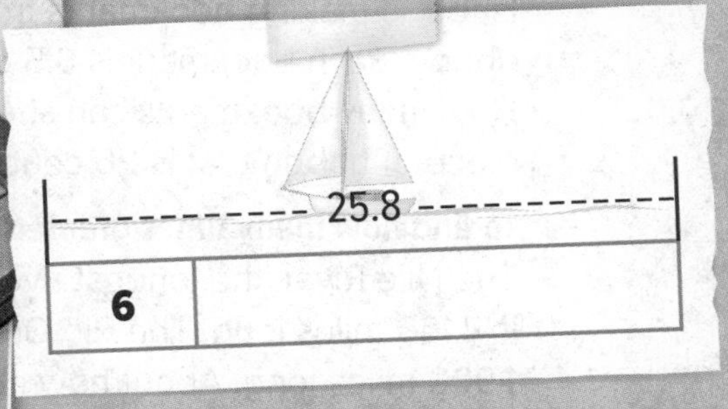

25.8	
6	

FOLDABLES® Study Organizer

Cut out the Foldable on page FL3 of this book.

Place your Foldable on page 506.

Use the Foldable throughout this chapter to help you learn about expressions.

What Tools Do You Need?

Vocabulary

algebra	defining the variable	like terms
algebraic expression	Distributive Property	numerical expression
Associative Properties	equivalent expressions	perfect square
base	evaluate	powers
coefficient	exponent	properties
Commutative Properties	factor the expression	term
constant	Identity Properties	variable

Study Skill: Reading Math

Meaning of Division Look for these other meanings when you are solving a word problem.

- **To share:**
  Zach and his friend are going to share 3 apples equally. How many apples will each boy have?

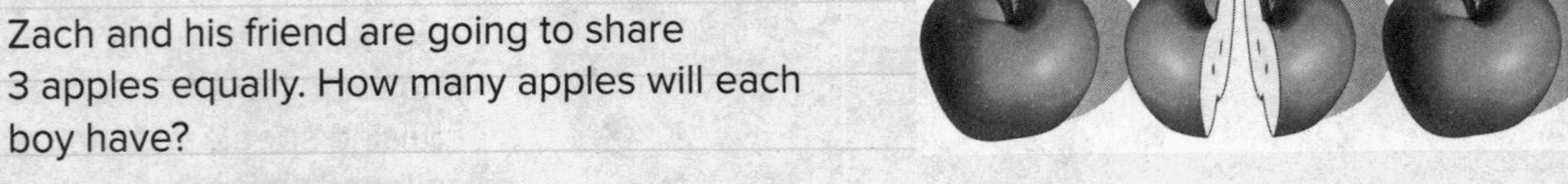

- **To take away equal amounts:**
  Isabel is making bookmarks from a piece of ribbon. Each bookmark is 6.5 centimeters long. How many bookmarks can she make from a piece of ribbon that is 26 centimeters long?

- **To find how many times greater:**
  The Nile River, the longest river on Earth, is 4,160 miles long. The Rio Grande River is 1,900 miles long. About how many times as long is the Nile as the Rio Grande?

Practice

Identify the meaning of division shown in each problem. Then solve the problem.

1. The Jackson family wants to buy a flat-screen television that costs $1,200. They plan to pay in six equal payments. What will be the amount of each payment?

2. A full-grown blue whale can weigh 150 tons. An adult African elephant weights about 5 tons. How many times as great does a blue whale weigh as an African elephant?

What Do You Already Know?

Read each statement. Decide whether you agree (A) or disagree (D). Place a checkmark in the appropriate column and then justify your reasoning.

Expressions			
Statement	A	D	Why?
You must follow the order of operations to find the value of a numerical expression.			
A variable is a symbol used to represent an operation.			
The phrase 4 less than x is written as $4 - x$.			
The multiplicative identity is 0.			
Subtraction is a commutative operation.			
The Distributive Property combines addition and multiplication.			

When Will You Use This?

Here are a few examples of how expressions are used in the real world.

Activity 1 Use the Internet to find the cost of admission to a science museum. How much will your admission ticket cost? Are there any other events at the museum that you would pay to see?

____

____

Activity 2 Go online at **connectED.mcgraw-hill.com** to read the graphic novel ***Science Center***. Suppose you were to go to the science center at noon. If you plan to watch the movie, how much will you pay? ______________________________

Try the Quick Check below.
Or, take the Online Readiness Quiz.

Quick Review

Common Core Review 4.NBT.5, 5.NF.1

Example 1

Multiply $5 \times 5 \times 5 \times 5$.

5 is used as a factor four times.

$5 \times 5 \times 5 \times 5 = 625$

Example 2

Find $3\frac{7}{8} - 1\frac{1}{2}$.

$$\begin{array}{rrl} 3\frac{7}{8} = & 3\frac{7}{8} & \text{Rename using the LCD, 8.} \\ -1\frac{1}{2} = & -1\frac{4}{8} & \\ \hline & 2\frac{3}{8} & \text{Subtract.} \end{array}$$

Quick Check

Number Patterns **Multiply.**

1. $7 \times 7 \times 7 =$ ________

2. $2 \times 2 \times 2 =$ ________

3. $9 \times 9 \times 9 \times 9 =$ ________

Fractions **Add or subtract. Write in simplest form.**

4. $\frac{4}{5} - \frac{1}{2} =$ ________

5. $\frac{8}{9} + \frac{2}{3} =$ ________

6. $3\frac{1}{10} - 2\frac{5}{6} =$ ________

7. What fraction more of the coupon books did Jabar sell than Guto?

____

Coupon Book Sales	
Student	**Fraction of Total Sales**
Guto	$\frac{1}{12}$
Holly	$\frac{3}{40}$
Jabar	$\frac{2}{15}$

Which problems did you answer correctly in the Quick Check? Shade those exercise numbers below.

1 2 3 4 5 6 7

Inquiry Lab

Structure of Expressions

HOW can you identify the parts of an expression using mathematical terms?

Content Standards
6.EE.2, 6.EE.2b

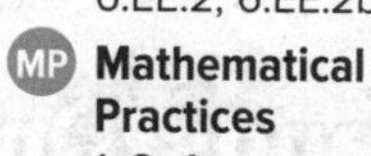

Mathematical Practices
1, 3, 4

Fitness Fortress recycles plastic water bottles. On Saturday, 8 bottles were placed in the bins. On Sunday, 8 more bottles were recycled.

Hands-On Activity 1

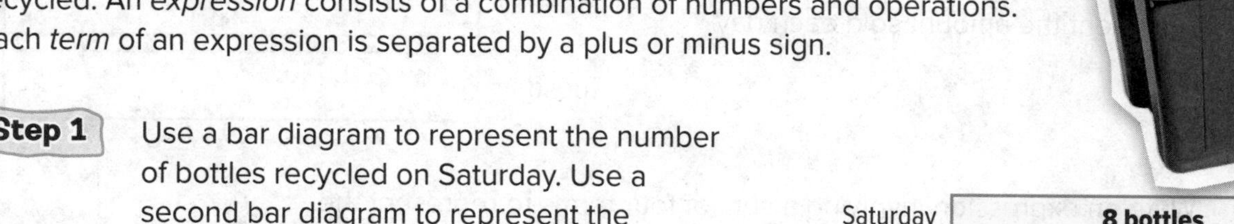

You can use an expression to represent the number of bottles that were recycled. An *expression* consists of a combination of numbers and operations. Each *term* of an expression is separated by a plus or minus sign.

Step 1 Use a bar diagram to represent the number of bottles recycled on Saturday. Use a second bar diagram to represent the number of bottles recycled on Sunday.

Saturday	**8 bottles**
Sunday	**8 bottles**

Step 2 The addition expression 8 + 8 represents the total. How many terms are in the expression? ☐

Does the expression represent a *sum*, *product*, or *quotient*?

____

Step 3 The multiplication expression 2 × 8 also represents the total.

How many terms are in the expression? ☐

Does the expression represent a *sum*, *product*, or *quotient*?

____

Investigate

Work with a partner. Rewrite each sum as a product. Then identify the number of terms in each expression.

1. 14 + 14 = ____________

Sum: ____________

Product: ____________

2. 92 + 92 + 92 = ____________

Sum: ____________

Product: ____________

Some expressions can be written as the product of a sum. For example, $2 \times (3 + 4)$ represents the product of 2 and the sum of 3 and 4. The expression $2 \times (3 + 4)$ can also be thought of as the product of two *factors*.

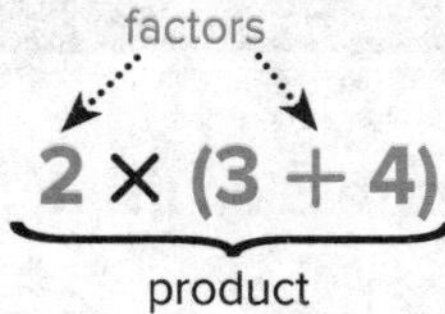

Hands-On Activity 2

Melina and Kendrick are selling tins of cashews for a school fundraiser. Melina sold 5 tins on Monday and 5 tins on Tuesday. Kendrick sold 4 tins Monday and 4 tins on Tuesday.

Step 1 Divide and label each bar diagram to represent the amount sold each day.

Monday

Tuesday

Step 2 Write an expression involving a sum of four terms to represent the total amount sold.

☐ + ☐ + ☐ + ☐

Step 3 Complete the expression below involving the product of a sum to represent the total amount sold.

$2 \times (\square + \square)$

In the expression above, what are the two factors? ______________

In the expression above, which factor can be thought of as both a single term and a sum of two terms? ______________

Investigate

Work with a partner. Rewrite each sum as the product of a sum. Then identify the factors.

3. $1 + 4 + 1 + 4 =$ ______________

   Factors: ______________

4. $32 + 32 + 2 + 2 =$ ______________

   Factors: ______________

5. $79 + 8 + 79 + 8 =$ ______________

   Factors: ______________

6. $19 + 56 + 56 + 19 =$ ______________

   Factors: ______________

Work with a partner. Represent each expression using bar diagrams.

7. 5 + 5

8. 9 + 9

Work with a partner. Represent each expression using bar diagrams. Then identify the factors.

9. 2 × (3 + 1)

Factors: ______________

Which factor is also a sum?

____

10. 2 × (5 + 2)

Factors: ______________

Which factor is also a sum?

____

Work with a partner. Represent each diagram as a sum.

11. ______________

17	17

12. ______________

74	74

Work with a partner. Represent each diagram as the product of a sum. Then identify the factors.

13. Product: ______________

Factors: ______________

Which factor is also a sum?

____

5	8
5	8

14. Product: ______________

Factors: ______________

Which factor is also a sum?

____

54	58
54	58

Analyze and Reflect

Work with a partner to match each description to the correct expression. The first one is already done for you.

	Description	Expression
15.	This expression is a sum of two terms.	a. $(1 + 2) \times 2$
16.	This expression can be thought of as a product of two factors. One of the factors is the sum of 6 and 4.	b. $6 + 6$
17.	This expression can be thought of as a product of two factors. One of the factors is the sum of 1 and 2.	c. $14 \div 7$
18.	This expression is the quotient of 14 and 7.	d. $(6 + 4) \times 2$

19. MP **Reason Inductively** Consuela wrote the expression $2 \times (31 + 47)$. She states that the expression is a product and that the expression $(31 + 47)$ is a factor. Marcus states that the expression $(31 + 47)$ is a sum of two terms. Who is correct? Explain. ______

Create

20. MP **Model with Mathematics** Write an expression and a real-world problem for the situation modeled to the right.

4 pounds	6 pounds
4 pounds	6 pounds

21. Inquiry HOW can you identify the parts of an expression using mathematical terms?

Lesson 1

Powers and Exponents

Vocabulary Start-Up 

A product of like factors can be written in exponential form using an exponent and a base. The **base** is the number used as a factor. The **exponent** tells how many times a base is used as a factor.

1. Fill in the boxes with the words *factors*, *exponent*, and *base*.

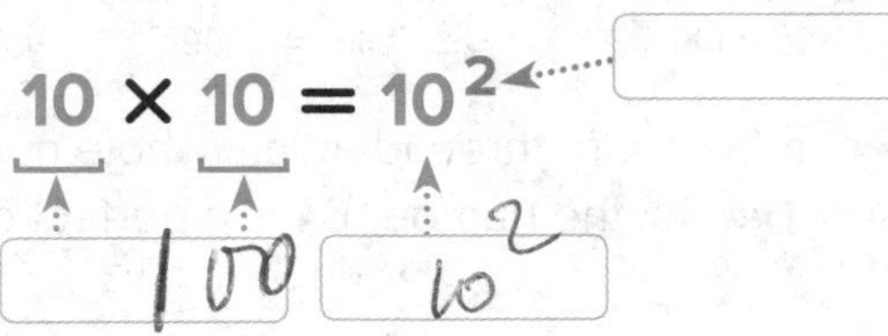

$$10 \times 10 = 10^2$$

2. Give an example of an exponent.

3. Write the definition of exponent in your own words.

Real-World Link

MP3 players come in different storage sizes, such as 2GB, 4GB, or 16GB, where GB means gigabyte. One gigabyte is equal to $10 \times 10 \times 10 \times 10 \times 10 \times 10 \times 10 \times 10 \times 10$ bytes.

What is this number written with exponents? ☐

Which MP Mathematical Practices did you use? Shade the circle(s) that applies.

① Persevere with Problems
② Reason Abstractly
③ Construct an Argument
④ Model with Mathematics
⑤ Use Math Tools
⑥ Attend to Precision
⑦ Make Use of Structure
⑧ Use Repeated Reasoning

 **Essential Question**

HOW is it helpful to write numbers in different ways?

 **Vocabulary**

base
exponent
powers
perfect square

 **Common Core State Standards**

Content Standards
6.EE.1, 6.NS.3

MP Mathematical Practices
1, 3, 4, 6, 8

Work Zone

Write Products as Powers

Numbers expressed using exponents are called **powers**. For example, 100 is a power of 10 because it can be written as 10^2. Numbers like 100 are **perfect squares** because they are the squares of whole numbers.

$10 \times 10 = 100$
$10^2 = 100$

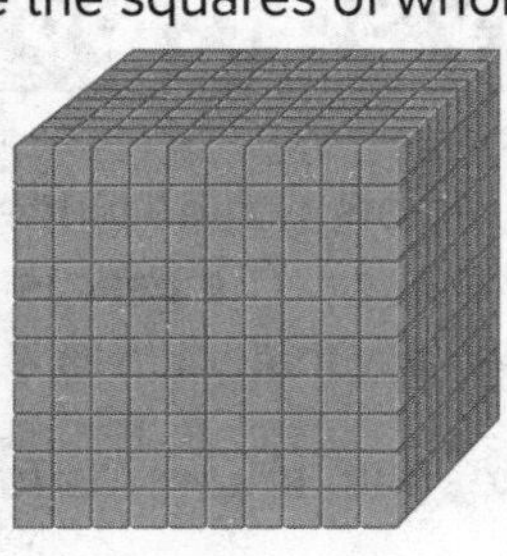

$10 \times 10 \times 10 = 1{,}000$
$10^3 = 1{,}000$

Perfect cubes are numbers with three identical whole number factors such as $4 \times 4 \times 4 = 64$. So, the number 64 is a perfect cube.

Examples

1. Write $6 \times 6 \times 6 \times 6$ using an exponent.

$6 \times 6 \times 6 \times 6 = 6^4$ — 6 is used as a factor four times.

2. Write $4 \times 4 \times 4$ using an exponent.

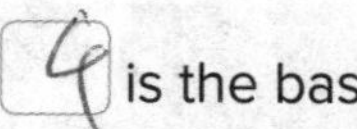

The factor [4] is the base.

The factor is multiplied [3] times.

The exponent is [3].

So, $4 \times 4 \times 4$ can be written as 4^3.

Show your work.

Got It? Do these problems to find out.

Write each product using an exponent.

a. $7 \times 7 \times 7 \times 7$

b. $9 \times 9 \times 9 \times 9 \times 9 \times 9 \times 9$

a. ________

b. ________

Write Powers as Products

To write powers as products, determine the base and the exponent. The base of 10^2 is 10 and the exponent is 2. To read powers, consider the exponent. The power 10^2 is read as *ten squared* and 10^3 is read as *ten cubed*.

Examples

Tutor

3. Write 5^2 as a product of the same factor. Then find the value.

The base is 5. The exponent is 2. So, 5 is used as a factor two times.

$5^2 = 5 \times 5$ Write 5^2 as a product.

$= 25$ Multiply 5 by itself.

4. Write 1.5^3 as a product of the same factor. Then find the value.

The base is 1.5. The exponent is 3. So, 1.5 is used as a factor three times.

$1.5^3 = 1.5 \times 1.5 \times 1.5$ Write 1.5^3 as a product.

$= 3.375$ Multiply.

5. Write $\left(\frac{1}{2}\right)^3$ as a product of the same factor. Then find the value.

The base is $\frac{1}{2}$. The exponent is 3. So $\frac{1}{2}$ is used as a factor three times.

$\left(\frac{1}{2}\right)^3 = \frac{1}{2} \times \frac{1}{2} \times \frac{1}{2}$ Write $\left(\frac{1}{2}\right)^3$ as a product.

$= \frac{1}{8}$ Multiply.

Notation

In Example 5, the fraction $\frac{1}{2}$ is set in parentheses to note that the entire fraction is the base

$\left(\frac{1}{2}\right)^3 = \frac{1}{2} \times \frac{1}{2} \times \frac{1}{2} = \frac{1}{8}$

Without the parentheses, it is understood that the base is only the numerator of the fraction.

$\frac{1^3}{2} = \frac{1 \times 1 \times 1}{2} = \frac{1}{2}$

Got It? Do these problems to find out.

Write each power as a product of the same factor. Then find the value.

c. 10^5 100000

d. 2.1^2 4.41

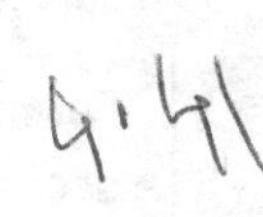

e. $\left(\frac{1}{4}\right)^2$ $\frac{1}{16}$

c. ________

d. ________

e. ________

Real World Example

6. **STEM** **The zoo has an aquarium that holds around 7^4 gallons of water. About how many gallons of water does the aquarium hold?**

$7^4 = 7 \times 7 \times 7 \times 7$ Write 7^4 as a product.

$= 2,401$ Multiply.

So, the aquarium holds about 2,401 gallons of water.

Got It? **Do this problem to find out.**

f. ______

f. **STEM** Michigan has more than 10^4 inland lakes. Find the value of 10^4.

Guided Practice

Write each product using an exponent. (Examples 1 and 2)

1. $8 \times 8 \times 8 =$ ______

2. $1 \times 1 \times 1 \times 1 \times 1 =$ ______

Write each power as a product of the same factor. Then find the value. (Examples 3–5)

3. $\left(\frac{1}{7}\right)^3 =$ ______

4. $2^5 =$ ______

5. $1.4^2 =$ ______

6. Coal mines have shafts that can be as much as 7^3 feet deep. About how many feet deep into Earth's crust are these shafts? (Example 6)

7. **Building on the Essential Question** How is using exponents helpful? ______

Rate Yourself!

How confident are you about powers and exponents? Shade the ring on the target.

For more help, go online to access a Personal Tutor.

Name ________________________ My Homework ________________________

Independent Practice

Go online for Step-by-Step Solutions

Write each product using an exponent. (Examples 1 and 2)

1. $6 \times 6 =$ 6^2

Show your work.

2. $1 \times 1 \times 1 =$ 1^3

3. $5 \times 5 \times 5 \times 5 \times 5 \times 5 =$ 5^6

4. $12 \times 12 =$ 12^2

5. $27 \times 27 \times 27 \times 27 =$ 27^4

6. $15 \times 15 \times 15 =$ 15^3

Write each power as a product of the same factor. Then find the value. (Examples 3–5)

7. $6^4 =$ ______

8. $0.5^3 =$ ______

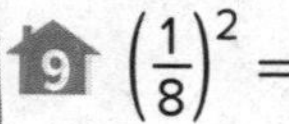

9. $\left(\frac{1}{8}\right)^2 =$ ______

10. **MP Identify Repeated Reasoning** A byte is a basic unit of measurement for information storage involving computers. (Example 6)

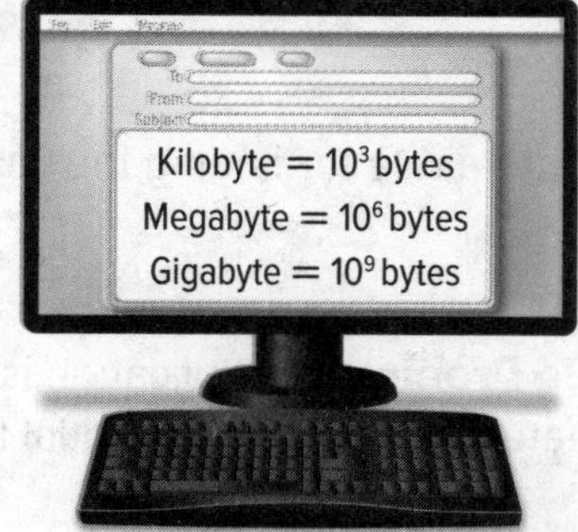

a. A kilobyte is equal to 10^3 bytes. Write 10^3 as a product of the same factor. Then find the value.

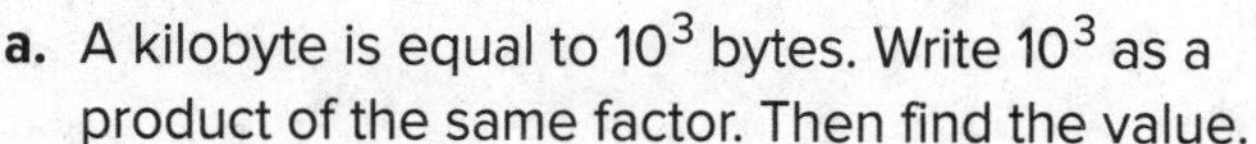

____

____

b. A megabyte is equal to 10^6 bytes. Write 10^6 as a product of the same factor. Then find the value.

____

____

c. How many more bytes of information are in a gigabyte than a megabyte? ______________

____

Find the value of each expression.

11. $0.5^4 + 1 =$ ______

12. $3.2^3 \times 10 =$ ______

13. $10.3^3 + 8 =$ ______

H.O.T. Problems Higher Order Thinking

14. **Model with Mathematics** Write a power whose value is greater than 1,000. ______

15. **Persevere with Problems** Use the table to solve.

Powers of 2	Powers of 4	Powers of 10
$2^4 = 16$	$4^4 = 256$	$10^4 = 10{,}000$
$2^3 = 8$	$4^3 = 64$	$10^3 = 1{,}000$
$2^2 = 4$	$4^2 = 16$	$10^2 = 100$
$2^1 =$	$4^1 =$	$10^1 =$
$2^0 =$	$4^0 =$	$10^0 =$

a. Describe the pattern for the powers of 2. Write the values of 2^1 and 2^0 in the table.

b. Describe the pattern for the powers of 4. Write the values of 4^1 and 4^0 in the table.

c. Describe the pattern for the powers of 10. Write the values of 10^1 and 10^0 in the table.

d. Write a rule for finding the value of any base with an exponent of 0.

16. **Be Precise** Multiplication is defined as repeated addition. Use the word repeated to define exponential form. Justify your reasoning.

17. **Reason Inductively** Suppose the population of the United States is about 230 million. Is this number closer to 10^7 or 10^8? Explain your reasoning.

Extra Practice

Write each product using an exponent.

18. $6 \times 6 \times 6 =$ 6^3

mework Help

The factor 6 is used 3 times.
The base is 6.
The exponent is 3.

19. $10 \times 10 \times 10 =$ ______

20. $32 \times 32 \times 32 \times 32 =$ ______

21. $9 \times 9 =$ ______

22. $7 \times 7 \times 7 \times 7 \times 7 \times 7 =$ ______

23. $13 \times 13 \times 13 \times 13 \times 13 =$ ______

Write each power as a product of the same factor. Then find the value.

24. $3^7 =$ ______

25. $0.06^2 =$ ______

26. $\left(\frac{1}{4}\right)^3 =$ ______

27. MP **Be Precise** The baseball infield at the right has an area of 90^2 square feet. What is the area of the infield?

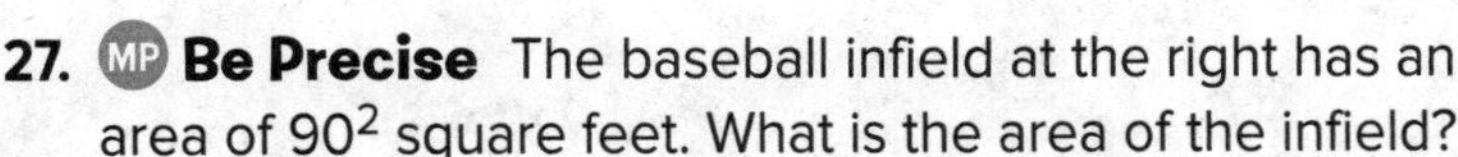

______

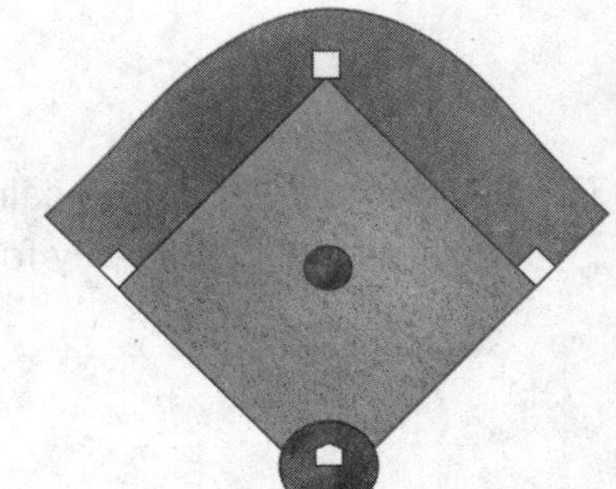

28. Last week Bakery Marvels baked 5^5 muffins. How many muffins did Bakery Marvels bake?

______

29. Luke ran 3.5^3 miles in the month of January. How many miles did Luke run in January? ______

Power Up! Common Core Test Practice

30. Mrs. Torrey traveled about 8 × 8 × 8 × 8 miles from Ohio to Hawaii. Select values to complete the model below to show the repeated multiplication as a power.

2	4
6	8

The base is ______.

The exponent is ______.

The repeated multiplication can be expressed as the power: ______.

About how many miles did Mrs. Torrey travel? ______

31. Claire used counters to make the pattern shown below.

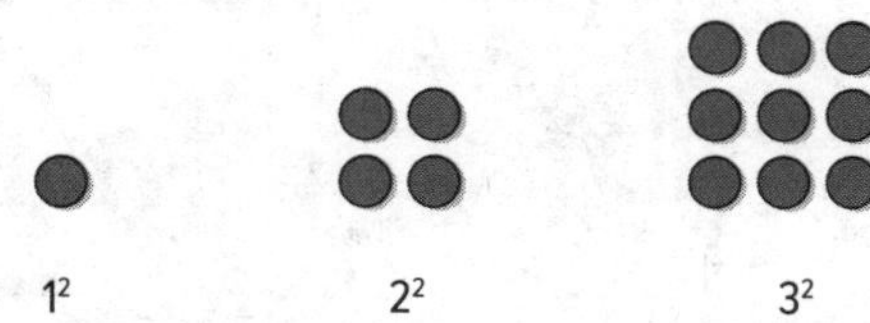

She continues the pattern for several more figures. Which of the following accurately describe how many counters she uses in different figures? Select all that apply.

- ☐ There are 25 counters in the 5th figure.
- ☐ There are 42 counters in the 6th figure.
- ☐ There are 81 counters in the 9th figure.
- ☐ There are 121 counters in the 11th figure.

Common Core Spiral Review

Multiply or divide. 4.NBT.5, 4.NBT.6

32. 6 × 8 = ______

33. 64 ÷ 8 = ______

34. 42 ÷ 7 = ______

35. All video games are on sale at The Game House for $29 each. How much will Bella pay for 3 video games? 4.NBT.5

______

______

36. Max and two of his friends carpooled on a visit to the zoo. The cost of admission was $12 per person. Parking cost $7 per car. How much did the group pay on their visit to the zoo? 4.OA.3 ______

 Need more practice? Download more Extra Practice at **connectED.mcgraw-hill.com.**

Lesson 2

Numerical Expressions

Real-World Link

Snacks The table shows the cost of different snacks at a concession stand at the school hockey game.

Item	Price ($)
Popcorn	2
Juice or Soda	1
Hot Dog	4

1.  = $ ☐

2. 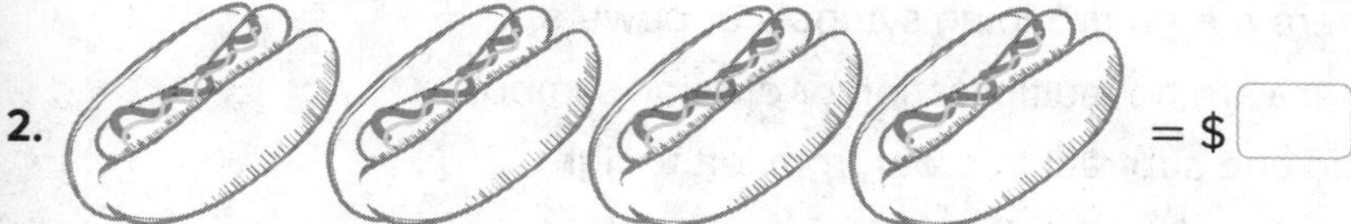 = $ ☐

3. Find the total cost of buying 3 boxes of popcorn and 4 hot dogs.

4. What operations could you use in Exercises 1–2? Explain how to find the answer to Exercise 3 using these operations.

Which MP **Mathematical Practices** did you use? Shade the circle(s) that applies.

① Persevere with Problems
② Reason Abstractly
③ Construct an Argument
④ Model with Mathematics
⑤ Use Math Tools
⑥ Attend to Precision
⑦ Make Use of Structure
⑧ Use Repeated Reasoning

Essential Question

HOW is it helpful to write numbers in different ways?

Vocabulary

numerical expression
order of operations

Common Core State Standards

Content Standards
6.EE.1

MP **Mathematical Practices**
1, 2, 3, 4, 5

Key Concept

Order of Operations

1. Simplify the expressions inside grouping symbols, like parentheses.
2. Find the value of all powers.
3. Multiply and divide in order from left to right.
4. Add and subtract in order from left to right.

Work Zone

A **numerical expression** like $3 \times 2 + 4 \times 4$ is a combination of numbers and operations. The **order of operations** tells you which operation to perform first so that everyone finds the same value for an expression.

Examples

Find the value of each expression.

1. $10 - 2 + 8$

There are no grouping symbols or powers.

There are no multiplication or division symbols.

Add and subtract in order from left to right.

$10 - 2 + 8 = 8 + 8$ Subtract 2 from 10 first.

$= 16$ Add 8 and 8.

2. $4 + 3 \times 5$

There are no grouping symbols or powers.

Multiply before adding.

$4 + 3 \times 5 = 4 + 15$ Multiply 3 and 5.

$= 19$ Add 4 and 15.

Got It? **Do these problems to find out.**

a. $10 + 2 \times 15$

b. $16 \div 2 \times 4$

a. ______

b. ______

Parentheses and Exponents

Expressions inside grouping symbols, such as parentheses are simplified first. Follow the order of operations inside parentheses. For example in the expression $3 + (4^2 + 5)$, you will need to find the value of the power, 4^2, before you can add the expression inside the parentheses.

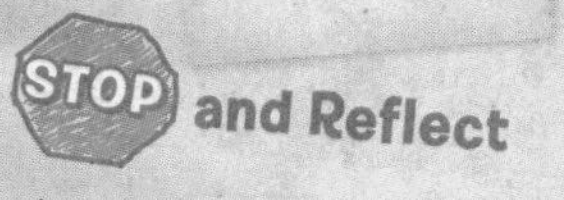

Why is it important to have the order of operations?

Examples

Find the value of each expression.

3. $20 \div 4 + 17 \times (9 - 6)$

$20 \div 4 + 17 \times (9 - 6) = 20 \div 4 + 17 \times 3$	Subtract 6 from 9.
$= 5 + 17 \times 3$	Divide 20 by 4.
$= 5 + 51$	Multiply 17 by 3.
$= 56$	Add 5 and 51.

4. $3 \times 6^2 + 4$

$3 \times 6^2 + 4 = 3 \times 36 + 4$	Find 6^2.
$= 108 + 4$	Multiply 3 and 36.
$= 112$	Add 108 and 4.

5. $5 + (8^2 - 2) \times 2$

$5 + (8^2 - 2) \times 2 = 5 + (\square - 2) \times 2$	Simplify the exponent.
$= 5 + \square \times 2$	Simplify inside parentheses.
$= 5 + \square$	Multiply.
$= \square$	Add.

Got It? **Do these problems to find out.**

c. $25 \times (5 - 2) \div 5 - 12$

d. $24 \div (2^3 + 4)$

c. _______

d. _______

Example

6. **Write an expression for the total cost of 5 lotions, 2 candles, and 4 lip balms. Find the total cost.**

Cost of Items			
Item	Lotion	Candle	Lip balm
Cost ($)	5	7	2

$5 \times \$5 + 2 \times \$7 + 4 \times \$2$

$= 5^2 + 2 \times 7 + 4 \times 2$

$= 25 + 2 \times 7 + 4 \times 2$ — Simplify 5^2 to find the cost of the lotions.

$= 25 + 14 + 4 \times 2$ — Multiply 2 and 7 to find the cost of the candles.

$= 25 + 14 + 8$ — Multiply 4 and 2 to find the cost of the lip balms.

$= 47$

The total cost of the items is $47.

Got It? **Do these problems to find out.**

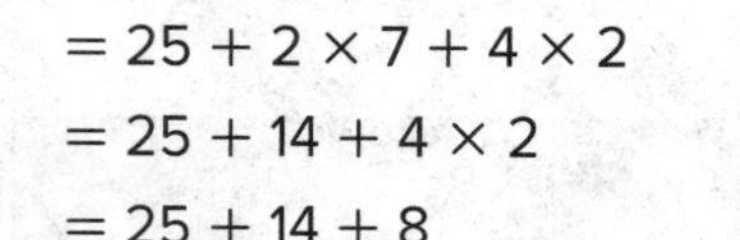

Show your work.

e. ____________

e. Alexis and 3 friends are at the mall. Each person buys a pretzel for $4, sauce for $1, and a drink for $2. Write an expression for the total and find the total cost.

Guided Practice

Find the value of each expression. (Examples 1–5)

Show your work.

1. $9 + 3 - 5 =$ ____________

2. $(26 + 5) \times 2 - 15 =$ ____________

3. $5^2 + 8 \div 2 =$ ____________

4. **Financial Literacy** Tickets to a play cost $10 for members and $24 for nonmembers. Write an expression to find the total cost of 4 nonmember tickets and 2 member tickets. Then find the total cost. (Example 6)

____

5. **Building on the Essential Question** How are grouping symbols helpful in simplifying expressions correctly?

____

____

____

Rate Yourself!

How well do you understand order of operations? Circle the image that applies.

Clear

Somewhat Clear

Not So Clear

For more help, go online to access a Personal Tutor.

Name ______ My Homework ______

Independent Practice

Go online for Step-by-Step Solutions 

Find the value of each expression. (Examples 1–5)

1. $8 + 4 - 3 =$ ______

2. $38 - 19 + 12 =$ ______

3. $7 + 9 \times (3 + 8) =$ ______

4. $15 - 2^3 \div 4 =$ ______

5. $55 \div 11 + 7 \times (2 + 14) =$ ______

6. $5^3 - 12 \div 3 =$ ______

7. $8 \times (2^4 - 3) + 8 =$ ______

8. $9 + 4^3 \times (20 - 8) \div 2 + 6 =$ ______

9. Financial Literacy Tyree and four friends go to the movies. Each person buys a movie ticket for \$7, a snack for \$5, and a drink for \$2. Write an expression for the total cost of the trip to the movies. Then find the total cost. (Example 6)

______

______

10. Financial Literacy The Molina family went to a concert together. They purchased 4 concert tickets for \$25 each, 3 T-shirts for \$15 each, and a poster for \$10. Write an expression for the total cost. Then find the total cost. (Example 6)

______

______

11. **Use Math Tools** A wholesaler sells rolls of fruit snacks in two sizes of bags. The table shows the number of rolls that come in each bag. Write an expression that could be used to determine the number of rolls in 3 large bags and 2 small bags. Then find the number of rolls.

Bag	Number of Rolls
Large	10
Small	5

H.O.T. Problems Higher Order Thinking

12. **Find the Error** Luis is finding $9 - 6 + 2$. Find his mistake and correct it.

13. **Reason Inductively** Use the expression $34 - 12 \div 2 + 7$.

a. Place parentheses in the expression so that the value of the expression is 18.

b. Place parentheses in the expression to find a value other than 18. Then find the value of the new expression.

14. **Persevere with Problems** Write an expression with a value of 12. It should contain four numbers and two different operations.

15. **Use Math Tools** Place parentheses in each equation, if needed, to make each equation true.

a. $7 + 3 \times 2 + 4 = 25$

b. $8^2 \div 4 \times 8 = 2$

c. $16 + 8 - 5 \times 2 = 14$

16. **Which One Doesn't Belong?** Which expression does not belong with the other three? Justify your response.

$6^2 - 9$ | 3^3 | $(5 + 4)^2 \div 3$ | $4 \times 5 + 9$

Name ______________________ My Homework ______________________

Extra Practice

Find the value of each expression.

17. $9 + 12 - 15 =$ 6

Homework Help: $9 + 12 - 15 = 21 - 15$
$= 6$

18. $22 - 17 + 8 =$ ______

19. $(9 + 2) \times 6 - 5 =$ ______

20. $27 \div (3 + 6) \times 5 - 12 =$ ______

21. $26 + 6^2 \div 4 =$ ______

22. $22 \div 2 \times 3^2 =$ ______

23. $12 \div 4 + (5^2 - 6) =$ ______

24. $96 \div 4^2 + (25 \times 2) - 15 - 3 =$ ______

25. **Financial Literacy** Admission to a circus is \$16 for adults and \$8 for children. Write an expression to find the total cost of 3 adult tickets and 8 children's tickets. Then find the total cost.

____

____

26. MP **Reason Inductively** Addison is making caramel apples. She has $2\frac{1}{2}$ bags of apples. One full bag has 8 apples, and each apple weighs 5 ounces. Write an expression that could be used to find the total number of ounces of apples Addison has. Then find the total number of ounces.

____

____

Power Up! Common Core Test Practice

27. Kailey wants to buy 4 pencils and 3 notebooks. Which of the following represent the total cost? Select all that apply.

Pencils	\$0.50
Notebooks	\$2.25

- ☐ 4(\$0.50) + 3(\$2.25)
- ☐ 7(\$0.50 + \$2.25)
- ☐ \$8.75
- ☐ \$19.25

28. Denzel has $3\frac{2}{5}$ boxes of party favors. One full box contains 15 bags of favors, and each bag has 3 favors in it. In addition to this he has 7 extra party favors that are not in bags or boxes.

Select the correct operation to model a numerical expression for the total number of party favors.

×

÷

+

−

$3\frac{2}{5}$ ☐ 15 ☐ 3 ☐ 7

How many party favors does Denzel have in all?

☐

Common Core Spiral Review

Find the missing number. **4.NBT.4**

29. 131 + ☐ = 140

30. ☐ − 6 = 354

31. ☐ + 210 = 224

32. Use skip counting and the number line to find the missing number. **4.OA.5**

3 × ☐ = 12

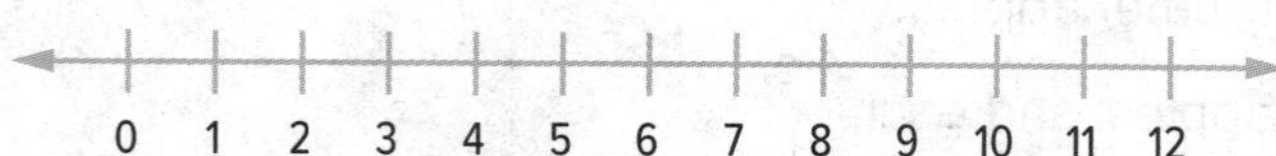

33. Sophie earns \$7 an hour babysitting and \$8 an hour for cleaning the house. Last week she babysat for 3 hours and cleaned for 2 hours.

How much did Sophie earn last week? **4.OA.3** ______

Algebra: Variables and Expressions

Vocabulary Start-Up

Algebra is a language of symbols including variables. A **variable** is a symbol, usually a letter, used to represent a number.

Scan the lesson to complete the graphic organizer.

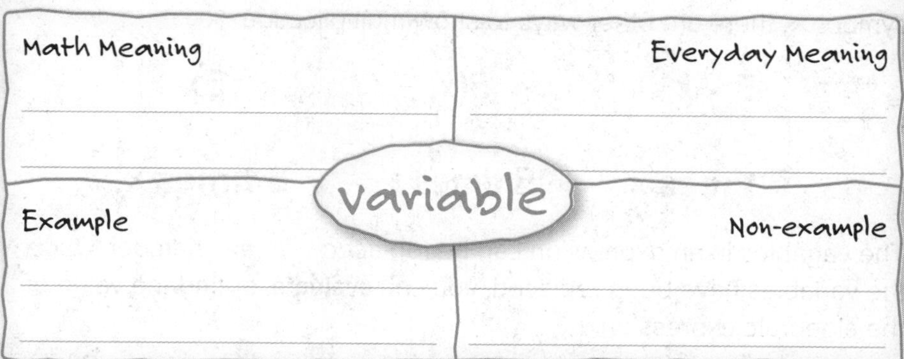

Essential Question

HOW is it helpful to write numbers in different ways?

Vocabulary

algebra
variable
algebraic expression
evaluate

Common Core State Standards

Content Standards
6.EE.2, 6.EE.2c, 6.EE.6

MP Mathematical Practices
1, 2, 3, 4, 6

Real-World Link

A box contains an unknown number of markers. There are 2 markers outside the box. The total number of markers is represented by the bar diagram below.

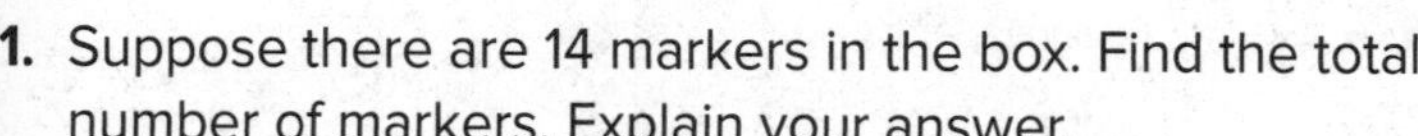

1. Suppose there are 14 markers in the box. Find the total number of markers. Explain your answer. ______

______

______

Which  **Mathematical Practices** did you use? Shade the circle(s) that applies.

① Persevere with Problems
② Reason Abstractly
③ Construct an Argument
④ Model with Mathematics
⑤ Use Math Tools
⑥ Attend to Precision
⑦ Make Use of Structure
⑧ Use Repeated Reasoning

Work Zone

Evaluate One-Step Expressions

Algebraic expressions contain at least one variable and at least one operation. For example, the expression $n + 2$ represents *the sum of an unknown number and two.*

Any letter can be used as a variable. → $n + 2$

The letter x is often used as a variable. To avoid confusion with the symbol ×, there are other ways to show multiplication.

$5 \cdot x$	$5(x)$	$5x$
↑	↑	↑
5 times x	5 times x	5 times x

The variables in an expression can be replaced with any number. Once the variables have been replaced, you can **evaluate**, or find the value of, the algebraic expression.

Examples

Tutor

1. Evaluate $16 + b$ if $b = 25$.

$16 + b = 16 + 25$ Replace b with 25.

$= 41$ Add 16 and 25.

2. Evaluate $x - y$ if $x = 64$ and $y = 27$.

$x - y = 64 - 27$ Replace x with 64 and y with 27.

$= 37$ Subtract 27 from 64.

3. Evaluate $6x$ if $x = \frac{1}{2}$.

$6x = 6 \cdot \frac{1}{2}$ Replace x with $\frac{1}{2}$.

$= 3$ Multiply 6 and $\frac{1}{2}$.

Got It? **Do these problems to find out.**

Evaluate each expression if $a = 6$, $b = 4$, and $c = \frac{1}{3}$.

a. $a + 8$ **b.** $a - b$ **c.** $a \cdot b$ **d.** $9c$

Show your work.

a. ____________

b. ____________

c. ____________

d. ____________

Evaluate Multi-Step Expressions

To evaluate multi-step expressions, replace each variable with the correct value and follow the order of operations.

Examples

4. Evaluate $5t + 4$ if $t = 3$.

$5t + 4 = 5 \cdot 3 + 4$ Replace t with 3.

$= 15 + 4$ Multiply 5 and 3.

$= 19$ Add 15 and 4.

5. Evaluate $4x^2$ if $x = \frac{1}{8}$.

$4x^2 = 4 \cdot \left(\frac{1}{8}\right)^2$ Replace x with $\frac{1}{8}$.

$= 4 \cdot \frac{1}{64}$ Simplify $\left(\frac{1}{8}\right)^2$.

$= \frac{1}{16}$ Multiply.

6. Evaluate $10a + 7$ if $a = \frac{1}{5}$.

$10a + 7 = 10\left(\frac{\square}{\square}\right) + 7$ Replace a with $\frac{1}{5}$.

$= \square + 7$ Multiply 10 and $\frac{1}{5}$.

$= \square$ Add.

Got It? Do these problems to find out.

Evaluate each expression if $d = 12$ and $e = \frac{1}{3}$.

e. $2d - 5$

f. $50 - 3d$

g. $9e^2$

e. ____________

f. ____________

g. ____________

Example

7. **Khalil is wrapping a gift for his brother's birthday. The box has side lengths that are $\frac{1}{2}$ foot. Use the expression $6s^2$, where s represents the length of a side, to find the surface area of the box he is wrapping. Write your answer in square feet.**

$6s^2 = 6 \cdot \left(\frac{1}{2}\right)^2$ Replace s with $\frac{1}{2}$.

$= 6 \cdot \frac{1}{4}$ Simplify $\left(\frac{1}{2}\right)^2$.

$= \frac{6}{4}$ or $1\frac{1}{2}$ Multiply.

So, the surface area of the box is $1\frac{1}{2}$ square feet.

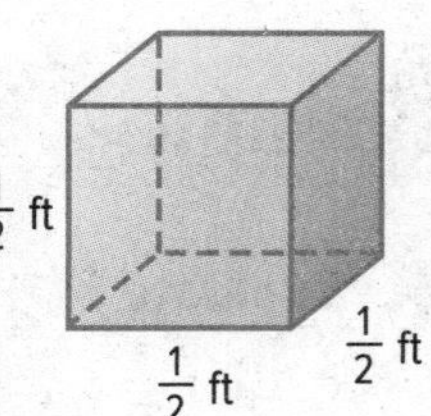

Guided Practice

Evaluate each expression if $m = 4$, $z = 9$, and $r = \frac{1}{6}$. (Examples 1–6)

1. $3 + m$ ______

Show your work.

2. $z - m$ ______

3. $12r$ ______

4. $4m - 2$ ______

5. $60r - 4$ ______

6. $3r^2$ ______

7. The amount of money that remains from a 20-dollar bill after Malina buys 4 party favors for p dollars each is $20 - 4p$. Find the amount remaining if each favor cost \$3. (Example 7)

______

8. **Building on the Essential Question** How are numerical expressions and algebraic expressions different?

______

______

______

Rate Yourself!

Are you ready to move on? Shade the section that applies.

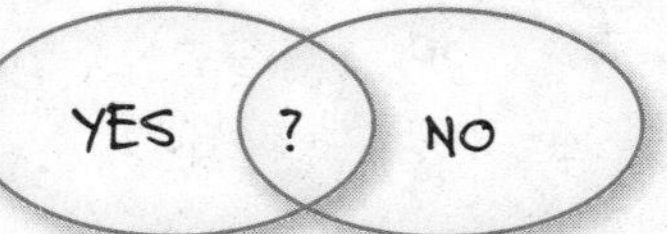

For more help, go online to access a Personal Tutor.

Independent Practice

Go online for Step-by-Step Solutions 

Evaluate each expression if $m = 2$, $n = 16$, and $p = \frac{1}{3}$. (Examples 1–6)

1. $m + 10$ ______

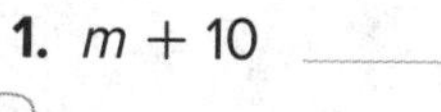

2. $n \div 4$ ______ 

3. $m + n$ ______

4. $6m - 1$ ______

5. $3p$ ______

6. $12p$ ______

 7. $12m - 4$ ______

8. $9p^2$ ______

9. A paper recycling bin has the dimensions shown. Use the expression s^3, where s represents the length of a side, to find the volume of the bin. Write your answer in cubic meters. (Example 7)

____

10. **MP Model with Mathematics** Refer to the graphic novel frame below for Exercises **a–b**.

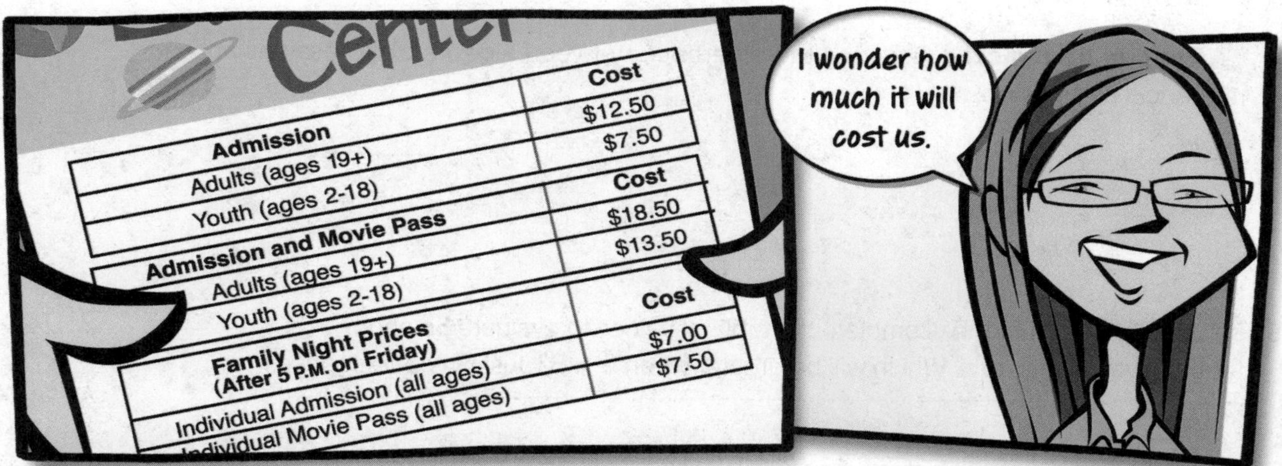

a. What is the total cost for one individual admission and one individual movie pass on Family Night? ______________________

b. The expression $14.50x$ can be used to find the total cost for x tickets on Family Night for admission and the movie. What is the cost for 3 tickets? ______________________

11. **Financial Literacy** Julian earns $13.50 per hour. His company deducts 23% of his pay each week for taxes. Julian uses the expression $0.77(13.50h)$ to compute his earnings after taxes for the hours h he works. What will be his earnings after taxes, if he works 40 hours? ______

Evaluate each expression if $x = 3$, $y = 12$, and $z = 8$.

12. $4z + 8 - 6$ ______

13. $7z \div 4 + 5x$ ______

14. $y^2 \div (3z)$ ______

15. **MP Be Precise** To find the area of a trapezoid, use the expression $\frac{1}{2}h(b_1 + b_2)$, where h represents the height, b_1 represents the length of the top base, and b_2 represents the length of the bottom base. What is the area of the trapezoidal table? ______

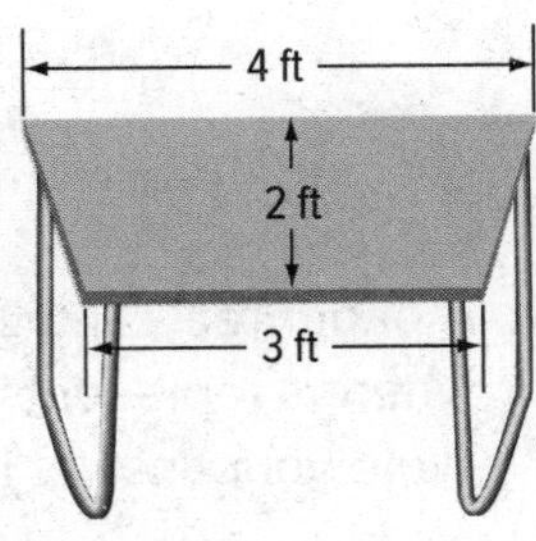

H.O.T. Problems Higher Order Thinking

16. **MP Persevere with Problems** Isandro and Yvette each have a calculator. Yvette starts at 100 and subtracts 7 each time. Isandro starts at zero and adds 3 each time. If they press the keys at the same time, will their displays ever show the same number? If so, what is the number? ______

17. **MP Reason Abstractly** Describe the difference between algebraic expressions and numerical expressions.

______

______

______

18. **MP Justify Conclusions** Complete the table of values to evaluate $5n$ and 5^n for the given values of n. Which will be greater when $n > 5$? Justify your response.

n	1	2	3	4
$5n$				
5^n				

______

______

______

Name _______________ My Homework _______________

Extra Practice

Evaluate each expression if $m = 2$, $n = 16$, and $g = \frac{1}{5}$.

19. $n + 8$ 24

$n + 8 = 16 + 8$

Homework Help $= 24$

20. $12 \div m$ ______

21. $n - m$ ______

22. $2n - 6$ ______

23. $15g$ ______

24. $45g$ ______

25. $7m + 8$ ______

26. $50g^2$ ______

27. **Financial Literacy** Colton earns \$7 per hour plus \$1.50 for each pizza delivery. The expression $7h + 1.50d$ can be used to find the total earnings after h hours and d deliveries have been made. How much money will Colton earn after working 15 hours and making 8 deliveries?

______

28. MP **Be Precise** As a member of a music club, you can order CDs for \$14.99 each. The music club also charges \$4.99 for each shipment. The expression $14.99n + 4.99$ represents the cost of n CDs. Find the total cost for ordering 3 CDs.

______

Evaluate each expression if $a = \frac{1}{2}$, $b = 15$, and $c = 9$.

29. $c^2 + a$ ______

30. $2ac$ ______

31. $b^2 - 5c$ ______

32. What is the value of $st \div (6r)$ if $r = 5$, $s = 32$, and $t = 45$?

______

Power Up! Common Core Test Practice

33. The height of the triangle shown can be found using the expression $48 \div b$, where b is the base of the triangle.

What is the height of the triangle? ☐

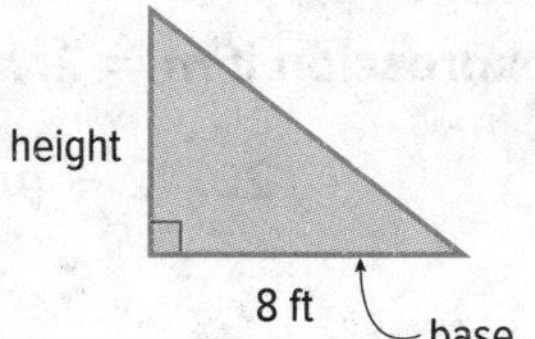

34. The table shows the total medal counts for the top 6 medal winning countries at the 2012 Summer Olympic Games. The top 6 countries earned a total of 421 medals at the Games. Based on the information in the table, write an expression that represents the total amount of medals earned by these countries.

Total Medal Count	
Country	**Number of Medals**
United States	104
China	88
Russia	82
Great Britain	x
Germany	44
Japan	38

☐

How could you use your expression and the information given to determine the number of medals Great Britain earned? Explain your reasoning.

☐

Common Core Spiral Review

Write the symbol <, >, or = for each description. 4.NF.2, 4.NF.7

35. equal to ______

36. greater than ______

37. less than ______

38. Write a number sentence to show that *two plus four equals six.* 5.OA.2

______

39. Write a number sentence to show *the sum of fourteen and eight is twenty-two.* 5.OA.2

______

40. Gianna skied three times farther than Xavier. Xavier skied four miles. How far did Gianna ski? 4.NBT.5

______

Inquiry Lab
Write Expressions

HOW can bar diagrams help you to write expressions in which letters stand for numbers?

Content Standards
6.EE.2, 6.EE.2a, 6.EE.2b

Mathematical Practices
1, 3, 4

Kevin has 6 more baseball cards than Elian. Write an algebraic expression to represent the number of baseball cards Kevin has.

What do you know? ______________________________

____

What do you need to know? ______________________________

____

Hands-On Activity 1

Algebraic expressions are similar to numerical expressions.

Step 1 Elian has an unknown number of baseball cards *c*. Use a bar diagram to show Elian's cards.

Step 2 Kevin has 6 more baseball cards than Elian. Complete the bar diagram below to show how many baseball cards Kevin has.

Kevin | ***c* cards** | ☐ **cards**

So, Kevin has ☐ + ☐ baseball cards.

Recall that the terms of an expression are separated by addition or subtraction signs.

How many terms are in the expression? ☐

Does the expression represent a *sum, difference, product,* or *quotient?*

____

Hands-On Activity 2

Sam sent 10 fewer messages in July than in August. Write an algebraic expression to represent the number of text messages Sam sent in July.

Step 1 Sam sent an unknown number of messages *m* in August. Label the bar diagram to represent the messages Sam sent in August.

August | ***m* messages**

Step 2 Sam sent 10 fewer messages in July. Label the bar diagram to show the messages Sam sent in July.

July | ☐ **messages** | ☐ **fewer**

So, Sam sent ☐ − 10 messages in July.

How many terms are in the expression? ☐

Does the expression represent a *sum, difference, product,* or *quotient?*

Hands-On Activity 3

A bottlenose dolphin can swim *d* miles per hour. Humans swim one-third as fast as dolphins. Write an algebraic expression that could be used to find out how fast humans can swim.

Step 1 Dolphins can swim an unknown number of miles per hour *d*. Use a bar diagram to represent the speed a dolphin swims.

Dolphins | ***d* miles per hour**

Step 2 Humans swim one-third as fast as dolphins. Divide and shade a second bar diagram to represent the speed humans can swim.

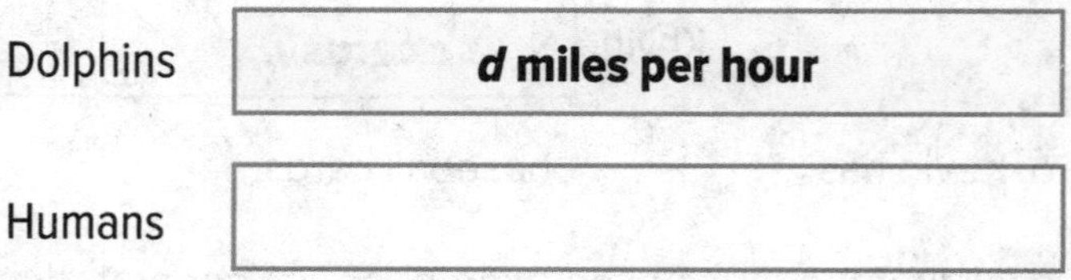

Dolphins | ***d* miles per hour**

Humans |

So, humans can swim ☐ ÷ ☐ miles per hour.

How many terms are in the expression? ☐

Does the expression represent a *sum, difference, product,* or *quotient?*

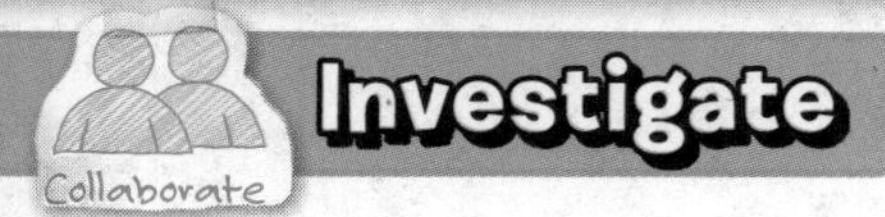

Work with a partner. Write a real-world problem and algebraic expression for each situation modeled.

Show your work.

1.

Year 1	*p* people	
Year 2	*p* people	43 people

2.

Bag of Apples	*p* pounds	
Bag of Oranges		

3.

Dasan	*b* baseball caps
Dion	*b* baseball caps
	2 caps

4.

Kent	*m* square miles
Ames	*m* square miles
	12

5.

Harry	*m* minutes			
Janice				

6.

Sixth Grade	*h* inches	
Seventh Grade	*h* inches	2 inches

Analyze and Reflect

Collaborate

Work with a partner to complete the table. The first one is done for you.

	Algebraic Expression	Word Phrase	Model
	$a + 8$	the sum of a number and 8	[*a* | 8]
7.	$r - 4$		
8.	$5w$		
9.	$\frac{c}{3}$		
10.	$7 + m$		

11. MP **Reason Inductively** Write an algebraic expression that represents *a number y divided by 10.* ______

Create

On Your Own

12. MP **Model with Mathematics** Write a real-world situation and an algebraic expression that is represented by the bar diagram.

w	*w*	*w*	3

______

______

13. Inquiry HOW can bar diagrams help you to write expressions in which letters stand for numbers?

______

______

Lesson 4

Algebra: Write Expressions

Real-World Link

Watch

Airports Missouri has 8 major commercial airports. California has 24 major commercial airports.

1. **Alabama has 4 fewer airports than Missouri.**

   a. Underline the key math word in the problem.

   b. Circle the operation you would use to determine how many airports are located in Alabama. Explain.

   $+$ $-$ $\times$ $\div$

2. **California has three times as many airports as Georgia.**

   a. Underline the key math word in the problem.

   b. Circle the operation you would use to find how many airports Georgia has. Explain.

   $+$ $-$ $\times$ $\div$

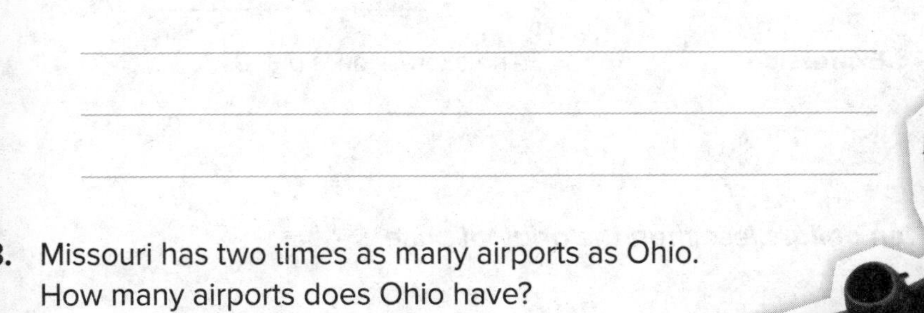

3. Missouri has two times as many airports as Ohio. How many airports does Ohio have?

   8 ◯ 2 = ______

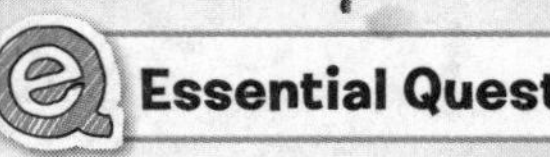

Essential Question

HOW is it helpful to write numbers in different ways?

Vocabulary

defining the variable

Common Core State Standards

Content Standards
6.EE.2, 6.EE.2a, 6.EE.2c, 6.EE.6

MP Mathematical Practices
1, 2, 3, 4, 6

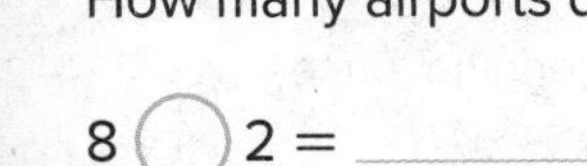

Which MP Mathematical Practices did you use?
Shade the circle(s) that applies.

① Persevere with Problems
② Reason Abstractly
③ Construct an Argument
④ Model with Mathematics
⑤ Use Math Tools
⑥ Attend to Precision
⑦ Make Use of Structure
⑧ Use Repeated Reasoning

Work Zone

Write Phrases as Algebraic Expressions

To write verbal phrases as algebraic expressions, follow the steps below. In the second step, **defining the variable**, choose a variable and decide what it represents.

Words	Describe the situation. Use only the most important words.
Variable	Choose a variable to represent the unknown quantity.
Expression	Translate your verbal phrase into an algebraic expression.

Examples

Write each phrase as an algebraic expression.

1. ***eight dollars more than Ryan earned***

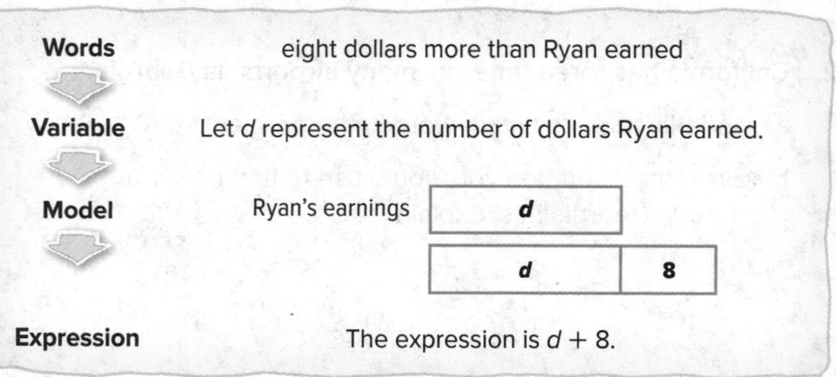

2. ***ten dollars less than the original price***

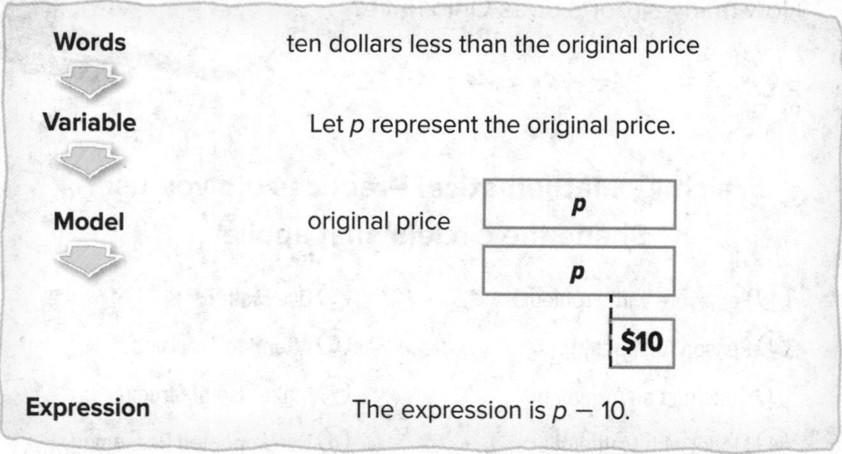

Less Than

You can write ten more than a number as either $10 + p$ or $p + 10$. But ten less than a number can only be written as $p - 10$.

3. ***four times the number of gallons***

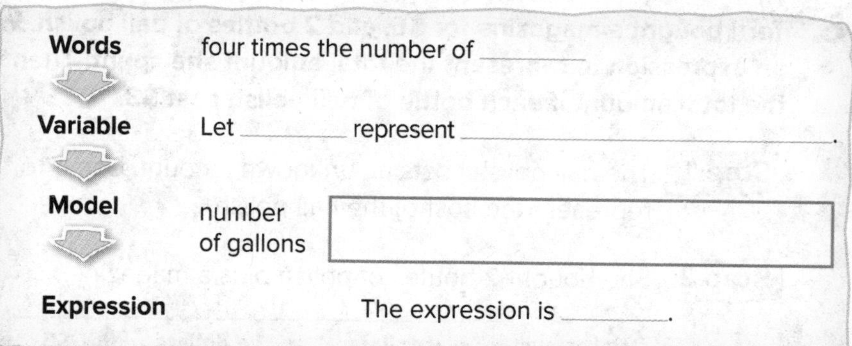

Got It? Do these problems to find out.

a. four points fewer than the Bulls scored

b. 12 times the number of feet

c. the total cost of a shirt and an \$8 pair of socks

a. ________

b. ________

c. ________

Write Two-Step Expressions

Two-step expressions contain two different operations.

Example

4. **Write the phrase *5 less than 3 times the number of points* as an algebraic expression.**

Words		5 less than 3 times the number of points
Variable		Let p represent the number of points.
Model	number of points	p p p ; 5
Expression		The expression is $3p - 5$.

Got It? Do this problem to find out.

d. Write the phrase *\$3 more than four times the cost of a pretzel* as an algebraic expression.

d. ________

Example

5. Terri bought a magazine for $5, and 2 bottles of nail polish. Write an expression to represent the total amount she spent. Then find the total amount if each bottle of nail polish cost $3.

Step 1 The nail polish costs an unknown amount. Use *d* to represent the cost of the nail polish.

Step 2 She bought 2 bottles of polish plus a magazine.

total amount	***d* dollars**	***d* dollars**	**$5**

The expression is $2 \times d + 5$ or $2d + 5$.

$2d + 5 = 2(3) + 5$ Replace *d* with 3.

$= 6 + 5$ Multiply.

$= 11$ Add.

So, the total amount is $11.

Guided Practice

Define a variable and write each phrase as an algebraic expression. (Examples 1–4)

1. four times more money than Elliot saved ______
2. half as many pages as George read ______
3. the width of a box that is 4 inches less than the length ______
4. the cost of 5 CDs and a $12 DVD ______
5. Shoko bought a box of popcorn for $3.50 and three medium drinks. Define a variable and write an expression to represent the total amount they spent. Then find the total amount if one drink costs $1.50. (Example 5)

______

6. **Building on the Essential Question** How can writing phrases as algebraic expressions help you solve problems?

______

______

______

Rate Yourself!

☐ I understand how to write algebraic expressions.

Great! You're ready to move on!

☐ I still have some questions about writing algebraic expressions.

No Problem! Go online to access a Personal Tutor.

Name ________________ My Homework ________________

Independent Practice

Go online for Step-by-Step Solutions

Define a variable and write each phrase as an algebraic expression. (Examples 1–4)

1. six feet less than the width

2. 6 hours more per week than Theodore studies

3. six years less than Tracey's age

4. 2 less than one-third of the points that the Panthers scored

5. The United States House of Representatives has 35 more members than four times the number of members in the United States Senate. Define a variable and write an expression to represent the number of members in the House of Representatives. Then find the number of members in the House of Representatives, if there are 100 members in the Senate. (Example 5)

6. MP **Multiple Representations** Dani uses the table to help her convert measurements when she is sewing.

Number of Inches	12	24	36	48
Number of Feet	1	2	3	4

   a. **Words** Describe the relationship between the number of inches and the number of feet.

   b. **Symbols** Write an expression for the number of feet in x inches.

   c. **Numbers** Find the number of feet in 252 inches.

7. MP **Be Precise** An inch is equal to about 2.54 centimeters. Write an expression which estimates the number of centimeters in x inches. Then estimate the number of centimeters in 12 inches.

8. **Financial Literacy** On a recent day, a Euro was equal to about 1.2 American dollars. Write an expression which estimates the number of dollars in x Euros. Then estimate the number of American dollars equal to 25 Euros.

9. Justin is 2 years older than one third Marcella's age. Aimee is four years younger than 2 times Justin's age. Define a variable and write an expression to represent Justin's age. Then find Justin's age and Aimee's age if Marcella is 63 years old.

H.O.T. Problems Higher Order Thinking

10. MP **Find the Error** Elisa is writing an algebraic expression for the phrase *5 less than a number.* Find her mistake and correct it.

11. MP **Persevere with Problems** Wendy earns $2 for every table she serves plus 20% of the total customer order. Define a variable and write an expression to represent the amount of money she earns for one table.

12. MP **Justify Conclusions** If n represents the amount of songs stored on an MP3 player, analyze the meaning of the expressions $n + 7$, $n - 2$, $4n$, and $n \div 2$.

13. MP **Justify Conclusions** Determine whether the statement below is *always*, *sometimes*, or *never* true. Justify your reasoning.

The expressions $x - 3$ and $y - 3$ represent the same value.

14. MP **Reason Inductively** Suppose x is an odd number. Write an expression to represent each of the following:

   a. The odd number immediately following x. __________

   b. The odd number immediately preceding x. __________

Extra Practice

Define a variable and write each phrase as an algebraic expression.

15. four times as many apples a = the number of apples; $4 \times a$ or $4a$

16. ten more shoes than Ruben

17. \$5 dollars less on dinner than James spent

18. 3 more than twice as many ringtones as Mary

19. Melinda goes bowling on Saturday afternoons. She bowls three games and pays for shoe rental. Define a variable and write an expression to represent the total cost Melinda pays. Then find the total cost if one game cost \$4.

Bowl-A-Rama	
One Game	■
Shoe Rental	\$2

20. Kiyo bought a pizza for \$12.75 and four medium drinks at Pauli's Pizza. Define a variable and write an expression to represent the total amount of money he spent. Then find the total cost if one drink costs \$3.

21. Moesha's music library has 17 more than two times the number of songs than Damian's music library. Define a variable and write an expression to represent the number of songs in Moesha's music library. Then find the number of songs in Moesha's library if Damian has 5 songs in his library.

22. MP **Reason Abstractly** Cierra has 3 more than one half as many purses as Aisha. Define a variable and write an expression to represent the number of purses in Cierra's collection. Then find the number of purses in Cierra's collection if Aisha has 12 purses.

CCSS Power Up! Common Core Test Practice

23. Marco and his friends bought game tokens for \$15 and three admission tickets to Fun Palace. Let *t* represent the cost of an admission ticket. Determine if each statement below is true or false.

a. The expression $3(t + 15)$ represents the total amount that Marco and his friends spent. ☐ True ☐ False

b. If each admission ticket costs \$2.50, then Marco and his friends spent \$22.50 in all. ☐ True ☐ False

24. The table below shows the relationship between feet and yards.

Number of Feet	3	6	9	12	15
Number of Yards	1	2	3	4	5

Fill in each box to write an algebraic expression to a represent the number of feet in any number of yards.

f	$\frac{1}{3}$
y	feet
3	yards

Words ☐ times the number of ☐.

Variable Let ☐ represent the number of ☐.

Model ☐

Expression The number of feet in ☐ yards is given by the expression ☐ ☐.

CCSS Common Core Spiral Review

Evaluate each expression. 5.NBT.7

25. $7 + 0.8 =$ ______

26. $8.3 \times 1 =$ ______

27. $3.5 + (4 + 7) =$ ______

28. Samantha ran five miles each day for seven days. Mariska ran seven miles each day for five days. Did the girls run the same distance? Explain. 4.OA.2

______

______

Problem-Solving Investigation
Act It Out

Case #1 Table Trouble

Ariana is arranging tables for her volleyball banquet. The rectangular tables can seat up to 6 people, two on each side and one on each end. She can push tables together to create a longer table that seats more people.

How many people can be seated using four tables?

Content Standards
6.EE.2

Mathematical Practices
1, 3, 4

Understand *What are the facts?*

Each rectangular table can seat up to 6 people.

Plan *What is your strategy to solve this problem?*

Use the rectangle to represent one table. Use counters to represent each seat. Draw an X to show where each counter was placed.

Solve *How can you apply the strategy?*

Act out the situation to find the number seats at four tables. Use counters to represent each seat. Draw an X to show where each counter was placed.

Four tables can seat ☐ people.

Check *Does the answer make sense?*

Use the expression $4x + 2$, where x represents the number of tables.

So, ☐ × ☐ + ☐ = ☐. ✓

Analyze the Strategy

Tutor

Reason Inductively Explain how the act it out strategy could help you check the reasonableness of answers. ______________________________

____

Case #2 Step It Up

Assume the pattern continues in the figures at the right.

Find the number of unit squares in Figure 5.

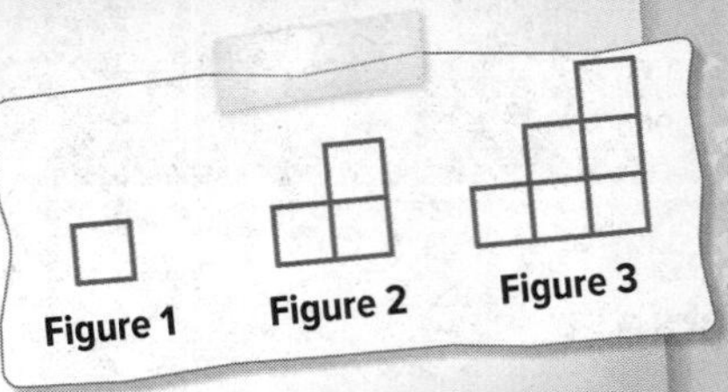

Understand

Read the problem. What are you being asked to find?

I need to find ______________________________.

Underline key words and values in the problem.
What information do you know?

Figure 1 has ☐ square. Figure 2 has ☐ squares. Figure 3 has ☐ squares.

Plan

Choose a problem-solving strategy.

I will use the ______________________ strategy.

Solve

Use your problem-solving strategy to solve the problem.

Use counters to recreate the figures.
Use 1 counter for Figure 1, 3 counters for Figure 2, and 6 counters for Figure 3.

☐ counters are added to Figure 1 to make Figure 2.

☐ counters are added to Figure 2 to make Figure 3.

Add ☐ counters to Figure 3 to make Figure 4.

Then add ☐ counters to Figure 4 to make Figure 5.

So, ______________________________.

Check

Use information from the problem to check your answer.
To check your answer, draw a model. Draw two additional squares for the first figure, three additional squares for the second figure, and so on.

Work with a small group to solve the following cases. Show your work on a separate piece of paper.

Case #3 Teams

Twenty-four students will be divided into four equal-size teams. Each student will count off, beginning with the number 1 as the first team.

If Nate is the eleventh student to count off, to which team number will he be assigned?

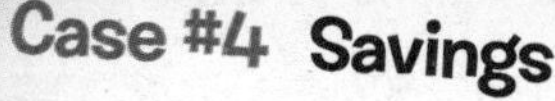

Case #4 Savings

Dakota has \$5.38 in her savings account. Each week she adds \$2.93.

How much money does Dakota have after 5 weeks? after n weeks?

Case #5 Vacations

The Florida tourism board surveyed people on their favorite vacation cities. Half of the people said Orlando, $\frac{1}{4}$ said Miami, $\frac{1}{8}$ said Kissimmee, $\frac{1}{16}$ responded Fort Lauderdale, $\frac{1}{32}$ said Key West, and the rest said Tampa.

If 22 people said Tampa, how many people responded Orlando?

Case #6 School

The birth months of the students in Miss Desimio's geography class are shown.

Birth Months		
June	July	April
March	July	June
October	May	August
June	April	October
May	October	April
September	December	January

What is the difference in the percentage of students born in June than in August? Round to the nearest whole percent.

Mid-Chapter Check

Vocabulary Check

1. **MP Be Precise** Define *powers.* Provide an example of power with an exponent of 2. (Lesson 1)

____

____

2. Fill in the blank in the sentence below with the correct term. (Lesson 2)

The ______________________________ tells you which operation to perform first so that everyone finds the same value for an expression.

Skills Check and Problem Solving

Write each power as a product of the same factor. Then find the value. (Lesson 1)

3. $7^2 =$ ______________

4. $5^5 =$ ______________

Evaluate each expression if $x = 6$. (Lesson 3)

5. $x + 11$ ________

6. $4(x - 5)$ ________

7. $2x \div 6$ ________

8. **MP Reason Abstractly** Tia is 8 years younger than her sister Annette. Annette is *y* years old. Write an algebraic expression that describes Tia's age. (Lesson 4)

____

9. **MP Reason Abstractly** The prices per pound of different types of nuts are shown. Write an expression that can be used to find the total cost of 2 pounds of peanuts, 3 pounds of cashews, and 1 pound of almonds, all for 20% off. (Lesson 2)

____

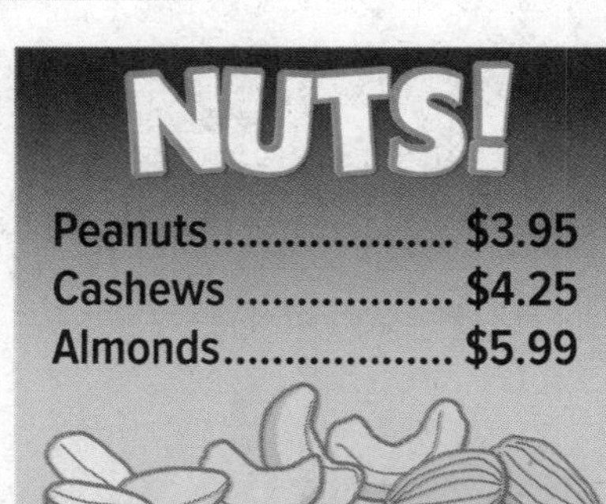

Lesson 5

Algebra: Properties

Real-World Link

Baking Angelica and Nari are baking cookies for a bake sale fundraiser. Angelica baked 6 sheets with 10 cookies each and Nari baked 10 sheets with 6 cookies each.

1. How many total cookies can Angelica bake?

   6 ◯ 10 = ☐

2. How many total cookies did Nari bake?

   10 ◯ 6 = ☐

3. What do you notice about your answers for Exercises 1 and 2?

   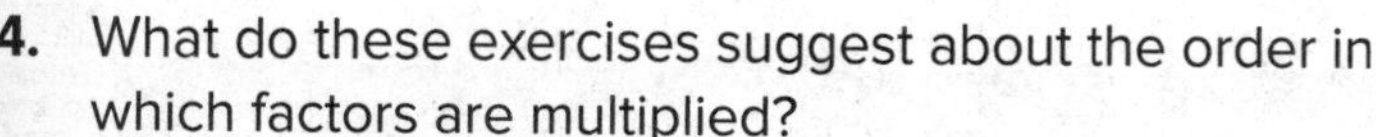

4. What do these exercises suggest about the order in which factors are multiplied?

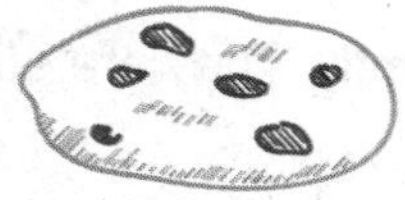

Essential Question

HOW is it helpful to write numbers in different ways?

Vocabulary

properties
Commutative Properties
Associative Properties
Identity Properties
equivalent expressions

Common Core State Standards

Content Standards
6.EE.3

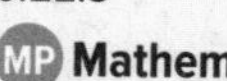

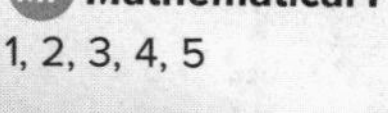

Mathematical Practices
1, 2, 3, 4, 5

Which MP **Mathematical Practices** did you use? Shade the circle(s) that applies.

① Persevere with Problems
② Reason Abstractly
③ Construct an Argument
④ Model with Mathematics
⑤ Use Math Tools
⑥ Attend to Precision
⑦ Make Use of Structure
⑧ Use Repeated Reasoning

Key Concept

Use Properties to Compare Expressions

Commutative Properties The order in which two numbers are added or multiplied does not change their sum or product.

$7 + 9 = 9 + 7$ $4 \cdot 6 = 6 \cdot 4$

$a + b = b + a$ $a \cdot b = b \cdot a$

Associative Properties The way in which three numbers are grouped when they are added or multiplied does not change their sum or product.

$3 + (9 + 4) = (3 + 9) + 4$ $8 \cdot (5 \cdot 7) = (8 \cdot 5) \cdot 7$

$a + (b + c) = (a + b) + c$ $a \cdot (b \cdot c) = (a \cdot b) \cdot c$

Identity Properties The sum of an addend and 0 is the addend. The product of a factor and 1 is the factor.

$13 + 0 = 13$ $7 \cdot 1 = 7$

$a + 0 = a$ $a \cdot 1 = a$

Work Zone

Properties are statements that are true for any number. The expressions 6×10 and 10×6 are called **equivalent expressions** because they have the same value. This illustrates the Commutative Property.

Examples

Determine whether the two expressions are equivalent. If so, tell what property is applied. If not, explain why.

1. **$15 + (5 + 8)$ and $(15 + 5) + 8$**

The numbers are grouped differently. They are equivalent by the Associative Property.

Use an = sign to compare the expressions.

So, $15 + (5 + 8) = (15 + 5) + 8$.

2. **$(20 - 12) - 3$ and $20 - (12 - 3)$**

The expressions are not equivalent because the Associative Property is not true for subtraction.

Use the $\neq$ sign to show the expressions are not equivalent.

So, $(20 - 12) - 3 \neq 20 - (12 - 3)$.

Determine whether the two expressions are equivalent. If so, tell what property is applied. If not, explain why.

3. **34 + 0 and 34**

The expressions are equivalent by the Identity Property.

So, 34 + 0 = 34.

4. **20 ÷ 5 and 5 ÷ 20**

The expressions are not equivalent because the Commutative Property does not hold for division.

So, 20 ÷ 5 ≠ 5 ÷ 20.

Division

The Commutative Property does not hold for division. To prove this, simplify the expressions in Example 4.

$20 \div 5 = 4$

$5 \div 20 = \frac{1}{4}$

Since 4 is not equal to $\frac{1}{4}$, expressions are not equivalent.

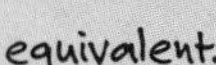

Show your work.

Got It? **Do these problems to find out.**

a. 5 × (6 × 3) and (5 × 6) × 3 **b.** 27 ÷ 3 and 3 ÷ 27

a. ________

b. ________

Use Properties to Solve Problems

Properties can also be used to write equivalent expressions and to solve problems.

Example

5. **In a recent season, the Kansas Jayhawks had 15 guards, 4 forwards, and 3 centers on their roster. Write two equivalent expressions using the Associative Property that can be used to find the total number of players on their roster.**

The Associative Property states that the grouping of numbers when they are added does not change the sum, so 15 + (4 + 3) is the same as (15 + 4) + 3.

Got It? **Do this problem to find out.**

c. **Financial Literacy** Brandi earned $7 babysitting and $12 cleaning out the garage. Write two equivalent expressions using the Commutative Property that can be used to find the total amount she earned.

c. ________

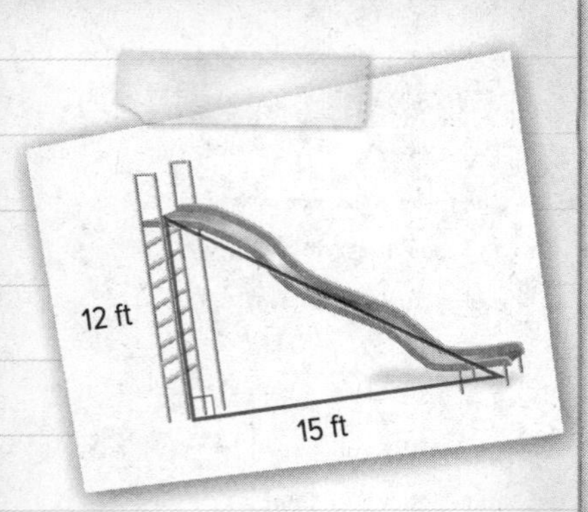

Example

6. The area of a triangle can be found using the expression $\frac{1}{2}bh$, where b is the base and h is the height. Find the area of the triangle shown at the left.

$\frac{1}{2}bh = \frac{1}{2}(15)(12)$ — Replace b with 15 and h with 12.

$= \frac{1}{2}(12)(15)$ — Commutative Property

$= 6(15)$ — Multiply. $\frac{1}{2} \times 12 = 6$

$= 90$ — Multiply.

The area of the triangle is 90 square feet.

Show your work.

Got It? **Do this problem to find out.**

d. Financial Literacy Vickie earned $6 an hour while working 11 hours over the weekend. She put $\frac{1}{3}$ of what she earned in a savings account. Find how much she put into the account.

d. ______

Guided Practice

Determine whether the two expressions are equivalent. If so, tell what property is applied. If not, explain why. (Examples 1–4)

1. (35 + 17) + 43 and 35 + (17 + 43) ______

2. (25 − 9) − 5 and 25 − (9 − 5) ______

3. 59 × 1 and 59 ______

4. At a gymnastics meet, a gymnast scored an 8.95 on the vault and a 9.2 on the uneven bars. Write two equivalent expressions that could be used to find her total score. (Example 5)

______

5. Nadia bought suntan lotion for $12, sunglasses for $15, and a towel for $18. Use the Associative Property to mentally find the total of her purchases. (Example 6) ______

______

6. **Building on the Essential Question** How can using properties help you to simplify expressions?

______

______

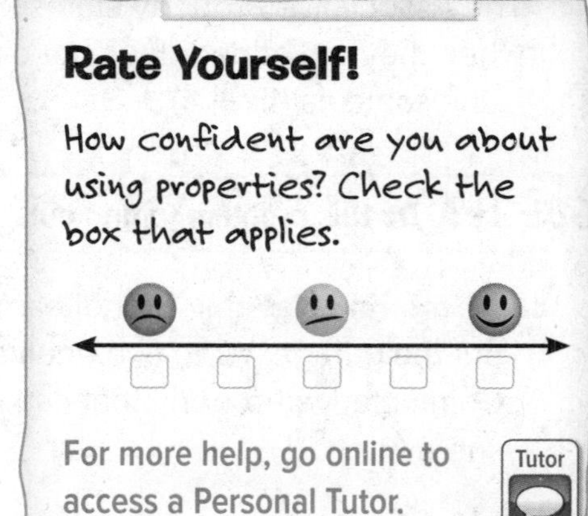

Independent Practice

Go online for Step-by-Step Solutions

Determine whether the two expressions are equivalent. If so, tell what property is applied. If not, explain why. (Examples 1–4)

1. $(8 + 27) + 52$ and $8 + (27 + 52)$ ______

2. $(3 \cdot 6) \cdot 9$ and $3 \cdot (6 \cdot 9)$ ______

3. $72 - (63 - 8)$ and $(72 - 63) - 8$ ______

4. $36 \div (12 \div 3)$ and $(36 \div 12) \div 3$ ______

5. $0 + 32$ and 0 ______

6. **STEM** Find the perimeter of the triangle shown. (Example 6)

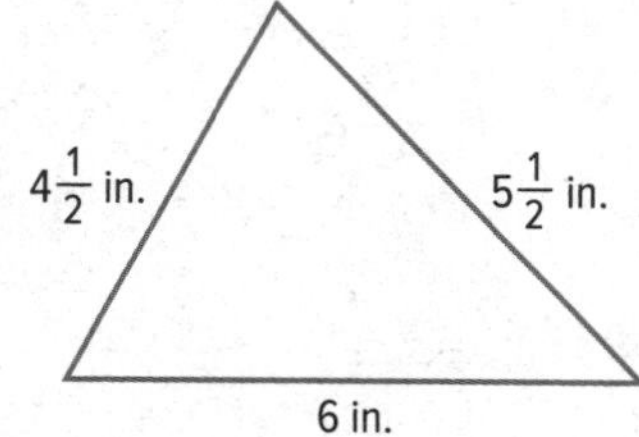

______

7. Each day, about 75,000 people visit Paris, France. Use the Commutative Property to write two equivalent expressions that could be used to find the number of people that visit over a 5-day period. (Example 5)

______

Use one or more properties to rewrite each expression as an expression that does not use parentheses.

8. $(y + 1) + 4 =$ ______

9. $(6 \cdot r) \cdot 7 =$ ______

Find the value of *x* that makes a true statement.

10. $24 + x = 24$ ______

11. $17 + x = 3 + 17$ ______

12. **Reason Abstractly** The graphic shows the driving distance between certain cities in Florida.

a. Write a number sentence that compares the mileage from Miami to Jacksonville to Tampa, and the mileage from Tampa to Jacksonville to Miami.

b. Refer to part a. Name the property that is illustrated by this sentence.

H.O.T. Problems Higher Order Thinking

13. **Reason Abstractly** Write two equivalent expressions that illustrate the Associative Property of Addition.

14. **Construct an Argument** Determine whether $(18 + 35) \times 4 = 18 + 35 \times 4$ is *true* or *false*. Explain.

15. **Persevere with Problems** A *counterexample* is an example showing that a statement is not true. Provide a counterexample to the following statement.

Division of whole numbers is commutative.

16. **Justify Conclusions** Do $(4 + 9) + 5 = (9 + 4) + 5$ and $(4 + 9) + 5 = 4 + (9 + 5)$ illustrate the same property? Justify your response.

17. **Reason Inductively** How can the Associative Property be used to mentally find $48 + 82$?

Extra Practice

Determine whether the two expressions are equivalent. If so, tell what property is applied. If not, explain why.

18. 64 + 0 and 64 yes; Identity Property

19. 23 • 1 and 23 ______

20. 8 ÷ 2 and 2 ÷ 8 ______

21. 46 + 15 and 15 + 46 ______

22. 13 • 1 and 1 ______

23. MP **Use Math Tools** Anita's mother hosted a party. The table shows the costs. Use the Associative Property to write two equivalent expressions that could be used to find the total amount spent.

______

______

Party Costs

Item	Cost ($)
Cake	12
Hot dogs and hamburgers	24
Drinks	6

24. Ella sold 37 necklaces for $20 each at the craft fair. She is going to donate half of the money she earned to charity. Use the Commutative Property to mentally find how much money she will donate. Explain the steps you used.

______

______

Use one or more properties to rewrite each expression as an expression that does not use parentheses.

25. $2 + (x + 4) =$ ______

26. $4 + (b + 0) =$ ______

27. $1 \cdot (n \cdot 8) =$ ______

28. $20 \cdot (6 \cdot y) =$ ______

29. $(6 + m) + 9$ ______

30. $(w \cdot 12) \cdot 3$ ______

CCSS Power Up! Common Core Test Practice

31. The table shows the number of desks in each classroom at three different schools. Which expressions represent the total number of desks? Select all that apply.

School	Number of Classrooms	Number of Desks per Classroom
Medina	12	25
Monroe	12	25
Yorktown	15	20

- ☐ $2 \times (12 \times 25) + 15 \times 20$
- ☐ $2 \times (12 + 12 + 15)$
- ☐ $(2 \times 12) \times 25 + 15 \times 20$
- ☐ $15 \times 20 + (2 \times 12) \times 25$

32. Determine if the two expressions in each pair are equivalent. If they are equivalent, select the property that is illustrated.

- Associative Property
- Commutative Property
- Identity Property

	Equivalent?	Property
$5 + 0$ and 5		
12×2 and 2×12		
$16 - 3$ and $3 - 16$		
$3 + (1 + 9)$ and $(3 + 1) + 9$		

CCSS Common Core Spiral Review

Write each number in expanded form. 4.NBT.2

33. 15 = ________

34. 37 = ________

35. 209 = ________

36. Lakisha had \$10 bills and \$1 bills in her wallet. She used seven bills to buy a pair of shoes for \$43. How many of each type of bill did she spend? 4.OA.3

____

____

37. Margo has 3 dimes. Justin has 5 dimes. They put their money into a donation box for a local pet shelter. What is the value of the money they added to the donation box? Explain. 4.OA.3

____

____

Inquiry Lab

The Distributive Property

HOW can you use models to evaluate and compare expressions?

Content Standards
6.EE.3

MP Mathematical Practices
1, 3, 5

Three friends are going to a concert at the fair. They will each pay for admission to the fair, which is $6.00, and admission to the concert, which is $22.00. What is the total that the three friends will spend?

Hands-On Activity 1

Watch

Write an expression to represent the amount spent in dollars.

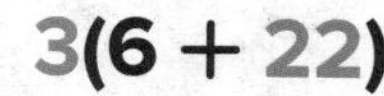

Step 2 Use area models to evaluate the expression.

Method 1 Add the lengths. Then multiply.

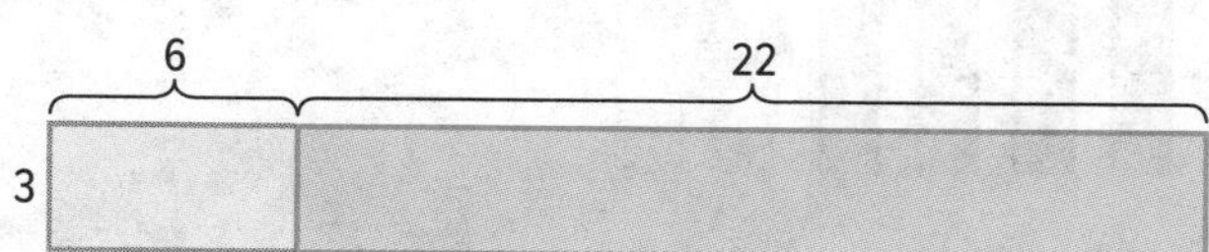

$3(6 + 22) = 3(28)$

$= \square$

Method 2 Find each area. Then add.

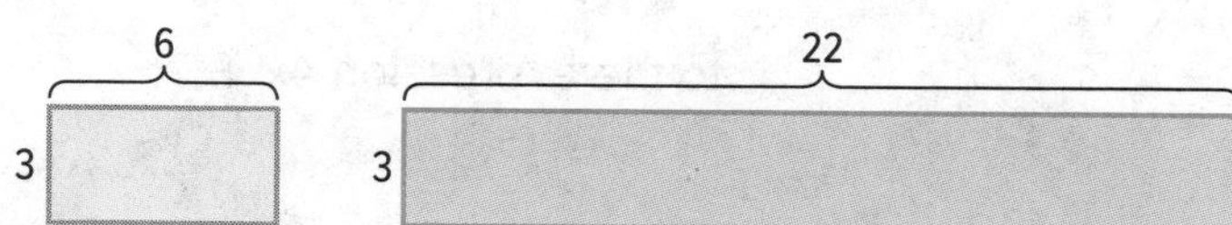

$3 \cdot 6 + 3 \cdot 22 = 18 + 66$

$= \square$

Since both expressions are equal to $\square$, they are equivalent.

So, $3(6 + 22) = 3 \cdot \square + 3 \cdot \square$.

Hands-On Activity 2

You can also use algebra tiles to model expressions with variables. Refer to the set of algebra tiles below.

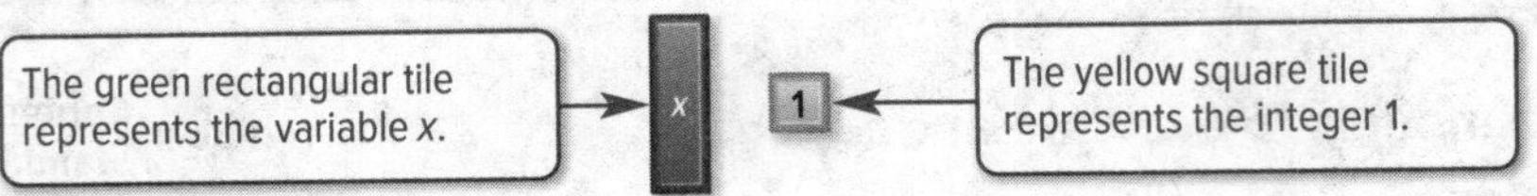

Just like 2(3) means 2 groups of 3, $2(x + 1)$ means 2 groups of $x + 1$.

Use algebra tiles to tell whether the expressions $2(2x + 1)$ and $4x + 2$ are equivalent.

Step 1 Model the expression $2(2x + 1)$.

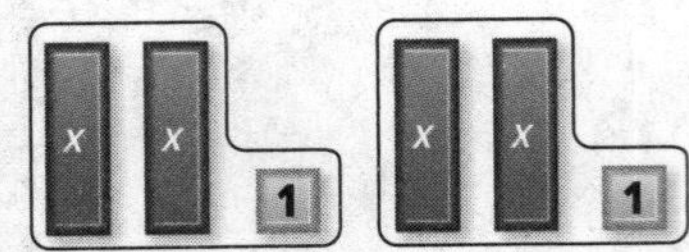

There are ☐ groups with $2x + 1$ in each group.

Step 2 Group like tiles together.

The model shows ☐ x-tiles and ☐ integer tiles.

Both models have the same number of x-tiles and the same number of integer tiles.

So, the expression $2(2x + 1)$ is ________ to the expression $4x + 2$.

Investigate

Work with a partner. Draw area models to show that the pairs of expressions are equivalent.

1. $2(4 + 6)$ and $(2 \cdot 4) + (2 \cdot 6)$

$2(4 + 6) = 2($ _______ $)$ $\quad$ $(2 \cdot 4) + (2 \cdot 6) =$ _______ $+$ _______

$=$ _______ $\quad$ $=$ _______

2. $4(3 + 2)$ and $(4 \cdot 3) + (4 \cdot 2)$

$4(3 + 2) = 4($ _______ $)$ $\quad$ $(4 \cdot 3) + (4 \cdot 2) =$ _______ $+$ _______

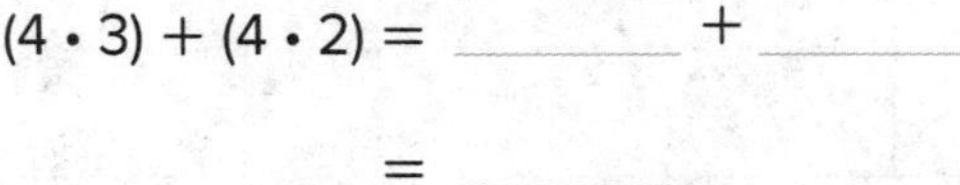

$=$ _______ $\quad$ $=$ _______

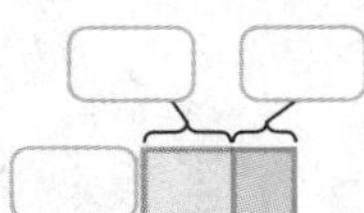

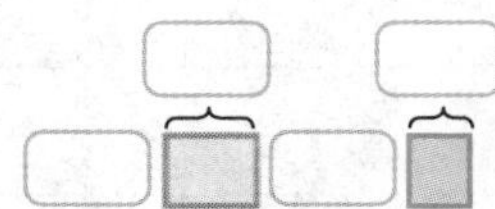

3. $6(20 + 3)$ and $(6 \cdot 20) + (6 \cdot 3)$

$6(20 + 3) = 6($ _______ $)$ $\quad$ $(6 \cdot 20) + (6 \cdot 3) =$ _______ $+$ _______

$=$ _______ $\quad$ $=$ _______

Use algebra tiles to tell whether the pairs of expressions are equivalent.

4. $3(x + 1)$ and $3x + 3$ _______

$3(x + 1)$: ____ x-tiles, ____ integer tiles $\quad$ $3x + 3$: ____ x-tiles, ____ integer tiles

5. $2(3x + 2)$ and $6x + 4$ _______

$2(3x + 2)$: ____ x-tiles, ____ integer tiles $\quad$ $6x + 4$: ____ x-tiles, ____ integer tiles

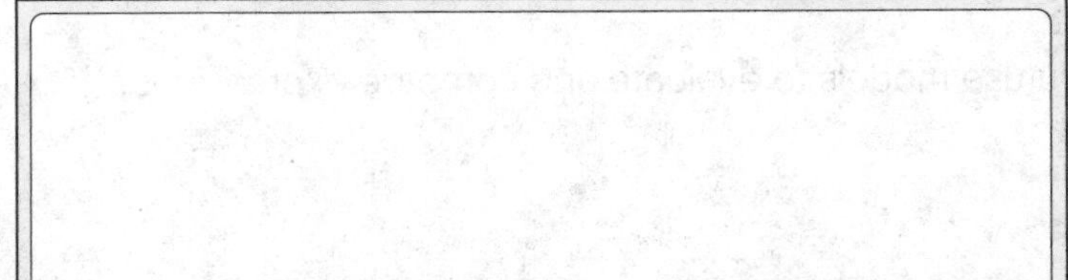

Analyze and Reflect

Collaborate

Work with a partner to complete the table. Use a model if needed. The first one is done for you.

	Expression	Rewrite the expression.	Evaluate.
	2(4 + 1)	2(4) + 2(1)	10
6.	7(8 + 4)		
7.	9(3 + 9)		
8.	5(3 + 5)		
9.	2(24 + 6)		
10.	3(16 + 5)		
11.	4(8 + 7)		
12.	6(22 + 9)		

13. **MP Use Math Tools** A friend decides that $4(x + 3) = 4x + 3$. Use the algebra tiles at the right to explain to your friend that $4(x + 3) = 4x + 12$.

Create

On Your Own

14. **MP Model with Mathematics** Write a real-world word problem that could be represented by the expression 3(23). Then explain how you could solve the problem mentally.

15. **Inquiry** HOW can you use models to evaluate and compare expressions?

Lesson 6

The Distributive Property

Real-World Link

Watch

Baseball Three friends went to a baseball game. Each ticket cost \$20 and all three friends bought a baseball hat for \$15 each.

1. What does the expression $3(20 + 15)$ represent?

   3 represents: ______________________

   20 represents: ______________________

   15 represents: ______________________

2. Evaluate the expression in Exercise 1.

   $(20 + 15) =$ ☐

   ☐ $\times$ ☐ $=$ ☐

3. What does the expression $3 \times 20 + 3 \times 15$ represent?

   3×20 represents: ______________________

   3×15 represents: ______________________

4. Evaluate the expression $3 \times 20 + 3 \times 15$.

   $3 \times 20 =$ ☐

   $3 \times 15 =$ ☐

   ☐ $+$ ☐ $=$ ☐

5. What do you notice about the answers to Exercises 2 and 4?

   ____

   ____

Essential Question

HOW is it helpful to write numbers in different ways?

Vocab

Vocabulary

Distributive Property
factor the expression

Common Core State Standards

Content Standards
6.EE.3, 6.NS.4

MP **Mathematical Practices**
1, 3, 4, 5, 6, 7, 8

Which MP **Mathematical Practices** did you use? Shade the circle(s) that applies.

① Persevere with Problems
② Reason Abstractly
③ Construct an Argument
④ Model with Mathematics
⑤ Use Math Tools
⑥ Attend to Precision
⑦ Make Use of Structure
⑧ Use Repeated Reasoning

Key Concept

Distributive Property

Work Zone

Words To multiply a sum by a number, multiply each addend by the number outside the parentheses.

Example

Numbers	Algebra
$2(7 + 4) = 2 \times 7 + 2 \times 4$	$a(b + c) = ab + ac$

The expressions $3(20 + 15)$ and $3 \times 20 + 3 \times 15$ show how the **Distributive Property** combines addition and multiplication.

Example

1. Find $9 \times 4\frac{1}{3}$ mentally using the Distributive Property.

$9 \times 4\frac{1}{3} = 9\left(4 + \frac{1}{3}\right)$ Write $4\frac{1}{3}$ as $4 + \frac{1}{3}$.

$= 9(4) + 9\left(\frac{1}{3}\right)$ Distributive Property

$= 36 + 3$ Multiply.

$= 39$ Add.

a. ______

b. ______

c. ______

Got It? Do these problems to find out.

Find each product mentally. Show the steps you used.

a. $5 \times 2\frac{3}{5}$ b. $12 \times 2\frac{1}{4}$ c. 2×3.6

Example

2. Use the Distributive Property to rewrite $2(x + 3)$.

$2(x + 3) = 2(x) + 2(3)$ Distributive Property

$= 2x + 6$ Multiply.

d. ______

e. ______

f. ______

Got It? Do these problems to find out.

Use the Distributive Property to rewrite each expression.

d. $8(x + 3)$ e. $5(9 + x)$ f. $4(x + 2)$

Example

3. Fran is making a pair of earrings and a bracelet for four friends. Each pair of earrings uses 4.5 centimeters of wire and each bracelet uses 13 centimeters. Write two equivalent expressions and then find how much total wire is needed.

Using the Distributive Property, $4(4.5) + 4(13)$ and $4(4.5 + 13)$ are equivalent expressions.

$4(4.5) + 4(13) = 18 + 52 = 70$ $\quad$ $4(4.5 + 13) = 4(17.5) = 70$

So, Fran needs 70 centimeters of wire.

Got It? **Do this problem to find out.**

g. Each day, Martin lifts weights for 10 minutes and runs on the treadmill for 25 minutes. Write two equivalent expressions and then find the total minutes that Martin exercises in 7 days.

g. ____________

Factor an Expression

When numeric or algebraic expressions are written as a product of their factors, the process is called **factoring the expression**.

Prime Factorization

The prime factorization of an algebraic expression contains both the prime factors and any variable factors. For example, the prime factorization of $6x$ is $2 \cdot 3 \cdot x$.

Example

4. Factor 12 + 8.

$12 = 2 \cdot 2 \cdot 3$ Write the prime factorization of 12 and 8.

$8 = 2 \cdot 2 \cdot 2$ Circle the common factors.

The GCF of 12 and 8 is $2 \cdot 2$ or 4.

Write each term as a product of the GCF and its remaining factor. Then use the Distributive Property to *factor out* the GCF.

$12 + 8 = \mathbf{4}(3) + \mathbf{4}(2)$ Rewrite each term using the GCF.

$= \mathbf{4}(3 + 2)$ Distributive Property

So, $12 + 8 = 4(3 + 2)$.

Got It? **Do these problems to find out.**

Factor each expression.

h. $9 + 21$ **i.** $14 + 28$ **j.** $80 + 56$

h. ____________

i. ____________

j. ____________

Example

5. Factor $3x + 15$.

$3x = 3 \cdot x$ Write the prime fractorization of 15 and $3x$.

$15 = 3 \cdot 5$ Circle the common factors.

The GCF of $3x$ and 15 is 3.

$3x + 15 = 3(x) + 3(5)$ Rewrite each term using the GCF.

$= 3(x + 5)$ Distributive Property

So, $3(x + 5) = 3x + 15$.

Show your work.

k. ______

l. ______

m. ______

Got It? Do these problems to find out.

Factor each expression.

k. $16 + 4x$ **l.** $7x + 42$ **m.** $36x + 30$

Guided Practice

1. Find $9 \times 8\frac{2}{3}$ mentally. Show the steps you used. (Example 1)

Use the Distributive Property to rewrite each algebraic expression. (Example 2)

2. $3(x + 1) =$ ______ **3.** $5(x + 8) =$ ______ **4.** $4(x + 6) =$ ______

Factor each expression. (Examples 4 and 5)

5. $25 + 60 =$ ______ **6.** $4x + 40 =$ ______

7. Financial Literacy Six friends are going to the state fair. The cost of one admission is \$9.50, and the cost for one ride on the Ferris wheel is \$1.50. Write two equivalent expressions and then find the total cost. (Example 3)

8. Building on the Essential Question How can the Distributive Property help you to rewrite expressions?

Rate Yourself!

How well do you understand the Distributive Property? Circle the image that applies.

Clear

Somewhat Clear

Not So Clear

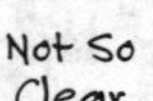

For more help, go online to access a Personal Tutor.

Name _______________ My Homework _______________

Independent Practice

Go online for Step-by-Step Solutions

Find each product mentally. Show the steps you used. (Example 1)

1. $9 \times 44 =$ ________

2. $4 \times 5\frac{1}{8} =$ ________

3. $7 \times 3.8 =$ ________

Use the Distributive Property to rewrite each algebraic expression. (Example 2)

4. $8(x + 7) =$ ________

5. $6(11 + x) =$ ________

6. $8(x + 1) =$ ________

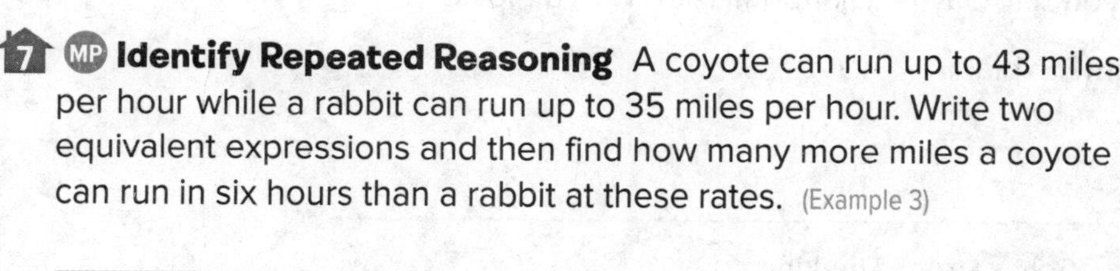

7. **MP Identify Repeated Reasoning** A coyote can run up to 43 miles per hour while a rabbit can run up to 35 miles per hour. Write two equivalent expressions and then find how many more miles a coyote can run in six hours than a rabbit at these rates. (Example 3)

____

____

Factor each expression. (Examples 4 and 5)

8. $8 + 16 =$ ________

9. $54 + 24 =$ ________

10. $63 + 81 =$ ________

11. $11x + 55 =$ ________

12. $32 + 16x =$ ________

13. $77x + 21 =$ ________

14. **MP Model with Mathematics** Refer to the graphic novel frame below for Exercises a–b.

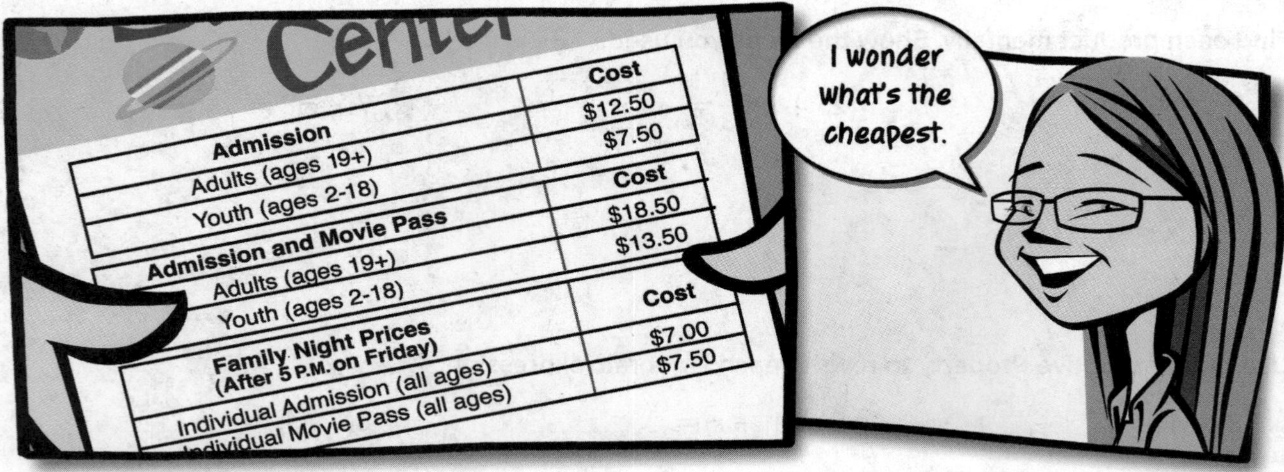

a. Write two equivalent expressions that demonstrate the Distributive Property for the cost of x tickets for admission and movie passes on Family Night. ______

b. Is it less expensive for a youth to pay regular admission with a movie pass or go on Family Night? Explain. ______

H.O.T. Problems Higher Order Thinking

15. **MP Persevere with Problems** Evaluate the expression 0.1(3.7) mentally. Justify your response using the Distributive Property. ______

16. **MP Identify Structure** Write two equivalent expressions involving decimals that illustrate the Distributive Property. ______

17. **MP Construct an Argument** A friend rewrote the expression $5(x + 2)$ as $5x + 2$. Write a few sentences to your friend explaining the error. Then, rewrite the expression $5(x + 2)$ correctly. ______

18. **MP Reason Inductively** Explain why $3(5x)$ is not equivalent to $(3 \cdot 5)(3 \cdot x)$.

Name ________________ My Homework ________________

Extra Practice

Find each product mentally. Show the steps you used.

19. $4 \times 38 =$ 152

Homework Help

$4(30) + 4(8)$
$= 120 + 32$
$= 152$

20. $11 \times 27 =$ ______

21. $3 \times 3.9 =$ ______

Use the Distributive Property to rewrite each algebraic expression.

22. $4(x + 2) =$ ______

23. $3(x + 7) =$ ______

24. $5(2x + 7) =$ ______

25. MP **Be Precise** Mrs. Singh bought 9 folders and 9 notebooks. The cost of each folder was \$2.50. Each notebook cost \$4. Write two equivalent expressions and then find the total cost.

____

26. MP **Be Precise** Five friends bought admission tickets to the museum and a box lunch. The cost of each admission ticket was \$11.75. Each box lunch cost \$5. Write two equivalent expressions and then find the total cost.

____

Factor each expression.

27. $27 + 12 =$ ______

28. $12 + 36 =$ ______

29. $16 + 20 =$ ______

30. $2x + 8 =$ ______

31. $30 + 12x =$ ______

32. $42x + 49 =$ ______

CCSS Power Up! Common Core Test Practice

33. Determine if each statement illustrates the Distributive Property. Select yes or no.

a. $7x + 1 = 7(x + 1)$ ☐ yes ☐ no

b. $3x + 6 = 3(x + 2)$ ☐ yes ☐ no

c. $5(x + 4) = 5x + 20$ ☐ yes ☐ no

d. $9(x + 4) = 9x + 4$ ☐ yes ☐ no

34. Toby and three of his friends ate lunch together at a deli. Each person ordered a sandwich and a drink.

Item	Cost ($)
Sandwich	2.75
Drink	1.25

Fill in each box to write an expression to represent the amount they spent altogether.

☐ × (☐ + ☐)

1.25	1.50	2
2.75	3	4

How much did Toby and his friends spend altogether? ☐

CCSS Common Core Spiral Review

Evaluate each expression. 5.NBT.7

35. $4 + 5.23 + 3 =$ ______

36. $4 \times 0 \times 9.17 =$ ______

37. $1.8 \times 1 \times 2 =$ ______

38. Elise and her sister Marta recorded the amount they saved each week for a month. How much did each person save? Use the information in the table to compare the total amount that Elise saved to the total amount Marta saved. 4.OA.3 ______

Week	Elise's Savings ($)	Marta's Savings ($)
1	20	15
2	15	20
3	10	10
4	20	20

39. Each bottle holds 16 fluid ounces of water. Bottles are packaged in 4 rows of 6 bottles. How many ounces of water are in each package? 4.NBT.4 ______

Inquiry Lab

Equivalent Expressions

HOW do you know that two expressions are equivalent?

Content Standards
6.EE.4

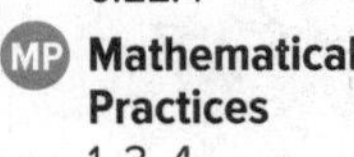

Mathematical Practices
1, 3, 4

Derrick and his friends bought tickets for the dirt bike rally. The cost of each ticket was x dollars. Derrick bought 2 tickets on Saturday and 3 tickets on Sunday. They paid $4 for parking. The expression $2x + 4 + 3x$ represents the total cost in dollars of the dirt bike rally.

Hands-On Activity

Simplify the expression $2x + 4 + 3x$ using algebra tiles.

Step 1 Choose tiles to represent each addend. Use ☐ x-tiles to model $2x$, ☐ 1 tiles to model 4, and ☐ x-tiles to model $3x$.

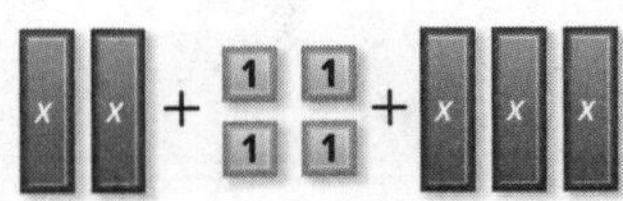

Step 2 Find the like terms. The like terms are ☐ and ☐ because they are both x-tiles. There are a total of ☐ x-tiles and four 1-tiles.

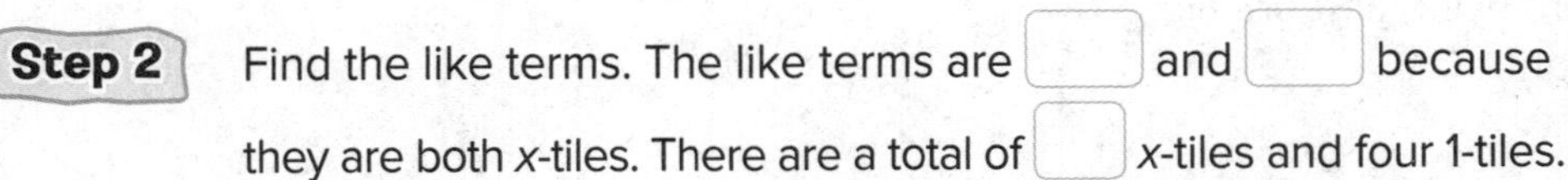

Step 3 Draw the algebra tiles in the space below, placing all like terms together.

Step 4 Rewrite the expression using addition to combine the like terms. Add $2x$ and $3x$.

So, $2x + 4 + 3x =$ ☐ + ☐.

Rearrange the algebra tiles to determine if $2x + 4 + 3x$ is equivalent to $4x + x + 4$. Are they equivalent expressions? ______

Investigate

Collaborate

Work with a partner. Simplify each expression using algebra tiles. Draw algebra tile models to represent each expression.

1. $x + 4x + x =$ ______

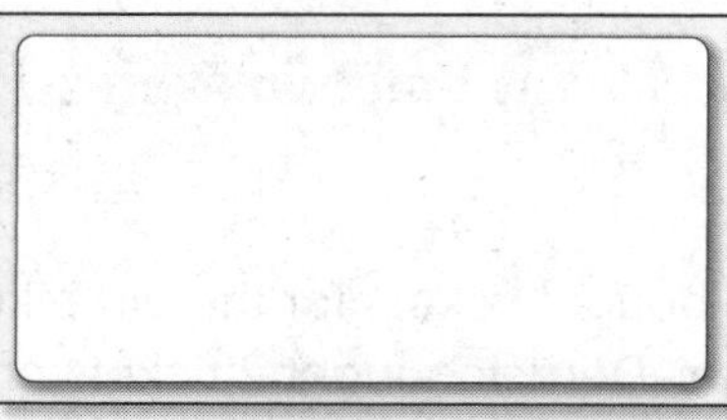

2. $4x + 7 + 2x =$ ______

3. $2(x + 2) =$ ______

4. Determine if the expressions $x + 1 + 3x$ and $4x + 1$ are equivalent using algebra tiles. Draw your tiles at the right.

______

Create

On Your Own

5. MP **Model with Mathematics** Maggie is x years old. Her brother Demarco is 4 years older than her. Anna is 3 times as old as Demarco. Write and simplify an expression that represents Anna's age. Explain.

______

______

6. Inquiry How do you know that two expressions are equivalent?

______

______

Lesson 7

Equivalent Expressions

Vocabulary Start-Up

When addition or subtraction signs separate an algebraic expression into parts, each part is called a **term**. The numerical factor of a term that contains a variable is called the **coefficient**. A term without a variable is called a **constant**. **Like terms** are terms that contain the same variables, such as x, $2x$, and $3x$.

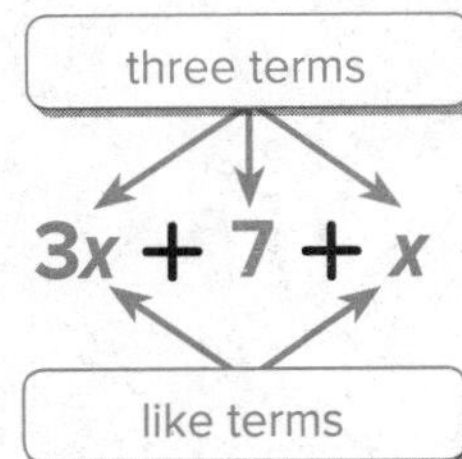

The three terms are $3x$, 7, and x.

The terms $3x$ and x are like terms because they have the same variable, x.

The constant is 7.

Label the graphic organizer below.

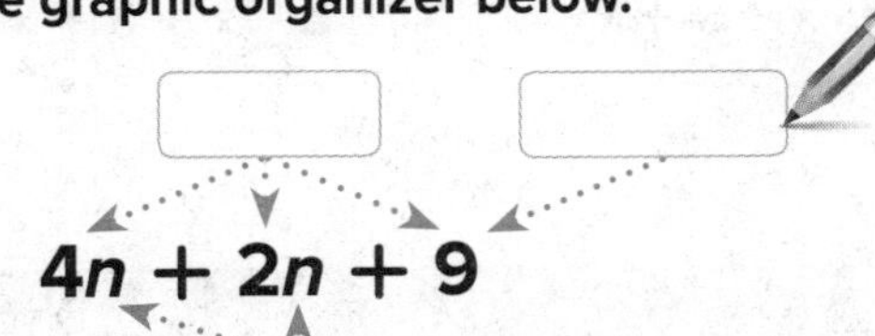

Essential Question

HOW is it helpful to write numbers in different ways?

Vocabulary

term
coefficient
constant
like terms

Common Core State Standards

Content Standards
6.EE.2, 6.EE.2b, 6.EE.3, 6.EE.4

MP Mathematical Practices
1, 3, 4, 5, 7

Real-World Link

Games Andrew's mother gave him a computer game and $10 for his birthday. His aunt gave him two computer games and $5. The expression $x + 10 + 2x + 5$, where x represents the cost of each game, can be used to represent Andrew's birthday gifts.

1. What is the coefficient of the term $2x$? ☐
2. How many terms are in the expression $x + 10 + 2x + 5$? ☐

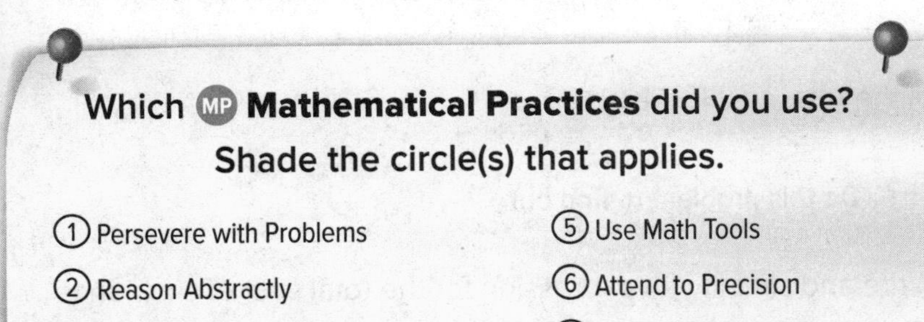

Which MP **Mathematical Practices** did you use? Shade the circle(s) that applies.

① Persevere with Problems
② Reason Abstractly
③ Construct an Argument
④ Model with Mathematics
⑤ Use Math Tools
⑥ Attend to Precision
⑦ Make Use of Structure
⑧ Use Repeated Reasoning

Work Zone

Equivalent Expressions
Two expressions are equivalent when the expressions have the same value, no matter what value is substituted for x. So, $24x$ is equivalent to $4(6x)$.

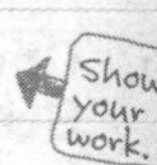

a. ________

b. ________

c. ________

d. ________

Simplify Expressions with One Variable

To simplify an algebraic expression, use properties to write an equivalent expression that has no like terms and no parentheses.

Numbers	Variables
$3 + 3 = 2(3)$ or 6	$x + x = 2x$

Example

1. Simplify the expression $4(6x)$.

$4(6x) = 4 \cdot (6 \cdot x)$ Parentheses indicate multiplication.
$= (4 \cdot 6) \cdot x$ Associative Property
$= 24x$ Multiply 4 and 6.

Got It? **Do these problems to find out.**

Simplify each expression.

a. $(3 \cdot x) \cdot 11$ **b.** $x + x + x$ **c.** $7x + 8 + x$

Example

2. Three friends will pay $\$x$ each for admission to the museum plus \$1 each to view the mummy exhibit. A fourth friend will pay admission but will not view the mummy exhibit. Write and simplify an expression that represents the total cost.

The expression $3(x + 1) + x$ represents the total cost.

cost of admission and exhibit for three friends → $3(x + 1)$
cost of admission for the fourth friend → x

$3(x + 1) + x = 3x + 3 + x$ Distributive Property
$= 3x + x + 3$ Commutative Property
$= 4x + 3$ Combine like terms.

So, the total cost is $\$4x + \$3$.

Got It? **Do this problem to find out.**

d. Write and simplify an expression for the total cost of six friends to go to the museum if only four friends view the mummy exhibit.

Simplify Expressions with Two Variables

Properties can be used to simplify or to factor expressions with two variables.

Compare the effects of operations on numbers to the effects of operations on variables.

Numbers	Variables
$3 + 3 + 4 = 2(3) + 4$	$x + x + y = 2x + y$

Examples

3. Simplify the expression $(14y + x) + 22y$.

$(14y + x) + 22y = (x + 14y) + 22y$ Commutative Property

$= x + (14y + 22y)$ Associative Property

$= x + 36y$ Combine like terms.

4. Simplify $4(2x + y)$ using the Distributive Property.

$4(2x + y) = 4(2x) + 4(y)$ Distributive Property

$= 8x + 4y$ Multiply.

5. Factor $27x + 18y$.

Step 1 Find the GCF of $27x$ and $18y$.

$27x = 3 \cdot 3 \cdot 3 \cdot x$ Write the prime factorization of $27x$ and $18y$.

$18y = 2 \cdot 3 \cdot 3 \cdot y$ Circle the common factors.

The GCF of $27x$ and $18y$ is $3 \cdot 3$ or 9.

Step 2 Write each term as a product of the GCF and its remaining factor. Then use the Distributive Property to *factor out* the GCF.

$27x + 18y = 9(3x) + 9(2y)$ Rewrite each term using the GCF.

$= 9(3x + 2y)$ Distributive Property

Got It? **Do these problems to find out.**

e. Simplify $3x + 9y + 2x$.

f. Simplify $7(3x + y)$.

g. Factor $12x + 8y$.

e. ____________

f. ____________

g. ____________

Example

6. The farmer's market sells fruit baskets. Each basket has 3 apples and 1 pear. Use a to represent the cost of each apple and p to represent the cost of each pear. Write and simplify an expression that represents the total cost of 5 baskets.

Use the expression $3a + p$ to represent the cost of each basket. Use $5(3a + p)$ to represent the cost of 5 baskets.

Use the Distributive Property to rewrite $5(3a + p)$.

$5(3a + p) = 5(3a) + 5(p)$ Distributive Property

$= 15a + 5p$ Multiply.

So, the total cost of five baskets is $15a + 5p$.

Guided Practice

Simplify each expression. (Examples 1, 3, and 4)

1. $5(6x) =$ ______

2. $2x + 5y + 7x =$ ______

3. $4(2x + 5y) =$ ______

4. Factor $35x + 28y$. (Example 5) ______

5. Mikayla bought five skirts at $\$x$ each. Three of the five skirts came with a matching top for an additional \$9 each. Write and simplify an expression that represents the total cost of her purchase. (Example 2)

______

6. The gift bag from Claire Cosmetics includes 5 bottles of nail polish and 2 tubes of lip gloss. Use p to represent the cost of each bottle of nail polish and g to represent the cost of each tube of lip gloss. Write and simplify an expression that represents the total cost of 8 gift bags. (Example 6)

______

7. **Building on the Essential Question** How can properties help to write equivalent algebraic expressions?

______

Rate Yourself!

Are you ready to move on? Shade the section that applies.

YES ? NO

For more help, go online to access a Personal Tutor.

Independent Practice

Go online for Step-by-Step Solutions 

Simplify each expression. (Examples 1, 3, and 4)

1. $x + 4x + 6x =$ ______

2. $3x + 4x + 5x =$ ______

3. $9(5x) =$ ______

4. $3x + 8y + 13x =$ ______

5. $7(3x + 5y) =$ ______

6. $3x + 6x + 2x =$ ______

Factor each expression. (Example 5)

7. $24x + 18y =$ ______

8. $16x + 40y =$ ______

9. Eight friends went to a hockey game. The price of admission per person was $\$x$. Four of the friends paid an extra \$6 each for a player guide book. Write and simplify an expression that represents the total cost. (Example 2)

______

10. Gabriella is x years old. Her sister, Felicia, is six years older than she is. Their mother is twice as old as Felicia. Their aunt, Tanya, is x years older than their mother. Write and simplify an expression that represents Tanya's age in years. (Example 2)

______

11. A DVD box set includes 3 thriller movies and 2 comedies. Use t to represent the cost of each thriller and c to represent the cost of each comedy. Write and simplify an expression that represents the total cost of 6 box sets. (Example 6)

______

12. A fall candle gift set has 4 vanilla candles and 6 pumpkin spice candles. Use v to represent the cost of each vanilla candle and p to represent the cost of each pumpkin candle. Write and simplify an expression that represents the total cost of 4 sets. (Example 6)

______

Find the value of y that makes each equation true for all values of x.

13. $3x + 6x = yx$ ______

14. $x + 5 + 11x = 12x + y$ ______

15. **Use Math Tools** Pizza Palace charges $x for a large cheese pizza and an additional fee based on the number of toppings ordered.

Pizza Palace Prices	
Pizza	**Price ($)**
large cheese	x
add 1 topping	add $0.75
add 2 toppings	add $1.50
add 3 toppings	add $2.25
add 4 toppings	add $3.00

a. Two large cheese pizzas and three large pepperoni pizzas are ordered. Write and simplify an expression that represents the total cost. ______

b. Write and simplify an expression that represents the total cost of eight large pizzas, if two are cheese and six have four toppings each.

______

c. Elsa orders three large cheese pizzas, a large pepperoni and mushroom pizza, and a large green pepper and onion pizza. Write and simplify an expression that represents the total cost.

______

H.O.T. Problems Higher Order Thinking

16. **Identify Structure** Write an expression that, when simplified, is equivalent to $15x + 7$. ______

17. **Reason Inductively** Explain why the expressions $y + y + y$ and $3y$ are equivalent.

______

Persevere with Problems For Exercises 18 and 19, simplify each expression.

18. $7x + 5(x + 3) + 4x + x + 2$ ______

19. $6 + 2(x + 8) + 3x + 11 + x$ ______

20. **Reason Abstractly** The algebraic expression shown below is missing two whole-number constants. Determine the constants so that the expression simplifies to $14x + 11$.

$$4x + 8(x + \square) + \square + 2x$$

Name ______________________ My Hom

Extra Practice

Simplify each expression.

21. $4x + 2x + 3x =$ 9x

$4x + 2x + 3x = (4x + 2x) + 3x$
$= 6x + 3x$
$= 9x$

Homework Help

22. $2x + 8x + 4x =$ ________

23. $7(3x) =$ ________

24. $8y + 4x + 6y =$ ________

25. $4(7x + 5y) =$ ________

26. $6x + 2x =$ ________

Factor each expression.

27. $10x + 15y =$ ________

28. $35x + 63y =$ ________

29. MP **Use Math Tools** Four friends went to see a movie. Each ticket cost $x. The table shows the prices of several items at the theater. They bought four large pretzels and four bottles of water. Write and simplify an expression that represents the total cost of tickets and snacks or beverages.

Snack or Beverage	Price
large popcorn	$4
large pretzel	$3
small soda	$2
bottle of water	$2

____

30. Seven friends have similar cell phone plans. The price of each plan is $x. Three of the seven friends pay an extra $4 per month for unlimited text messaging. Write and simplify an expression that represents the total cost of the seven plans.

____

31. A set of glassware includes 5 tall glasses and 3 juice glasses. Use t to represent the cost of each tall glass and j to represent the cost of each juice glass. Write and simplify an expression that represents the total of cost of 4 sets.

____

Identify the terms, like terms, coefficients, and constants in each expression.

32. $4y + 5 + 3y$

____

____

33. $2x + 3y + x + 7$

____

____

...n Core Test Practice

...ression?

...s to ship a package that weighs up to ... on the additional weight of the package.

Shipping Prices	
Weight	**Price ($)**
Up to 1 lb	x
Up to 2 lbs	add $1.50
Up to 3 lbs	add $3.00
Up to 4 lbs	add $4.50
Up to 5 lbs	add $6.00

Write and simplify an expression that represents the total cost of shipping 2 packages that each weigh 0.75 pound, 3 packages that each weigh 2.5 pounds, and 1 package that weighs 4.2 pounds.

Common Core Spiral Review

Find the missing number that makes the sentence true. **4.NF.3b**

36. $\frac{3}{8} = \frac{1}{8} + \frac{1}{8} + \frac{\square}{\square}$

37. $\frac{4}{7} = \frac{2}{7} + \frac{\square}{\square}$

38. $2\frac{5}{9} = 2 + \frac{\square}{\square}$

39. Find the missing number in the pattern below. **4.OA.5**

14, 21, ☐, 35, 42, ...

40. Soccer balls cost $18 each. Complete the table and use a pattern to find the cost of 2, 3, and 4 soccer balls. **4.OA.5**

Number of Soccer Balls	Addition Pattern	Total Cost ($)
1	18	$18
2	18 + 18	
3	18 + ☐ + ☐	
4	18 + ☐ + ☐ + ☐	

21ST CENTURY CAREER in Engineering

Water Slide Engineer

Do you love riding the twisting, turning, plunging slides at water parks? Do you have ideas that would make them more fun and exciting? If so, you should think about a career designing water slides! Water slide engineers apply engineering principles, the newest technology, and their creativity to design state-of-the-art water slides that are both innovative and safe. These engineers are responsible for designing not only the winding flumes that riders slide down, but also the pumping systems that allow the slides to have the appropriate flow of water.

College & Career READINESS

Is This the Career for You?

Are you interested in a career as a water slide engineer? Take some of the following courses in high school.

- Algebra
- Computer-Aided Drafting
- Engineering Calculus
- Engineering Technology
- Physics

Find out how math relates to a career in Engineering.

It's a Slippery Ride!

Use the information in the table to solve each problem.

1. The table shows the relationship between the number of minutes and the gallons of water pumped out on The Black Hole. Write an expression to determine the number of gallons pumped out for any number of minutes.

Number of Minutes (m)	Water Pumped Out (g)
3	3,000
6	6,000
9	9,000

2. Refer to the fact about Big Thunder. Define a variable. Then write an expression that could be used to find the number of feet that riders travel in any number of seconds.

3. Write two equivalent expressions that could be used to find the number of gallons of water pumped out of the Crush 'n' Gusher after 90 seconds. Then determine the number of gallons pumped in 90 seconds.

4. Explain how you could use the Distributive Property to find how many gallons of water are pumped out of The Black Hole in $2\frac{1}{2}$ minutes.

Water Slides	
Water Slide, Park	**Fact**
Big Thunder, Rapids Water Park	At the steepest drop, riders travel about 30 feet per second.
The Black Hole, Wet 'n Wild	Riders plummet 500 feet as water is pumped out at 1,000 gallons per minute.
Crush 'n' Gusher, Typhoon Lagoon	The water jet nozzle on each slide pumps out about 23 gallons of water per second.
Gulf Scream, Adventure Island	Riders hurl down a 210-foot slide at 25 miles per hour.

Career Project

It's time to update your career portfolio!
Find three water slides in your state. Use a spreadsheet to compare several features of the slides, such as the longest drop, total length, and gallons of water pumped. Describe how you, as a water slide engineer, would have designed the slides differently.

List several challenges associated with this career.

-
-
-
-
-

Chapter Review 

Vocabulary Check 

Complete the crossword puzzle using the vocabulary list at the beginning of the chapter.

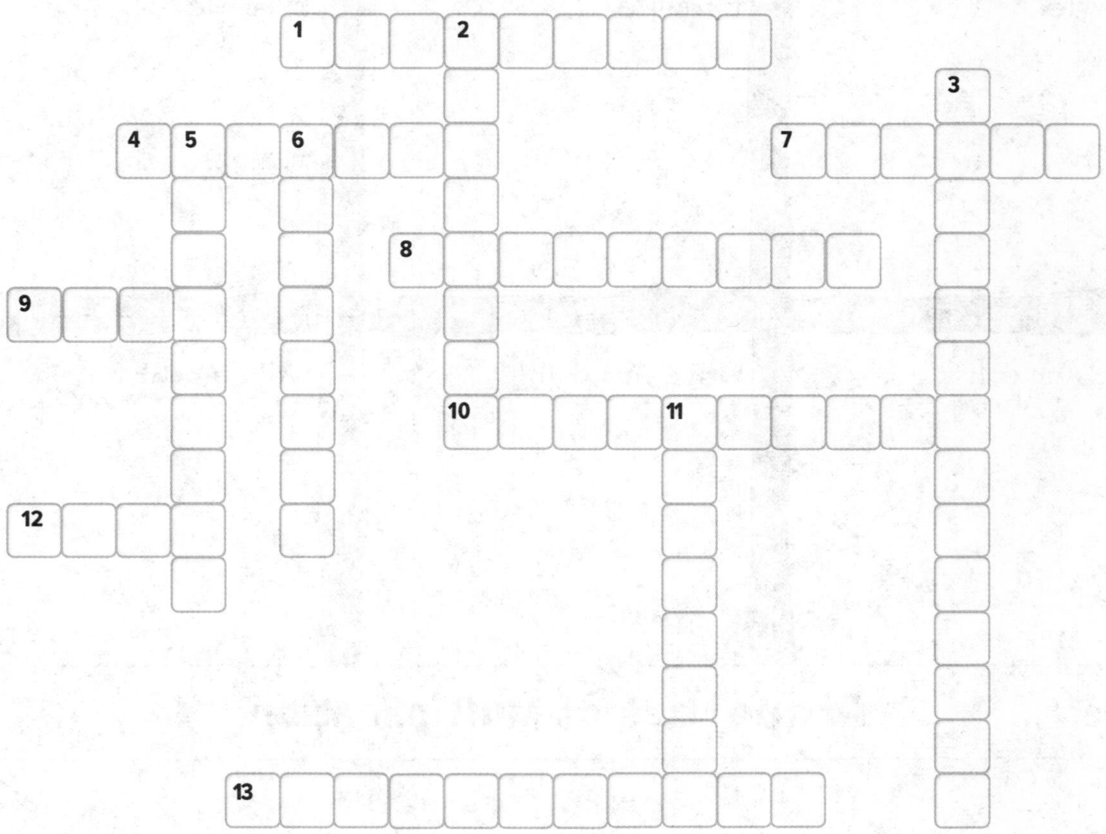

Across

1. an expression which combines variables, numbers, and at least one operation
4. a mathematical language of symbols, including variables
7. numbers expressed using exponents
8. an expression which combines numbers and operations
9. in a power, the number used as a factor
10. expressions that have the same value
12. each part of an algebraic expression separated by a plus or minus sign
13. the numerical factor of a term that contains a variable

Down

2. to find the value of an algebraic expression
3. numbers with square roots that are whole numbers
5. terms that contain the same variables to the same power
6. in a power, the number that tells how many times the base is used as a factor
11. a symbol used to represent a number

Key Concept Check

Use Your FOLDABLES®

Use your Foldable to help review the chapter.

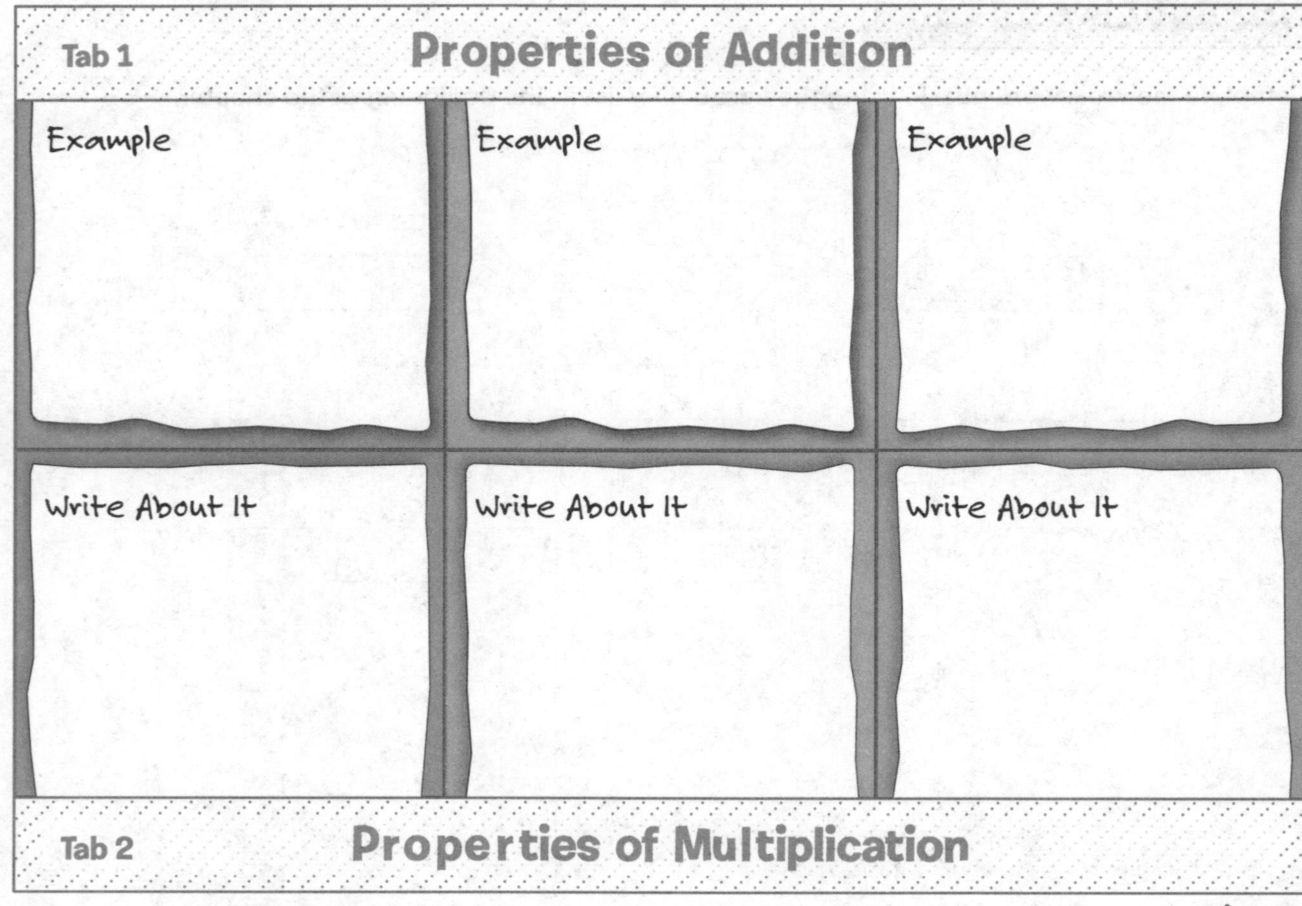

Got it?

Match each expression with the equivalent expression.

1. $2(6x + 6)$	**a.** $2(x + 3)$
2. $16x - 8$	**b.** $4x + 12$
3. $3(x - 2)$	**c.** $12x + 12$
4. $3(4x + 4)$	**d.** $3x - 6$
5. $2x + 6$	**e.** $8(2x - 1)$
6. $4(x + 3)$	**f.** $2x + 8$

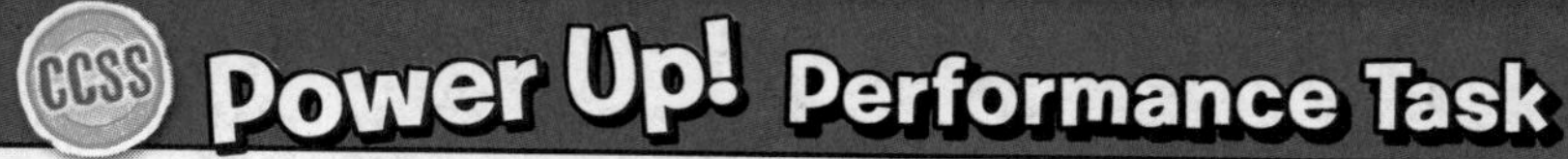

Cross Country Tryouts

The local high school is having tryouts for the cross country team. The school does not have a track, so the runners run around the school's football field. The cross country coach determines that the width of the field is seventy yards shorter than the length.

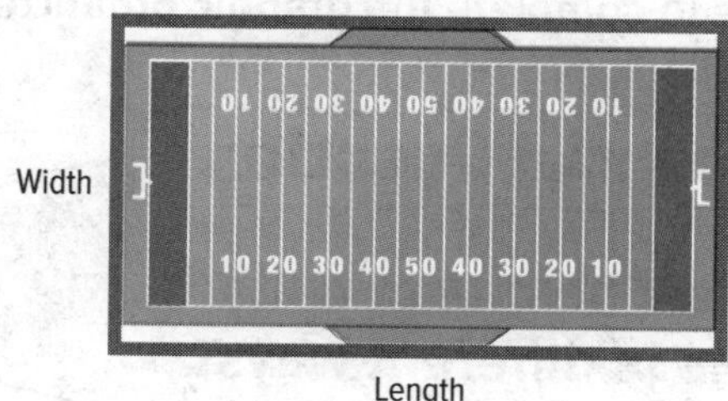

Write your answers on another piece of paper. Show all of your work to receive full credit.

Part A
Write an expression that represents the perimeter of the football field. Let x represent the length of the football field. Include parentheses in your expression. Next, write an equivalent expression that does not include parentheses. What property or properties did you use to simplify? Explain.

Part B
The cross-country coach later determines that the length of the football field is 120 yards. All students must run five laps. Using your answer from Part A, determine the actual number of yards that each athlete must run in the tryouts. In order to make the team, students must complete the laps in 6 minutes. How quickly must they run each lap?

Part C
Rita is the manager of the football team, and she has been assigned the task of painting a mascot in the middle of the field. The painting fits neatly in the shape of a square with one side of the painting equal to five yards. The area of square is given by the formula $A = s^2$, where s is the length of a side. What is the area of the painting in square yards?

Reflect

Answering the Essential Question

Use what you learned about expressions to complete the graphic organizer.

Essential Question

HOW is it helpful to write numbers in different ways?

Expression	Variable	Write a real-world example. What does the variable represent?
$7x$	x	Each ticket to the school play costs $7. The variable x represents the number of tickets purchased.
$9 + y$		
$23 - p$		
$\frac{d}{4}$		
$\frac{3}{5}c$		

Hi mama!

Answer the Essential Question. HOW is it helpful to write numbers in different ways?

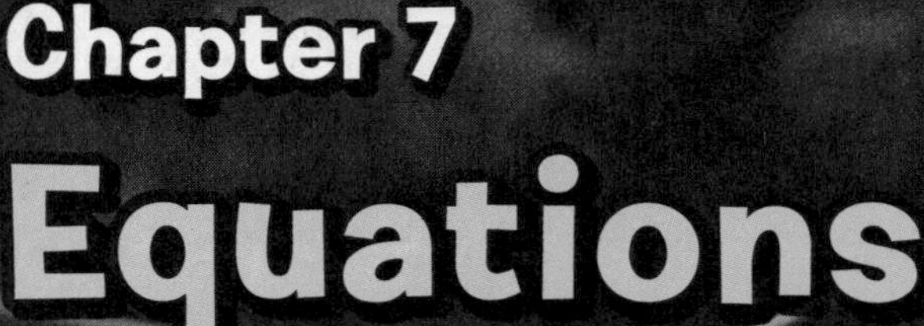

Chapter 7 Equations

Essential Question

HOW do you determine if two numbers or expressions are equal?

Common Core State Standards

Content Standards
6.EE.5, 6.EE.7, 6.RP.3

MP Mathematical Practices
1, 2, 3, 4, 5, 7

Math in the Real World

Zip lines can be used for entertainment or to access remote areas such as a rainforest canopy.

The speed differs based on the angle of the cable. On one zip line, the average speed is 44 ft/s. It takes 8 seconds to travel the length of the zip line. Fill in the table to find the distance.

Rate (ft/s)	×	Time (s)	=	Distance (ft)
44	×	1	=	
44	×	2	=	
44	×	3	=	
44	×	4	=	
44	×	5	=	
44	×	6	=	
44	×	7	=	
44	×	8	=	

FOLDABLES® Study Organizer

Cut out the Foldable on page FL5 of this book.

Place your Foldable on page 572.

Use the Foldable throughout this chapter to help you learn about equations.

What Tools Do You Need?

Vocabulary

Addition Property of Equality
Division Property of Equality
equals sign
equation
expressions
inverse operations
Multiplication Property of Equality
solution
solve
Subtraction Property of Equality

Study Skill: Studying Math

Simplify the Problem Read the problem carefully to determine what information is needed to solve the problem.

Step 1 **Read the problem.**

Kylie wants to order several pairs of running shorts from an online store. They cost $14 each, and there is a one-time shipping fee of $7. What is the total cost of buying any number of pairs of shorts?

Step 2 **Rewrite the problem to make it simpler. Keep all of the important information but use fewer words.**

Kylie wants to buy some ________ that cost ________ each plus a shipping fee of ________. What is the total cost for any number of pairs of shorts.

Step 3 **Rewrite the problem using even fewer words. Write a variable for the unknown.**

The total cost of *x* shorts is ________ + ________.

Step 4 **Translate the words into an expression.**

____

Use the method above to write an expression for each problem.

1. Akira is saving money to buy a bicycle. He has already saved $80 and plans to save an additional $5 each week. Find the total amount he has saved after any number of weeks.

____

2. A taxi company charges $1.50 per mile plus a $10 fee. What is the total cost of a taxi ride for any number of miles?

____

What Do You Already Know?

Place a checkmark below the face that expresses how much you know about each concept. Then scan the chapter to find a definition or example of it.

☹ I have no clue. 😐 I've heard of it. ☺ I know it!

Equations				
Concept	☹	😐	☺	Definition or Example
inverse operations				
solving addition equations				
solving division equations				
solving multiplication equations				
solving subtraction equations				
writing equations				

When Will You Use This?

Here are a few examples of how equations are used in the real world.

Activity 1 Describe a rewards system that you would use to earn a party for your class. What type of party would you want to have? How would you earn points? How long would you have to earn points? ______________________

____

____

Activity 2 Go online at **connectED.mcgraw-hill.com** to read the graphic novel ***Pizza Party Challenge***. How many points do you earn for reading a book? a magazine? ______________________

Try the Quick Check below.
Or, take the Online Readiness Quiz.

Quick Review

Common Core Review 5.NBT.7, 5.NF.1

Example 1

Find 1.37 − 0.75.

$$\begin{array}{r} \overset{0\,1}{\cancel{1}.37} \\ -\ 0.75 \\ \hline 0.62 \end{array}$$

Line up the decimal points.
Subtract.

Example 2

Find $\frac{3}{4} - \frac{5}{9}$.

The LCD of $\frac{3}{4}$ and $\frac{5}{9}$ is 36.

Write the problem. → Rename using the LCD, 36. → Subtract the numerators.

$$\frac{3}{4} \rightarrow \frac{3 \times 9}{4 \times 9} = \frac{27}{36} \rightarrow \frac{27}{36}$$

$$-\frac{5}{9} \rightarrow \frac{5 \times 4}{9 \times 4} = -\frac{20}{36} \rightarrow -\frac{20}{36}$$

$$\frac{7}{36}$$

Quick Check

Subtract Decimals **Find each difference.**

1. 2.34 − 1.23 = ______
2. 1.26 − 0.78 = ______
3. 3.65 − 0.96 = ______

Subtract Fractions **Find each difference. Write in simplest form.**

4. $\frac{7}{8} - \frac{1}{4} =$
5. $\frac{5}{6} - \frac{1}{2} =$
6. $\frac{3}{5} - \frac{2}{7} =$

7. Pamela ran $\frac{7}{10}$ mile on Tuesday and $\frac{3}{8}$ mile on Thursday. How much farther did she run on Tuesday?

______

Which problems did you answer correctly in the Quick Check? Shade those exercise numbers below.

1 2 3 4   7

Lesson 1
Equations

Vocabulary Start-Up

An **equation** is a mathematical sentence showing two expressions are equal. An equation contains an **equals sign**, =.

Equation	Expression
Definition	Definition
Example	Example

How are an equation and an expression similar?

How are an equation and an expression different?

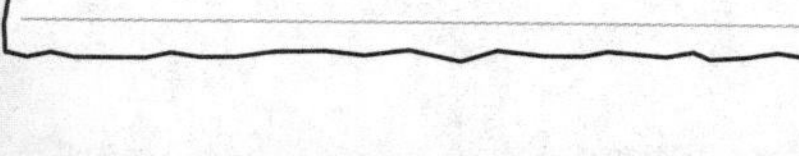

Real-World Link

Shopping Anna bought a package of 6 pair of socks. She writes the equation below to find how much she paid per pair. Circle the *solution* of the equation.

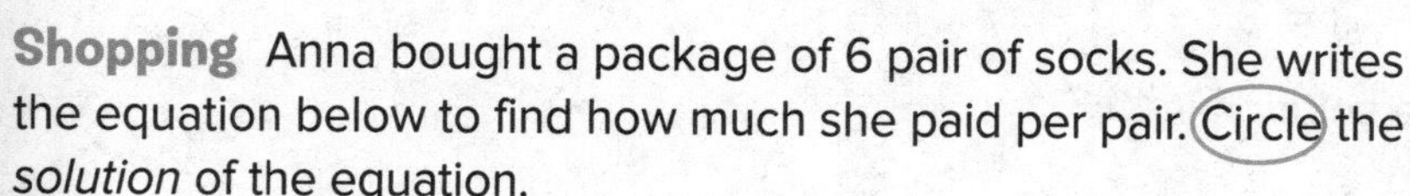

$$6x = \$9$$

\$0.50 \$1.50 \$2.00

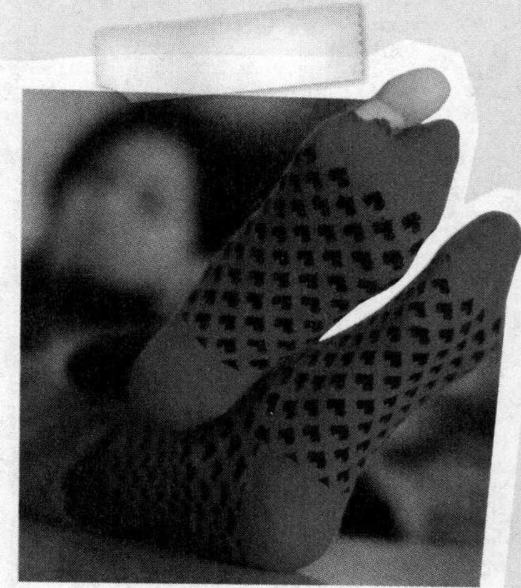

Which MP Mathematical Practices did you use? Shade the circle(s) that applies.

① Persevere with Problems
② Reason Abstractly
③ Construct an Argument
④ Model with Mathematics
⑤ Use Math Tools
⑥ Attend to Precision
⑦ Make Use of Structure
⑧ Use Repeated Reasoning

Essential Question

HOW do you determine if two numbers or expressions are equal?

Vocabulary

equation
equals sign
solve
solution

Common Core State Standards

Content Standards
6.EE.5

MP Mathematical Practices
1, 2, 3, 4, 7

Work Zone

Solve Addition and Subtraction Equations Mentally

When you replace a variable with a value that results in a true sentence, you **solve** the equation. That value for the variable is the **solution** of the equation.

$$2 + x = 9$$
$$2 + 7 = 9$$
$$9 = 9$$

The value for the variable that results in a true sentence is 7. So, 7 is the solution.

This sentence is true.

Examples

1. Is 3, 4, or 5 the solution of the equation $a + 7 = 11$?

Value of a	$a + 7 \stackrel{?}{=} 11$	Are Both Sides Equal?
3	$3 + 7 \stackrel{?}{=} 11$ $10 \neq 11$	no
4	$4 + 7 \stackrel{?}{=} 11$ $11 = 11$	yes ✓
5	$5 + 7 \stackrel{?}{=} 11$ $12 \neq 11$	no

The solution is 4.

STOP and Reflect

How can you check if your solution to an equation is correct?

2. Solve $g - 7 = 3$ mentally.

$g - 7 = 3$ Think What number minus 7 equals 3?

$10 - 7 = 3$ You know that $10 - 7 = 3$.

$3 = 3$

The solution is 10.

3. The total cost of a pair of skates and kneepads is \$63. The skates cost \$45. Use the *guess, check, and revise* strategy to solve the equation $45 + k = 63$ to find k, the cost of the kneepads.

Use the *guess, check, and revise* strategy.

Try 14.

$45 + k = 63$

$45 + 14 \stackrel{?}{=} 63$

$59 \neq 63$

Try 16.

$45 + k = 63$

$45 + 16 \stackrel{?}{=} 63$

$61 \neq 63$

Try 18.

$45 + k = 63$

$45 + 18 \stackrel{?}{=} 63$

$63 = 63$ ✓

So, the kneepads cost \$18.

Got it? **Do these problems to find out.**

a. Is 4, 5, or 6 the solution of the equation $c + 8 = 13$?

b. Solve $9 - x = 2$ mentally.

c. The difference between an ostrich's speed and a chicken's speed is 31 miles per hour. An ostrich can run at a speed of 40 miles per hour. Use mental math or the *guess, check, and revise* strategy to solve the equation $40 - c = 31$ to find c, the speed a chicken can run.

a. ________

b. ________

c. ________

Solve Multiplication and Division Equations Mentally

Multiplication and division equations are solved in a similar way to addition and subtraction equations.

Examples

Tutor

4. Is 3, 4, or 5 the solution of the equation $18 = 6z$?

Value of z	$18 \stackrel{?}{=} 6z$	Are Both Sides Equal?
3	$18 \stackrel{?}{=} 6 \cdot 3$ $18 = 18$	yes ✓
4	$18 \stackrel{?}{=} 6 \cdot 4$ $18 \neq 24$	no
5	$18 \stackrel{?}{=} 6 \cdot 5$ $18 \neq 30$	no

The solution is 3.

5. Solve $16 \div s = 8$ mentally.

$16 \div s = 8$ Think 16 divided by what number equals 8?

$16 \div 2 = 8$ You know that $16 \div 2 = 8$.

$8 = 8$

The solution is 2.

Got it? **Do these problems to find out.**

d. Is 2, 3, or 4 the solution of the equation $4n = 16$?

e. Solve $24 \div w = 8$ mentally.

d. ________

e. ________

Example

6. **Mason bought 72 sticks of gum. There are 8 sticks of gum in each package. Use the *guess, check, and revise* strategy to solve the equation $8 \cdot p = 72$ to find *p*, the number of packages Mason bought.**

Use the *guess, check, and revise* strategy.

Try 7.	Try 8.	Try 9.
$8 \cdot p = 72$	$8 \cdot p = 72$	$8 \cdot p = 72$
$8 \cdot 7 \stackrel{?}{=} 72$	$8 \cdot 8 \stackrel{?}{=} 72$	$8 \cdot 9 \stackrel{?}{=} 72$
$56 \neq 72$	$64 \neq 72$	$72 = 72$ ✓

So, Mason bought 9 packages of gum.

Guided Practice

Identify the solution of each equation from the list given. (Examples 1 and 4)

1. $9 + w = 17$; 7, 8, 9 ______

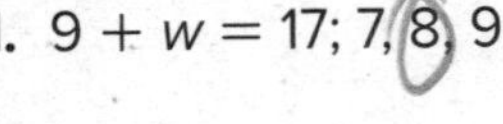

2. $8 \div c = 8$; 0, 1, 2 ______

Solve each equation mentally. (Examples 2 and 5)

3. $x - 11 = 23$

4. $4x = 32$

5. Mississippi and Georgia have a total of 21 electoral votes. Mississippi has 6 electoral votes. Use mental math or the *guess, check, and revise* strategy to solve the equation $6 + g = 21$ to find *g*, the number of electoral votes Georgia has. (Example 3)

______

6. Riley and her sister collect stickers. Riley has 220 stickers in her sticker collection. Her sister has 55 stickers in her collection. Riley has how many times as many stickers as her sister? Use mental math or the *guess, check, and revise* strategy to solve the equation $55x = 220$. (Example 6)

______

7. **Building on the Essential Question** How do you solve an equation? ______

______

Rate Yourself!

☐ I understand how to solve equations.

Great! You're ready to move on!

☐ I still have some questions about solve equations.

No Problem! Go online to access a Personal Tutor.

FOLDABLES Time to update your Foldable!

Name ______________________ My Homework ______________________

Extra Practice

Identify the solution of each equation from the list given.

21. $a + 15 = 23$; 6, 7, 8 ___8___

Try 6. $6 + 15 \neq 23$ Try 7. $7 + 15 \neq 23$ Try 8. $8 + 15 = 23$ ✓

Homework Help

22. $19 = p - 12$; 29, 30, 31 ________

23. $63 = 9k$; 6, 7, 8 ________

24. $36 \div s = 4$; 9, 10, 11 ________

Solve each equation mentally.

25. $j + 7 = 13$ =6

13−7=6=j ✓

26. $22 = 30 - m$ 8

m=30−18=
30−22=8

27. $25 - k = 20$

k=25−20=5 ✓

28. $5m = 25$

m=25÷5=5 ✓

29. $d \div 3 = 6$

d=6×3=18 ✓

30. $24 = 12k$

24÷12=2=k ✓

MP Identify Structure For Exercises 31–33, solve using mental math or the *guess, check, and revise* strategy.

31. Gabriella made 36 cookies. She gave away 28 cookies. Use the equation $28 + c = 36$ to find c, the number of cookies she kept.

____

32. The Lee family ate a total of 12 hotdogs at a cookout. Each family member ate 2 hotdogs. Use the equation $2m = 12$ to find m, the number of members in the Lee family. ______________________

33. A bottlenose dolphin is 96 inches long. There are 12 inches in 1 foot. Use the equation $12d = 96$ to find d, the length of the bottlenose dolphin in feet.

____

Power Up! Common Core Test Practice

34. Select the correct solution for each equation.

a. Mike bought a box of 12 golf balls for $18. Solve the equation $12x = \$18$ to find the price of each golf ball.
- ☐ $1.25
- ☐ $1.50
- ☐ $1.75

b. Tonya is 5 years older than Raul. Tonya is 16 years old. Solve the equation $r + 5 = 16$ to find Raul's age.
- ☐ 11 years
- ☐ 16 years
- ☐ 21 years

c. Mr. Caldwell divides 72 students into 12 equal groups. Solve the equation $\frac{72}{s} = 12$ to find the number of students in each group.
- ☐ 6 students
- ☐ 8 students
- ☐ 84 students

35. The graph shows the life expectancy of certain mammals. Write and solve an equation to find the difference *d* in the number of years a blue whale lives and the number of years a gorilla lives.

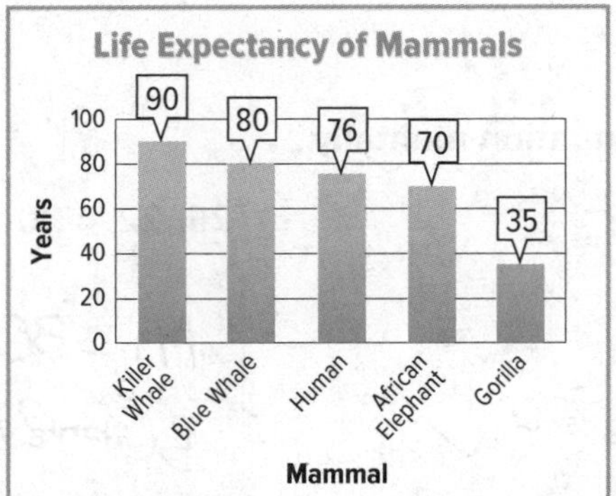

Common Core Spiral Review

Add. 4.NBT.4

36. $56 + 89 =$ ______

37. $37 + 26 =$ ______

38. $95 + 48 =$ ______

39. $29 + 86 =$ ______

40. $64 + 48 =$ ______

41. $31 + 62 =$ ______

42. The table shows the number of raffle tickets the art club sold during the beginning of the week. On Friday, the art club sold what they sold on Monday and Wednesday together. How many tickets did they sell on Friday? 4.NBT.4 ______

Day	Tickets Sold
Monday	42
Tuesday	67
Wednesday	54

Inquiry Lab

Solve and Write Addition Equations

HOW do you solve addition equations using models?

Content Standards
6.EE.5, 6.EE.7

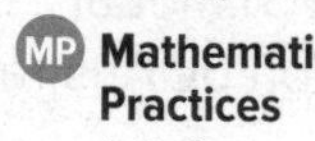

Mathematical Practices
1, 3, 4,

Bryan played two baseball games last weekend. He got 7 hits in all. He had 3 hits in the first game. How many hits did he get in the second game?

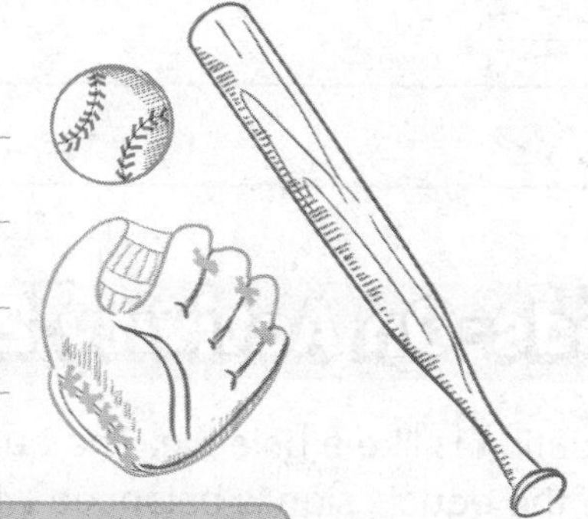

What do you know? ______________________

____

What do you need to find? ______________________

____

Hands-On Activity 1

Step 1 Define a variable. Use the variable s to represent the number of hits Bryan had in the second game.

Step 2 Use a bar diagram to help write the equation.

7	
hits in second game, *s*	**3**

The total length of the diagram represents ______________.

The 3 represents ______________.

☐ + ☐ = ☐

Step 3 Work backward. Rewrite the equation as a subtraction sentence and solve.

☐ – ☐ = ☐

So, Bryan had ☐ hits in the second game.

Investigate

Collaborate

Work with a partner. Write and solve an addition equation using a bar diagram.

1. In the 2008 Summer Olympics, the United States won 11 more medals in swimming than Australia. The United States won a total of 31 medals. Find the number of medals won by Australia.

2. A lion can run 50 miles per hour. This is 20 miles per hour faster than a house cat. Find the speed of a house cat.

Hands-On Activity 2

An equation is like a balance. The quantity on the left side of the equals sign is balanced with the quantity on the right.

To solve an addition equation using cups and counters, subtract the same number of counters from each side of the mat so that the equation remains balanced.

Solve $x + 1 = 5$ using cups and counters.

Step 1 Model the equation. Use a cup to represent x.

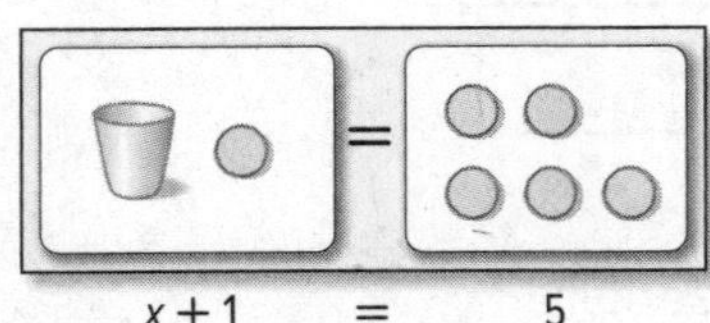

$x + 1 \quad = \quad 5$

Step 2 Use the model above. Cross out 1 counter from each side so that the cup is by itself.

Step 3 There are ☐ counters remaining on the right side, so $x =$ ☐.

So, the solution is ☐.

Check $x + 1 = 5$ Write the original equation.

☐ $+ 1 \stackrel{?}{=} 5$ Replace x with your solution.

☐ $= 5$ Is the sentence true? ______

Investigate

Work with a partner. Solve each equation using cups and counters. Draw cups and counters to show your work.

3. $1 + x = 8$

$x =$ ______

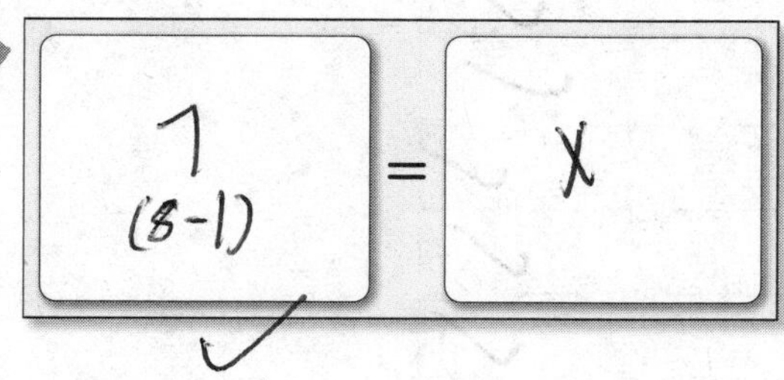

4. $x + 2 = 7$

$x =$ 5

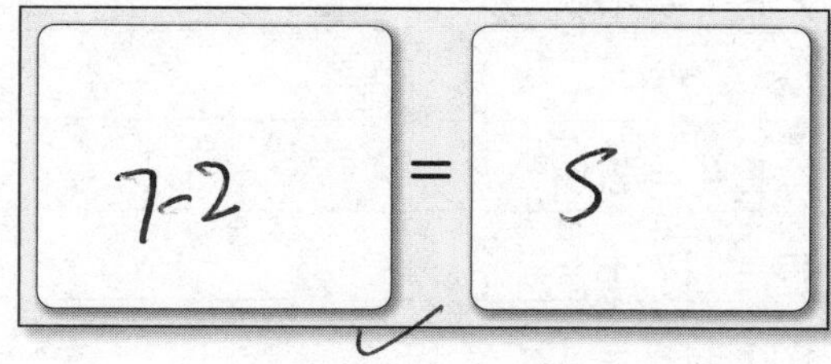

5. $3 + x = 6$

$x =$ 2

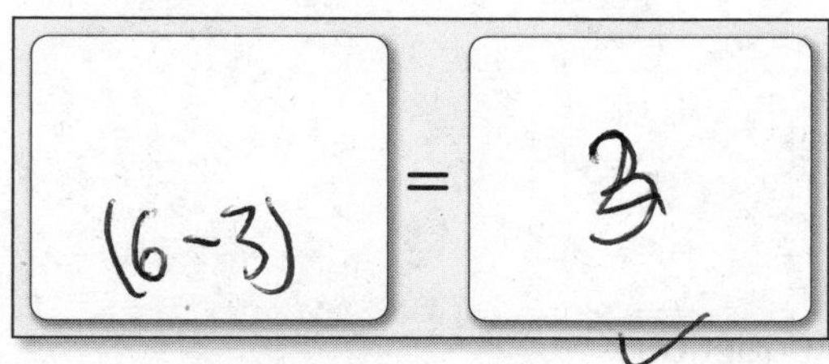

6. $x + 5 = 7$

$x =$ 2

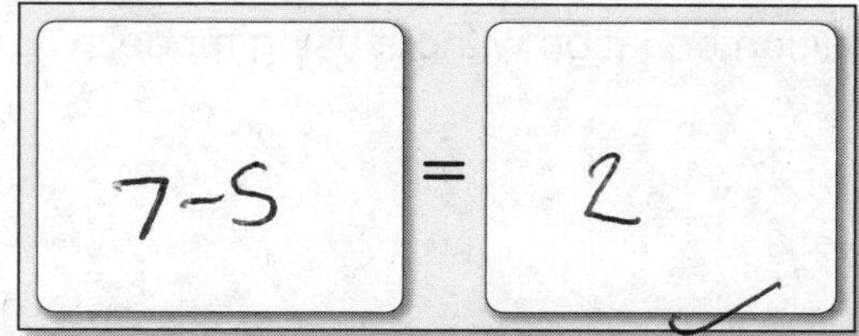

Work with a partner. Write a real-world problem that can be represented by the equation. Then solve each addition equation using the model of your choice.

7. $9 = x + 3$

x = 6

8. $4 + x = 6$

x = 2

9. Terrell bought an MP3 player. He spent the rest of his money on an Internet music subscription for $25.95. If he started with $135, how much was the MP3 player? Write and solve an equation using a bar diagram.

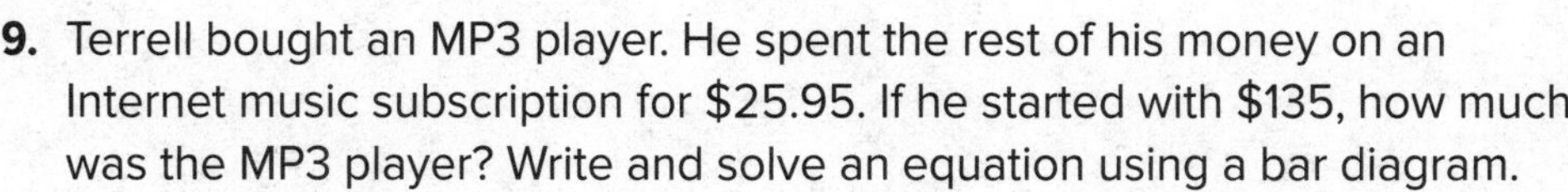

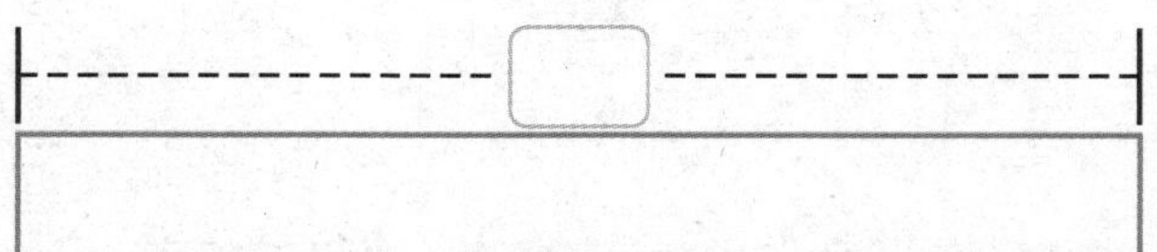

Analyze and Reflect

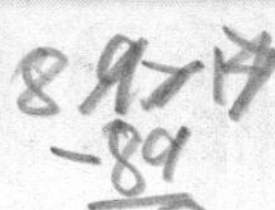

Work with a partner to complete the table. The first one is done for you.

	Addition Equation	Subtraction Sentence	Solution
	$x + 1 = 3$	$3 - 1 = x$	$x = 2$
10.	$y + 9 = 12$	$12 - 9 = x$	$x = 3$
11.	$14 = 7 + m$	$14 - 7 = x$	$x = 7$
12.	$8 + f = 20$	$20 - 8 = x$	$x = 12$
13.	$47 = 17 + v$	$47 - 17 = x$	$x = 30$
14.	$100 + c = 129$	$129 - 100 = x$	$x = 29$
15.	$h + 89.4 = 97.4$	$97.4 - 89.4 = x$	$x = 8$

16. MP **Reason Inductively** Write a rule that you can use to solve an addition equation without using models.

17. How can the number family 3, 4, 7 help you to solve the equation $3 + x = 7$?

Create

18. MP **Model with Mathematics** Write a real-world problem for the equation modeled below. Then write the equation and solve.

6 weeks	
length of vacation, *v*	**2 weeks**

19. Inquiry HOW do you solve addition equations using models?

Solve and Write Addition Equations

Real-World Link

Tools

Miniature Golf On the second hole of miniature golf, it took Anne 3 putts to sink the golf ball. Her score is now 5. She represents this situation with cups and counters.

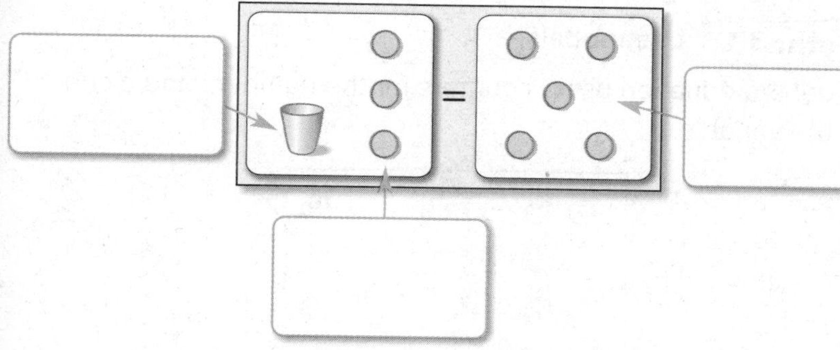

1. Fill in the boxes above using the phrases below:
   - Her score on the first hole is unknown.
   - Her score is now 5.
   - She scored a 3 on the second hole.

2. Write the addition equation shown in the figure.

3. Explain how to solve the equation.

4. What was Anne's score on the first hole? ☐

Essential Question

HOW do you determine if two numbers or expressions are equal?

Vocabulary

inverse operations
Subtraction Property of Equality

Common Core State Standards

Content Standards
6.EE.5, 6.EE.7

MP Mathematical Practices
1, 2, 3, 4, 5

Which MP Mathematical Practices did you use?
Shade the circle(s) that applies.

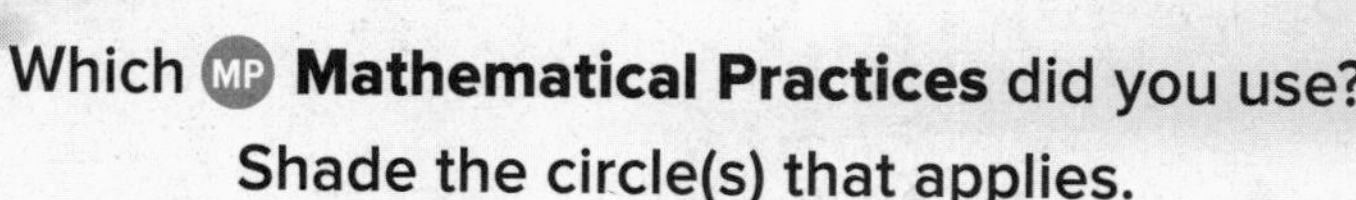

① Persevere with Problems
② Reason Abstractly
③ Construct an Argument
④ Model with Mathematics
⑤ Use Math Tools
⑥ Attend to Precision
⑦ Make Use of Structure
⑧ Use Repeated Reasoning

Work Zone

Solve an Equation By Subtracting

In Lesson 1, you mentally solved equations. Another way is to use **inverse operations**, which *undo* each other. For example, to solve an addition equation, use subtraction.

Example

1. Solve $8 = x + 3$. Check your solution.

Method 1 **Use models.**

Model the equation using counters for the numbers and a cup for the variable.

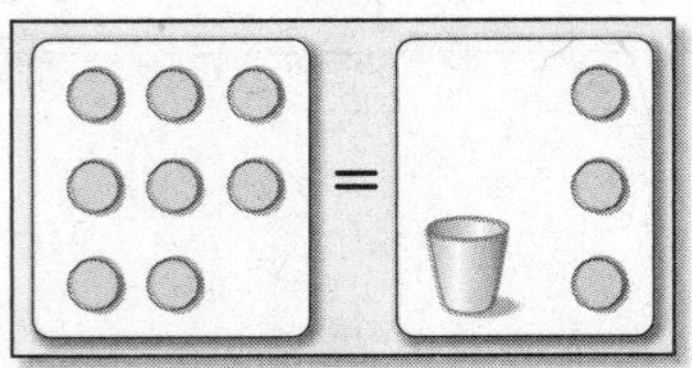

Remove 3 counters from each side.

There are 5 counters remaining.

Method 2 **Use symbols.**

$8 = x + 3$	Write the equation.
$-3 = \quad -3$	Subtract 3 from each side to "undo" the addition of 3 on the right.
$5 = x$	

Check

$8 = x + 3$	Write the equation.
$8 \stackrel{?}{=} 5 + 3$	Replace x with 5.
$8 = 8$ ✓	This sentence is true.

Using either method, the solution is 5.

Got it? **Do these problems to find out.**

Solve each equation. Check your solution.

a. $c + 2 = 5$ **b.** $6 = x + 5$ **c.** $3.5 + y = 12.75$

Show your work.

a. ____________

b. ____________

c. ____________

Subtraction Property of Equality

Key Concept

Words If you subtract the same number from each side of an equation, the two sides remain equal.

Examples

Numbers	Algebra
$\begin{array}{r} 5 = 5 \\ -3 = -3 \\ \hline 2 = 2 \end{array}$	$\begin{array}{r} x + 2 = 3 \\ -2 = -2 \\ \hline x = 1 \end{array}$

When you solve an equation by subtracting the same number from each side of the equation, you are using the **Subtraction Property of Equality**.

Example

2. Ruben and Tariq have 245.5 downloaded minutes of music. If Ruben has 132 minutes, how many belong to Tariq? Write and solve an addition equation to find how many minutes belong to Tariq.

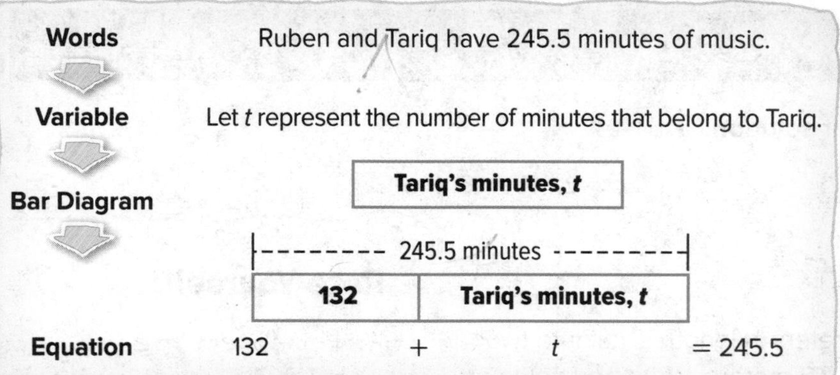

$$\begin{array}{rl} 132 + t = 245.5 & \text{Write the equation.} \\ -132 \quad = -132 & \text{Subtract 132 from each side.} \\ \hline t = 113.5 & \text{Simplify.} \end{array}$$

So, 113.5 minutes belong to Tariq.

Check $132 + 113.5 = 245.5$ ✓

Checking Solutions

You should always check your solution. You will know immediately whether your solution is correct or not.

Got it? **Do this problem to find out.**

d. Suppose Ruben had 147.5 minutes of the 245.5 that were downloaded. Write and solve an addition equation to find how many minutes belong to Tariq.

d. ________

Example

3. A male gorilla weighs 379 pounds on average. This is 181 pounds more than the weight of the average female gorilla. Write and solve an addition equation to find the weight of an average female gorilla.

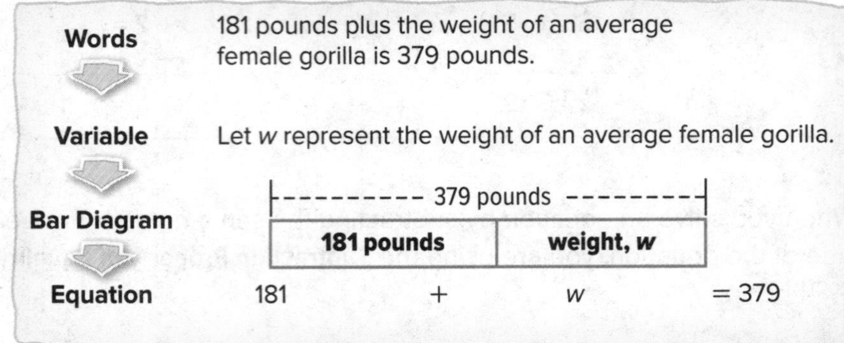

Words	181 pounds plus the weight of an average female gorilla is 379 pounds.
Variable	Let w represent the weight of an average female gorilla.
Bar Diagram	379 pounds: **181 pounds** | **weight, *w***
Equation	$181 + w = 379$

$$\begin{aligned} 181 + w &= 379 \\ -181 \quad &= -181 \\ w &= 198 \end{aligned}$$

Write the equation.

Subtract 181 from each side.

$379 - 181 = 198$

So, an average female gorilla weighs 198 pounds.

Check $181 + 198 = 379$ ✓

Guided Practice

Solve each equation. Check your solution. (Example 1)

1. $y + 7 = 10$

   y=3 ✓

2. $10 = 6 + e$

   e=4 ✓

3. A board that measures 19.5 meters in length is cut into two pieces. One piece measures 7.2 meters. Write and solve an equation to find the length of the other piece. (Example 2)

4. It takes 43 facial muscles to frown. This is 26 more muscles than it takes to smile. Write and solve an equation to find the number of muscles it takes to smile. (Example 3)

5. **Building on the Essential Question** How can the Subtraction Property of Equality be used to solve addition equations?

Rate Yourself!

How confident are you about writing and solving addition equations? Shade the ring on the target.

For more help, go online to access a Personal Tutor.

FOLDABLES Time to update your Foldable!

Name Arjun Mahesh My Homework 9-14-19

Independent Practice

Go online for Step-by-Step Solutions

eHelp

Solve each equation. Check your solution. (Example 1)

1. $c + 3 = 6$

   c = 3

   Show your work.

2. $9 = 2 + x$

   9-2=x

   7=x

3. $7 + a = 9$

   a = 2

4. Zacarias and Paz together have $756.80. If Zacarias has $489.50, how much does Paz have? Write and solve an addition equation to find how much money belongs to Paz. (Example 2) ______

5. The average length of a King Cobra is 118 inches, which is 22 inches longer than a Black Mamba. Write and solve an addition equation to find the average length of a Black Mamba. (Example 3) ______

6. MP **Model with Mathematics** Refer to the graphic novel frame below for Exercises a–b.

READING REWARD
50 Points = Pizza Party

ITEM READ	POINTS
Book	5
Magazine	1
Newspaper	1

Remember, I need 50 points for the pizza party.

a. If Mei has already earned 30 points, write and solve an addition equation to find the number of points she still needs.

______

b. Suppose Julie has already earned 36 points. Write and solve an addition equation to find the number of points she still needs to earn the pizza party. ______

Solve each equation. Check your solution.

7. $a + \frac{1}{10} = \frac{5}{10}$

8. $m + \frac{1}{3} = \frac{2}{3}$

9. $\frac{3}{4} = x + \frac{1}{2}$

10. $\frac{7}{8} = y + \frac{1}{4}$

H.O.T. Problems Higher Order Thinking

11. **MP Reason Inductively** Write two different addition equations that have 12 as the solution.

12. **MP Persevere with Problems** In the equation $x + y = 5$, the value for x is a whole number greater than 2 but less than 6. Determine the possible solutions for y.

13. **MP Which One Doesn't Belong?** Identify the equation that does not belong with the other three. Explain your reasoning.

$6 + x = 9$	$15 = x + 12$	$x + 9 = 11$	$7 + x = 10$

14. **MP Find the Error** Melody is solving the equation $x + 12 = 31$. Find her mistake and correct it.

$$\begin{aligned} x + 12 &= 31 \\ +12 &= +12 \\ \hline x &= 43 \end{aligned}$$

15. **MP Reason Abstractly** Suppose $x + y = 13$ and the value of x increases by 4. If their sum remains the same, what must happen to the value of y?

Name ____________________ My Homework ____________________

Extra Practice

Solve each equation. Check your solution.

16. $x + 5 = 11$

Homework Help

17. $7 = 4 + y$

18. $5 + g = 6$

19. $d + 3 = 8$

20. $x + 4 = 6$

21. $3 + f = 8$

22. Enrique and Levi together have 386 trading cards. If Enrique has 221 trading cards, how many does Levi have? Write and solve an addition equation to find how many trading cards are Levi's.

23. Eliott is 63 inches tall, which is 9 inches taller than his cousin, Jackson. Write and solve an addition equation to find Jackson's height.

24. MP **Use Math Tools** The table shows the heights of three monster trucks. Bigfoot 5 is 4.9 feet taller than Bigfoot 2. Write and solve an addition equation to find the height of Bigfoot 2. ____________

Truck	Height (ft)
Bigfoot 5	15.4
Swamp Thing	12.2
Bigfoot 2	■

Solve each equation. Check your solution.

25. $t + \frac{8}{10} = \frac{9}{10}$

26. $\frac{5}{8} + n = \frac{7}{8}$

27. $t + \frac{1}{4} = \frac{3}{4}$

Power Up! Common Core Test Practice

28. Nathan has scored 174 points this basketball season. This is 29 points more than Will has scored. Select the correct items to complete the bar diagram representing the number of points Will has scored this season.

Will's points, x
29 points
174 points

What equation is modeled by the bar diagram? ______

How many points has Will scored? ______

29. Niko wants to buy a skateboard that costs $85. He has already saved $15. Fill in the box to complete each statement.

a. The equation ______ can be used to find the amount of money Niko still needs to save to buy the skateboard.

b. Niko still needs to save ______ to buy the skateboard.

Common Core Spiral Review

Subtract. 4.NBT.4

30. $22 - 8 =$ ______

31. $72 - 34 =$ ______

32. $34 - 19 =$ ______

33. $51 - 32 =$ ______

34. $66 - 14 =$ ______

35. $49 - 32 =$ ______

36. The table shows the distances three friends hiked. How much farther did Isabella hike than Devon? 5.NBT.7 ______

Name	Distance Hiked (mi)
Devon	1.85
Franco	2.55
Isabella	2.25

Inquiry Lab

Solve and Write Subtraction Equations

HOW do you solve subtraction equations using models?

Content Standards
6.EE.5, 6.EE.7

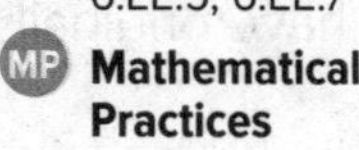

Mathematical Practices
1, 3, 4

Zack gave 5 trading cards to his sister. Now he has 41 cards. How many cards did he have originally?

What do you know? ____

____

What do you need to find? ____________________________________

____

Hands-On Activity

Step 1 Define a variable. Use the variable c to represent the number of cards Zack had originally.

Step 2 Use a bar diagram to help write the equation.

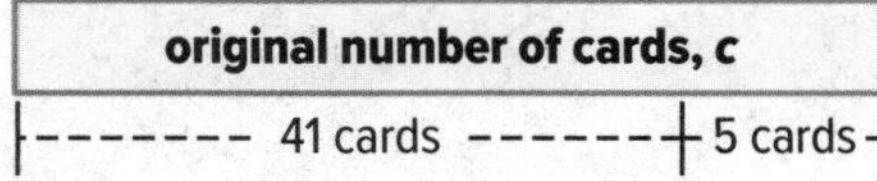

The total length of the diagram shows __________________________.

The number 41 represents ____________________________________.

The number 5 represents _____________________________________.

☐ − ☐ = ☐

Step 3 Work backward. Rewrite the equation as an addition sentence and solve.

☐ + ☐ = ☐

So, Zack originally had ☐ trading cards.

Investigate

Work with a partner. Write and solve a subtraction equation using a bar diagram.

1. Mariska gave her friend Elise 8 beads and was left with 37 beads. How many did she have originally?

2. Clinton has $12 after buying a snack at the mall. The snack cost $5. How much money did Clinton have originally?

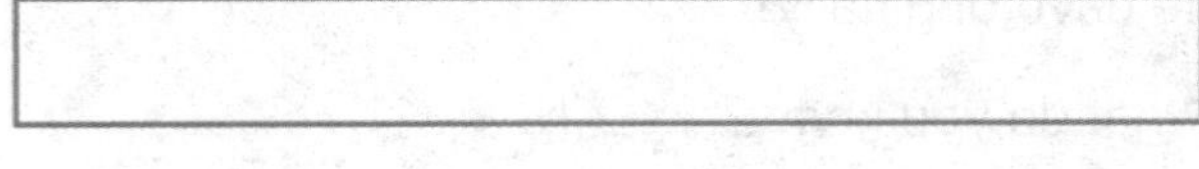

3. The Martin County Cat Shelter placed 8 cats with new owners on Monday. On Tuesday, 31 cats remained at the shelter. How many cats were at the shelter originally?

Create

4. **MP Reason Inductively** Write a rule for solving equations like $x - 4 = 7$.

5. **MP Model with Mathematics** Write a real-world subtraction problem for the equation modeled below. Then write the equation and solve.

miles driven, *m*

128 miles | 67 miles

6. **Inquiry** HOW do you solve subtraction equations using models?

Lesson 3

Solve and Write Subtraction Equations

Real-World Link

Bowling Meghan's bowling score was 39 points less than Charmaine's. Meghan's score was 109.

1. Let s represent Charmaine's score. Write an equation for *39 points less than Charmaine's score is equal to 109.*

2. Use the number line to find Charmaine's score by counting forward.

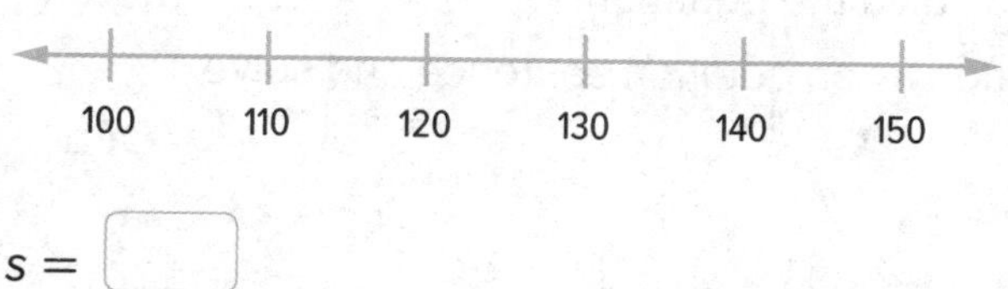

$s =$ ☐

3. What operation does counting forward suggest?

4. Would it be reasonable to use cups and counters to solve this equation? Explain.

Essential Question

HOW do you determine if two numbers or expressions are equal?

Vocabulary

Addition Property of Equality

Common Core State Standards

Content Standards
6.EE.5, 6.EE.7

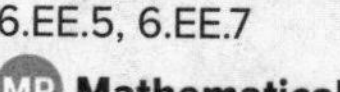

Mathematical Practices
1, 3, 4, 5

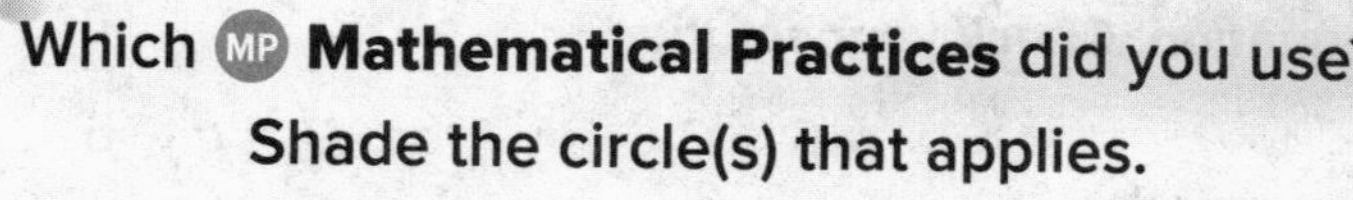

Which MP **Mathematical Practices** did you use? Shade the circle(s) that applies.

① Persevere with Problems
② Reason Abstractly
③ Construct an Argument
④ Model with Mathematics
⑤ Use Math Tools
⑥ Attend to Precision
⑦ Make Use of Structure
⑧ Use Repeated Reasoning

Work Zone

Solve an Equation by Adding

Because addition and subtraction are inverse operations, you can solve a subtraction equation by adding.

Example

1. **Solve $x - 2 = 3$. Check your solution.**

Method 1 **Use models.**

Model the equation.

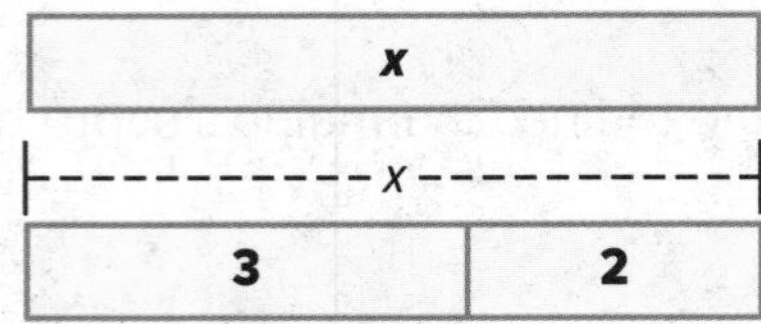

Work backward to solve the equation.

Rewrite the equation as an addition sentence and solve.

$3 + 2 = 5$

Method 2 **Use symbols.**

$x - 2 = 3$	Write the equation.
$+2 = +2$	Add 2 to each side.
$x = 5$	Simplify.

Check

$x - 2 = 3$	Write the equation.
$5 - 2 \stackrel{?}{=} 3$	Replace x with 5.
$3 = 3$ ✓	This sentence is true.

Using either method, the solution is 5.

Show your work.

a. __________

b. __________

c. __________

Got it? **Do these problems to find out.**

Solve each equation. Check your solution.

a. $x - 7 = 4$ **b.** $y - 6 = 8$ **c.** $9 = a - 5$

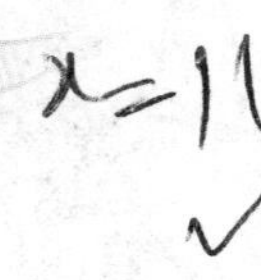

x=11 ✓ y=14 ✓ a=14

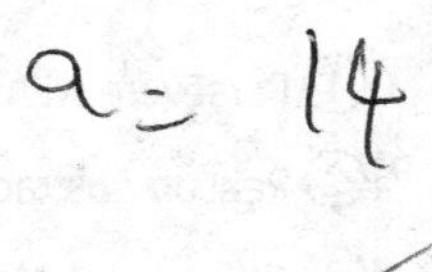

Addition Property of Equality

Key Concept

Words If you add the same number to each side of an equation, the two sides remain equal.

Examples

Numbers	Algebra
$\begin{array}{r} 5 = 5 \\ +3 = +3 \\ \hline 8 = 8 \end{array}$	$\begin{array}{r} x - 2 = 3 \\ +2 = +2 \\ \hline x = 5 \end{array}$

When you solve an equation by adding the same number to each side of the equation, you are using the **Addition Property of Equality**.

Example

2. **STEM At age 25, Gherman Titov of Russia was the youngest person to travel into space. This is 52 years less than the oldest person to travel in space, John Glenn. How old was John Glenn? Write and solve a subtraction equation.**

Words Oldest age minus youngest age is 52 years.

Variable Let a represent the oldest age in space.

Bar Diagram

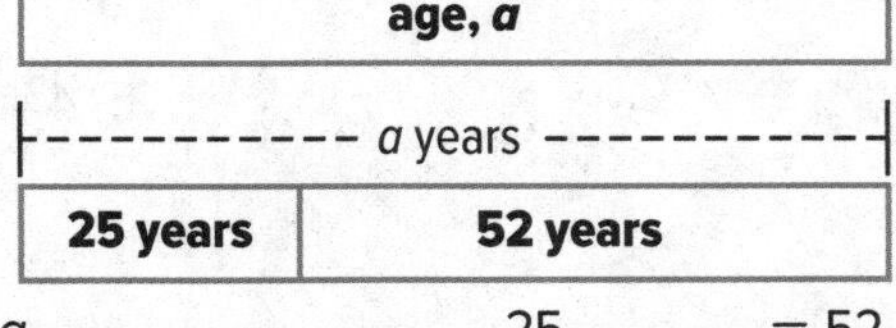

Equation $a \quad - \quad 25 \quad = 52$

$$\begin{array}{rl} a - 25 = 52 & \text{Write the equation.} \\ +25 = +25 & \text{Add 25 to each side.} \\ \hline a = 77 & \text{Simplify.} \end{array}$$

John Glenn was 77 years old.

Check $77 - 25 = 52$ ✓

Got it? **Do this problem to find out.**

d. Georgia's height is 4 inches less than Sienna's height. Georgia is 58 inches tall. Write and solve a subtraction equation to find Sienna's height.

d. ____________

Example

3. **Raheem's rollerblades cost \$70.25 less than his bicycle. His rollerblades cost \$43.50. How much did his bicycle cost? Write and solve a subtraction equation.**

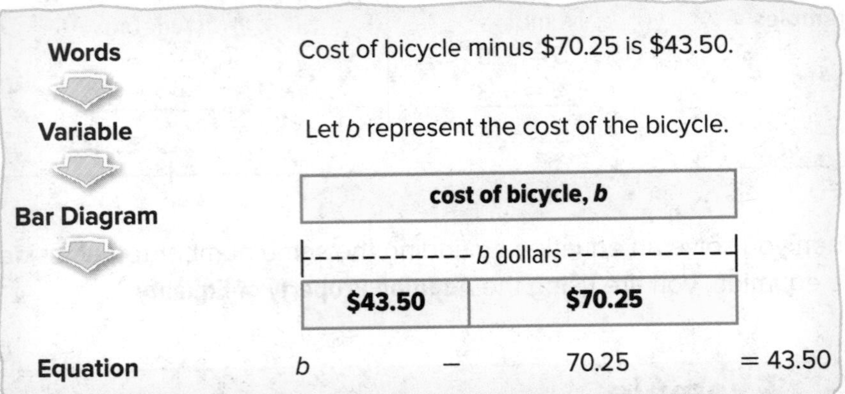

STOP and Reflect

How is solving an addition equation different from solving a subtraction equation? Explain below.

$$
\begin{aligned}
b - 70.25 &= 43.50 && \text{Write the equation.}\\
\underline{+\,70.25} &= \underline{+\,70.25} && \text{Add 70.25 to each side.}\\
b &= 113.75 && \text{Simplify.}
\end{aligned}
$$

The bicycle cost \$113.75.

Check $113.75 - 70.25 = 43.50$ ✓

Guided Practice

Solve each equation. Check your solution. (Example 1)

1. $a - 5 = 9$

2. $b - 3 = 7$

3. $4 = y - 8$

4. Catherine studied 1.25 hours for her science test. This was 0.5 hour less than she studied for her algebra test. Write and solve a subtraction equation to find how long she studied for her algebra test. (Examples 2 and 3)

5. **Building on the Essential Question** How can the Addition Property of Equality be used to solve subtraction equations?

Rate Yourself!

Are you ready to move on? Shade the section that applies.

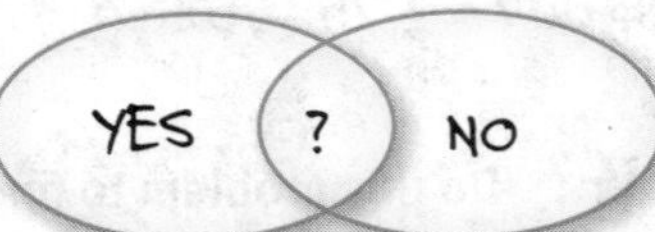

For more help, go online to access a Personal Tutor.

FOLDABLES Time to update your Foldable!

Name ______________________ My Homework ______________________

Independent Practice

Go online for Step-by-Step Solutions — eHelp

Solve each equation. Check your solution. (Examples 1 and 3)

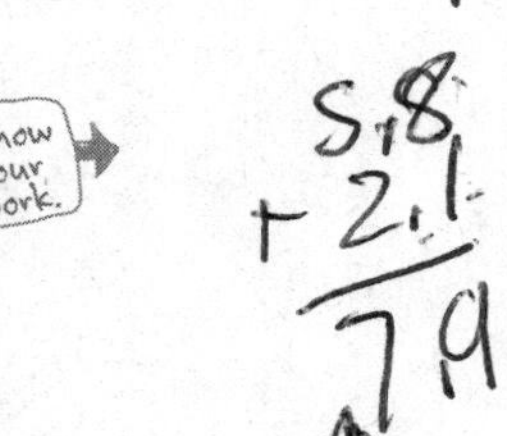

1. $c - 1 = 8$

2. $t - 7 = 2$

3. $1 = g - 3$

4. $a - 2.1 = 5.8$

5. $a - 1.1 = 2.3$

6. $4.6 = e - 3.2$

7. Pete is 15 years old. This is 6 years younger than his sister Victoria. Write and solve a subtraction equation to find Victoria's age. (Example 2)

____

8. A CD costs \$14.95. This is \$7.55 less than the cost of a DVD. Write and solve a subtraction equation to find the cost of the DVD. (Example 3)

____

9. If $b - 10 = 5$, what is the value of $b + 6$? ____________

Solve each equation. Check your solution.

10. $m - \frac{1}{3} = \frac{2}{3}$

11. $n - \frac{1}{4} = \frac{3}{4}$

12. $s - \frac{1}{3} = \frac{7}{9}$

13. Alejandra spent her birthday money on a video game that cost \$24, a controller for \$13, and a memory card for \$16. The total tax was \$3. Write and solve a subtraction equation to find how much money Alejandra gave the cashier if she received \$4 in change.

____

14. **MP Multiple Representations** The bar diagram represents a subtraction equation.

x°F	
74°F	13°F

a. **Words** Write a real-world problem that can be represented by the bar diagram. ______________________

____

____

b. **Algebra** Write a subtraction equation that can be represented by the bar diagram. ______________________

c. **Numbers** Solve the equation you wrote in part b. ______________________

H.O.T. Problems Higher Order Thinking

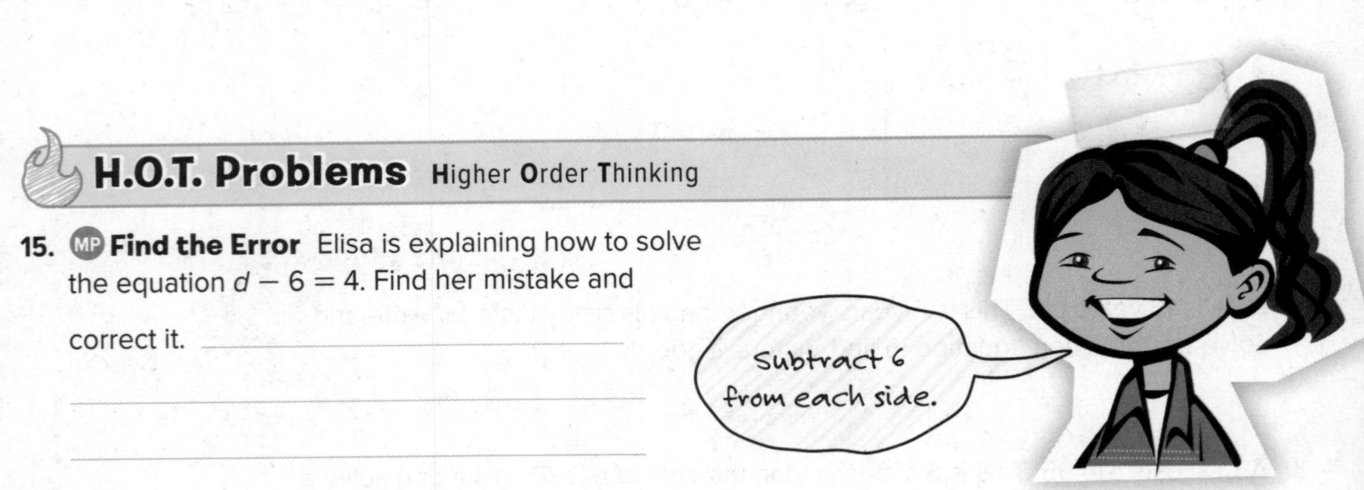

15. **MP Find the Error** Elisa is explaining how to solve the equation $d - 6 = 4$. Find her mistake and correct it. ______________________

____

____

16. **MP Model with Mathematics** Write a real-world problem that could be represented by $d - 32 = 64$. ______________________

____

17. **MP Persevere with Problems** Another type of subtraction equation is $16 - b = 7$. Explain how you would solve this equation then solve it.

____

____

18. **MP Reason Inductively** Which of the following is true concerning $x - 5 = 13$? ______________________

I To find the value of x, add 5 to each side.
II To find the value of x, subtract 5 from each side.
III To find the value of x, add 13 to each side.
IV To find the value of x, subtract 13 from each side.

Name _______________ My Homework _______________

Extra Practice

Solve each equation. Check your solution.

19. $f - 1 = 5$

Homework Help

$f - 1 = 5$
$+1 = +1$
$f = 6$

20. $2 = e - 1$

21. $r - 3 = 1$

22. $z - 6.3 = 2.1$

23. $t - 9.25 = 5.45$

24. $k - 32.9 = 16.5$

25. MP **Use Math Tools** North Carolina has 12 less electoral votes than Florida. Write and solve a subtraction equation to find the number of electoral votes for Florida. _______________

____

Electoral Votes	
State	**Number of Votes**
Florida	■
North Carolina	15

26. Marty's cat weighs 10.4 pounds. This is 24.4 pounds less than the weight of his dog. Write and solve a subtraction equation to find the weight of Marty's dog. _______________

____

27. Find the value of t if $t - 7 = 12$. _______________

____

Solve each equation. Check your solution.

28. $s - \frac{1}{2} = \frac{1}{2}$

29. $h - \frac{1}{4} = \frac{1}{4}$

30. $c - 1 = \frac{3}{4}$

31. At a movie, Angelo bought a medium popcorn for \$4, a small drink for \$3, and a box of fruit snacks for \$5. Write and solve a subtraction equation to find how much money Angelo gave the cashier if he received \$3 in change.

____

CCSS Power Up! Common Core Test Practice

32. Xavier's age is 3 years less than Paula's age. Xavier is 11 years old. Select the correct items to complete the bar diagram below representing Paula's age.

Paula's age, p
3
11

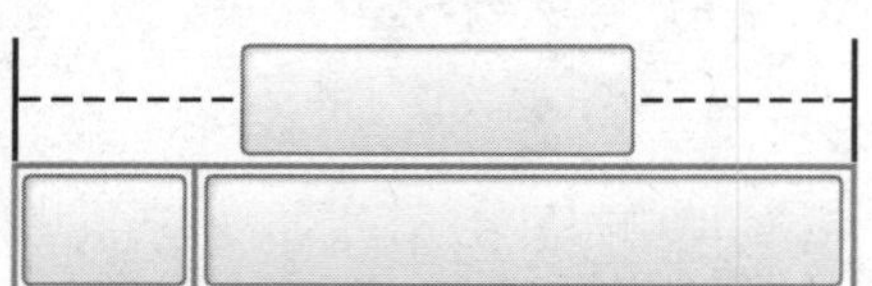

What equation is modeled by the bar diagram?

How old is Paula?

33. Owen bought a pair of shoes and the shirt shown. The cost of the shirt was \$42 less than the price of the shoes. Let s represent the price of the shoes. Determine if each statement is true or false.

a. The equation $s - 22 = 42$ models the situation. ☐ True ☐ False

b. The equation $42 - s = 22$ models the situation. ☐ True ☐ False

c. The cost of the shoes was \$64. ☐ True ☐ False

CCSS Common Core Spiral Review

Multiply. 4.NBT.5

34. $63 \times 8 =$ ______

35. $19 \times 6 =$ ______

36. $27 \times 5 =$ ______

37. $13 \times 8 =$ ______

38. $36 \times 4 =$ ______

39. $21 \times 3 =$ ______

40. The Cozy Cat Shop has 3 calico cats for every gray cat. If they have 9 calico cats available, how many gray cats do they have? 4.NBT.6

______

 Problem-Solving Investigation

Guess, Check, and Revise

Content Standards
6.EE.7

Mathematical Practices
1, 3, 4

Case #1 Smart Money

Damian used $20 bills and $10 bills to pay for his $100 guitar lesson.

If he paid with 8 bills, how many of each bill did he use?

Understand *What are the facts?*

- Damian paid with 8 bills that add to $100.
- The money was in $20 bills and $10 bills.

Plan *What is your strategy to solve this problem?*

Make a guess until you find an answer that makes sense for the problem.

Solve *How can you apply the strategy?*

Use addends that have a sum of 8 to find the number of $20 and $10 bills.

Number of $20 bills	Number of $10 bills	Total Amount	Compare to $100
1	7	1($20) + 7($10) = $	
2	6	2($20) + 6($10) = $	
3	5	3($20) + 5($10) = $	
4	4	4($20) + 4($10) = $	

Damian paid with ☐ $20 bills and ☐ $10 bills.

Check *Does the answer make sense?*

The other combinations are either less than or greater than $100.

Analyze the Strategy

Reason Inductively Monique received $100 in $10 and $5 bills, including eight $10 bills. Use the equation $x + 80 = 100$ to find how much money x was given to her in $5 bills. How many $5 bills did she receive?

Case #2 Anime Adventure

A book store sells used graphic novels in packages of 5 and new graphic novels in packages of 3.

If Amy buys a total of 16 graphic novels, how many packages of new and used graphic novels did she buy?

1 Understand

Read the problem. What are you being asked to find?

I need to find ____

____.

Underline key words and values in the problem. What information do you know?

The ________ novels come in packages of ☐ and the ________ novels come in packages of ☐. Amy buys ☐ graphic novels.

Is there any information that you do *not* need to know?

I do not need to know ____.

2 Plan

Choose a problem-solving strategy.

I will use the ______________________________ strategy.

3 Solve

Use your problem-solving strategy to solve the problem. Make a guess.

2 used packages and 1 new package ☐(5) + ☐(3); ☐ < 16

3 used packages and 2 new packages ☐(5) + ☐(3); ☐ > 16

2 used packages and 2 new packages ☐(5) + ☐(3); ☐ = 16

So, ____.

4 Check

Use information from the problem to check your answer.

Make a list of multiples of 3 and a list of multiples of 5. Look for a combination of these multiples that add up to 16.

____

____

Work with a small group to solve the following cases. Show your work on a separate piece of paper.

Case #3 Quizzes

On a science quiz, Ivan earned 18 points. There are six problems worth 2 points each and two problems worth 4 points each.

Find the number of problems of each type Ivan answered correctly.

Case #4 Numbers

Kathryn is thinking of four numbers from 1 through 9 with a sum of 18. Each number is used only once.

Find the numbers.

Case #5 Equations

Use the symbols +, −, ×, or ÷ to make the following equation true. Use each symbol only once.

3 ■ 4 ■ 6 ■ 1 = 18

Case #6 Money

Nathaniel is saving money to buy a new graphics card for his computer that costs $260.

If he is saving $18 a month and already has $134, in how many more months will he have enough money for the graphics card?

Mid-Chapter Check

Vocabulary Check

1. Define *equation*. Give an example of an equation and an example of an expression. Use a variable in each example. (Lesson 1)

2. Fill in the blank in the sentence below with the correct term. (Lesson 2)

   You can solve equations using ______________, which undo each other.

Skills Check and Problem Solving

Circle the solution of the equation from the list given. (Lesson 1)

3. $x + 22 = 27$; 5, 6, 7

4. $17 + n = 24$; 6, 7, 8

Solve each equation. Check your solution. (Lessons 2 and 3)

5. $63 + d = 105$

6. $h + 7.9 = 13$

7. $a + 1.6 = 2.1$

8. $p - 13 = 29$

9. $y - 9 = 26$

10. $r - 5\frac{1}{6} = 10$

11. **MP Use Math Tools** The difference between the water levels for high and low tide was 3.6 feet. Write and solve an equation to find the water level at high tide. (Lesson 3)

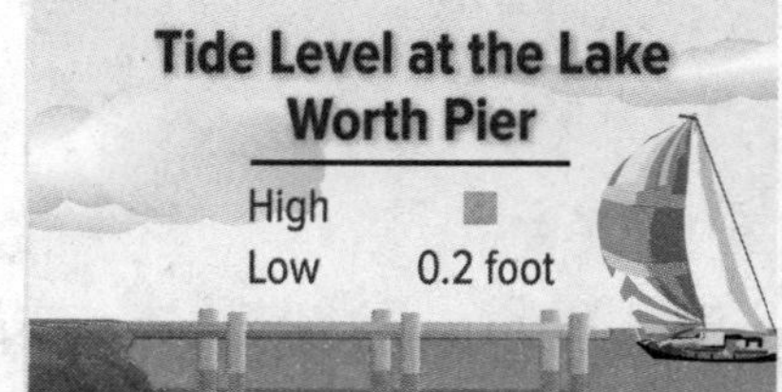

Tide Level at the Lake Worth Pier

High	■
Low	0.2 foot

12. **MP Persevere with Problems** If $x + 9.8 = 14.7$, what is the value of $8(x - 3.7)$? (Lesson 2)

Inquiry Lab

Solve and Write Multiplication Equations

HOW do you solve multiplication equations using models?

Content Standards
6.EE.5, 6.EE.7

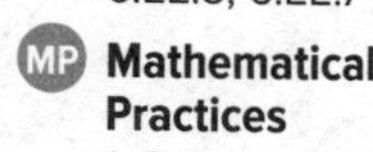

Mathematical Practices
1, 3, 4

In 5 days, Nicole ran a total of 10 miles. She ran the same amount each day. How much did she run each day?

What do you know? ______________________________

____

What do you need to find? ______________________________

____

Hands-On Activity 1

Step 1 Define a variable. Use the variable d to represent the distance run in one day.

Step 2 Use a bar diagram to help write the equation.

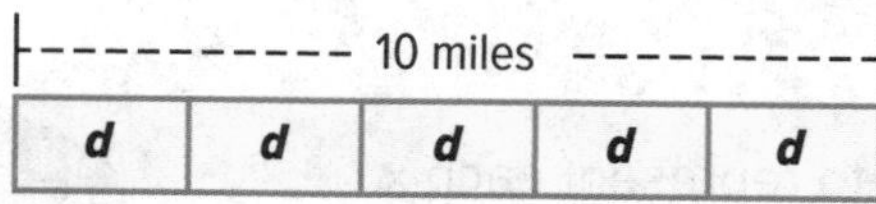

The total length of the diagram shows ______________.

The variable d appears in the diagram ☐ times.

☐ $d =$ ☐

Step 3 Work backward. Rewrite the equation as a division sentence and solve.

☐ $\div$ ☐ $= d$

So, Nicole ran ☐ miles each day.

Investigate

Collaborate

Work with a partner. Define the variable. Then write and solve a multiplication equation using a bar diagram.

1. Suppose Nicole ran 12 miles in four days. If she ran the same distance *d* each day, how many miles did she run in one day?

2. Krista has owned her cell phone for 8 months, which is twice as long as her sister Allie has owned her cell phone. How many months m has Allie had her cell phone?

Hands-On Activity 2

Solve $3x = 12$. Check your solution.

Step 1 Model the equation. Use one cup to represent each *x*.

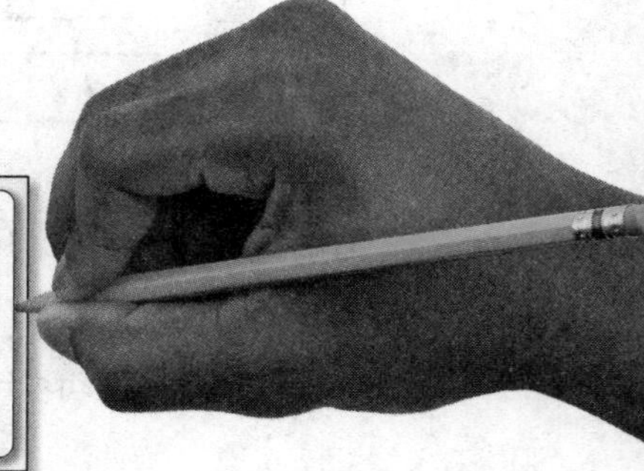

Step 2 Use the model above. Divide the 12 counters equally by circling 3 groups. There are ☐ counters in each group.

So, the solution is ☐.

Check $3\square = 12$ Write the original equation.

$3(\square) \stackrel{?}{=} 12$ Replace *x* with your solution.

$\square = 12$ Is the sentence true? ______

Investigate

Work with a partner. Solve each equation using cups and counters.

3. $4n = 8$

$n =$ ______

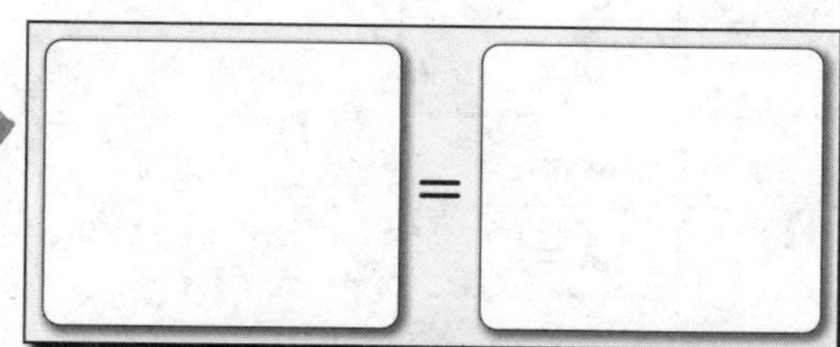

4. $3x = 9$

$x =$ ______

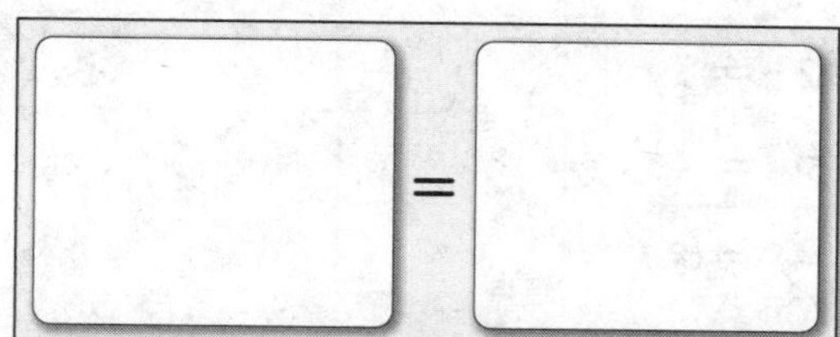

5. $10 = 5x$

$x =$ ______

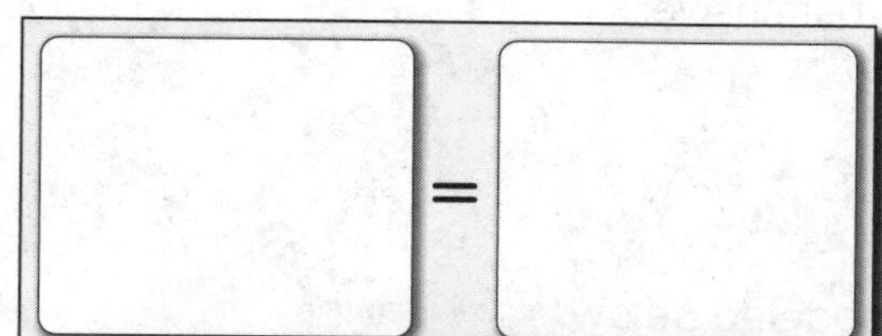

6. $6x = 12$

$x =$ ______

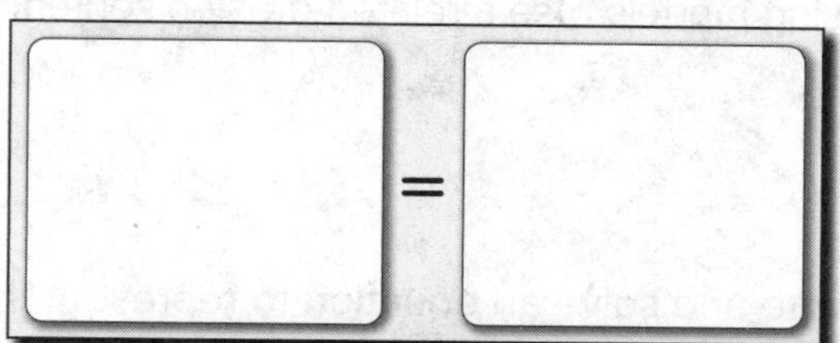

Define a variable. Then write and solve a multiplication equation using a bar diagram.

7. The average lifespan of a horse is 40 years, which is five times longer than the average lifespan of a guinea pig. Use the bar diagram below to find the average lifespan of a guinea pig. Label each section of the diagram. ______

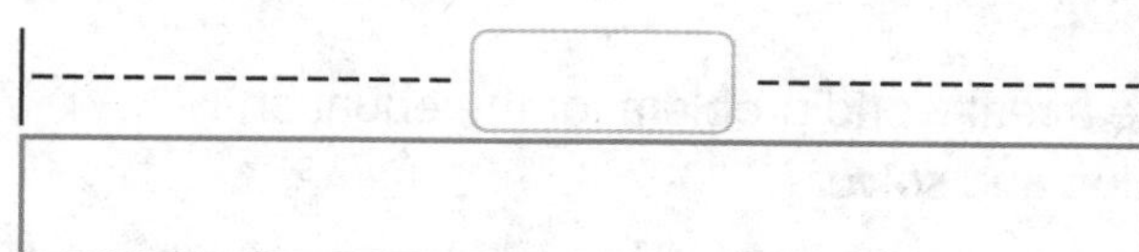

8. Kosumi is saving an equal amount each week for 4 weeks to buy a $40 video game. Use the bar diagram below to find how much he is saving each week. Label each section of the diagram.

______

Analyze and Reflect

Work with a partner to complete the table. The first one is done for you.

	Multiplication Equation	Coefficient	Variable	Product	Division Sentence	Solution
	$7g = 14$	7	g	14	$14 \div 7 = g$	$g = 2$
9.	$21 = 3y$					$y =$
10.	$5m = 45$					$m =$
11.	$48 = 8d$					$d =$
12.	$16f = 32$					$f =$
13.	$39 = 13b$					$b =$

14. MP **Reason Inductively** Write a rule for solving equations like $2x = 24$ without using models. Use a related division sentence to explain your answer.

15. Write and solve an equation to represent the situation modeled below.

Create

16. MP **Model with Mathematics** Write a real-world problem for the equation modeled below. Then write the equation and solve.

$12			
c	c	c	c

17. Inquiry HOW do you solve multiplication equations using models?

Solve and Write Multiplication Equations

Vocabulary Start-Up

The equation $3x = 9$ is a multiplication equation. In $3x$, 3 is the coefficient of x because it is the number by which x is multiplied.

Fill in the table. The first one is done for you.

Prefix	Root Word	New Word	Meaning
co-	pilot	copilot	the second pilot that flies with the primary pilot of the plane
co-	author		
co-	operate		
co-	efficient		

Essential Question

HOW do you determine if two numbers or expressions are equal?

Vocabulary

Division Property of Equality

Common Core State Standards

Content Standards
6.EE.5, 6.EE.7, 6.RP.3

MP Mathematical Practices
1, 2, 3, 4, 5

Real-World Link

Ringtones Matthew is downloading ringtones. The cost to download each ringtone is $2. When Matthew is finished he has spent a total of $10. Let x represent the number of ringtones. What does the expression $2x$ represent?

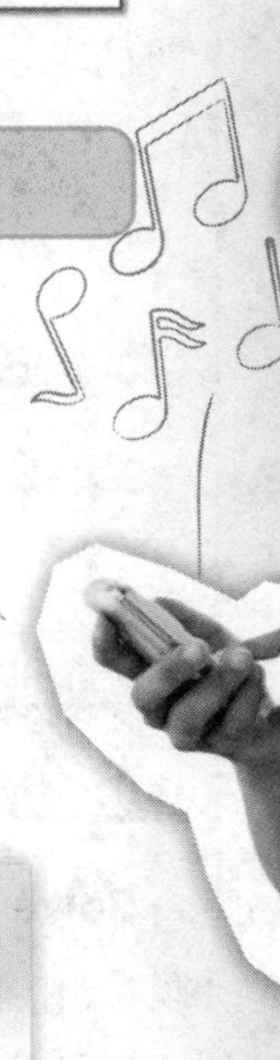

Which MP **Mathematical Practices** did you use? Shade the circle(s) that applies.

① Persevere with Problems
② Reason Abstractly
③ Construct an Argument
④ Model with Mathematics
⑤ Use Math Tools
⑥ Attend to Precision
⑦ Make Use of Structure
⑧ Use Repeated Reasoning

Work Zone

Solve a Multiplication Equation

A multiplication equation is an equation like $2x = 10$ because the variable x is multiplied by 2. Multiplication and division are inverse operations. So, to solve a multiplication equation, use division.

Examples

1. Solve $2x = 10$. Check your solution.

$2x = 10$ Write the equation.

$\frac{2x}{2} = \frac{10}{2}$ Divide each side by the coefficient 2.

$x = 5$

Check $2x = 10$ Write the original equation.

$2(5) \stackrel{?}{=} 10$ Replace x with 5.

$10 = 10$ This sentence is true. ✓

2. Solve $3x = 6$. Check your solution.

Fill in the boxes below.

$3x = 6$ Write the equation.

$\frac{3x}{\square} = \frac{6}{\square}$ Divide each side by the coefficient ☐.

$x = \square$

Check $3\square = 6$ Write the original equation.

$3(\square) \stackrel{?}{=} 6$ Replace x with ☐.

$\square = 6$ This sentence is ☐. ✓

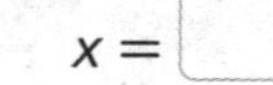

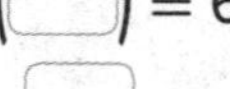

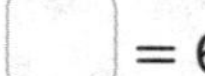

Got it? **Do these problems to find out.**

Solve each equation. Check your solution.

a. $3x = 15$ **b.** $8 = 4x$ **c.** $2x = 14$

a. ____________

b. ____________

c. ____________

Division Property of Equality

Key Concept

Words If you divide each side of an equation by the same nonzero number, the two sides remain equal.

Examples

Numbers	Algebra
$18 = 18$	$3x = 12$
$\frac{18}{6} = \frac{18}{6}$	$\frac{3x}{3} = \frac{12}{3}$
$3 = 3$	$x = 4$

When you solve an equation by dividing both sides of the equation by the same number, you are using the **Division Property of Equality**.

Example

3. Vicente and some friends shared the cost of a package of blank CDs. The package cost $24 and each person contributed $6. How many people shared the cost of the CDs?

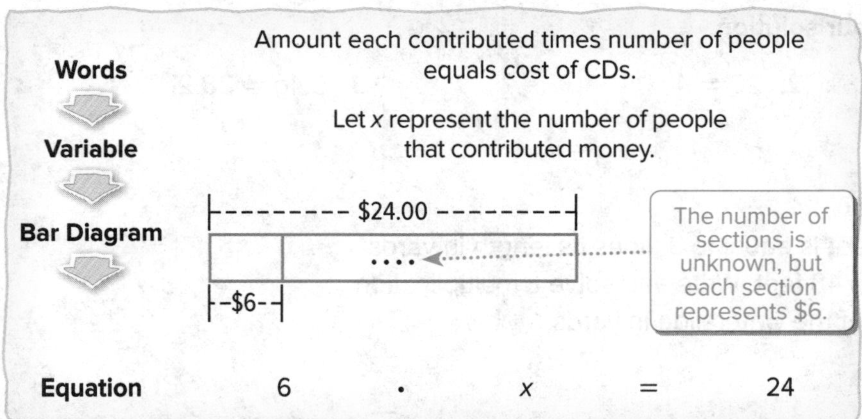

$6x = 24$ Write the equation.

$\frac{6x}{6} = \frac{24}{6}$ Divide each side by 6.

$x = 4$ Simplify.

Check $6 \times 4 = 24$ ✓

There were 4 people who split the cost of the CDs.

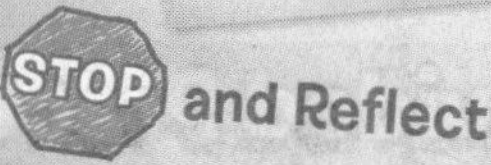

What is the coeffienct in the equation in Example 3?

Got it? Do this problem to find out.

d. In 2004, Pen Hadow and Simon Murray walked 680 miles to the South Pole. The trip took 58 days. Suppose they traveled the same distance each day. Write and solve a multiplication equation to find about how many miles they traveled each day.

d. ________

e. ______

f. ______

g. ______

Example

4. **Solve $3.28x = 19.68$. Check your solution.**

$3.28x = 19.68$ Write the equation.

$\frac{3.28x}{3.28} = \frac{19.68}{3.28}$ Divide each side by 3.28.

$x = 6$

Check $3.28x = 19.68$ Write the original equation.

$3.28(6) \stackrel{?}{=} 19.68$ Replace *x* with 6.

$19.68 = 19.68$ This sentence is true. ✓

Got it? **Do these problems to find out.**

Solve each equation. Check your solution.

e. $2.25n = 6.75$ **f.** $1.7b = 8.5$ **g.** $6.15y = 55.35$

Guided Practice

Solve each equation. Check your solution. (Examples 1, 2, and 4)

1. $2a = 6$

2. $20 = 4c$

3. $9.4g = 28.2$

4. The length of an object in feet is equal to 3 times its length in yards. The length of a waterslide is 48 feet. Write and solve a multiplication equation to find the length of the waterslide in yards. (Example 3)

______

5. The total time to burn a CD is 18 minutes. Last weekend, Demitri spent 90 minutes burning CDs. Write and solve a multiplication equation to find the number of CDs Demitri burned last weekend. Explain how you can check your solution. (Example 3) ______

______

6. **Building on the Essential Question** How can the Division Property of Equality be used to solve multiplication equations? ______

______

______

Rate Yourself!

How well do you understand solving and writing multiplication equations? Circle the image that applies.

Clear Somewhat Clear Not So Clear

For more help, go online to access a Personal Tutor.

FOLDABLES Time to update your Foldable!

Name ______________________ My Homework ______________________

Independent Practice

Go online for Step-by-Step Solutions

Solve each equation. Check your solution. (Examples 1, 2, and 4)

1. $4g = 24$

2. $5d = 30$

3. $36 = 6e$

4. $1.5x = 3$

5. $2.5y = 5$

6. $8.1 = 0.9a$

7. A jewelry store is selling a set of 4 pairs of gemstone earrings for $58, including tax. Neva and three of her friends want to buy the set so each could have one pair of earrings. Write and solve a multiplication equation to find how much each person should pay. (Example 3)

____

Solve each equation. Check your solution.

8. $39 = 1\frac{3}{10}b$

9. $\frac{1}{2}e = \frac{1}{4}$

10. $\frac{2}{5}g = \frac{3}{5}$

11. MP **Use Math Tools** Use the table that shows football data.

a. Morten Andersen played in the NFL for 25 years. Write and solve an equation to find how many points he averaged each year.

____

b. Jason Hanson played in the NFL for 20 years. Write and solve an equation to find how many points he averaged each year.

____

Top NFL Kickers	
Player	**Career Points**
Morten Andersen	2,544
Gary Anderson	2,434
Jason Hanson	2,150
John Carney	2,062
Adam Vinatieri	2,006

12. **STEM** An average person's heart beats about 103,680 times a day. Write and solve an equation to find about how many times the average person's heart beats in one minute.

____

13. **Model with Mathematics** Problems involving constant speed can be solved by the formula distance = rate × time. Fernando's family traveled 272 miles on a road trip last weekend. They drove for 4 hours. What was the rate at which Fernando's family traveled? Write and solve a multiplication equation.

distance	=	rate	×	time

Fernando's family traveled an average rate of ______ miles per hour.

H.O.T. Problems Higher Order Thinking

14. **Find the Error** Noah is solving $5x = 75$. Find his mistake and correct it.

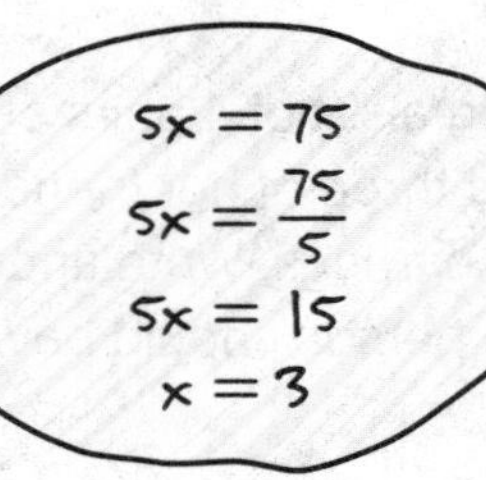

15. **Which One Doesn't Belong?** Identify the equation that does not belong with the other three. Explain your reasoning.

$5x = 20$ | $4b = 7$ | $8w = 32$ | $12y = 48$

16. **Persevere with Problems** Explain how you know that the equations $\frac{1}{4} = 2x$ and $\frac{1}{4} \div x = 2$ have the same solution. Then, find the solution.

17. **Model with Mathematics** Write a real-world problem that could be represented by the equation $4r = 240$. Then solve the equation and interpret the solution.

Extra Practice

Solve each equation. Check your solution.

18. $4c = 16$

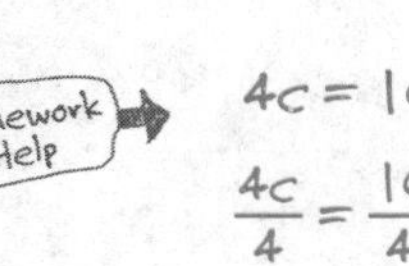

$4c = 16$

$\frac{4c}{4} = \frac{16}{4}$

$c = 4$

19. $5t = 25$

20. $5a = 15$

21. $3f = 12$

22. $21 = 3g$

23. $6x = 12$

24. $5.9q = 23.6$

25. $2.55d = 17.85$

26. $6.5a = 32.5$

27. The Raimonde family drove 1,764 miles across the United States on their vacation. If it took a total of 28 hours, write and solve a multiplication equation to find their average speed in miles per hour.

______

28. MP **Reason Abstractly** Four friends went bowling one afternoon. Use the table that shows the bowling data.

Player	Score
Bryan	320
Carson	366
Jana	522
Pilar	488

a. Carson bowled 3 games. Write and solve an equation to find how many points he averaged each game. ______

b. Jana bowled 5 games. Write and solve an equation to find how many points she averaged each game. ______

Copy and Solve **Solve each equation. Show your work on a separate piece of paper.**

29. $1\frac{2}{5}x = 7$

30. $3\frac{1}{2}r = 28$

31. $2\frac{1}{4}w = 6\frac{3}{4}$

32. $2\frac{3}{4}a = 19\frac{1}{4}$

33. $1\frac{1}{2}c = 6$

34. $3\frac{3}{4}m = 33\frac{3}{4}$

Power Up! Common Core Test Practice

35. Mr. Solomon bikes at a constant speed of 12 miles per hour. He wants to find the number of hours it will take him to bike 54 miles. Determine if each statement is true or false.

a. To find the number of hours, subtract 12 from 54. ☐ True ☐ False

b. To find the number of hours, divide 54 by 12. ☐ True ☐ False

c. It will take Mr. Solomon 5 hours to bike 54 miles. ☐ True ☐ False

36. The table shows some of the nutritional information for a bottle of iced tea. Marguerite wants to determine how many grams of sugar are in each serving. Let *s* represent the grams of sugar in each serving. Select the correct values to model the situation with a multiplication equation.

Nutritional Facts (2 Servings)
Calories: 80
Total Fat: 0 grams
Sodium: 50 milligrams
Sugars: 64 grams

☐ × ☐ = ☐

0	50	80
2	64	*s*

How many grams of sugar are in each serving? ☐

Common Core Spiral Review

Divide. 5.NTB.6

37. $138 \div 6 =$ ______

38. $80 \div 5 =$ ______

39. $208 \div 4 =$ ______

40. $217 \div 7 =$ ______

41. $216 \div 24 =$ ______

42. $378 \div 6 =$ ______

43. The table shows the cost of concessions at a concert. Evan spent $31.50 buying popcorn for his class. How many bags of popcorn did Evan buy? 5.NTB.7

______

Item	Cost ($)
Nachos	$3.00
Popcorn	$1.50
Water	$2.00

44. After dinner, $\frac{3}{4}$ of a pie remains. If Tasha eats $\frac{1}{6}$ of the remaining pie, how much of the total pie does Tasha eat? 6.NS.1 ______

Inquiry Lab

Solve and Write Division Equations

HOW do you solve division equations using models?

Content Standards
6.EE.5, 6.EE.7

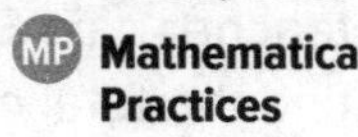

Mathematical Practices
1, 3, 4

Four friends decided to split the cost of season concert tickets equally. Each person paid $35. Find the total cost of the season concert tickets.

What do you know? ______________________________

What do you need to find? ______________________________

Hands-On Activity

Step 1 Define a variable. Use the variable c to represent the total cost of the tickets.

Step 2 Use a bar diagram to help write the equation.

total cost, c

amount each person pays	amount each person pays	amount each person pays	amount each person pays

$35 (one section)

The total length of the diagram shows ______________.

The number 35 represents ______________.

There are four equal sections because ______________.

☐ ÷ ☐ = ☐

Step 3 Work backward. Rewrite the equation as a multiplication sentence and solve.

☐ × ☐ = c

So, the total cost of the season tickets was $☐.

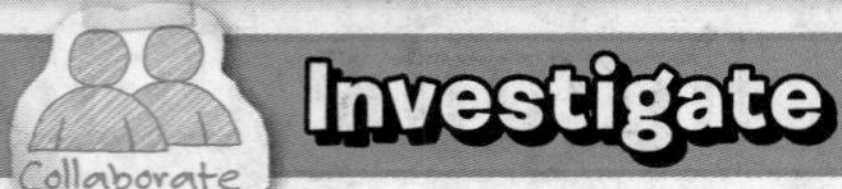

MP Model with Mathematics **Work with a partner. Write and solve a division equation using a bar diagram.**

1. Three teachers went to a conference. They shared the cost of gasoline *g* equally. Each teacher paid $38.50. Draw a bar diagram to find the total cost of gasoline.

2. Silvia has completed 8 math exercises *e*. This is one fourth of the assignment. How many exercises were assigned?

3. Antonio bought a shirt for $\frac{1}{2}$ off. He paid $21.75 for the shirt *s*. Draw a bar diagram to find the original cost of the shirt.

4. Six friends are sharing the cost for a pizza party *p* equally. Each person paid $15.25. Find the total cost of the pizza party.

Create

On Your Own

5. **MP Model with Mathematics** Write a real-world division problem for the equation modeled below. Then write the equation and solve.

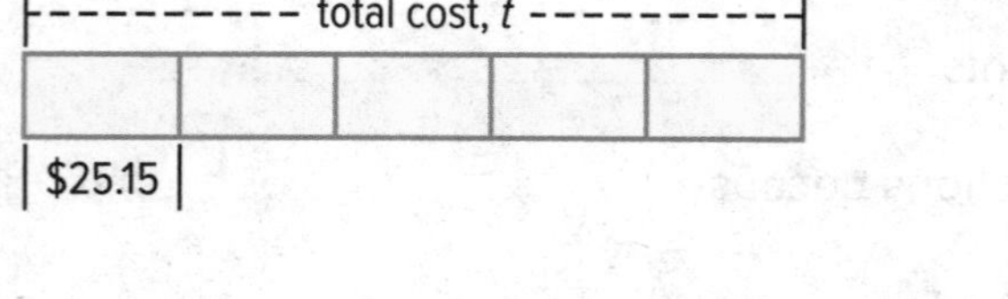

6. Inquiry HOW do you solve division equations using models?

Solve and Write Division Equations

Real-World Link

Allowances Leslie spends $5 a month on snacks at school, which is one fourth of her monthly allowance. Complete the questions below to find Leslie's monthly allowance.

1. Draw a bar diagram to represent $5 as one fourth of Leslie's monthly allowance.

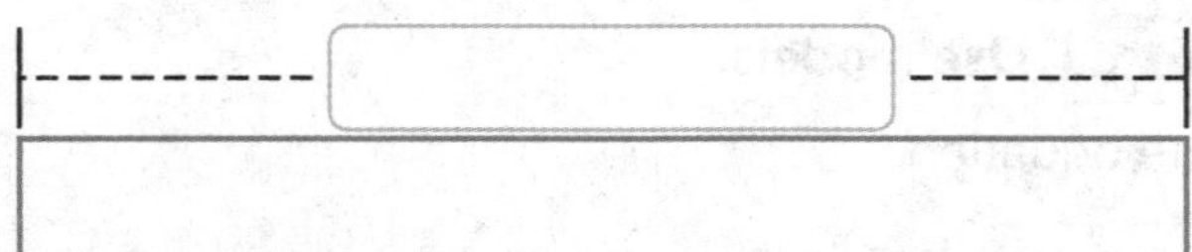

2. What is Leslie's monthly allowance?

3. What operation did you use to find Leslie's allowance?

4. How can you check your answer to determine if it is accurate?

Essential Question

HOW do you determine if two numbers or expressions are equal?

Vocabulary

Multiplication Property of Equality

Common Core State Standards

Content Standards
6.EE.5, 6.EE.7

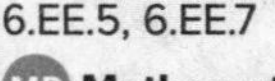

1, 2, 3, 4, 7

Which MP **Mathematical Practices** did you use? Shade the circle(s) that applies.

① Persevere with Problems
② Reason Abstractly
③ Construct an Argument
④ Model with Mathematics
⑤ Use Math Tools
⑥ Attend to Precision
⑦ Make Use of Structure
⑧ Use Repeated Reasoning

Work Zone

Solve Division Equations

In the situation on the previous page, equation $\frac{a}{4} = 5$, where a represents the monthly allowance, means the monthly allowance *divided by* 4 equals \$5. Since multiplication and division are inverse operations, use multiplication to solve division equations.

Example

1. Solve $\frac{a}{3} = 7$. Check your solution.

Method 1 **Use models.**

Model the equation.

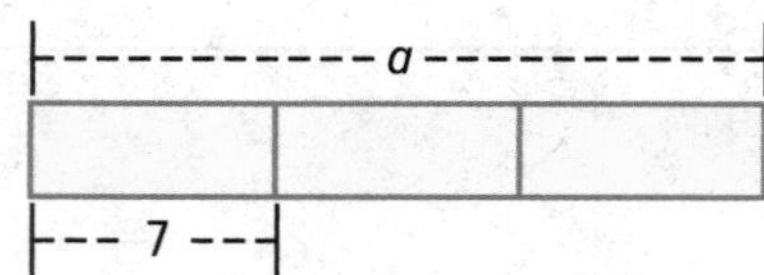

Solve the equation. Work backward.

Since $\frac{a}{3} = 7$, $7 \times 3 = a$. So, $a = 21$.

Method 2 **Use symbols.**

$\frac{a}{3} = 7$	Write the equation.
$\frac{a}{3}(3) = 7(3)$	Multiply each side by 3.
$a = 21$	Simplify.
Check $\frac{a}{3} = 7$	Write the original equation.
$\frac{21}{3} \stackrel{?}{=} 7$	Replace a with 21.
$7 = 7$	This is a true sentence. ✓

Using either method, the solution is 21.

Got it? **Do these problems to find out.**

Solve each equation. Check your solution.

a. $\frac{x}{8} = 9$

b. $\frac{y}{4} = 8$

c. $\frac{m}{5} = 9$

d. $30 = \frac{b}{2}$

Show your work.

a. ____________

b. ____________

c. ____________

d. ____________

Multiplication Property of Equality

Words If you multiply each side of an equation by the same nonzero number, the two sides remain equal.

Examples

Numbers	Algebra
$3 = 3$	$\frac{x}{4} = 7$
$3(6) = 3(6)$	$\frac{x}{4}(4) = 7(4)$
$18 = 18$	$x = 28$

and Reflect

How is solving a multiplication equation similar to solving a division equation? How is it different? Explain below.

When you solve an equation by multiplying each side of the equation by the same number, you are using the **Multiplication Property of Equality**.

Real World Example

Tutor

2. The weight of an object on the Moon is one sixth that of its weight on Earth. If an object weighs 35 pounds on the Moon, write and solve a division equation to find its weight on Earth.

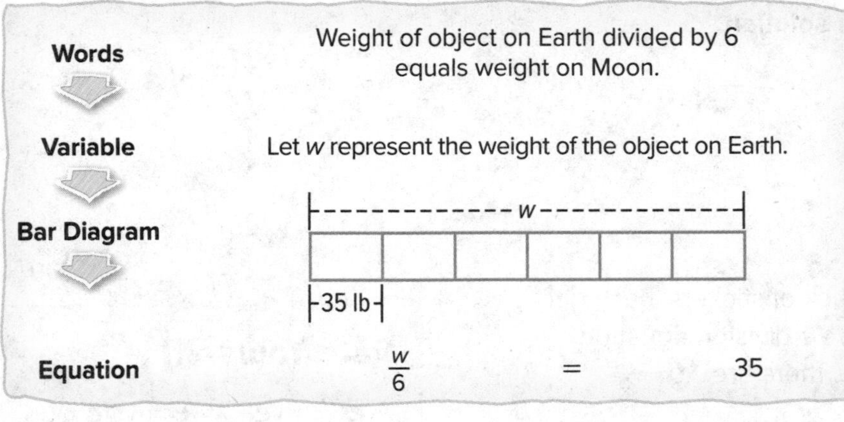

$\frac{w}{6} = 35$ Write the equation.

$\frac{w}{6}(6) = 35(6)$ Multiply each side by 6.

$w = 210$ $6 \times 35 = 210$

The object weighs 210 pounds on Earth.

Got it? Do this problem to find out.

e. Nathan picked a total of 60 apples in $\frac{1}{3}$ hour. Write and solve a division equation to find how many apples Nathan could pick in 1 hour.

e. ______

Example

3. Carla is buying ribbon for costumes. She wants to divide the ribbon into 8.5 inch pieces for 16 costumes. Write and solve a division equation to find the length of ribbon Carla should buy.

Let r represent the length of ribbon Carla should buy.

$\frac{r}{8.5} = 16$ Write the equation.

$\frac{r}{8.5}(8.5) = 16(8.5)$ Multiply each side by 8.5.

$r = 136$ $8.5 \times 16 = 136$

Carla should buy 136 inches of ribbon.

Got it? Do this problem to find out.

f. Allison is baking a pie. She wants 4.5 strawberries in each serving for 8 people. Write and solve a division equation to find how many strawberries Alison will need.

f. _______________

Guided Practice

Solve each equation. Check your solution. (Example 1)

1. $\frac{m}{6} = 10$

2. $\frac{k}{5} = 11$

3. $\frac{v}{13} = 14$

4. Kerry and Tya are sharing a pack of stickers. Each girl gets 11 stickers. Write and solve a division equation to find how many total stickers there are. (Example 2)

____

5. Chen is buying a ham. He wants to divide it into 6.5-ounce servings for 12 people. Write and solve a division equation to find what size ham Chen should buy. (Example 3) _______________

6. **Building on the Essential Question** When solving an equation, why is it necessary to perform the same operation on each side of the equals sign?

____

____

Rate Yourself!

Are you ready to move on? Shade the section that applies.

I have a few questions. | I'm ready to move on. | I have a lot of questions.

For more help, go online to access a Personal Tutor.

FOLDABLES Time to update your Foldable!

Name ______________________ My Homework ______________________

Independent Practice

Go online for Step-by-Step Solutions

Solve each equation. Check your solution. (Examples 1 and 3)

1. $5 = \frac{p}{4}$

2. $17 = \frac{w}{6}$

3. $4.7 = \frac{g}{3.2}$

Write and solve a division equation to solve each problem. (Examples 2 and 3)

4. Sophia is buying party favors. She has a budget of $2.75 a person for 6 people. How much can Sophia spend on party favors?

____

5. Caroline baked 3 dozen oatmeal raisin cookies for the bake sale at school. This is one fourth the number of dozens of cookies she baked in all. How many dozens of cookies did she bake in all?

____

6. **MP Model with Mathematics** Refer to the graphic novel frame below for Exercises a–b.

ITEM READ	POINTS
Book	5
Magazine	1
Newspaper	1

a. If Mei has earned 30 points, write and solve a multiplication equation to find how many books she needs to read. _______________

b. Suppose Mei has read 7 books. Write and solve a division equation to find the number of points she has earned. _______________

7. **MP Identify Structure** Write the property used to solve each type of equation.

H.O.T. Problems Higher Order Thinking

8. **MP Reason Abstractly** Write a division equation that has a solution of 42.

9. **MP Reason Inductively** *True* or *false:* $\frac{x}{3}$ is equivalent to $\frac{1}{3}x$. Explain your reasoning.

10. **MP Persevere with Problems** Explain how you would solve $\frac{16}{c} = 8$. Then solve the equation.

11. **MP Multiple Representations** Every autumn, the North American Monarch butterfly migrates up to 3,000 miles to California and Mexico where it hibernates until early spring. The butterfly travels on average 50 miles per day.

a. **Algebra** Write an equation that represents the distance d a butterfly will travel in t days. ____________

b. **Tables** Use the equation to complete the table.

c. **Words** Use the pattern in the table to determine how many days it will take the butterfly to travel 2,500 miles. ____________

Time (days)	1	2	3	4	5
Distance (miles)					

Extra Practice

Solve each equation. Check your solution.

12. $4 = \frac{r}{8}$

Homework Help

$4 = \frac{r}{8}$

$4(8) = \frac{r}{8}(8)$

$32 = r$

13. $12 = \frac{q}{7}$

14. $18 = \frac{r}{2}$

15. $\frac{h}{13} = 13$

16. $\frac{j}{12} = 11$

17. $\frac{z}{7} = 8$

18. $\frac{c}{0.2} = 7$

19. $\frac{d}{12} = 0.25$

20. $\frac{m}{16} = 0.5$

MP **Identify Structure** **Write and solve a division equation to solve each problem.**

21. One third of a bird's eggs hatched. If 2 eggs hatched, how many eggs did the bird lay? ______

22. Marcel is purchasing a board to build a bookcase. He wants to divide the board into 1.75-foot sections. He needs 6 sections. What size board does Marcel need? ______

23. Blake is cutting a piece of rope into fourths. If each piece is 16 inches long, what is the length of the entire rope? ______

24. MP **Justify Conclusions** A model plane is $\frac{1}{48}$ the size of the actual plane. If the model plane is 28 inches long, how long is the actual plane? Explain your reasoning to a classmate. ______

Power Up! Common Core Test Practice

25. Alfred does chores to earn money in the summer. The table shows the amount he earns per chore. Alfred weeded the garden 6 times over the summer. Write and solve a division equation to find how much he earned weeding the garden.

Chore	Amount Earned ($)
mow lawn	$10
wash car	$ 5
weed garden	$ 8

26. Shana ran 6 miles in 1 week. This was one third of what she ran in the month. Let m represent the number of miles Shana ran in the month. Select the correct values to model the situation with a division equation.

$\frac{\square}{\square} = \square$

1
3
6
7
m

How many miles did Shana run in the month? Explain how you can check your answer.

Common Core Spiral Review

Fill in each ◯ with <, >, or = to make a true sentence. 4.NF.7

27. 6.5 ◯ 5.2

28. 1.9 ◯ 1.7

29. 2.2 ◯ 2.2

30. 5.6 ◯ 6.5

31. 4.2 ◯ 3.9

32. 5.5 ◯ 5.7

33. The table shows the number of inches in different number of feet.

How many inches are in 5 feet? 4.OA.5 ____________________

Feet	Inches
1	12
2	24
3	36
4	48

34. Describe the pattern shown below. Then find the next number in the pattern. 4.OA.5

4, 8, 12, 16, 20, 24, . . .

____

21ST CENTURY CAREER in Music

Sound Engineer

Do you enjoy using electronics to make music sound better? If so, you might want to explore a career in sound engineering. Sound engineers, or audio technicians, prepare the sound equipment for recording sessions and live concert performances. They are responsible for operating consoles and other equipment to control, replay, and mix sound from various sources. Sound engineers adjust the microphones, amplifiers, and levels of various instrument and voice tones so that everything sounds great together.

Is This the Career for You?

Are you interested in a career as a sound engineer? Take some of the following courses in high school.

- Algebra
- Electronic Technology
- Music and Computers
- Physics
- Sound Engineering

Find out how math relates to a career in Music.

MP Amping the Band!

Use the information in the table and the diagram to solve each problem.

1. In the diagram, the distance between the microphones is 6 feet. This is 3 times the distance d from each microphone to the sound source. Write an equation that represents this situation. ______

2. Solve the equation that you wrote in Exercise 1. Explain the solution. ______

3. The distance from the microphone to the acoustic guitar sound hole is about 11 inches less than what it should be. Write an equation that models this situation. ______

4. Solve the equation that you wrote in Exercise 3. Explain the solution. ______

5. The microphone is about 9 times farther from the electric guitar amplifier than it should be to produce a natural, well-balanced sound. Write and solve an equation to find how far from the amplifier the microphone should be placed.

Microphone Mistakes		
Sound Source	**Location of Microphone**	**Resulting Sound**
Acoustic guitar	3 inches from sound hole	very bassy
Electric guitar amplifier	36 inches from amp	thin, reduced bass

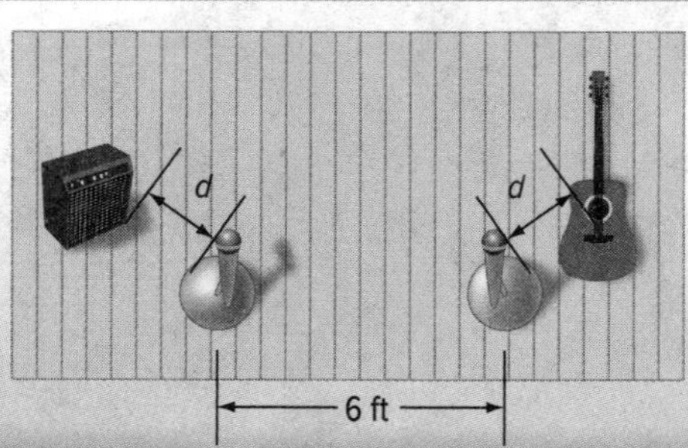

MP Career Project

It's time to update your career portfolio! Go to the *Occupational Outlook Handbook* online and research careers in sound engineering. Make a list of the advantages and disadvantages of working in that field.

List several challenges associated with this career.

-
-
-
-
-

Chapter Review

Vocabulary Check

Write the correct term for each clue in the crossword puzzle.

Across

1. property of equality used to solve multiplication equations
4. replace a variable with a value that results in a true sentence
5. the value of a variable that makes an equation true
6. mathematical sentence showing two expressions are equal
7. property of equality used to solve subtraction equations
8. a combination of numbers, variables, and at least one operation

Down

2. a symbol of equality
3. operations which undo each other

Use Your FOLDABLES

Use your Foldable to help review the chapter.

Tape here

Tab 4

Tab 3

Tab 2

Tab 1

Models

Symbols

division (÷)

Got it?

Match each equation with its solution.

1. $8x = 128$	**a.** $x = 68$
2. $13 + x = 29$	**b.** $x = 39$
3. $72 = 3x$	**c.** $x = 18$
4. $x - 22 = 17$	**d.** $x = 16$
5. $\frac{x}{4} = 17$	**e.** $x = 24$
6. $x - 18 = 33$	**f.** $x = 51$

Study Buddies

Lesha and Maria spend the weekend studying for upcoming tests. They start with math, since that is their favorite subject. The table lists their scores for the first three math tests of the semester.

Student	Test #1	Test #2	Test #3
Lesha	75	100	100
Maria	92	x	88

Write your answers on another piece of paper. Show all of your work to receive full credit.

Part A

Maria cannot remember what she scored on the second test, but she knows that the sum of the three tests is 270. Write and solve an addition equation to determine what she scored on her second test.

Part B

An A grade will be given to students having at least 450 total test points. There are two more tests to take before the semester is over. Lesha wants to know what she needs to score on the next two tests to finish with an A. Write and solve an equation to determine what score she needs to average on the next two tests if each question is worth 1 point. Explain your reasoning.

Part C

Consider the equation $5x = 8$. Write a scenario pertaining to the girls' studying that is represented by this equation. Solve the equation and explain what the answer represents.

Reflect

Answering the Essential Question

Use what you learned about expressions and equations to complete the graphic organizer.

Essential Question

HOW do you determine if two numbers or expressions are equal?

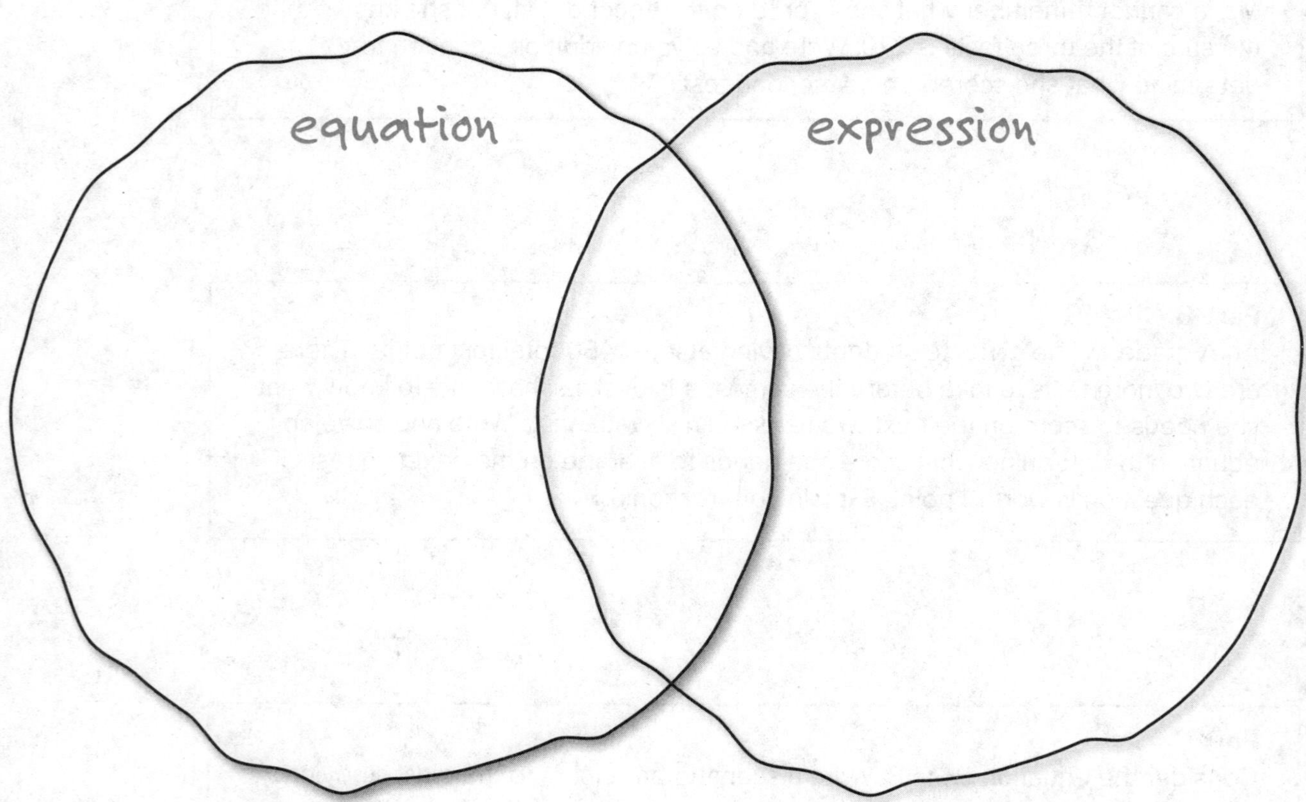

Answer the Essential Question. HOW do you determine if two numbers or expressions are equal?

Chapter 8
Functions and Inequalities

Essential Question

HOW are symbols, such as <, >, and =, useful?

Common Core State Standards

Content Standards
6.EE.2, 6.EE.2c, 6.EE.5, 6.EE.6, 6.EE.8, 6.EE.9

MP Mathematical Practices
1, 2, 3, 4, 5, 6, 7, 8

Math in the Real World

Ocean Life In the ocean, clownfish and sea anemones benefit one another. Clownfish chase away different species of fish that eat the sea anemone. Sea anemones have tentacles that are coated in poison. These tentacles protect the clownfish from predators.

A clownfish can be up to 3.5 inches in length. Some species of sea anemones can be up to 39 inches wide. Compare 3.5 inches and 39 inches.

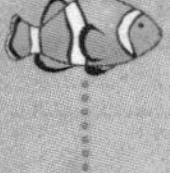

☐ < ☐

1. Cut out the Foldable on page FL7 of this book.
2. Place your Foldable on page 646.
3. Use the Foldable throughout this chapter to help you learn about functions and inequalities.

What Tools Do You Need?

Vocabulary

- arithmetic sequence
- dependent variable
- function
- function rule
- function table
- geometric sequence
- independent variable
- inequality
- linear function
- sequence
- term

Study Skill: Writing Math

Describe Data

When you *describe* something, you represent it in words.

Mark surveyed his class to find their favorite flavor of sugarless gum. Describe the data.

- Eight more people favor peppermint gum over cinnamon gum.
- The total number of people surveyed is 40.

These statements describe the data. What other ways can you describe the data? ______

Favorite Flavor of Sugarless Gum	
Flavor	**Number**
Cinnamon	10
Peppermint	18
Watermelon	12

Describe the data below.

1.

Least Favorite "Bug"	
Kind	**Number**
Centipede	2
Cockroach	18
Spider	30

2.

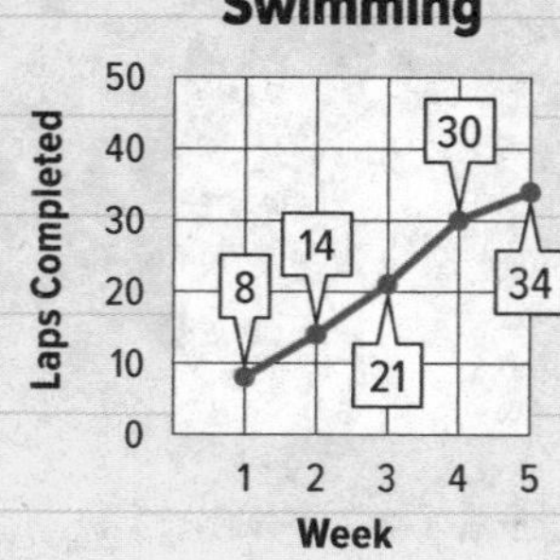

What Do You Already Know?

List three things you already know about functions and inequalities in the first section. Then list three things you would like to learn about functions and inequalities in the second section.

Functions and Inequalities

What I know	What I want to find out

When Will You Use This?

Here are a few examples of how inequalities are used in the real world.

Activity 1 Ask your parents to help you research the cost of an upcoming concert in your region. Give the concert and the cost of one ticket. Are there any additional fees? If so, how much are they?

Activity 2 Go online at **connectED.mcgraw-hill.com** to read the graphic novel ***Concert Ticket Tally***. What is the total cost of two tickets to the concert?

Try the Quick Check below.
Or, take the Online Readiness Quiz.

CCSS Quick Review

Common Core Review 4.NBT.2, 6.EE.7

Example 1

Fill in the ◯ with $<$, $>$, or $=$ to make a true statement.

71,238 ◯ 71,832

71,238 Use place value. Line up the digits.
71,832 Compare the hundreds place. $2 < 8$

So, $71{,}238 < 71{,}832$.

Example 2

Solve $54 + x = 180$.

$$\begin{aligned} 54 + x &= 180 && \text{Write the equation.} \\ -54 \quad &= -54 && \text{Subtract.} \\ x &= 126 \end{aligned}$$

Check $54 + 126 \stackrel{?}{=} 180$
$180 = 180$ ✓

Quick Check

Compare Numbers **Fill in each ◯ with $<$, $>$, or $=$, to make the inequality true.**

1. 302,788 ◯ 203,788

2. 54,300 ◯ 543,000

3. 892,341 ◯ 892,431

4. The table shows the number of bones in humans. Compare 300 and 206. ____________

Bones in Humans	
Baby	300
Adult	206

Solve Equations **Solve each equation.**

5. $x + 44 = 90$ ________

6. $x - 7 = 18$ ________

7. $16m = 48$ ________

8. In the first two basketball games, Lee scored a total of 40 points. If he scored 21 points in the second game, how many points did he score in the first game?

____

Which problems did you answer correctly in the Quick Check? Shade those exercise numbers below.

① ② ③ ④ ⑤ 

⑧

Lesson 1
Function Tables

Real-World Link

Science A ruby-throated hummingbird beats its wings about 52 beats per second.

1. Make a table showing show many times this bird beats its wings in 2 seconds.

Number of Seconds (s)	s • 52	Wing Beats
2	2 • 52	

2. Make a table to show how many times it beats its wings in 6 seconds.

Number of Seconds (s)	s • 52	Wing Beats
6		

3. Make a table to show how many times it beats its wings in 20 seconds.

Number of Seconds (s)	s • 52	Wing Beats

4. A giant hummingbird beats its wings about 10 times per second. Make a table to show how many times the Giant Hummingbird beats its wings in 3 seconds.

Number of Seconds (s)	s • 10	Wing Beats

Which MP Mathematical Practices did you use? Shade the circle(s) that applies.

① Persevere with Problems
② Reason Abstractly
③ Construct an Argument
④ Model with Mathematics
⑤ Use Math Tools
⑥ Attend to Precision
⑦ Make Use of Structure
⑧ Use Repeated Reasoning

Essential Question

HOW are symbols, such as <, >, and =, useful?

Vocabulary

function
function rule
function table
independent variable
dependent variable

Common Core State Standards

Content Standards
6.EE.2, 6.EE.2c, 6.EE.9

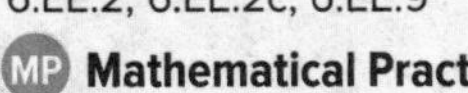

Mathematical Practices
1, 3, 4, 5

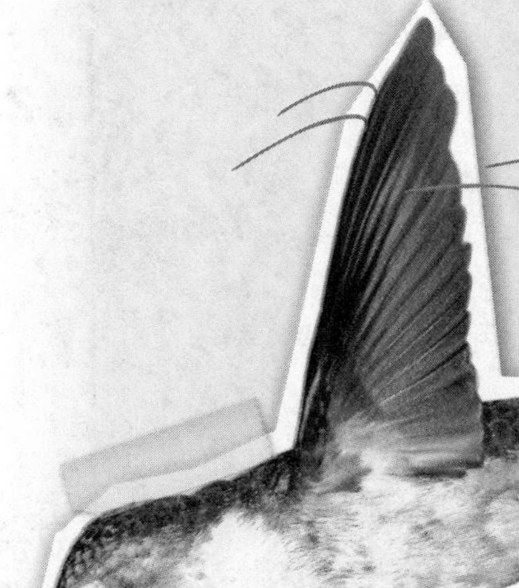

Work Zone

Find the Output for a Function Table

A **function** is a relation that assigns exactly one output value to one input value. The number of wing beats (output) depends on the number of seconds (input). The **function rule** describes the relationship between each input and output. You can organize the input-output values and the function rule in a **function table**.

In a function, the input value is also known as the **independent variable**, since it can be any number you choose. The value of the output depends upon the input value, so the output value is known as the **dependent variable**.

and Reflect

What values were used for the independent variable in Example 1? Answer below.

Examples

1. The output is 7 more than the input. Complete a function table for this relation.

The function rule is $x + 7$. Add 7 to each input.

Input (x)	$x + 7$	Output
10		
12		
14		

→

Input (x)	$x + 7$	Output
10	$10 + 7$	17
12	$12 + 7$	19
14	$14 + 7$	21

2. The output is 5 times the input. Complete a function table for this relation.

The function rule is $5x$. Multiply each input by 5.

Input (x)	$5x$	Output
8		
10		
12		

→

Input (x)	$5x$	Output
8	5 •	
10	5 •	
12	5 •	

Got it? Do these problems to find out.

a.

Input (x)	$x - 4$	Output
4		
7		
10		

b.

Input (x)	$3x$	Output
0		
2		
5		

Find the Input for a Function Table

The input and output of a function table can be represented as a set of ordered pairs, or a *relation*. In this lesson, the *x*-values represent the input and the *y*-values represent the output.

Example

3. Find the input for the function table.

Input (x)	$3x$	Output
		6
		15
		21

Use the *work backward* strategy to determine the input. If the output is found by multiplying by 3, then the input is found by dividing by 3.

The input values are $6 \div 3$ or 2, $15 \div 3$ or 5, and $21 \div 3$ or 7.

Got it? **Do these problems to find out.**

c.

Input (x)	$2x - 1$	Output
		1
		3
		5

d.

Input (x)	$3x + 2$	Output
		17
		20
		29

Real World

Example

4. The Gomez family is traveling at a rate of 70 miles per hour. The function rule that represents this situation is $70x$, where x is the number of hours. Make a table to find how many hours they have driven at 140 miles, 280 miles, and 350 miles. Then graph the function.

Input (x)	$70x$	Output (y)
2	70(2)	140
4	70(4)	280
5	70(5)	350

Use the *work backward* strategy. Divide each output by 70.

The missing input values are $140 \div 70$ or 2, $280 \div 70$ or 4, and $350 \div 70$ or 5.

The input and output values are the ordered pairs (x, y). Plot each ordered pair on the graph.

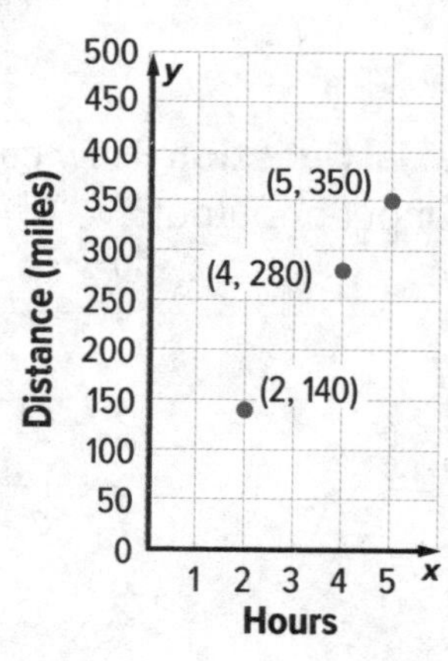

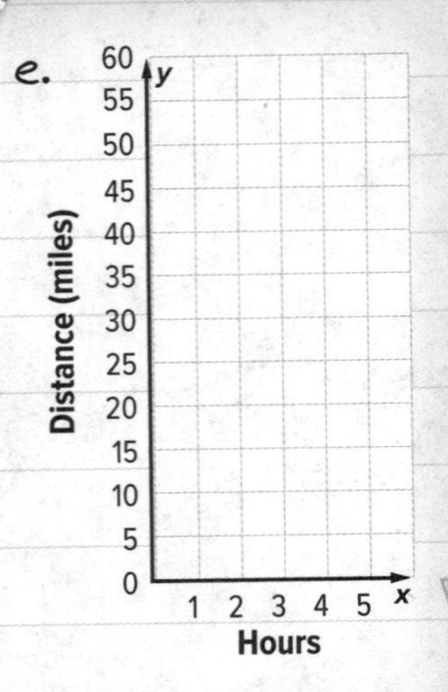

Got it? **Do this problem to find out.**

e. Briana bikes 12 miles per hour. The function rule that represents this situation is $12x$, where x is the number of hours. Make a table to find how many hours she has biked when she has gone 12, 36, and 48 miles. Then graph the function.

Input (x)	$12x$	Output (y)

Guided Practice

Check

1. Isaiah is buying jelly beans. In bulk, they cost \$3 per pound, and a candy dish costs \$2. The function rule, $3x + 2$ where x is the number of pounds, can be used to find the total cost of x pounds of jelly beans and 1 dish. Make a table that shows the total cost of buying 2, 3, or 4 pounds of jelly beans and 1 dish. (Examples 1 and 2)

Pounds (x)	$3x + 2$	Cost (\$) (y)

2. Jasper hikes 4 miles per hour. The function rule that represents this situation is $4x$, where x is the number of hours. Make a table to find how many hours he has hiked when he has gone 8, 12, and 20 miles. Then graph the function. (Examples 3 and 4)

Hours (x)	$4x$	Miles (y)

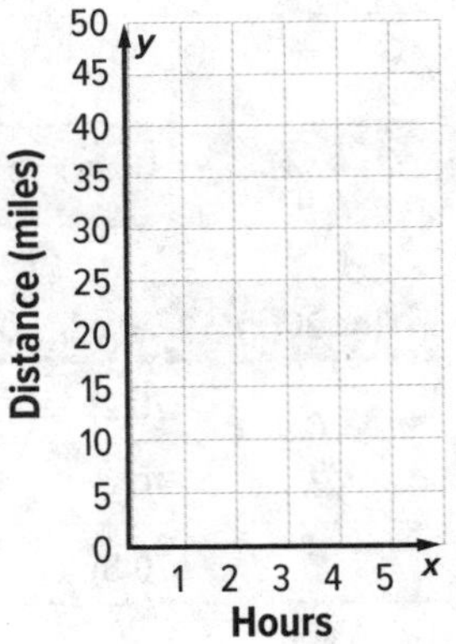

3. **Building on the Essential Question** How can a function table help you find input or output?

Rate Yourself!

Are you ready to move on? Shade the section that applies.

I have a few questions.

I'm ready to move on.

I have a lot of questions.

For more help, go online to access a Personal Tutor.

Tutor

FOLDABLES Time to update your Foldable!

Name ______________________ My Homework ______________________

Independent Practice

Go online for Step-by-Step Solutions 

MP Use Math Tools Complete each function table. (Examples 1–3)

1.

Input (x)	3x + 5	Output
0		
3		
9		

2.

Input (x)	x − 4	Output
4		
8		
11		

3.

Input (x)	x + 2	Output
		2
		3
		8

4.

Input (x)	2x + 4	Output
		18
		22
		34

5. Whitney has a total of 30 cupcakes for her guests. The function rule, $30 \div x$ where x is the number of guests, can be used to find the number of cupcakes per guest. Make a table of values that shows the number of cupcakes each guest will get if there are 6, 10, or 15 guests. Then graph the function. (Examples 1 and 2)

Number of Guests (x)	30 ÷ x	Cupcakes per Guest (y)

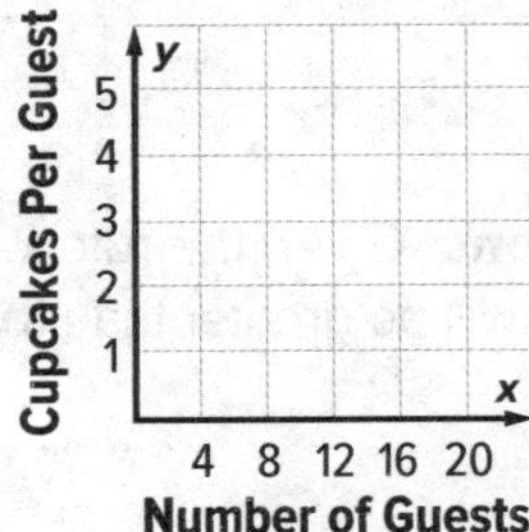

6. Bella rollerblades 8 miles in one hour. The function rule that represents this situation is $8x$, where x is the number of hours. Make a table to find how many hours she has skated when she has traveled 16, 24, and 32 miles. Then graph the function. (Examples 3 and 4)

Hours (x)	8x	Miles (y)

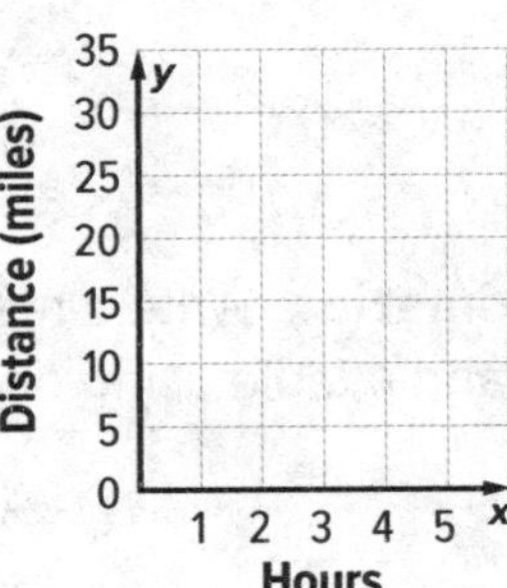

7. Refer to Exercise 6. How many miles would Bella travel if she skated for 7 hours? ______________________

H.O.T. Problems Higher Order Thinking

8. **MP Find the Error** Daniella is finding the output when the function rule is $10 \div x$ and the input is 2. Find her mistake and correct it.

9. **MP Persevere with Problems** Around 223 million Americans keep containers filled with coins in their home. Suppose each of the 223 million people started putting their coins back into circulation at a rate of $10 per year. Create a function table that shows the amount of money that would be recirculated in 1, 2, and 3 years.

10. **MP Reason Inductively** Explain how to find the input given a function rule and output.

11. **MP Justify Conclusions** Given the rule $x \div n$, describe the values of n for which the output value will be greater than the input value. Justify your response.

12. **MP Reason Inductively** Compare and contrast the tables used in this lesson to ratio tables.

13. **MP Model with Mathematics** Write a real-world problem that can be represented by a rule and a table using division.

Extra Practice

Use Math Tools Complete each function table.

14.

Input (x)	x + 3	Output
0	0 + 3	3
2	2 + 3	5
4	4 + 3	7

15.

Input (x)	4x + 2	Output
1		
3		
6		

16.

Input (x)	x − 1	Output
		0
		2
		4

17.

Input (x)	2x − 6	Output
		0
		6
		12

18. Ricardo weighs 2 pounds more than twice his sister's weight. The function rule, $2x + 2$ where x is his sister's weight, can be used to find Ricardo's weight. Make a table of values that show Ricardo's weight when his sister is 20, 30, and 40 pounds. Then graph the function.

Ricardo's Sister's Weight (x)	2x + 2	Ricardo's Weight (y)

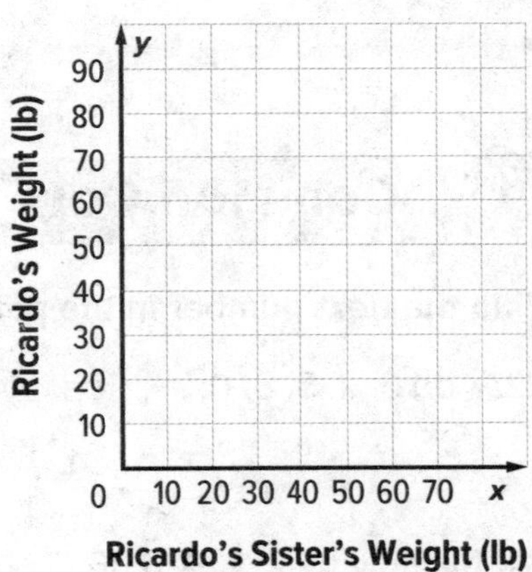

19. The Quinn family drove at a rate of 55 miles per hour. The function rule that represents this situation is $55x$, where x is the number of hours. Make a table to find how many hours they have traveled when they have driven 165, 220, and 275 miles. Then graph the function.

Hours (x)	55x	Miles (y)

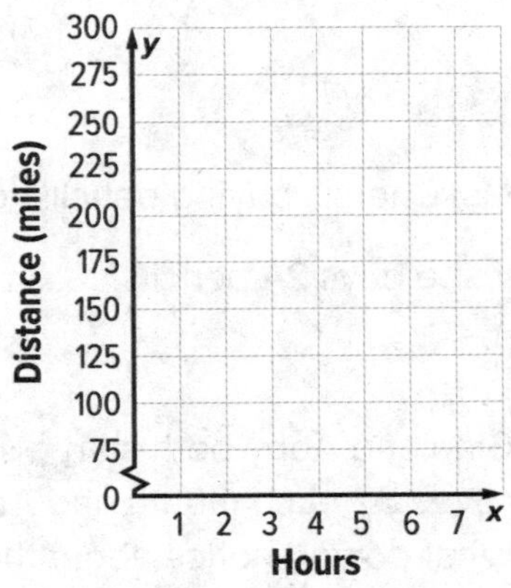

CCSS Power Up! Common Core Test Practice

20. In football, a touchdown is worth 6 points. Complete the table that shows the points earned for scoring 1, 2, 3, 4, and 5 touchdowns.

Number of Touchdowns (x)	$6x$	Points (y)
1		
2		
3		
4		
5		

How many points would a team earn for scoring 8 touchdowns?

[ ]

21. Refer to the function table at the right. Determine if each statement is true or false.

a. The output value when $x = 5$ is 3. ☐ True ☐ False

b. The output value when $x = 6$ is 13. ☐ True ☐ False

c. The output value when $x = 7$ is 16. ☐ True ☐ False

Input (x)	$3x - 5$	Output (y)
5	$3(5) - 5$	■
6	$3(6) - 5$	■
7	$3(7) - 5$	■

CCSS Common Core Spiral Review

Find the next number in the pattern using the given rule. 5.OA.3

22. Add 3: 2, 5, 8, 11, . . . ______

23. Subtract 2: 10, 8, 6, 4, . . . ______

24. Multiply by 2: 2, 4, 8, 16 . . . ______

25. Subtract 7: 84, 77, 70, 63, . . . ______

26. Multiply by 2: 3, 6, 12, 24, . . . ______

27. Add 15: 12, 27, 42, 57, . . . ______

28. Ms. Chen is buying pencils for her class. What is the cost if she buys 24 pencils? 5.NBT.7 ______

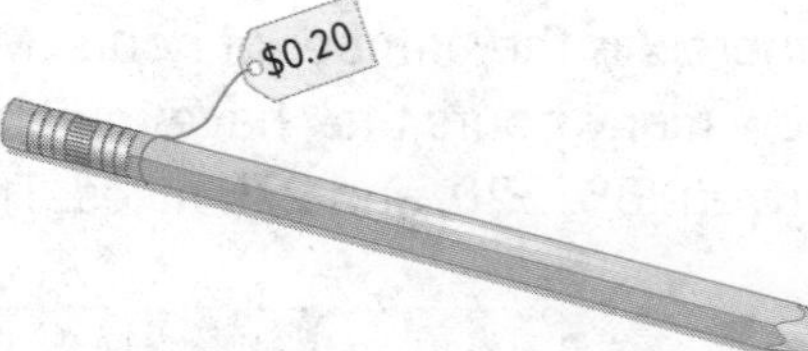

29. Gino and Abby both start a saving account in May. Gino saves \$2 each month and Abby saves \$4 each month. What do you notice about the amount in each account each month? 5.OA.3

______

Month	Gino's Account ($)	Abby's Account ($)
May	2	4
June	4	8
July	6	12

Lesson 2

Function Rules

Vocabulary Start-Up 

A **sequence** is a list of numbers in a specific order. Each number in the list is called a **term** of the sequence.

Arithmetic sequences can be found by adding the same number to the previous term. In a **geometric sequence**, each term is found by multiplying the previous term by the same number.

Compare arithmetic sequences and geometric sequences.

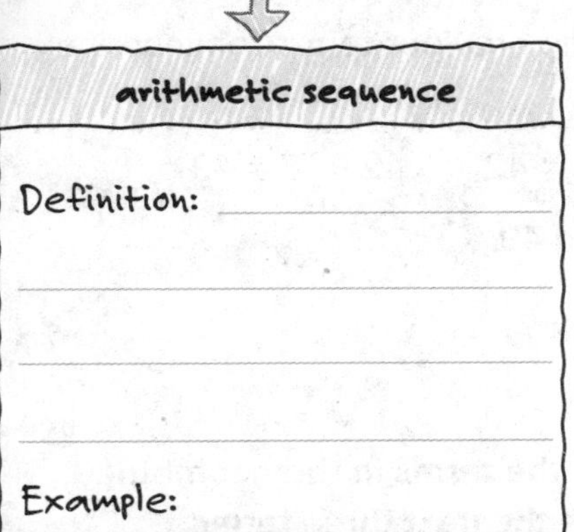

geometric sequence

Definition:

Example:

 **Essential Question**

HOW are symbols, such as <, >, and =, useful?

 **Vocabulary**

sequence
term
arithmetic sequence
geometric sequence

 **Common Core State Standards**

Content Standards
6.EE.2, 6.EE.2c, 6.EE.6, 6.EE.9

MP Mathematical Practices
1, 3, 4, 7

 Real-World Link

Delivery The China Palace sells lunch specials for $6 with a delivery charge of $5 per order. Fill in the table with the next three numbers in the sequence.

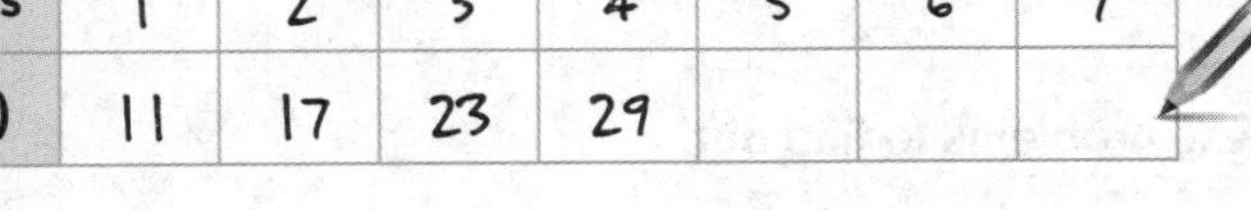

Specials	1	2	3	4	5	6	7
Cost ($)	11	17	23	29			

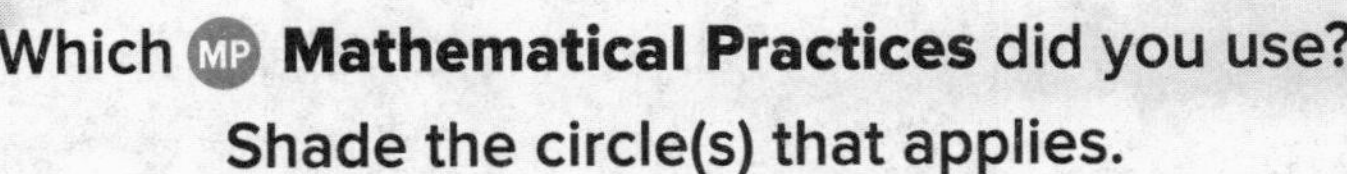

Which MP **Mathematical Practices** did you use? Shade the circle(s) that applies.

① Persevere with Problems
② Reason Abstractly
③ Construct an Argument
④ Model with Mathematics
⑤ Use Math Tools
⑥ Attend to Precision
⑦ Make Use of Structure
⑧ Use Repeated Reasoning

Work Zone

Arithmetic and Geometric Sequences

Determining if a sequence is arithmetic or geometric can help you find the pattern. When you know the pattern, you can continue the sequence to find missing terms.

Examples

1. Describe the relationship between the terms in the arithmetic sequence 7, 14, 21, 28, ... Then write the next three terms.

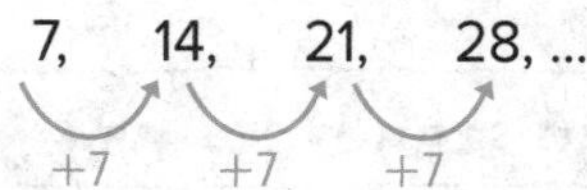

Each term is found by adding 7 to the previous term. Continue the pattern to find the next three terms.

$28 + 7 = 35$ $\quad 35 + 7 = 42$ $\quad 42 + 7 = 49$

The next three terms are 35, 42, and 49.

2. Describe the relationship between the terms in the geometric sequence 2, 4, 8, 16, ... Then write the next three terms.

Show your work.

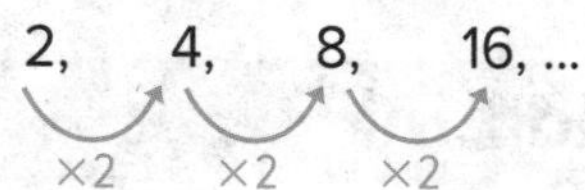

Each term is found by multiplying the previous term by two. Continue the pattern to find the next three terms.

$16 \times 2 = 32$ $\quad 32 \times 2 = 64$ $\quad 64 \times 2 = 128$

The next three terms are 32, 64, and 128.

Got it? Do these problems to find out.

a. 0, 15, 30, 45, ...

b. 4.5, 4, 3.5, 3, ...

c. 1, 3, 9, 27, ...

d. 3, 6, 12, 24, ...

a. ________

b. ________

c. ________

d. ________

Find a Rule

A sequence can also be shown in a table. The table gives both the position of each term in the list and the value of the term.

List

8, 16, 24, 32, ...

Table

Position	1	2	3	4
Value of Term	8	16	24	32

You can write an algebraic expression to describe a sequence. The value of each term can be described as a function of its position in the sequence.

In the table above, the position can be considered the input, and the value of the term as the output.

Example

3. Use words and symbols to describe the value of each term as a function of its position. Then find the value of the tenth term.

Position	1	2	3	4	n
Value of Term	3	6	9	12	■

Notice that the value of each term is 3 times its position number. So, the value of the term in position n is $3n$.

Position	Multiply by 3	Value of Term
1	1×3	3
2	2×3	6
3	3×3	9
4	4×3	12
n	$n \times 3$	$3n$

Now find the value of the tenth term.

$3n = 3 \cdot 10$ Replace n with 10.

$= 30$ Multiply.

The value of the tenth term in the sequence is 30.

You can check your rule by working backward. Divide each term by 3 to check the position.

Got it? Do these problems to find out.

Use words and symbols to describe the value of each term as a function of its position. Then find the value of the eighth term.

e.

Position	2	3	4	5	n
Value of Term	12	18	24	30	■

f.

Position	3	4	5	6	n
Value of Term	7	8	9	10	■

e. ____________

f. ____________

Example

4. The table shows the number of necklaces Ari can make, based on the number of hours she works. Write a function rule to find the number of necklaces she can make in x hours.

To find the rule, determine the function.

Hours (x)	Number of Necklaces
1	5
2	7
3	9
x	■

Notice that the values 5, 7, 9, ... increase by 2, so the rule includes $2x$. If the rule were simply $2x$, then the number of necklaces in 1 hour would be 2. But this value is 5, which is three more than $2x$.

To test the rule $2x + 3$, use the *guess, check, and revise* strategy.

Row 1: $2x + 3 = 2(1) + 3 = 2 + 3$ or 5

Row 3: $2x + 3 = 2(3) + 3 = 6 + 3$ or 9

The rule $2x + 3$ represents the function table.

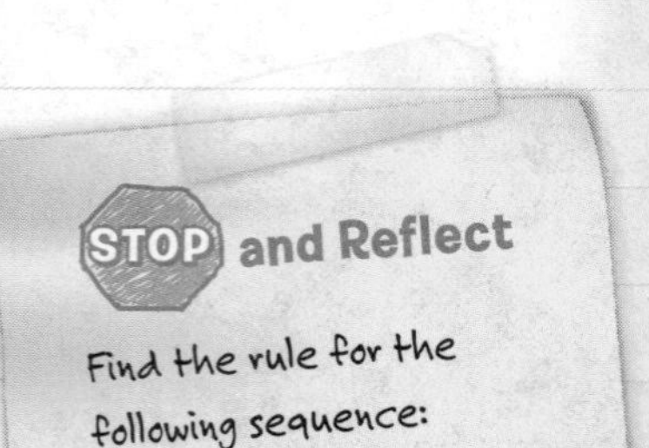

Guided Practice

1. Describe the relationship between the terms in the sequence 13, 26, 52, 104, ... Then write the next three terms in the sequence. (Examples 1 and 2)

2. Use words and symbols to describe the value of each term as a function of its position. Then find the value of the fifteenth term in the sequence. (Example 3)

Position	1	2	3	4	n
Value of Term	2	4	6	8	■

3. The table at the right shows the fee for overdue books at a library, based on the number of weeks the book is overdue. Write a function rule to find the fee for a book that is x weeks overdue. (Example 4)

Weeks Overdue (x)	Fee ($)
1	3
2	5
3	7
4	9
x	■

4. **Building on the Essential Question** What is the difference between an arithmetic sequence and a geometric sequence?

Rate Yourself!

Are you ready to move on? Shade the section that applies.

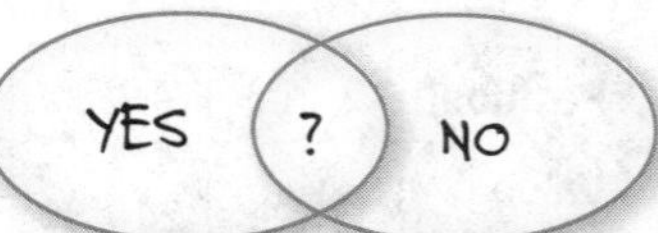

For more help, go online to access a Personal Tutor.

Name ______________________ My Homework ______________________

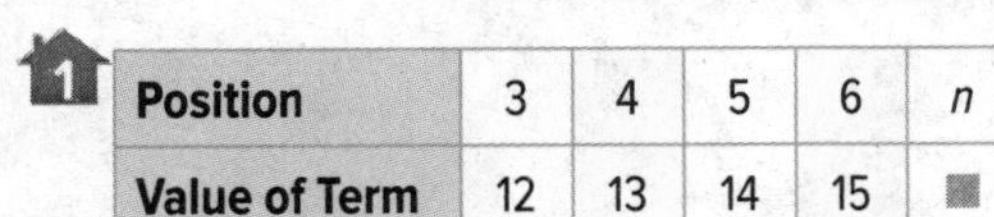

Independent Practice

Go online for Step-by-Step Solutions

Use words and symbols to describe the value of each term as a function of its position. Then find the value of the twelfth term in the sequence. (Examples 1–3)

1.

Position	3	4	5	6	n
Value of Term	12	13	14	15	■

2.

Position	2	3	4	5	n
Value of Term	24	36	48	60	■

3. Describe the relationship between the terms in the sequence 6, 18, 54, 162, ... Then write the next three terms in the sequence. (Example 2)

4. The table shows the amount it costs to rock climb at an indoor rock climbing facility, based on the number of hours. What is the rule to find the amount charged to rock climb for x hours? (Example 4)

Time (x)	Amount ($)
1	13
2	21
3	29
4	37
x	■

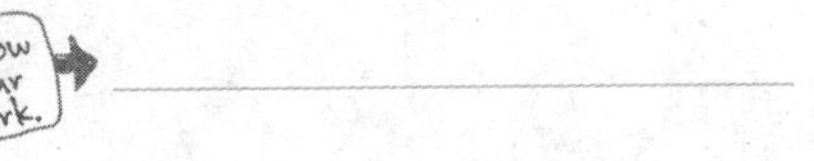

MP Identify Structure Determine how the next term in each sequence can be found. Then find the next two terms in the sequence.

5. 4, 16, 28, 40, ...

show your work.

6. 1.5, 3.9, 6.3, 8.7, ...

7. $2\frac{1}{4}, 2\frac{3}{4}, 3\frac{1}{4}, 3\frac{3}{4}, ...$

Find the missing number in each sequence.

8. 30, ______, 19, $13\frac{1}{2}$, ...

9. 43.8, 36.7, ______, 22.5, ...

State whether each sequence is arithmetic or geometric. Then find the next two terms in the sequence.

10. 1, 6, 36, 216

11. 0.75, 1.75, 2.75, 3.75

12. 0, 13, 26, 39

13 Jay is stacking cereal boxes to create a store display. The number of boxes in each row is shown in the table. Is the pattern an example of an arithmetic sequence or a geometric sequence? Explain. How many boxes will be in row 5?

Row	Number of Boxes
1	4
2	6
3	8
4	10
5	■

H.O.T. Problems Higher Order Thinking

14. MP **Reason Inductively** Create a sequence in which $1\frac{1}{4}$ is added to each number.

15. MP **Persevere with Problems** Refer to the table below. Use words and symbols to generalize the relationship of each term as a function of its position. Then determine the value of the term when $n = 100$.

Position	1	2	3	4	5	n
Value of Term	1	4	9	16	25	■

16. MP **Justify Conclusions** What is the rule to find the value of the missing term in the sequence in the table at the right? Justify your response.

Position, x	Value of Term
1	1
2	5
3	9
4	13
5	17
x	■

Name ______________________________ My Homework ______________

Extra Practice

Use words and symbols to describe the value of each term as a function of its position. Then find the value of the twelfth term in the sequence.

17.

Position	6	7	8	9	*n*
Value of Term	2	3	4	5	■

Homework Help

Look at position 6 and the value of the term. 2 is 4 less than 6, so try subtracting 4 from the other position numbers listed. The function rule is $n - 4$. $12 - 4 = 8$

subtract 4 from the position number; $n - 4$; 8

18.

Position	1	2	3	4	*n*
Value of Term	5	10	15	20	■

19. Describe the relationship between the terms in the sequence 4, 12, 36, 108, ... Then write the next three terms in the sequence.

20. The table shows the cost of a pizza based on the number of toppings. Write a function rule to find the cost for a pizza with *x* toppings.

Number of Toppings (*x*)	Cost ($)
1	12
2	14
3	16
4	18

MP **Identify Structure Determine how the next term in each sequence can be found. Then find the next two terms in the sequence.**

21. 1, 4, 7, 10, ...

22. 2.3, 3.2, 4.1, 5.0, ...

23. $1\frac{1}{2}$, 3, $4\frac{1}{2}$, 6, ...

Find the missing number in each sequence.

24. 7, ________, 16, $20\frac{1}{2}$, ...

25. 14.6, ________, 24, 28.7, ...

Power Up! Common Core Test Practice

26. Which of the following statements is true about the sequence below? Select all that apply.

3, 21, 39, 57, ...

- ☐ This is a geometric sequence.
- ☐ This is an arithmetic sequence.
- ☐ The fifth term of the sequence is 71.
- ☐ Each term is found by adding 18 to the previous term.

27. The table shows the number of cans of soup in each level of a display at a grocery store.

Level (n)	Number of Cans
1	3
2	6
3	12
4	24
n	■

Select the correct values to complete each statement.

2	3	4	6
48	64	72	96

To find additional terms of the sequence, multiply the previous term by ______.

There will be ______ cans of soup in the sixth level of the display.

The sequence of numbers represents a(n) ______ sequence.

Common Core Spiral Review

Multiply. 5.NBT.5

28. $62 \times 3 =$ ______

29. $12 \times 7 =$ ______

30. $16 \times 8 =$ ______

31. The table shows the cost to rent from Ray's Rentals. How much would it cost to rent a video game for 3 weeks? 5.NBT.7 ______

Rental	Cost per Week ($)
Movie	3.50
Video Game	4.50
Game System	20

32. Plot and label points $K(3, 4)$, $A(1, 3)$, and $J(4, 2)$ on the graph. 5.G.2

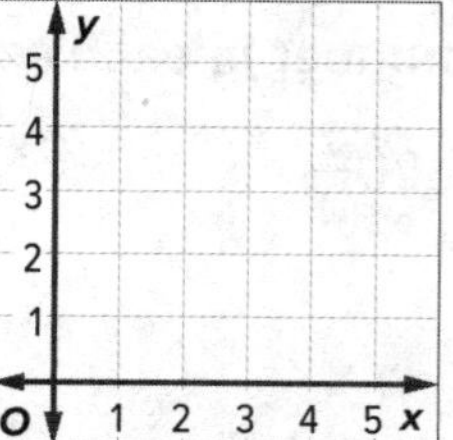

Lesson 3

Functions and Equations

Vocabulary Start-Up

A **linear function** is a function whose graph is a line.

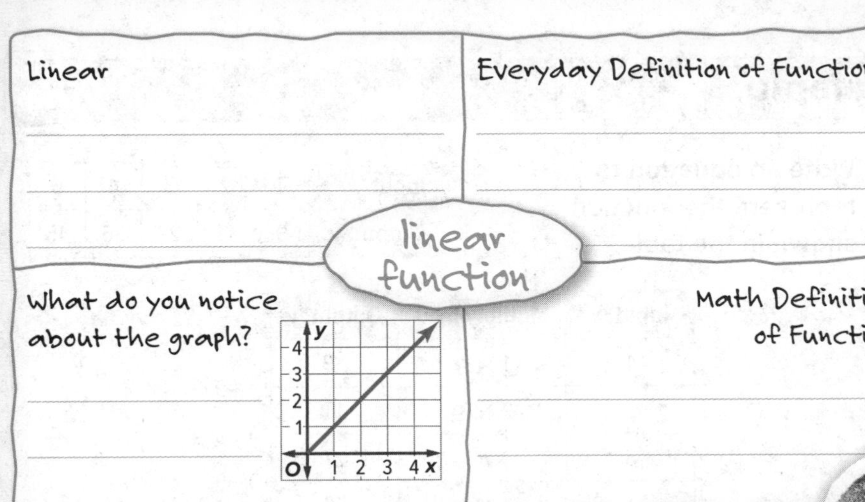

Essential Question

HOW are symbols, such as <, >, and =, useful?

Vocabulary

linear function

Common Core State Standards

Content Standards
6.EE.9

MP Mathematical Practices
1, 3, 4, 8

Real-World Link

Babysitting The table shows the amount of money Carli earns based on the number of hours she babysits.

1. Write a sentence that describes the relationship between the number of hours she babysits and her earnings.

2. Does she earn the same amount each hour?

   Explain.

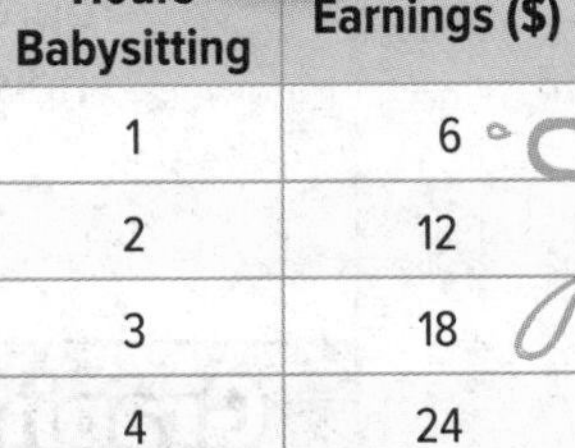

Hours Babysitting	Earnings ($)
1	6
2	12
3	18
4	24

Which MP Mathematical Practices did you use? Shade the circle(s) that applies.

① Persevere with Problems
② Reason Abstractly
③ Construct an Argument
④ Model with Mathematics
⑤ Use Math Tools
⑥ Attend to Precision
⑦ Make Use of Structure
⑧ Use Repeated Reasoning

Work Zone

STOP and Reflect

In the equation $d = 36t$, where d is the distance traveled and t is the time, which variable is independent and which is dependent? Explain below.

Write an Equation to Represent a Function

You can use an equation to represent a function. The input, or independent variable, represents the *x*-value, and the output, or dependent variable, represents the *y*-value. An equation expresses the dependent variable in terms of the independent variable.

Example

1. Write an equation to represent the function shown in the table.

Input, x	1	2	3	4	5
Output, y	9	18	27	36	45

Input, x	Multiply by 9	Output, y
1	1×9	9
2	2×9	18
3	3×9	27
4	4×9	36
5	5×9	45

+9, +9, +9, +9

The value of y is equal to 9 times the value of x. So, the equation that represents the function is $y = 9x$.

Got it? **Do this problem to find out.**

a. Write an equation to represent the function shown in the table.

Input, x	1	2	3	4	5
Output, y	16	32	48	64	80

a. ____________

Graph Linear Functions

You can also graph a function. If the graph is a line, the function is then called a *linear equation*. When graphing the function, the input is the *x*-coordinate and the output is the *y*-coordinate.

$$(\textit{input}, \textit{output}) \longrightarrow (x, y)$$

Example

2. Graph $y = 2x$.

Step 1 Make a table of ordered pairs. Select any three values for x. Substitute these values for x to find y.

x	$2x$	y	(x, y)
0	2(0)	0	(0, 0)
1	2(1)	2	(1, 2)
2	2(2)	4	(2, 4)

Step 2 Graph each ordered pair. Draw a line through each point.

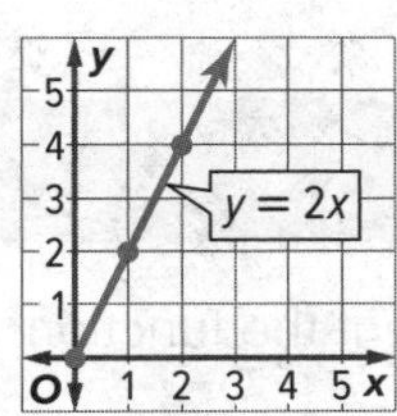

Got it? Do these problems to find out.

b. $y = x + 1$ **c.** $y = 3x + 2$

Show your work.

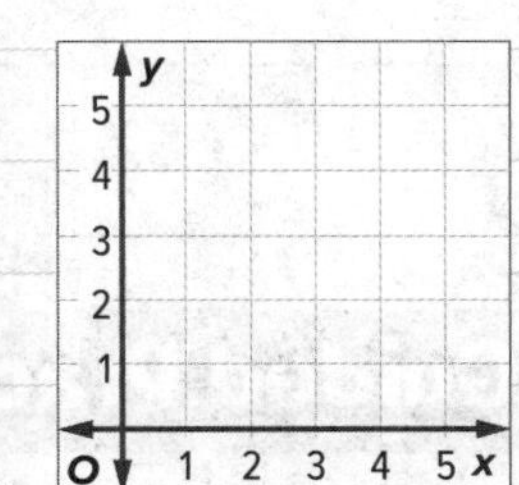

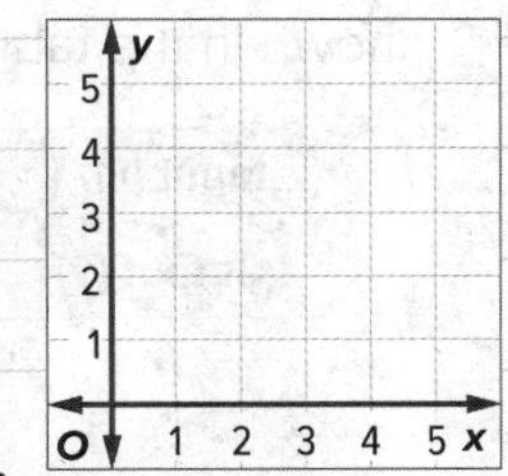

Examples

Martino constructed the graph shown, which shows the height of his cactus after several years of growth.

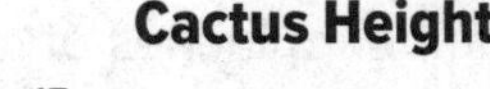

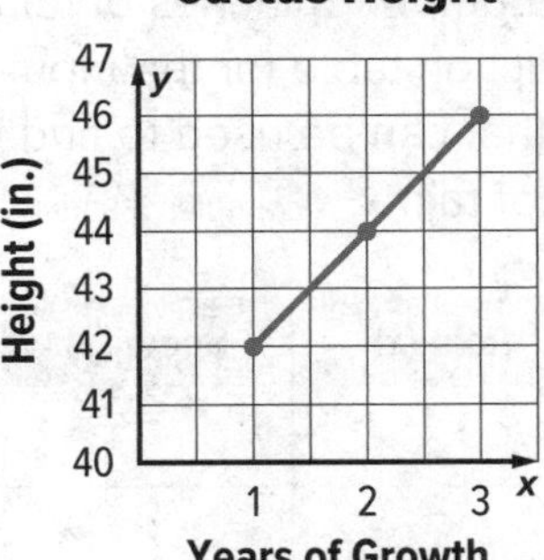

3. Make a function table for the input-output values.

The three input values are 1, 2, and 3. The corresponding output values are 42, 44, and 46.

Input (x)	Output (y)
1	42
2	44
3	46

4. Write an equation from the graph that could be used to find the height y of the cactus after x years.

Since the output values increase by 2, the equation includes $2x$. Each output value is 40 more than twice the input. So, the equation is $y = 2x + 40$.

Show your work.

Magazines (x)	Total (y)

d. ____________

Got it? **Do this problem to find out.**

d. The graph shows the total amount y that you spend if you buy one book and x magazines. Make a function table for the input-output values. Write an equation from the graph that could be used to find the total amount y if you buy one book and x magazines.

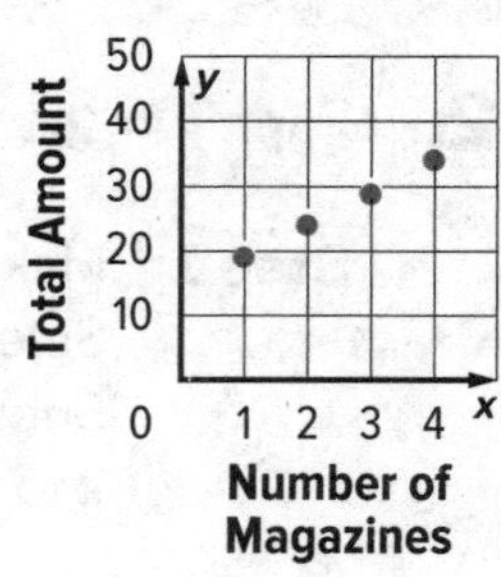

Guided Practice

1. Write an equation to represent the function shown in the table. (Example 1)

Input (x)	0	1	2	3	4
Output (y)	0	4	8	12	16

2. Graph the function $y = x + 3$. (Example 2)

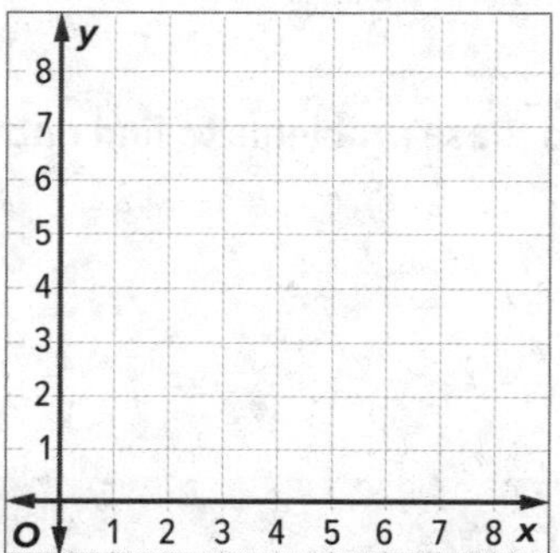

3. The graph below shows the number of inches of rainfall x equivalent to inches of snow y. Make a function table for the input-output values. Write an equation from the graph that can be used to find the total inches of snow y equivalent to inches of rain x. (Examples 3 and 4)

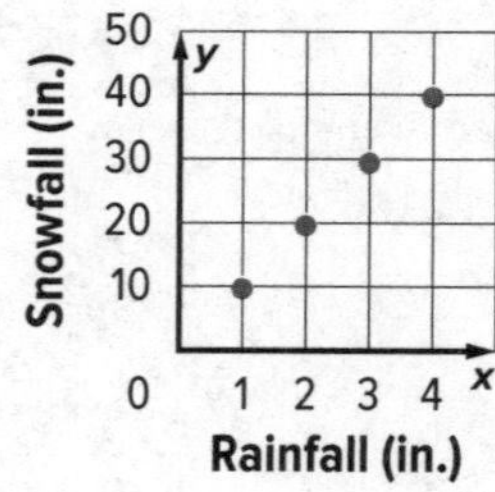

Rain (x)	Snow (y)

4. **Building on the Essential Question** How are ordered pairs of a function used to create the graph of the function?

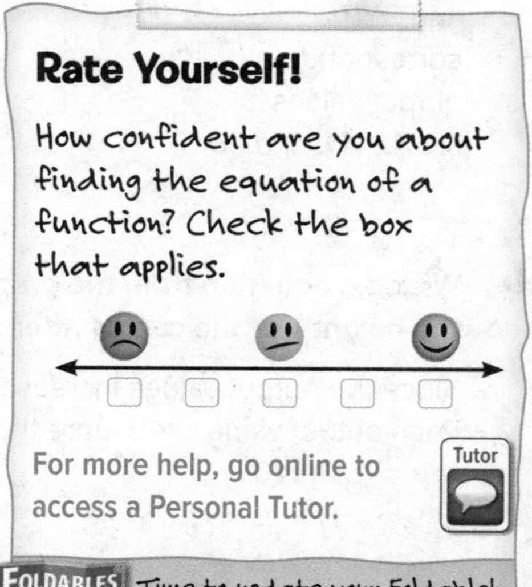

Name ______________________ My Homework ______________________

Independent Practice

Go online for Step-by-Step Solutions 

eHelp

Write an equation to represent each function. (Example 1)

1.

Input (x)	1	2	3	4	5
Output (y)	6	12	18	24	30

2.

Input (x)	0	1	2	3	4
Output (y)	0	15	30	45	60

Graph each equation. (Example 2)

 **3.** $y = x + 4$

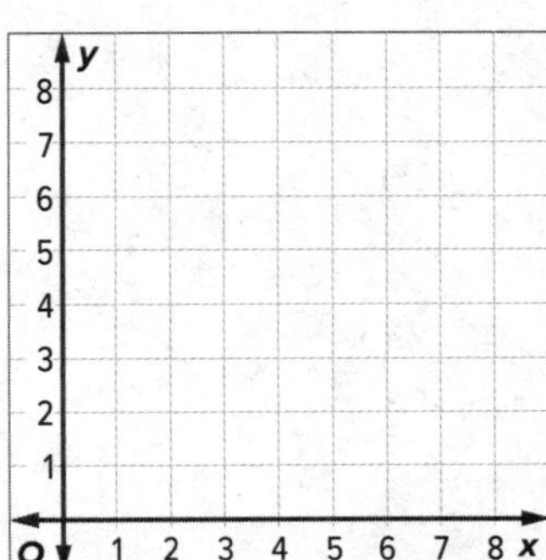

4. $y = 2x + 0.5$

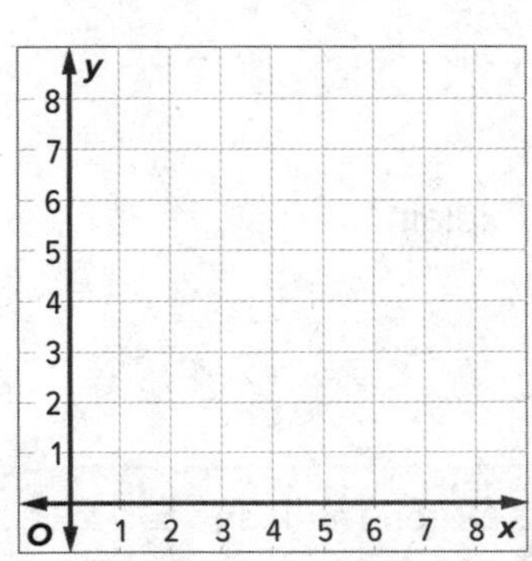

5. $y = 0.5x + 1$

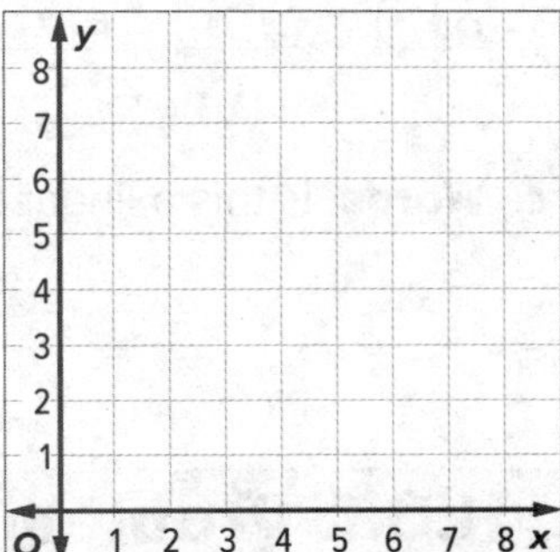

6. The graph shows the charges for a health club in a month. Make a function table for the input-output values. Write an equation that can be used to find the total charge y for the number of x classes. (Examples 3 and 4)

Input (x)				
Output (y)				

7. The graph shows the amount of money Pasha spent on lunch. Make a function table for the input-output values. Write an equation that can be used to find the money spent y for any number of days x. (Examples 3 and 4)

Input (x)				
Output (y)				

8. **Multiple Representations** The table shows the area of a square with the given side length.

Side Length (x)	Area of Square (y)
0	0
1	1
2	4
3	9

a. **Variables** Write an equation that could represent the function table.

b. **Graphs** Graph the function.

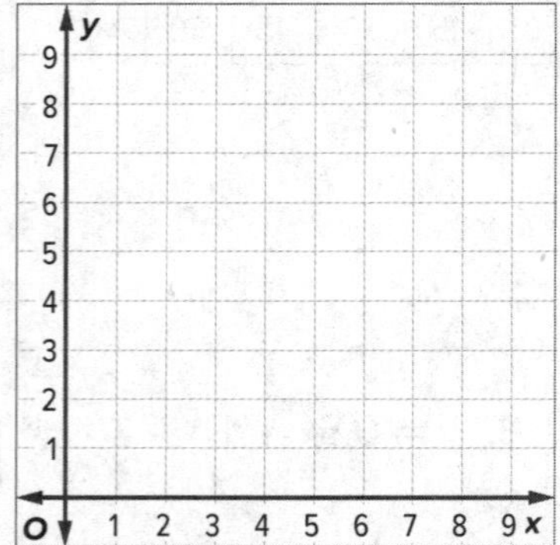

c. **Words** Is this a linear function? Explain.

H.O.T. Problems Higher Order Thinking

9. **Model with Mathematics** Write about a real-world situation that can be represented by the equation $y = 7x$. Be sure to explain what the variables represent in the situation.

10. **Persevere with Problems** Write an equation to represent the function in the table shown below.

Input (x)	6	8	10	12	14	16
Output (y)	0	1	2	3	4	5

11. **Persevere with Problems** The inverse of a relationship can be found by switching the coordinates in each ordered pair. Complete the table for three input and output values of $y = x + 3$ and its inverse. Then use the table to write an equation of the inverse of $y = x + 3$.

$y = x + 3$			
Input (x)			
Output (y)			

Inverse of $y = x + 3$			
Input (x)			
Output (y)			

Name ________________ My Homework ________________

Extra Practice

MP Identify Repeated Reasoning **Write an equation to represent each function.**

12.

Input (x)	0	1	2	3	4
Output (y)	0	11	22	33	44

$y = 11x$

Each output y is 11 times each input x.

13.

Input (x)	1	2	3	4	5
Output (y)	10	20	30	40	50

Graph each equation.

14. $y = 4x$

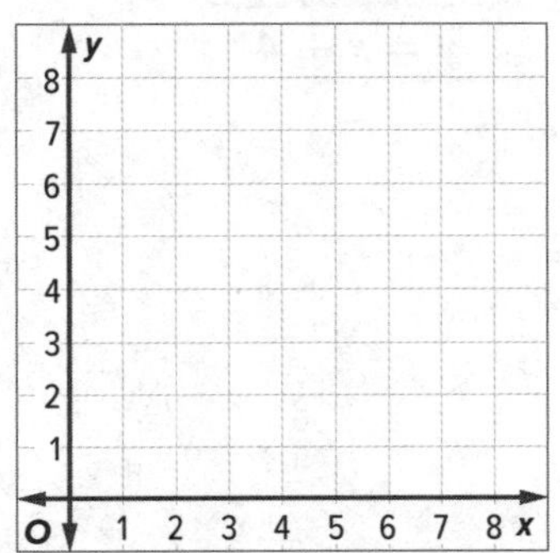

15. $y = 0.5x$

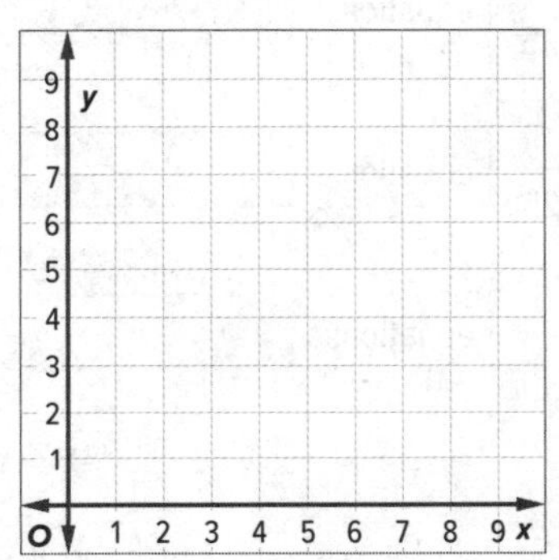

16. $y = x + 0.5$

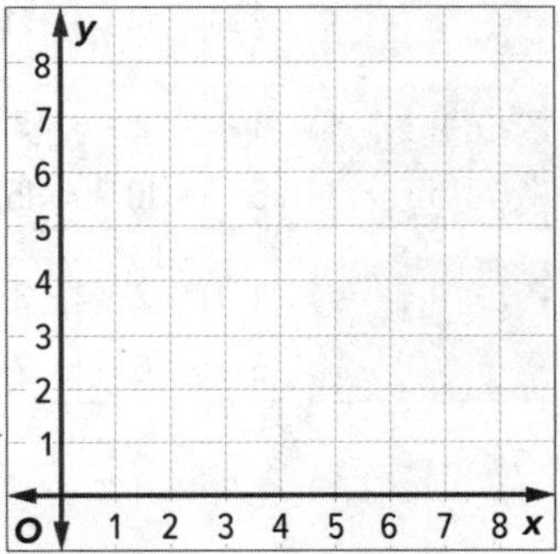

17. A company charges $50 per month for satellite television service plus an additional $5 for each movie ordered. The equation $y = 50 + 5x$ describes the total amount y a customer will pay if they order x movies. Graph the function.

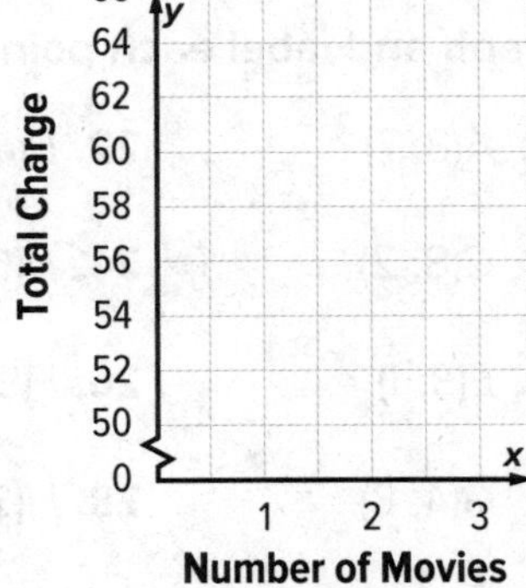

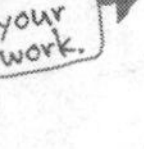

18. A fair charges an admission fee of $8. Each ride is an additional $2. The equation $y = 8 + 2x$ describes the total charge y for the number of rides x. Graph the function.

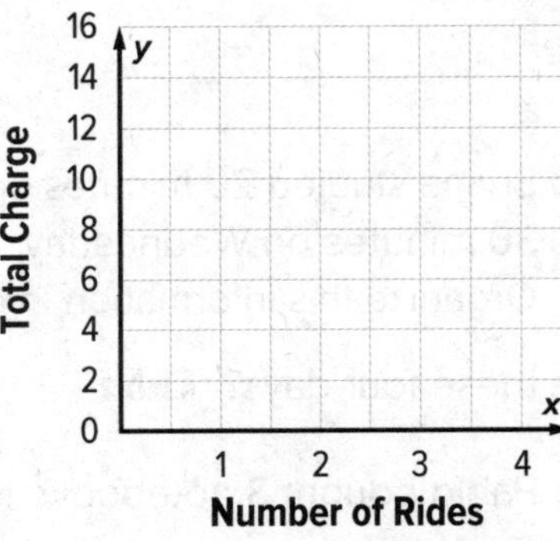

Power Up! Common Core Test Practice

19. The table shows the total cost of admission to a zoo for different numbers of guests. Determine if each statement is true or false.

Number of Guests, x	Total Cost ($), y
1	7
2	14
3	21
4	28

a. The total admission for 12 guests is $84. ☐ True ☐ False

b. The equation $y = 7x$ can be used to find the total admission for x guests. ☐ True ☐ False

c. The total admission for 10 guests is $63. ☐ True ☐ False

20. Match each function table to the correct equation.

Input (x)	1	2	3	4	5
Output (y)	9	10	11	12	13

equation: ________

Input (x)	1	2	3	4	5
Output (y)	7	14	21	28	35

equation: ________

Input (x)	1	2	3	4	5
Output (y)	5	10	15	20	25

equation: ________

Input (x)	1	2	3	4	5
Output (y)	5	6	7	8	9

equation: ________

$y = 5x$

$y = 7x$

$y = x + 8$

$y = x + 4$

Common Core Spiral Review

Graph and label each point. 5.G.2

21. $A(3, 7)$ **22.** $B(4, 3)$

23. $C(8, 2)$ **24.** $D(6, 5)$

25. $E(3, 1)$ **26.** $F(9, 4)$

27. $G(4, 8)$ **28.** $H(2, 6)$

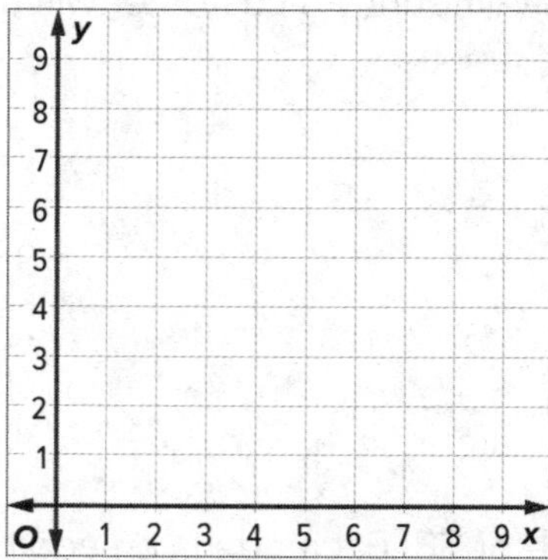

29. Shana studied 20 minutes on Monday, 45 minutes on Tuesday, 30 minutes on Wednesday, and 45 minutes on Thursday. Organize this information in the table. How long did she study these four days? **4.MD.2** ________

Day	Time Studied (min)

30. Pablo bought 3 notebooks for $5.85. How much did each notebook cost? **5.NBT.7** ________

Multiple Representations of Functions

Real-World Link

Museum A group of friends are going to the museum. Each friend must pay an admission price of $9.

Total Cost of Admission	
Number of Friends, *x*	Total Cost ($), *y*
1	9
2	
3	
4	

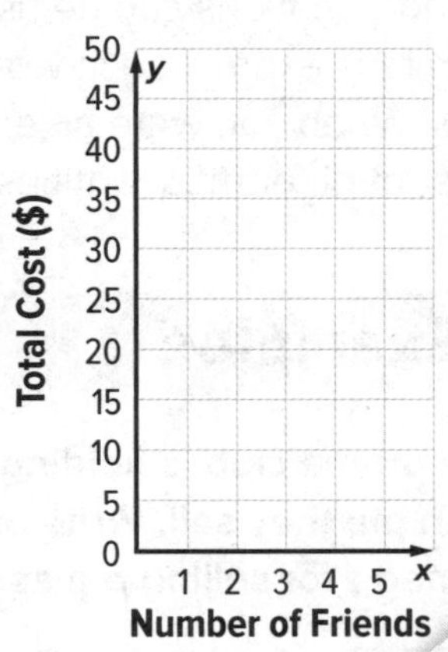

Essential Question

HOW are symbols, such as <, >, and =, useful?

Common Core State Standards

Content Standards
6.EE.9

MP Mathematical Practices
1, 2, 3, 4

1. Complete the table and graph the ordered pairs (number of friends, total cost).

2. Describe the graph.

3. Write an equation to find the cost of *n* tickets.

4. List the ordered pair for the cost when 5 friends go to the museum. Describe the location.

Which MP Mathematical Practices did you use? Shade the circle(s) that applies.

① Persevere with Problems
② Reason Abstractly
③ Construct an Argument
④ Model with Mathematics
⑤ Use Math Tools
⑥ Attend to Precision
⑦ Make Use of Structure
⑧ Use Repeated Reasoning

Key Concept

Represent Functions Using Words and Equations

Work Zone

Words A runner's distance in a marathon is equal to 8 miles per hour times the number of hours.

Equation $d = 8t$

Words and equations can be used to describe functions. For example, when a rate is expressed in words, it can be written as an equation with variables. When you write an equation, determine what variables to use to represent different quantities.

Examples

1. The drama club is holding a bake sale. They are charging $5 for each pie they sell. Write an equation to find the total amount earned *t* for selling *p* pies.

Variables
You can use any letter as a variable in an equation. If you graph the equation, make sure to label the axes with the correct variable.

Words	Total earned equals $5 times the number of pies sold.
Variable	Let t represent the total earned and p represent the number of pies sold.
Equation	$t \quad = \quad 5 \quad \cdot \quad p$

So, the equation is $t = 5p$.

2. In a science report, Mia finds that the average adult breathes 14 times each minute when not active. Write an equation to find the total breaths *b* a non-active person takes in *m* minutes.

Let b represent the total breaths and m represent the number of minutes.

The number of total breaths equals 14 times the number of minutes.

So, the equation is $b = 14m$.

Got it? **Do these problems to find out.**

a. __________

b. __________

a. A mouse can travel 8 miles per hour. Write an equation to find the total distance d a mouse can travel in h hours.

b. Samantha can make 36 cookies each hour. Write an equation to find the total number of cookies c that she can make in h hours.

Represent Functions Using Tables and Graphs

Key Concept

Table

Time (h), t	Distance (mi), d
0	0
1	8
2	16

Graph

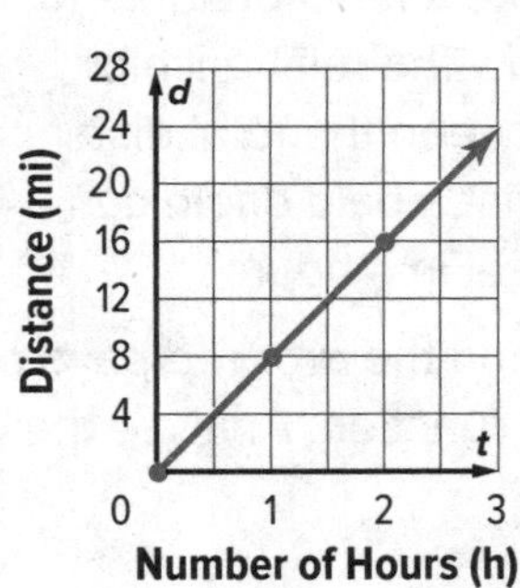

and Reflect

What are the independent and dependent variables in Example 3? Explain below.

Tables and graphs can also be used to represent functions.

Examples

The Student Council is holding a car wash to raise money. They are charging $7 for each car they wash.

3. Write an equation and make a function table to show the relationship between the number of cars washed c and the total amount earned t.

Cars Washed, c	$7c$	Total Earned ($), t
1	1×7	7
2	2×7	14
3	3×7	21
4	4×7	28

Using the assigned variables, the total earned t equals $7 times the number of cars washed c. So, the equation is $t = 7c$.

The total earned (output) is equal to $7 times the number of cars washed (input).

Write $7c$ in the middle column of the table.

4. Graph the ordered pairs. Analyze the graph.

Find the ordered pairs (c, t). The ordered pairs are (1, 7), (2, 14), (3, 21), and (4, 28). Now graph the ordered pairs.

The graph is linear because the amount earned increases by $7 for each car washed.

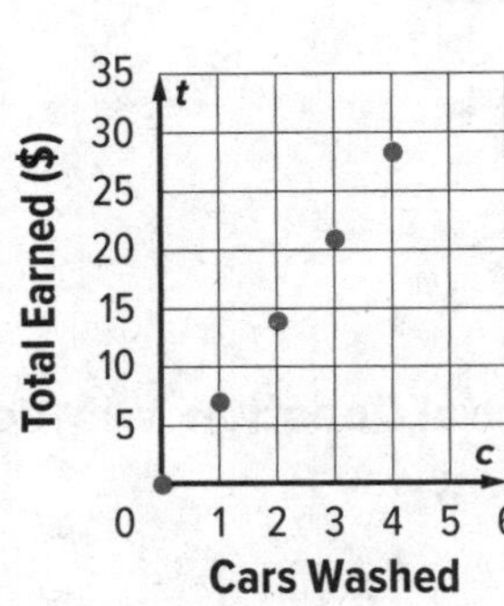

c. ____________

d. ____________

Got it? **Do these problems to find out.**

While in normal flight, a bald eagle flies at an average speed of 30 miles per hour.

c. Write an equation and make a function table to show the relationship between the total distance *d* that a bald eagle can travel in *h* hours.

d. Graph the ordered pairs of the function. Analyze the graph.

Time (h), *h*			
Distance (mi), *d*			

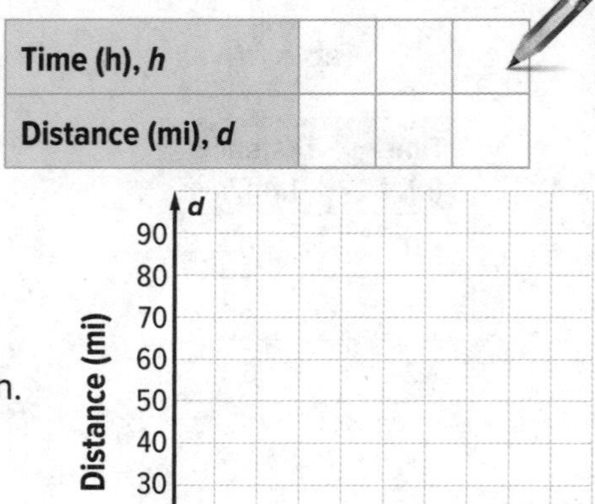

Guided Practice

1. The school cafeteria sells lunch passes that allow a student to purchase any number of lunches in advance for \$3 per lunch. (Examples 1–4)

a. Write an equation to find *t*, the total cost in dollars for a lunch pass with *n* lunches. ____________________

b. Make a function table to show the relationship between the number of lunches *n* and the cost *t*.

c. Graph the ordered pairs. Analyze the graph.

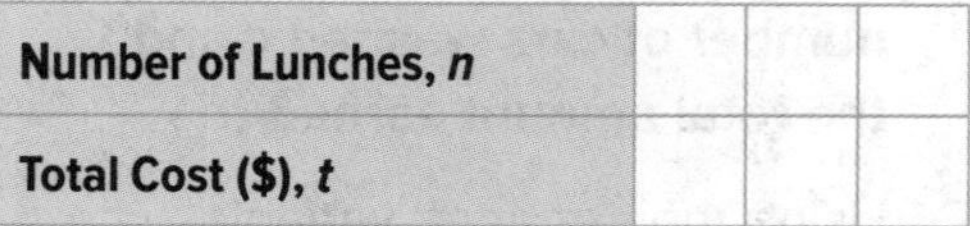

Number of Lunches, *n*			
Total Cost (\$), *t*			

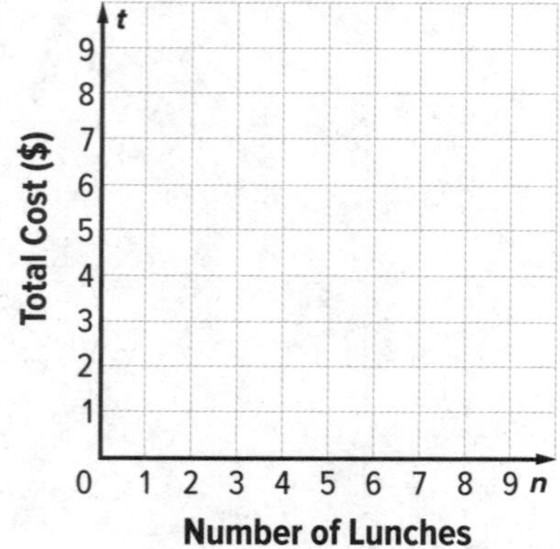

____

2. **Building on the Essential Question** Why do you represent functions in different ways?

____

____

Rate Yourself!

How well do you understand the different ways to represent functions? Circle the image that applies.

Clear

Somewhat Clear

Not So Clear

For more help, go online to access a Personal Tutor.

FOLDABLES Time to update your Foldable!

Independent Practice

Go online for Step-by-Step Solutions

1 An African elephant eats 400 pounds of vegetation each day. (Examples 1–4)

a. Write an equation to find *v*, the number of pounds of vegetation an African elephant eats in *d* days. ________________

b. Make a table to show the relationship between the number of pounds *v* an African elephant eats in days *d*.

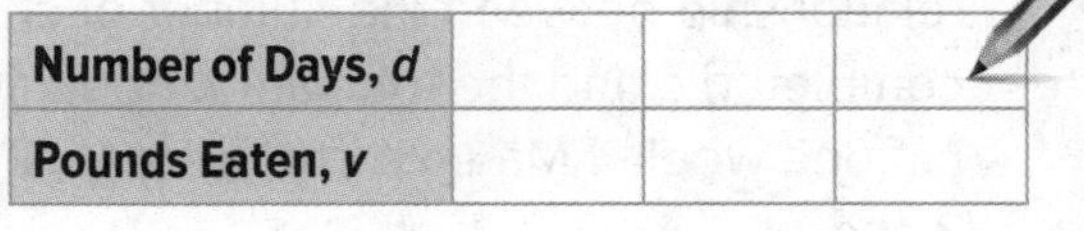

Number of Days, *d*			
Pounds Eaten, *v*			

c. Graph the ordered pairs. Analyze the graph.

____

____

____

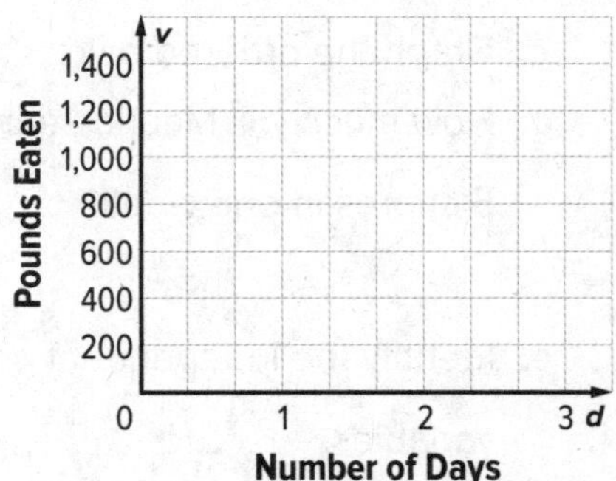

2. MP **Model with Mathematics** Refer to the graphic novel frame below for Exercises a–c.

a. Let *f* represent the cost of ordering each ticket online. Write an equation that could be used to find the cost of ordering each ticket online.

____

b. Solve the equation from part a. ________________

c. Another friend wants to go to the concert. What is the total cost of ordering three tickets online?

____

3 Maurice receives \$3 per week for allowance and earns an additional \$1.75 for each chore he completes.

a. Write an equation to find t, the total amount earned for c chores in one week. ______________________

b. Make a function table to show the relationship between the number of chores completed c and the total amount earned t in one week if Maurice completes 1, 2, or 3 chores.

Number of Chores, c			
Total Earned (\$), t			

c. Graph the ordered pairs.

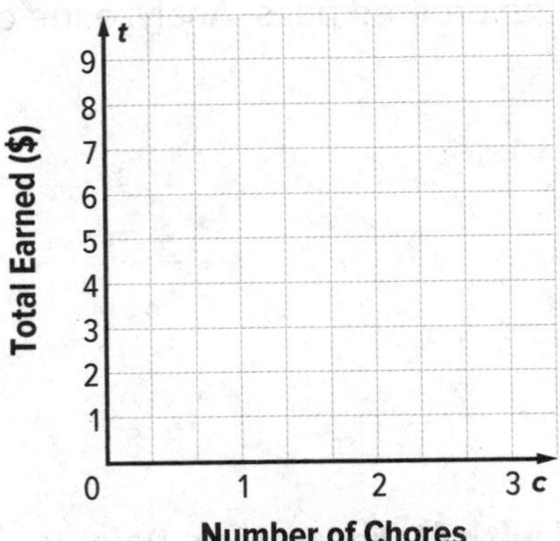

d. How much will Maurice earn if he completes 5 chores in one week? ______________________

e. Identify the independent and dependent variables. ______________________

H.O.T. Problems Higher Order Thinking

4. MP **Reason Abstractly** What would the graph of $y = x$ look like? Name three ordered pairs that lie on the line. ______________________

5. MP **Persevere with Problems** Boards 4 U charges \$10 per hour to rent a snowboard while Slopes charges \$12 per hour. Will the cost to rent snowboards at each place ever be the same for the same number of hours after zero hours? If so, for what number of hours? ______________________

6. MP **Model with Mathematics** Write a real-world problem in which you could graph a function. ______________________

7. MP **Reason Abstractly** A movie rental club charges a one-time fee of \$25 to join and \$2 for every movie rented. Write an equation that represents the cost of joining the club and renting any number of movies. ______________________

Extra Practice

8. In a video game, each player earns 5 points for reaching the next level and 15 points for each coin collected.

a. Write an equation to find p, the total points for collecting c coins after reaching the next level. $p = 5 + 15c$

Total points p equals 15 times the number of coins c collected plus 5 points for reaching the next level. So, the equation is $p = 5 + 15c$.

b. Make a table to show the relationship between the number of coins collected c and the total points p.

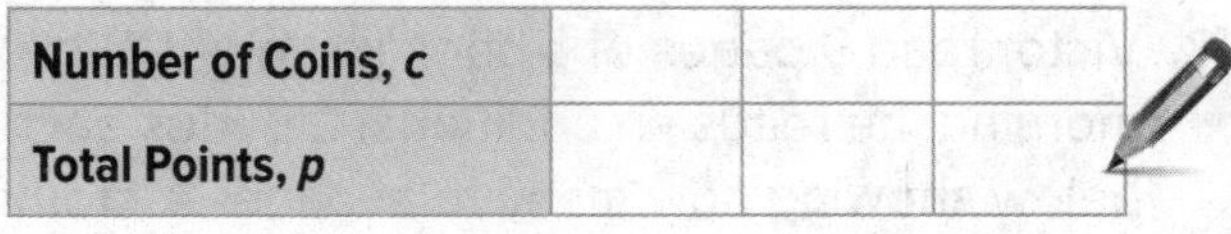

Number of Coins, c			
Total Points, p			

c. Graph the ordered pairs. Analyze the graph.

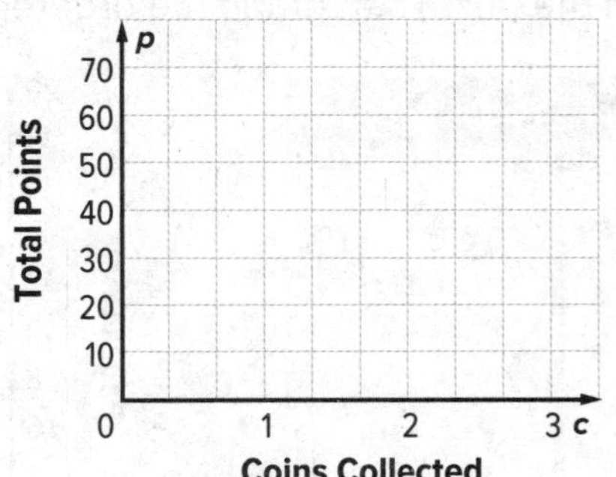

9. Two disc jockeys charge different rates. The Music Man charges \$45 per hour and Road Tunes charges \$35 per hour. Write equations to represent the total cost t of hiring either disc jockey for any number of hours n. ________________

140
120
100
80
60
40
20
0
t
The Music Man
Road Tunes
n
1 2 3 4
Total Fee (\$)
Number of Hours (h)

Copy and Solve **For Exercise 10, show your work on a separate piece of paper.**

10. MP **Construct an Argument** A catering service offers lasagna and chicken parmesan. Each pan of lasagna serves 24 people.

a. Write an equation to represent the number of people n served by any number of pans p of lasagna.

b. Make a function table to show the relationship between the number of pans p and the number of people served n.

c. Graph the ordered pairs.

d. The same catering company offers chicken parmesan that serves 16 people per pan. How many more people would 5 pans of lasagna serve than 5 pans of chicken parmesan? Explain your reasoning to a classmate.

Power Up! Common Core Test Practice

11. For each table Ariella waits on at a restaurant, she is paid \$4.00 plus 18% of the total bill. Let b represent the amount of the total bill and let m represent the total amount of money Ariella earns.

Write an equation that could be used to find the total amount of money Ariella earns per table. ______

If a table has a total bill of \$35, how much will Ariella earn? ______

12. Victor read 9 pages of a book last night. While riding the bus to school this morning, he reads an additional 2 pages each minute. Complete the table below showing how many total pages Victor will have read after m minutes of reading on the bus. Then graph the ordered pairs on the coordinate plane.

Minutes (m)	Pages Read (p)
0	
1	
2	
3	
4	

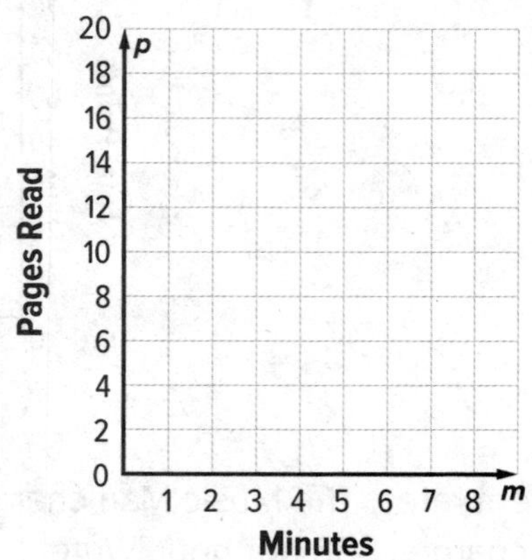

Write an equation to represent the situation. ______

Common Core Spiral Review

Fill in each ◯ with < or > to make a true statement. 4.NBT.2, 5.NBT.3b

13. 116 ◯ 161

14. 63 ◯ 61

15. 105 ◯ 115

16. 50 ◯ 500

17. 12 ◯ 1.2

18. 44 ◯ 49

19. Albert swam 13 laps on Monday, 12 laps on Tuesday, 16 laps on Wednesday, 15 laps on Thursday, and 10 laps on Friday. Graph each of these numbers on the number line. Which day did he swim the greatest number of laps? 6.NS.6c ______

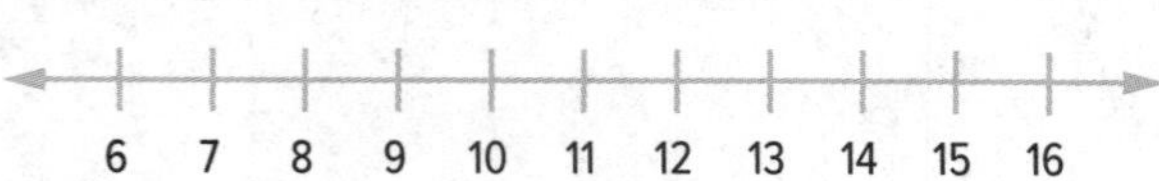

Problem-Solving Investigation
Make a Table

Content Standards
6.EE.9

Mathematical Practices
1, 3, 4

Case #1 Splitting Up

Blue-green algae is a type of bacteria that can double its population by splitting up to four times in one day.

If it grows at this rate, how many bacteria will be formed at the end of one day?

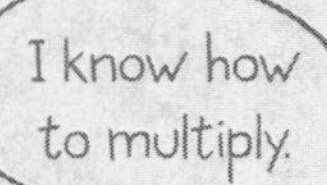

Understand *What are the facts?*

- Blue-green algae can double its population up to four times in one day.

Plan *What is your strategy to solve this problem?*

Make a table to display and organize the information.

Solve *How can you apply the strategy?*

Follow the pattern to find the total number of bacteria after 1 day.

Day Number	Number of Times Split	Total Number of Bacteria	
1	0	1	← ×2
1	1	2 	← ×2
1	2		← ×2
1	3		← ×2
1	4		← ×2

Check *Does the answer make sense?*

Use the equation $t = 2^n$ where n represents the number of times the bacteria split and t represents the total number of bacteria. $2^4 = 16$

Analyze the Strategy

Tutor

Justify Conclusions If the bacteria continue to grow at this rate, would the number of bacteria be over 1,000 within a week? Explain. ____________________

____

Case #2 Game On!

Miguel and Lauren are testing two versions of a new video game. In Miguel's version he receives 25 points at the start of the game, plus 1 point for each level he completes. In Lauren's version she receives 20 points at the start of the game and 2 points for each level she completes.

At what level will they both have the same number of points?

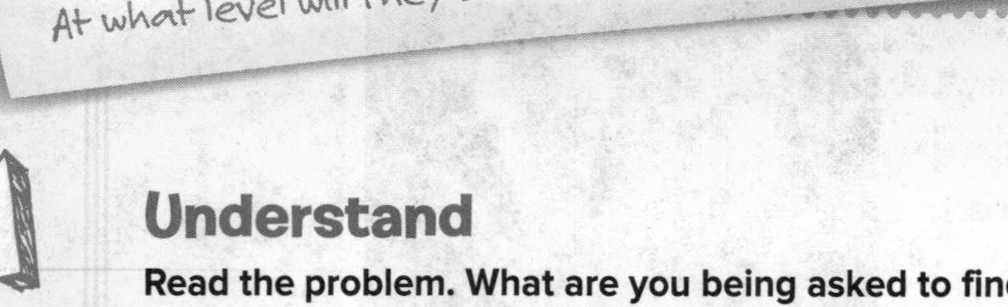

1 Understand

Read the problem. What are you being asked to find?

I need to find ______________________________.

Underline key words and values in the problem.
What information do you know?

Miguel starts with ☐ points and earns ☐ point for each level.

Lauren starts with ☐ points and earns ☐ points for each level.

2 Plan

Choose a problem-solving strategy.

I will use the ______________________________ strategy.

3 Solve

Use your problem-solving strategy to solve the problem.

	Start	Level 1	Level 2	Level 3	Level 4	Level 5
Miguel						
Lauren						

So, Miguel and Lauren will have the same score after completing Level ☐.

4 Check

Place the level number answer in each box and evaluate to check your answer.

Miguel: $25 + (1 \times \square) =$ __________

Lauren: $20 + (2 \times \square) =$ __________

Work with a small group to solve the following cases. Show your work on a separate piece of paper.

Case #3 Geometry

Determine how many cubes are used in each step.

Make a table to find the number of cubes in the seventh step.

Case #4 Car Rental

Anne Marie needs to rent a car for 9 days to take on vacation. The cost of renting a car is \$66 per day, \$15.99 for insurance, and \$42.50 to fill up the gas tank.

Find the total cost of her rental car.

Case #5 Numbers

The difference between two whole numbers is 14. Their product is 1,800.

What are the two numbers?

Case #6 Money

The admission for a fair is \$6 for adults, \$4 for children, and \$3 for senior citizens. Twelve people paid a total of \$50 for admission.

If 8 children attended, how many adults and senior citizens attended?

Mid-Chapter Check

Vocabulary Check

1. Define *sequence*. Give an example of an arithmetic and a geometric sequence. (Lesson 2)

____

____

2. Fill in the blank in the sentence below with the correct term. (Lesson 1)

A ______________ is a relation that assigns exactly one output value to one input value.

Skills Check and Problem Solving

Complete each function table. (Lesson 1)

3.

Input (x)	$2x + 6$	Output
0		
1		
2		

4.

Input (x)	$3x + 1$	Output
0		
1		
2		

MP **Identify Structure** **Find the rule for each function table.** (Lesson 2)

5.

Input (x)	Output
3	6
4	8
5	10

____

6.

Input (x)	Output
1	3
2	7
3	11

____

7.

Input (x)	Output
2	8
3	11
4	14

____

8. Arnold reads an average of 21 pages each day. Write an equation to represent the number of pages read after any number of days. (Lesson 4)

____

9. MP **Reason Abstractly** The table shows the cost of renting an inner tube to use at the Wave-a-Rama Water Park. Explain how to write an equation to represent the data in the table. Then give the equation for the data. (Lesson 3)

Input (x)	Cost (y)
2	$11.00
3	$16.50
4	$22.00

____

____

Inquiry Lab

Inequalities

HOW can bar diagrams help you to compare quantities?

Content Standards
6.EE.5, 6.EE.8

Mathematical Practices
1, 3, 4

In saltwater fishing, any flounder that is caught may be kept if it is greater than or equal to 12 inches long. Any flounder shorter than that must be released back into the water. Pat caught a flounder that is 14 inches long. He wants to know if he can keep the fish.

An *inequality* is a mathematical sentence that compares quantities. An inequality like $x < 7$ or $x > 5$ can be written to express how a variable compares to a number.

Step 1 Label the minimum length of flounders that may be kept.

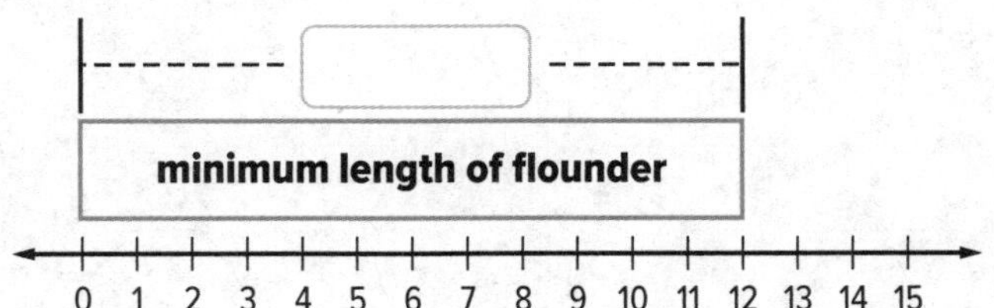

Step 2 Label the length of the flounder Pat caught on the top bar diagram.

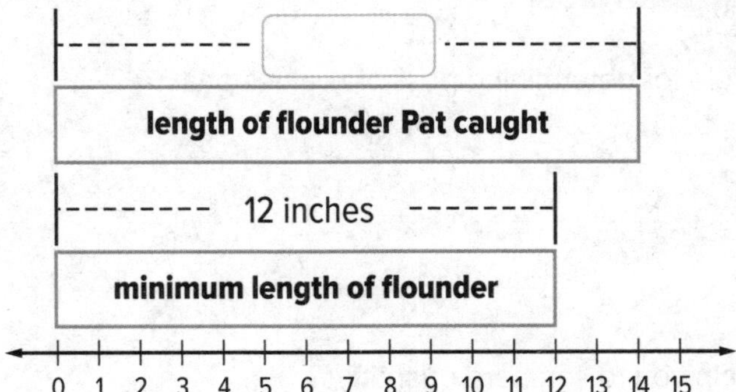

The bar representing Pat's fish is ______________ than the bar representing the minimum length that can be kept.

So, Pat ______________ keep the fish.

Investigate

MP Model with Mathematics Work with a partner. Draw bar diagrams to solve each problem.

1. For flights within the United States, luggage must be no more than 50 pounds. Imelda's luggage weighs 53 pounds. Can she take the luggage on her flight? ______

2. Byron needs at least 20 minutes between the end of his soccer practice and the start of his dentist appointment. His practice ends at 4:30 and his appointment is at 5:00. Does he have enough time? ______

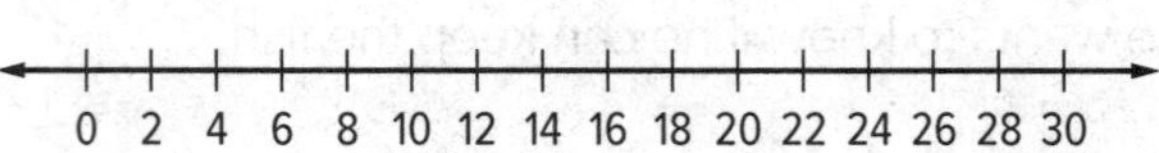

3. **MP Reason Inductively** Which inequality is used when the situation involves a "minimum"? Explain. ______

4. **MP Reason Inductively** Which inequality is used when the situation involves a "maximum"? Explain. ______

Create

5. **MP Reason Inductively** Write a rule for determining possible values of a variable in an inequality. ______

6. **Inquiry** HOW can bar diagrams help you to compare quantities?

Lesson 5
Inequalities

Vocabulary Start-Up

An **inequality** is a mathematical sentence that compares quantities.

Essential Question

HOW are symbols, such as <, >, and =, useful?

Vocabulary

inequality

Common Core State Standards

Content Standards
6.EE.5, 6.EE.8

MP Mathematical Practices
1, 2, 3, 4, 6, 7

Real-World Link

Compare the following using < or >.

1. the score after 2 goals is ◯ the score after 3 goals
2. the cost to download 10 songs is ◯ the cost to download 2 songs
3. the outside temperature in summer is ◯ the outside temperature in winter
4. the height of a 1st grade student is ◯ the height of a 6th grade student
5. the time to eat lunch is ◯ the time to brush your teeth

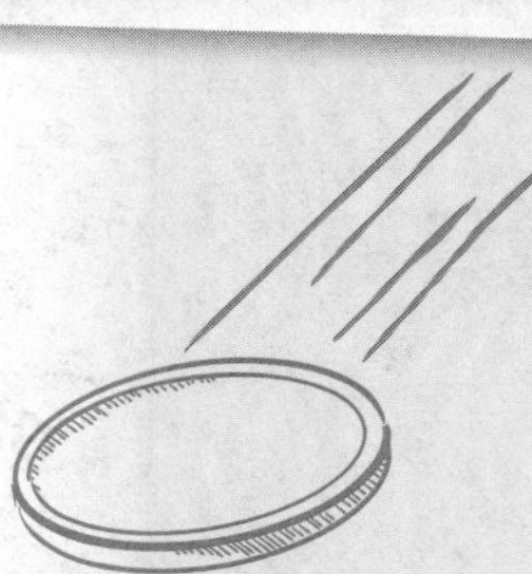

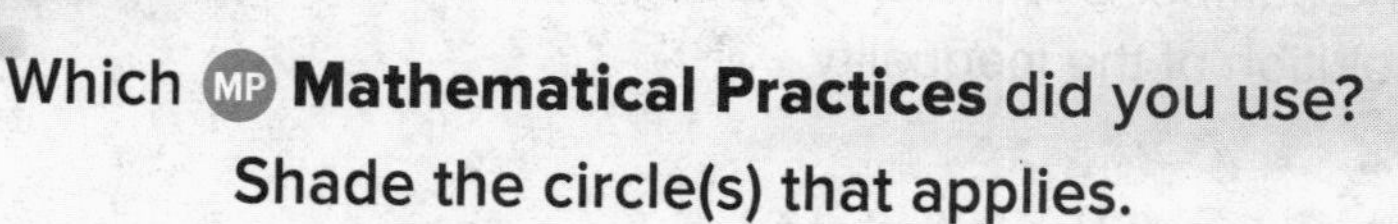

Which MP **Mathematical Practices** did you use? Shade the circle(s) that applies.

① Persevere with Problems
② Reason Abstractly
③ Construct an Argument
④ Model with Mathematics
⑤ Use Math Tools
⑥ Attend to Precision
⑦ Make Use of Structure
⑧ Use Repeated Reasoning

Key Concept

Inequalities

Symbols	$<$	$>$	$\leq$	$\geq$
Words	• is less than • is fewer than	• is greater than • is more than	• is less than or equal to • is at most	• is greater than or equal to • is at least
Examples	$3 < 5$	$8 > 4$	$7 \leq 10$	$12 \geq 9$

Work Zone

Inequalities can be solved by finding values of the variables that make the inequality true.

Example

1. Of the numbers 6, 7, or 8, which is a solution of the inequality $f + 2 < 9$?

Replace *f* with each of the numbers.

$f + 2 < 9$ Write the inequality.

$6 + 2 \overset{?}{<} 9$ Replace *f* with 6.

$8 < 9$ ✓ This is a true statement.

$f + 2 < 9$ Write the inequality.

$7 + 2 \overset{?}{<} 9$ Replace *f* with 7.

$9 < 9$ ✗ This is not a true statement.

$f + 2 < 9$ Write the inequality.

$8 + 2 \overset{?}{<} 9$ Replace *f* with 8.

$10 < 9$ ✗ This is not a true statement.

Since the number 6 is the only value that makes a true statement, 6 is a solution of the inequality.

Show your work.

Got it? **Do this problem to find out.**

a. Of the numbers 8, 9, or 10, which is a solution of the inequality $n - 3 > 6$?

a. ________

Determine Solutions of an Inequality

Since an inequality uses greater than and less than symbols, one-variable inequalities have infinitely many solutions. For example, any rational number greater than 4 will make the inequality $x > 4$ true.

Examples

Is the given value a solution of the inequality?

2. $x + 3 > 9, x = 4$

$x + 3 > 9$ Write the inequality.

$4 + 3 \overset{?}{>} 9$ Replace x with 4.

$7 \not> 9$ Simplify.

Since 7 is not greater than 9, 4 is not a solution.

3. $12 \leq 18 - y, y = 6$

$12 \leq 18 - y$ Write the inequality.

$12 \overset{?}{\leq} 18 - 6$ Replace y with 6.

$12 \leq 12$ Simplify.

Since 12 = 12, 6 is a solution.

4. $17 \geq 11 + x, x = 8$

$17 \geq 11 + x$ Write the inequality.

$17 \overset{?}{\geq} 11 + \square$ Replace x with $\square$.

$17 \not\geq \square$ Simplify.

Since $\square$ is not greater than or equal to $\square$, $\square$ is not a solution.

Name two solutions of the inequality $12 > 6 + y$.

Got it? Do these problems to find out.

b. $a + 7 > 15, a = 9$

c. $22 \leq 15 + b, b = 6$

d. $n - 4 < 6, n = 10$

e. $12 \geq 5 + g, g = 7$

b. ________

c. ________

d. ________

e. ________

Example

5. **Luisa works at a gift shop. She receives a bonus if she makes more than 20 balloon bouquets in a month. Which months did Luisa receive a bonus? Use the inequality $b > 20$, where b represents the number of balloon bouquets made each month, to solve.**

Balloon Sales	
Month	**Number Sold**
July	25
August	12
September	18
October	32

Use the *guess, check, and revise* strategy.

Try 25.	Try 12.	Try 18.	Try 32.
$b > 20$	$b > 20$	$b > 20$	$b > 20$
$25 > 20$ Yes	$12 > 20$ No	$18 > 20$ No	$32 > 20$ Yes

So, Luisa received a bonus in July and October.

Guided Practice

Determine which number is a solution of the inequality. (Example 1)

1. $9 + a < 17$; 7, 8, 9 ________

2. $b - 10 > 5$; 14, 15, 16 ________

Is the given value a solution of the inequality? (Examples 2–4)

3. $x - 5 < 5, x = 15$

____

4. $32 \geq 8n, n = 3$

____

5. If the bakery sells more than 45 bagels in a day, they make a profit. Use the inequality $b > 45$ to determine which days the bakery makes a profit. (Example 5)

Day	Number of Bagels Sold
Monday	18
Tuesday	25
Wednesday	21
Thursday	36
Friday	50
Saturday	48
Sunday	40

____

6. **Building on the Essential Question** How can mental math help you find solutions to inequalities?

____

Rate Yourself!

☐ I understand how to solve inequalities.

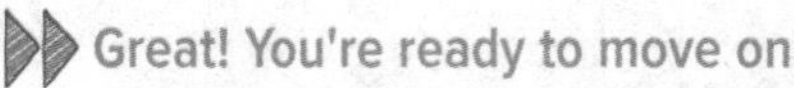

Great! You're ready to move on!

☐ I still have some questions about solving inequalities.

No Problem! Go online to access a Personal Tutor.

Name ______________________ My Homework ______________________

Independent Practice

Go online for Step-by-Step Solutions

Determine which number is a solution of the inequality. (Example 1)

1. $1 + f < 7$; 5, 6, 7 ________

2. $g - 3 > 4$; 6, 7, 8 ________

Is the given value a solution of the inequality? (Examples 2–4)

3. $q - 2 > 16$, $q = 20$ ________

4. $t - 7 < 10$, $t = 28$ ________

5. The table shows the number of different types of roller coasters in the United States. An amusement park wants to build a new roller coaster. They will only build a roller coaster if there are less than 10 of that type in the United States. Use the inequality $r < 10$, where r is the number of a certain type of roller coaster, to determine which type(s) can be built. (Example 5)

Type	Number
Sit down (steel)	530
Sit down (wood)	112
Inverted	43
Flying	10
Stand up	8
Suspended	5

6. The table shows the number of different types of movies in Lavar's collection. He wants to buy a new movie to add to his collection. He only wants to buy a movie if he already has more than 15 movies of that type. Use the inequality $m > 15$, where m is the number of the type of movie, to determine which type(s) he can buy. (Example 5)

Movie Type	Number
Action	18
Comedy	24
Drama	12
Thriller	15

7. The number of text messages Lelah sent each month is shown in the table. She can send no more than 55 messages each month without being charged. Use the inequality $t \leq 55$, where t is the number of text messages in a month, to determine in which months she exceeded her limit. If each additional text costs $0.25, how much was Lelah charged from January to April?

Month	Text Messages
January	56
February	57
March	55
April	51

8. **Identify Structure** Use one-variable equations and inequalities to fill in the graphic organizer.

	Equation	Inequality
Example		
Number of Solutions		

H.O.T. Problems Higher Order Thinking

9. **Reason Inductively** State three numbers that are solutions to the inequality $x + 1 \leq 5$. ____________

10. **Persevere with Problems** If $x = 2$, is the following inequality *true* or *false*? Explain.

$$\frac{112}{8} + x \geq 15 + 4x - 7$$

11. **Reason Abstractly** If $a > b$ and $b > c$, what is true about the relationship between a and c? Explain your reasoning.

12. **Construct an Argument** Explain why inequalities of the form $x > c$ or $x < c$, where c is any rational number, have infinitely many solutions.

13. **Persevere with Problems** Analyze the relationship between the inequalities in each pair of inequalities below. Then write the integers that are solutions to each pair of inequalities.

a. $y > 4$ and $y \leq 6$ ____________

b. $x \geq -3$ and $x < 0$ ____________

c. $m < 5$ and $m > 3$ ____________

d. $r < -1$ and $r > 0$ ____________

Name ______ My Homework ______

Extra Practice

Determine which number is a solution of the inequality.

14. $5 - h \geq 2$; 3, 4, 5 3

Try 3. $5 - 3 \stackrel{?}{\geq} 2$, $2 \geq 2$ ✓

Try 4. $5 - 4 \stackrel{?}{\geq} 2$, $1 \geq 2$ X

Try 5. $5 - 5 \stackrel{?}{\geq} 2$, $0 \geq 2$ X

15. $j + 8 \leq 8$; 0, 1, 2 ______

Is the given value a solution of the inequality?

16. $25 \geq 5u$, $u = 5$ ______

17. $13 \leq 4v$, $v = 3$ ______

18. Mrs. Crane recorded the number of sandwiches sold in her deli on one day. If she sells more than 25 of a type of sandwich, she orders more meat from the butcher. Use the inequality $s > 25$, where s is the number of sandwiches sold, to determine which meats she needs to order. ______

Sandwich	Number Sold
Club	25
Ham	30
Roast beef	22
Turkey	28

19. The height of each member of a family is listed in the table. In order to ride a certain roller coaster at an amusement park, you must be at least 54 inches tall. Use the inequality $h \geq 54$, where h is a family member's height, to determine who can ride the roller coaster.

Name	Height (in.)
Carmen	66
Eliot	54
Isabella	49
Jackson	52
Ryan	71

20. MP **Be Precise** Pedro subscribes to a service where he can download up to five free ringtones each month. Each ringtone after that costs \$3.50 each. During which months did Pedro exceed the plan? How much is Pedro's additional cost in 6 months? ______

Month	Ringtones
January	5
February	6
March	4
April	8
May	5
June	4

Power Up! Common Core Test Practice

21. The number of moons for some of the planets are shown in the table.

Planets	Moons	Planets	Moons
Earth	1	Uranus	27
Mars	2	Saturn	47
Neptune	13	Jupiter	63

Let m represent the number of moons for a planet. Which of the following planets have moons that represent solutions of the inequality $m > 27$? Select all that apply.

- ☐ Jupiter
- ☐ Earth
- ☐ Saturn
- ☐ Uranus

22. The inequality $h \geq 48$, where h is a person's height in inches, can be used to determine who can ride the Screaming Eagle roller coaster. The table shows the heights of some friends who want to ride the roller coaster.

Name	Height (in.)
Chris	49
Gregorio	56
Heather	53
Jason	48
Molly	47
Tito	44

Complete the chart to show who is and who is not able to ride the roller coaster.

Able to Ride	Not Able to Ride

Common Core Spiral Review

Write an expression to represent each situation. 5.OA.2

23. Alexis had 5 stickers and her sister gave her 3 stickers. ____________

24. There were 7 lemons on the lemon tree. Then 2 fell off the tree. ____________

25. Gavin had 5 packages of hotdogs that each contained 8 hotdogs. ____________

26. The distance 4 friends walked is shown in the table. Graph the numbers on the number line. Who walked the shortest distance? 4.NBT2, 5.NBT.3b

____

Name	Miles Walked
Corrine	2.5
Makenna	1.5
Noah	3
Tristan	2

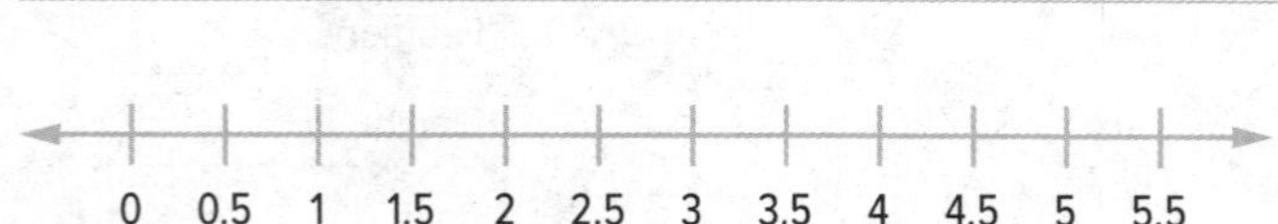

27. In one week, Carson read 4 books and Henry read 6 books. Fill in the blanks to compare the number of books they read. 4.NBT.2

________ > ________

Lesson 6

Write and Graph Inequalities

Real-World Link

Fair Look at the situations below. Circle the numbers that are possible answers in each situation.

1. Jessica spent more than $5 at the arcade.

   1 2 3 4 5 6 7 8 9 10 11 12 13 14 15

2. Less than 6 people rang the bell on the mallet game.

   1 2 3 4 5 6 7 8 9 10 11 12 13 14 15

3. There were less than 10 people in line for the Ferris wheel.

   1 2 3 4 5 6 7 8 9 10 11 12 13 14 15

4. It costs more than 6 tokens to ride the bumper cars.

   1 2 3 4 5 6 7 8 9 10 11 12 13 14 15

5. There are less than 8 lemonade stands.

   1 2 3 4 5 6 7 8 9 10 11 12 13 14 15

6. There are more than 12 different flavors of taffy.

   1 2 3 4 5 6 7 8 9 10 11 12 13 14 15

7. Describe any patterns you see in Exercises 1–6.

____

Essential Question

HOW are symbols, such as <, >, and =, useful?

Common Core State Standards

Content Standards
6.EE.6, 6.EE.8

MP Mathematical Practices
1, 3, 4, 5, 6

Which MP Mathematical Practices did you use? Shade the circle(s) that applies.

① Persevere with Problems
② Reason Abstractly
③ Construct an Argument
④ Model with Mathematics
⑤ Use Math Tools
⑥ Attend to Precision
⑦ Make Use of Structure
⑧ Use Repeated Reasoning

Work Zone

Write Inequalities

You can write an inequality to represent a situation.

Examples

Write an inequality for each sentence.

1. You must be over 12 years old to ride the go-karts.

Words	Your age	is over	12.
Variable		Let a = your age.	
Inequality	a	$>$	12

The inequality is $a > 12$.

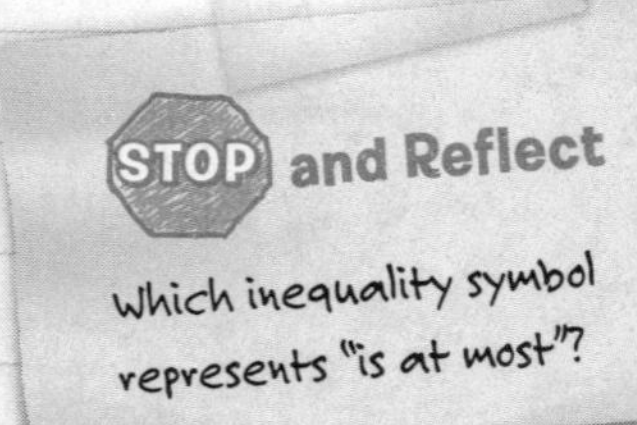

2. A pony is less than 14.2 hands tall.

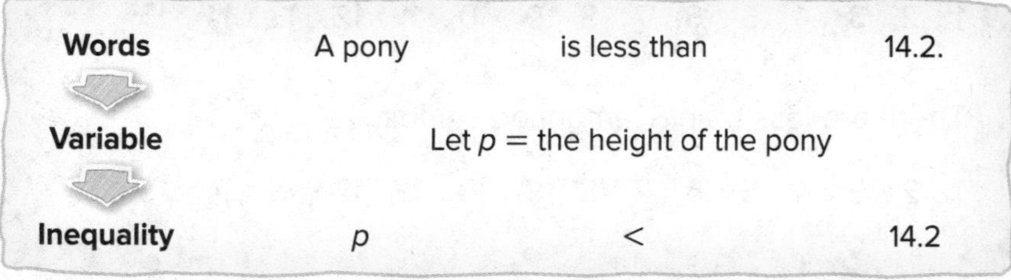

Words	A pony	is less than	14.2.
Variable		Let p = the height of the pony	
Inequality	p	$<$	14.2

The inequality is $p < 14.2$.

3. You must be at least 16 years old to have a driver's license.

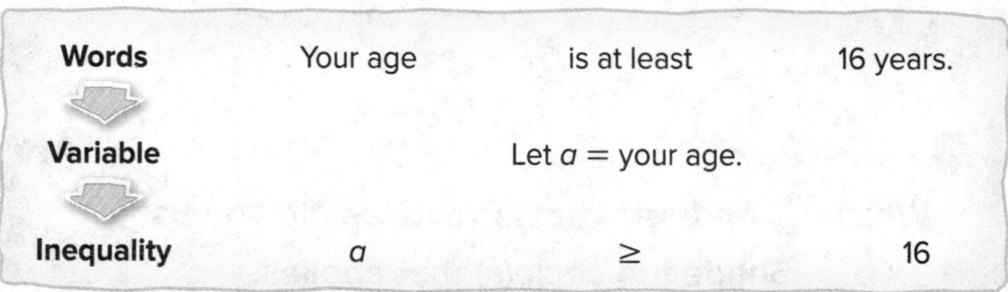

Words	Your age	is at least	16 years.
Variable		Let a = your age.	
Inequality	a	$\geq$	16

The inequality is $a \geq 16$.

Got it? **Do these problems to find out.**

Write an inequality for each sentence.

a. You must be older than 13 to play in the basketball league.

b. To use one stamp, your domestic letter must weigh under 3.5 ounces.

c. You must be over 48 inches tall to ride the roller coaster.

d. You must be at least 18 years old to vote.

a. ______

b. ______

c. ______

d. ______

Graph an Inequality

Inequalities can be graphed on a number line. Sometimes, it is impossible to show all the values that make an inequality true. The graph helps you see the values that make the inequality true.

Examples

Graph each inequality on a number line.

4. $n > 9$

Place an open dot at 9. Then draw a line and an arrow to the right.

The open dot means the number 9 is *not* included in the graph.

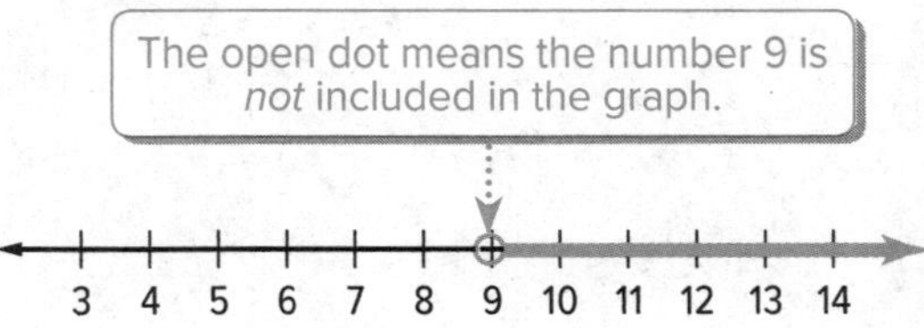

The values that lie on the line make the sentence true. All numbers greater than 9 make the sentence true.

5. $n \leq 10$

Place a closed dot at 10. Then draw a line and an arrow to the left.

The closed dot means the number 10 *is* included in the graph.

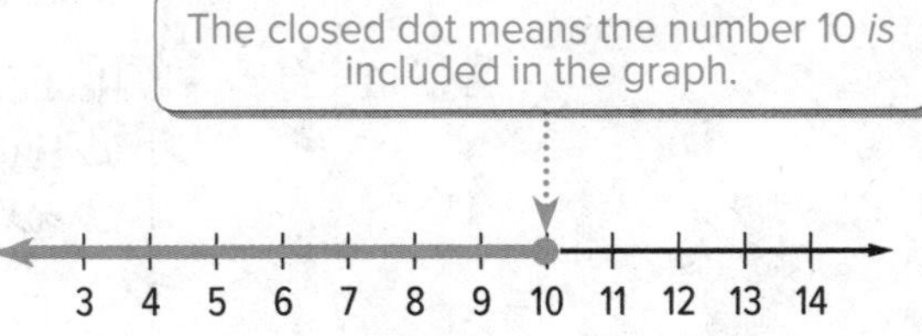

All numbers 10 and less make the sentence true.

Graphing Inequalities

When inequalities are graphed, an open dot means the number is not included (< or >) and a closed dot means it is included (≤ or ≥).

Got it? **Do these problems to find out.**

e. $a < 15$

f. $b \geq 7$

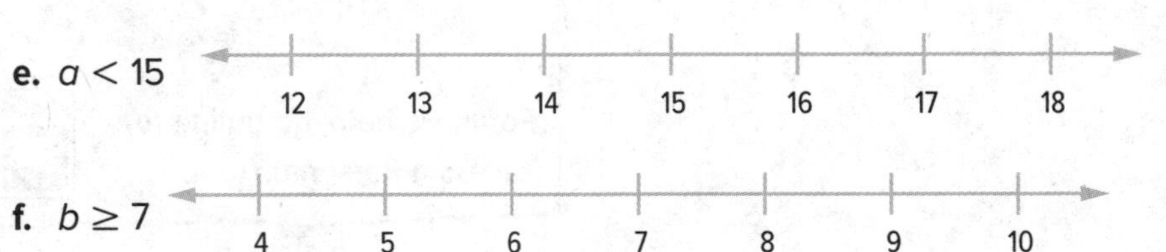

Example

6. Traffic on a residential street can travel at speeds of no more than 25 miles per hour. Write and graph an inequality to describe the possible speeds on the street.

Let s represent the speed on the street.

The inequality is $s \leq 25$.

Place a closed dot at 25. Then draw a line and an arrow to the left. All numbers 25 and less make the sentence true.

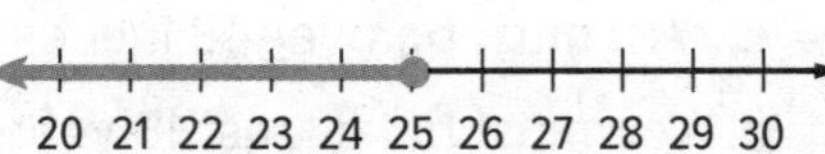

Guided Practice

Write an inequality for each sentence. (Examples 1–3)

1. The movie will be no more than 90 minutes in length. ___________

2. The mountain is at least 985 feet tall. ___________

Graph each inequality on a number line. (Examples 4 and 5)

3. $a \leq 6$

4. $b > 4$

5. Tasha can spend no more than $40 on new boots. Write and graph an inequality to describe how much she can spend. (Example 6) ___________

6. **Building on the Essential Question** How can graphing an inequality help to solve it? ___________

Rate Yourself!

How confident are you about writing and graphing inequalities? Shade the ring on the target.

For more help, go online to access a Personal Tutor.

Independent Practice

Go online for Step-by-Step Solutions eHelp

Write an inequality for each sentence. (Examples 1–3)

1. Swim practice will be no more than 35 laps. ______________

2. Kevin ran for less than 5 miles. ______________

3. The occupancy of the room must be less than 437 people. ______________

Graph each inequality on a number line. (Examples 4 and 5)

4. $f > 1$

5. $x \leq 5$

6. $y \geq 4$

7. A rewritable compact disc must have less than 20 songs on it. Write and graph an inequality to describe how many songs can be on the disc. (Example 6)

____

8. **Be Precise** Fill in the information in the table. The first is done for you.

Symbol	Words	Open or closed dot on number line?
>	greater than	open dot
	greater than or equal to	
	less than	
≤		

H.O.T. Problems Higher Order Thinking

9. **MP Find the Error** Mei is writing an inequality for the expression *at least 10 hours of community service*. Find her mistake and correct it.

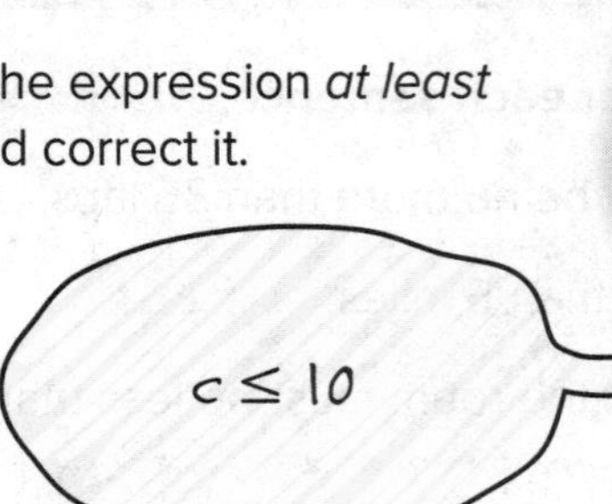

10. **MP Persevere with Problems** Name three solutions of the inequality $w \leq \frac{4}{5}$. Then justify your response using a number line.

11. **MP Justify Conclusions** Explain the difference between graphing an inequality with a closed dot and one with an open dot. Use examples to support your reasoning.

12. **MP Model with Mathematics** Graph the solution to each set of inequalities on a number line.

   a. $x > 5$ and $x < 8$

   b. $y \geq -2$ and $y < 7$

   c. $t < 3$ or $t \geq 6$

   d. $w \leq -5$ or $w \geq 0$

Name ______________________ My Homework ______________________

Extra Practice

Write an inequality for each sentence.

13. You cannot spend more than 50 dollars. $s \leq 50$

Let s represent what you can spend. Cannot spend more means you can spend less than or equal to 50 dollars.

14. More than 800 fans attended the opening soccer game. ______

15. The heavyweight division is greater than 200 pounds. ______

Graph each inequality on a number line.

16. $g < 6$

17. $z > 18$

18. $h \geq 3$

19. On a certain day, the temperature in Bismarck, North Dakota, was below 4°F. Write and graph an inequality to describe the possible temperatures.

______

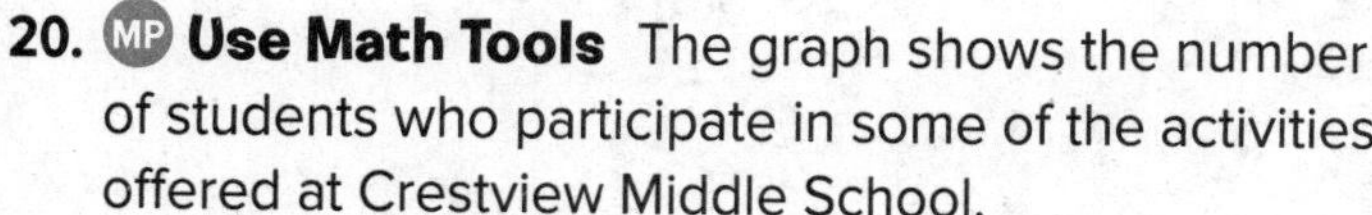

20. MP **Use Math Tools** The graph shows the number of students who participate in some of the activities offered at Crestview Middle School.

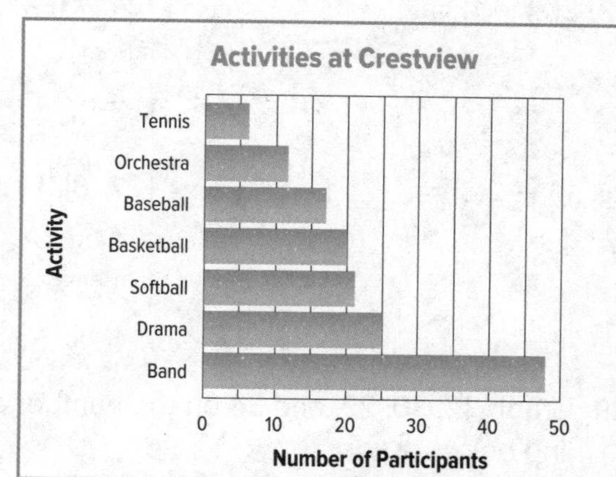

a. Which activities have more than 20 participants? at least 20? fewer than 19?

______

______

b. Write an inequality comparing the number of orchestra participants and the number of tennis participants.

______

______

CCSS Power Up! Common Core Test Practice

21. The table shows the number of different kinds of sports equipment sold at a sporting goods store.

Type	Number Sold in Store
Baseball	33
Basketball	n
Football	8
Hockey puck	3
Softball	21

The number of basketballs sold n is greater than the number of softballs sold. Determine if each statement is true or false.

a. The inequality $n > 21$ represents the situation. ☐ True ☐ False

b. The store sold more footballs than basketballs. ☐ True ☐ False

c. The store could have sold 22 basketballs. ☐ True ☐ False

22. Jason has less than 65 pages of his book left to read. Let p represent the number of pages left to read.

Write an inequality to represent this situation.

Graph the inequality on the number line.

Did you use a closed dot or an open dot at 65 on the number line? Explain your reasoning.

CCSS Common Core Spiral Review

Evaluate each expression. 5.OA.1

23. $8(2) - 11 =$ ______

24. $7 + 2(2) =$ ______

25. $3(5) - 7 =$ ______

26. $19 - 2(3) =$ ______

27. $3(4) - 7 =$ ______

28. $28 - 4(4) =$ ______

29. Graph 32, 30, 29, and 34 on the number line below. **6.NS.6c**

30. Graph 13, 15, 9, and 11 on the number line below. **6.NS.6c**

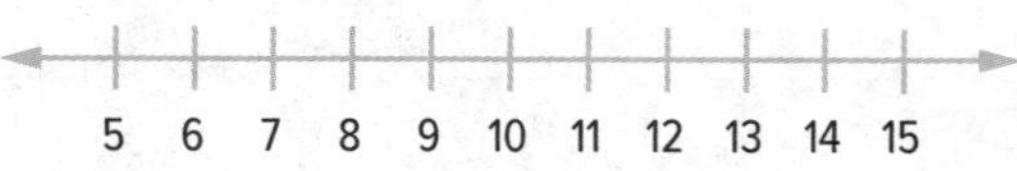

Inquiry Lab

Solve One-Step Inequalities

HOW can you use bar diagrams to solve one-step inequalities?

Content Standards
6.EE.5, 6.EE.8

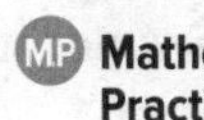

Mathematical Practices
1, 3, 4

In a recent Kentucky Derby, the total weight a horse could carry was less than 126 pounds. A jockey weighs a certain number of pounds and his equipment weighs 9 pounds. How much could the jockey weigh?

What do you know? ______________________________

____

What do you need to find? ______________________________

____

Hands-On Activity

You already learned that you can add or subtract the same quantity to each side of an equation when solving it. This is also true for inequalities.

Step 1 Model and solve the inequality $x + 9 < 126$ using a bar diagram. Place a dashed line on 126.

Step 2 The symbol is $<$, so a box is drawn to the left of 126.

Step 3 The bar represents $x +$ ☐. Label the bar diagram below.

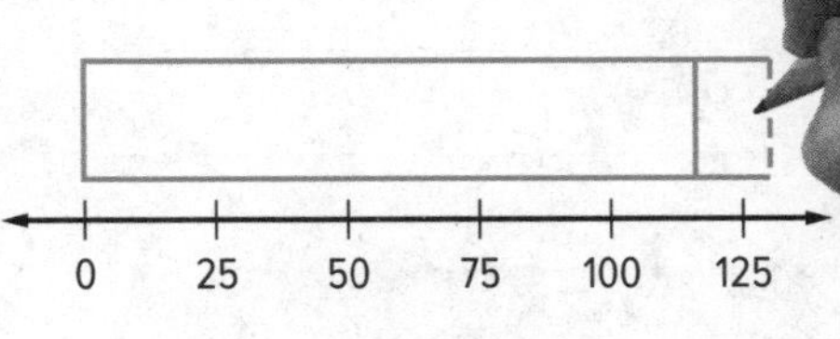

The section of the bar labeled x must be less than ☐ for the inequality to be true. So, $x <$ ☐.

Investigate

Work with a partner to solve each problem by using a model.

1. Regina sent x text messages before lunch. She sent another 4 text messages after lunch. She sent less than 7 text messages today. How many text messages could she have sent before lunch? Write your answer as an inequality. ______

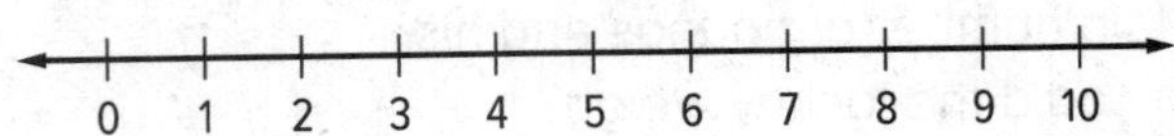

2. A player with five personal fouls cannot stay in the game. Dylan has already earned two personal fouls. How many more personal fouls x could he earn and still stay in the game? Write your answer as an inequality.

______

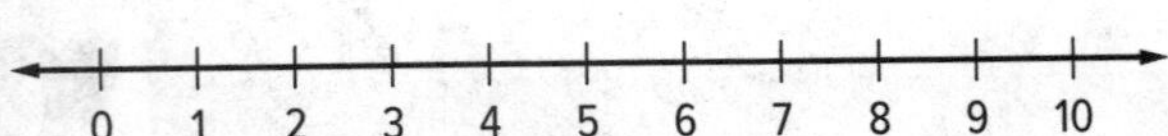

Work with a partner to solve by using the *guess, check, and revise* strategy. Find the least or greatest number that makes the inequality true.

3. $x - 5 \leq 1$ ______

4. $x + 3 \geq 8$ ______

Analyze and Reflect

5. **MP Reason Inductively** Explain how you could solve the inequality $x + 7 \leq 12$ using the *guess, check, and revise* strategy. Then solve. ______

Create

6. **MP Model with Mathematics** Write and solve a word problem using the inequality $x + 6 \leq 25$. ______

7. **Inquiry** HOW can you use bar diagrams to solve one-step inequalities?

Lesson 7

Solve One-Step Inequalities

Real-World Link

Baseball The graph shows the number of home runs that the top hitters on the baseball team hit last season.

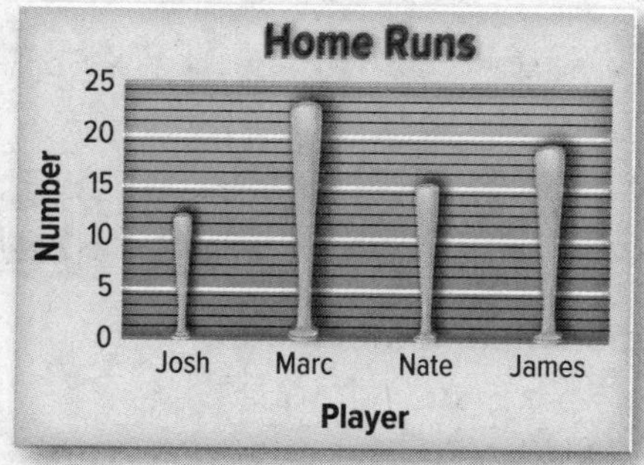

1. Write an inequality that compares the number of home runs Nate hit to the number of home runs Josh hit.

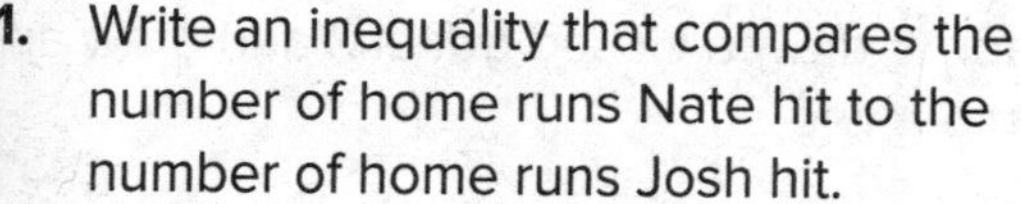

_____ > _____

2. Write an inequality that compares the number of home runs James hit to the number of home runs Marc hit.

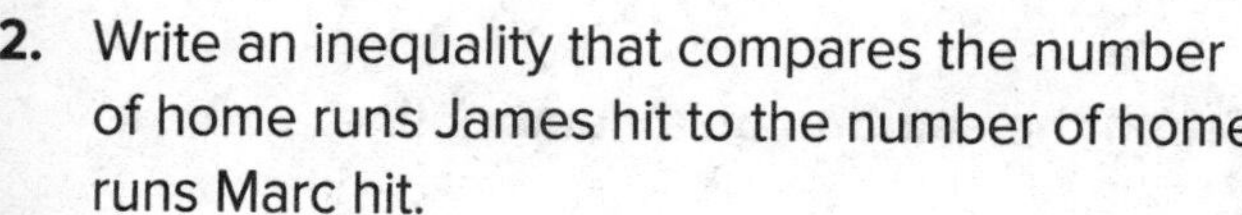

_____ < _____

3. Suppose James and Marc each hit 3 more home runs. Write a new inequality that compares the number of home runs James and Marc hit.

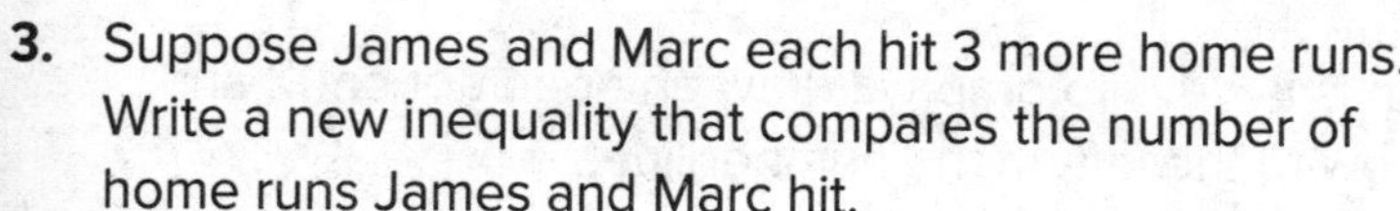

_____ < _____

Essential Question

HOW are symbols, such as <, >, and =, useful?

Common Core State Standards

Content Standards
6.EE.5, 6.EE.6, 6.EE.8

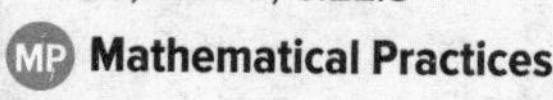

MP Mathematical Practices
1, 3, 4

Which MP **Mathematical Practices** did you use? Shade the circle(s) that applies.

① Persevere with Problems
② Reason Abstractly
③ Construct an Argument
④ Model with Mathematics
⑤ Use Math Tools
⑥ Attend to Precision
⑦ Make Use of Structure
⑧ Use Repeated Reasoning

Key Concept

Use Addition and Subtraction Properties to Solve Inequalities

Work Zone

Words When you add or subtract the same number from each side of an inequality, the inequality remains true.

Example

$$\begin{array}{r} 5 < 9 \\ +4 \quad +4 \\ \hline 9 < 13 \end{array} \qquad \begin{array}{r} 11 > 6 \\ -3 \quad -3 \\ \hline 8 > 3 \end{array}$$

These properties are also true for ≤ and ≥.

Examples

1. Solve $x + 7 \geq 10$. Graph the solution on a number line.

$x + 7 \geq 10$ Write the inequality.

$-7 \quad -7$ Subtract 7 from each side.

$x \geq 3$ Simplify.

The solution is $x \geq 3$. To graph it, draw a closed dot at 3 and draw an arrow to the right on the number line.

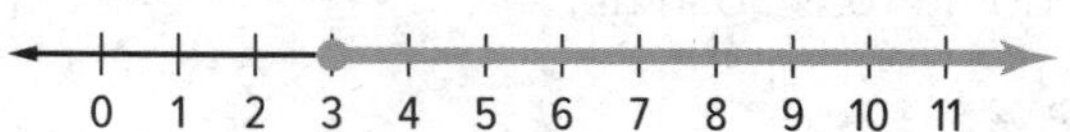

2. Solve $x - 3 < 9$. Graph the solution on a number line.

$x - 3 < 9$ Write the inequality.

$+3 \quad +3$ Add 3 to each side.

$x < 12$ Simplify.

The solution is $x < 12$. To graph it, draw an open dot on 12 and draw an arrow to the left on the number line.

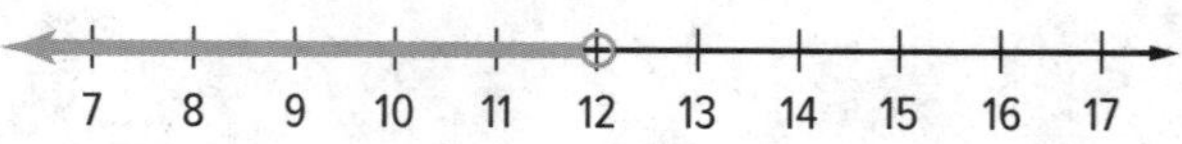

Got it? **Do these problems to find out.**

Show your work.

a. $n + 2 \leq 5$

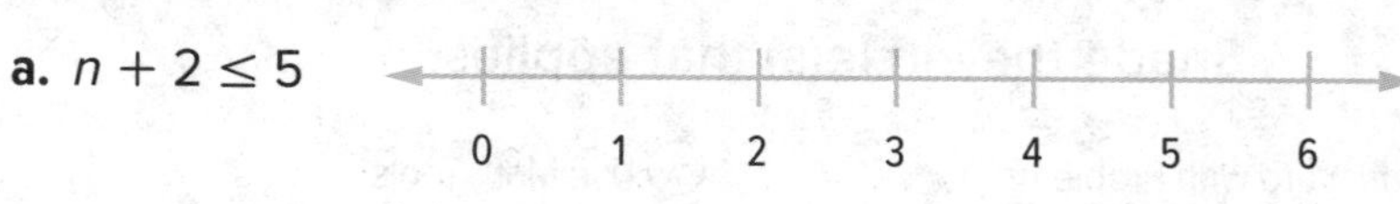

b. $y - 3 > 9$

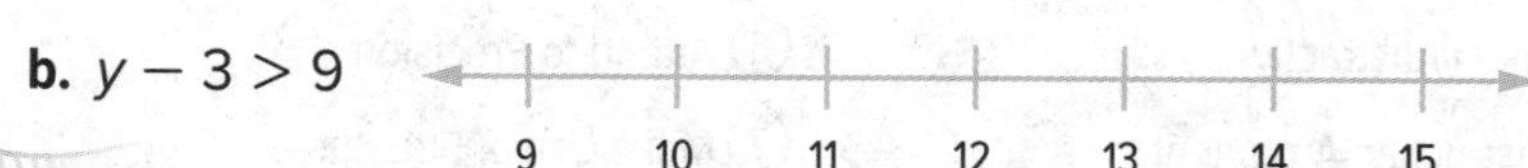

a. ________

b. ________

Use Multiplication and Division Properties to Solve Inequalities

Key Concept

Words When you multiply or divide each side of an inequality by the same *positive* number, the inequality remains true.

Example

$5 < 10$

$5 \times 2 < 10 \times 2$

$10 < 20$

$16 > 12$

$\frac{16}{2} > \frac{12}{2}$

$8 > 6$

These properties are also true for ≤ and ≥.

Examples

3. Solve $5x \leq 45$. Graph the solution on a number line.

$5x \leq 45$ Write the inequality.

$\frac{5x}{5} \leq \frac{45}{5}$ Divide each side by 5.

$x \leq 9$ Simplify.

The solution is $x \leq 9$.

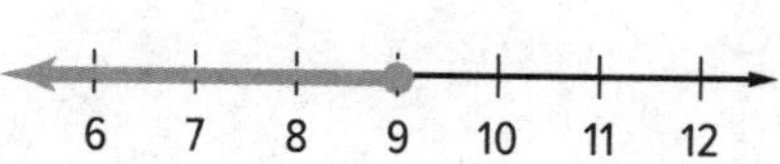

Checking Solutions You can check your solutions by substituting numbers into the inequality and testing to verify that it holds true.

4. Solve $\frac{x}{8} > 3$. Graph the solution on a number line.

$\frac{x}{8} > 3$ Write the inequality.

$\frac{x}{8}(8) > 3(8)$ Multiply each side by 8.

$x > 24$ Simplify.

The solution is $x > 24$.

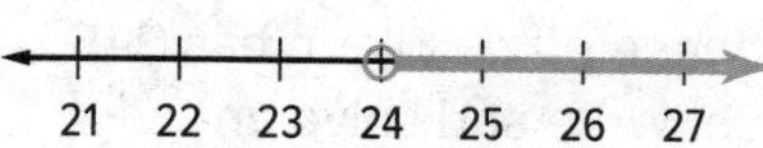

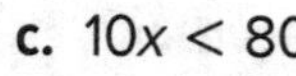

Got it? Do these problems to find out.

c. $10x < 80$

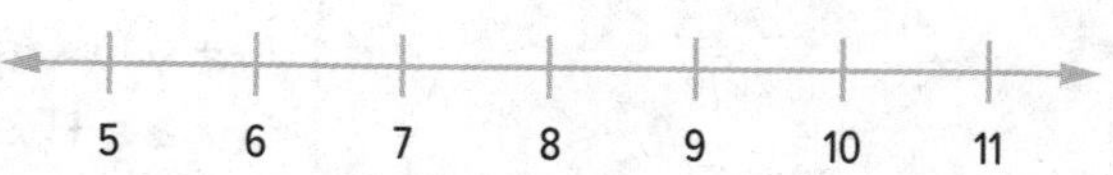

d. $\frac{x}{6} \geq 7$

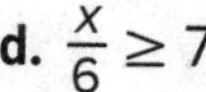

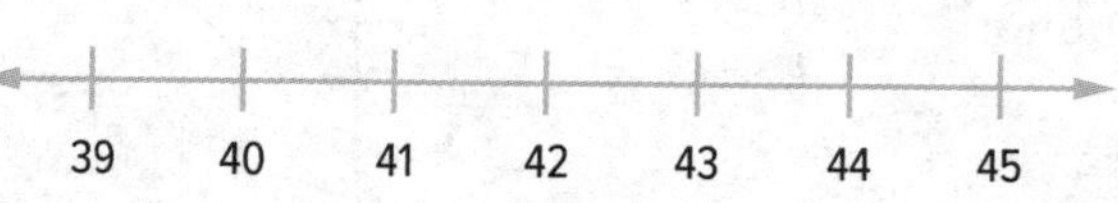

c. ____________

d. ____________

Example

5. Laverne is making bags of party favors for each of the 7 friends attending her birthday party. She does not want to spend more than $42 on the party favors. Write and solve an inequality to find the maximum cost for each party favor bag.

Words to Symbols
Remember, at most translates to $\leq$, while at least translates to $\geq$.

Let c represent the cost for each bag of party favors.

7 times the cost of each bag must be no more than $42.

$7c \leq 42$ Write the inequality.

$\frac{7c}{7} \leq \frac{42}{7}$ Divide each side by 7.

$c \leq 6$ Simplify.

Laverne can spend a maximum of $6 on each party favor bag.

Guided Practice

Solve each inequality. Graph the solution on a number line. (Examples 1–4)

1. $h - 6 \geq 13$ ______

2. $5y > 30$ ______

3. Johanna's parents give her $10 per week for lunch money. She cannot decide whether she wants to buy or pack her lunch. If a hot lunch at school costs $2, write and solve an inequality to find the maximum number of times per week Johanna can buy her lunch. (Example 5)

______

4. Tino's Pizza charges $9 for a cheese pizza. Eileen has $45 to buy pizza for the Spanish Club. Write and solve an inequality to find the maximum number of pizzas that Eileen can buy. (Example 5) ______

5. **Building on the Essential Question** How is solving an inequality similar to solving an equation?

Rate Yourself!

Are you ready to move on? Shade the section that applies.

YES ? NO

For more help, go online to access a Personal Tutor.

Tutor

Independent Practice

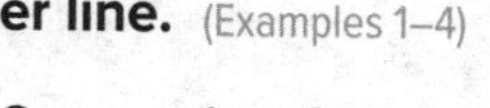

Solve each inequality. Graph the solution on a number line. (Examples 1–4)

1. $2 + y \leq 3$ ______

2. $w - 1 < 4$ ______

3. $7x > 56$ ______

4. $\frac{d}{3} \leq 2$ ______

5. A company charges $0.10 for each letter engraved. Bobby plans to spend no more than $5.00 on the engraving on a jewelry box. Write and solve an inequality to find the maximum number of letters he can have engraved. (Example 5)

______

6. **MP Model with Mathematics** Refer to the graphic novel frame below for Exercises a–b.

a. Suppose David has $65 to spend on his ticket and some shirts. He already spent $32.25 on his ticket and fee. Write an inequality that could be used to find the maximum number of shirts he can buy.

______

b. What is the maximum number of shirts he can buy?

______

Solve each inequality. Graph the solution on a number line.

7. $p - \frac{7}{12} > \frac{3}{10}$ ________

8. $f + 0.3 < 1.7$ ________

H.O.T. Problems Higher Order Thinking

9. **MP Model with Mathematics** Write a word problem that would have the solution $p \leq 21$.

10. **MP Persevere with Problems** In three math tests, you have scored 91, 95, and 88 points. You are about to take your next test. Suppose you want to have an average score of at least 90 points after all four tests. Explain a method you could use to find the score you must receive in order to average at least 90 points. Then find the least score.

11. **MP Construct an Argument** Does the order of the quantities in an inequality matter? Explain.

12. **MP Model with Mathematics** Write a real-world problem and an inequality that can be represented by the number line below.

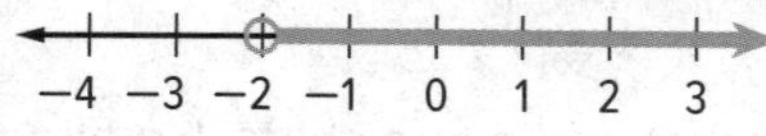

Extra Practice

Solve each inequality. Graph the solution on a number line.

13. $a + 4 < 9$ ______

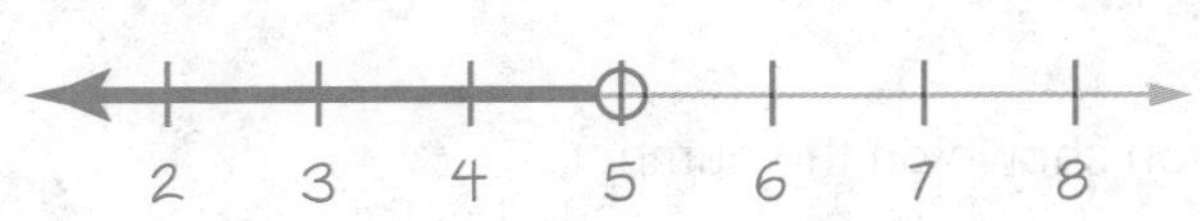

14. $x - 8 \geq 13$ ______

15. $d + 13 \geq 22$ ______

16. $25t \leq 100$ ______

17. $\frac{g}{2} < 6$ ______

18. $\frac{r}{9} > 8$ ______

19. A community needs to raise at least $5,000 to build a new skateboarding park. They are selling backpacks for $25 each to raise the money. Write and solve an inequality to determine the minimum number of backpacks they need to sell in order to reach this goal.

______

20. A sales associate at a computer store receives a bonus of $100 for every computer he sells. He wants to make $2,500 in bonuses next month. Write and solve an inequality to find the minimum number of computers he must sell. ______

MP Model with Mathematics Solve each inequality. Graph the solution on a number line.

21. $n + \frac{2}{7} \geq \frac{1}{2}$ ______

22. $0.2g > 1.8$ ______

Power Up! Common Core Test Practice

23. Use the graph of the inequality shown below.

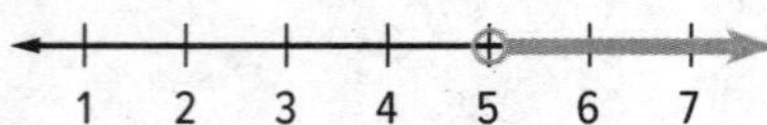

Which of the following inequalities have the solution shown on the number line? Select all that apply.

- ☐ $n + 3 < 8$
- ☐ $y + 1 > 6$
- ☐ $z - 4 > 1$
- ☐ $c - 7 > 12$

24. The table shows a gym class's average results for boys and girls participating in the long jump. Susan could jump no farther than 4 inches more than the average distance for females. Let *j* represent the distance that Susan could jump.

Gender	Distance
Male	10 ft 6 in.
Female	8 ft 4 in.

Write an inequality to represent the situation.

How far could Susan jump?

Common Core Spiral Review

Multiply. **4.NBT.5, 5.NBT.7**

25. $12 \times 12 =$ ______

26. $9 \times 13 =$ ______

27. $16 \times 12 =$ ______

28. $8.5 \times 6 =$ ______

29. $13.2 \times 5 =$ ______

30. $7 \times 11.5 =$ ______

31. Mitchell is painting several boards for scenery for the school play.

What is the area of the board shown? **4.MD.3** ______

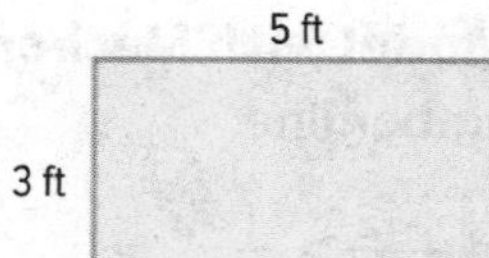

32. Daphne is painting her room. She knows that three of her bedroom walls are a total of 305 square feet. The fourth wall in her room measures 8 feet wide and 10 feet tall. How much total area will Daphne need to paint? **4.MD.3** ______

21ST CENTURY CAREER in Atmospheric Science

Meteorologist

Have you ever wondered how forecasters can predict severe storms such as hurricanes before they occur? Keeping track of changes in air pressure is one method that they use. Meteorologists study Earth's air pressure, temperature, humidity, and wind velocity. They use complex computer models to process and analyze weather data and to make accurate forecasts. In addition to understanding the processes of Earth's atmosphere, meteorologists must have a solid background in mathematics, computer science, and physics.

Is This the Career for You?

Are you interested in a career as a meteorologist? Take some of the following courses in high school.

- Algebra
- Calculus
- Earth and Its Environment
- Environmental Science
- Physics

Turn the page to find out how math relates to a career in Atmospheric Science.

MP The Pressure is On!

Use the information in the diagram and the table to solve each problem.

1. Write an inequality representing the temperature t of the ocean water during the formation of a hurricane. __________

2. Write an inequality representing the depth d of the water that must be greater than 80°F in order for a hurricane to form. __________

3. The air needs to be humid up to about 18,000 feet for a hurricane to form. Write an inequality to represent this altitude a of the air above the ocean. __________

4. Air pressure decreases during a storm. The difference between the normal air pressure n and the air pressure during the 1935 Florida Keys hurricane was greater than 121 millibars. Write and solve an inequality to find the normal air pressure in the Florida Keys before the hurricane.

____

5. The air pressure of Hurricane Katrina at landfall was greater than 17 millibars plus the air pressure p before landfall. Write and solve an inequality to find the air pressure of the storm before landfall.

____

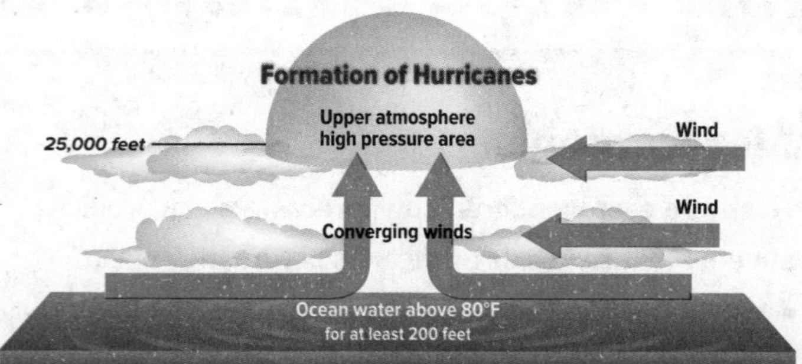

Top 5 Most Intense Hurricanes at Landfall in the U.S.

Rank	Hurricane	Pressure (millibars)
1	Florida Keys, (Labor Day), 1935	892
2	Hurricane Camille, 1969	909
3	Hurricane Katrina, 2005	920
4	Hurricane Andrew, 1992	922
5	Texas (Indianola), 1886	925

MP Career Project

It's time to update your career portfolio! Interview a meteorologist at a local television station. Be sure to ask what he or she likes most about being a meteorologist and what is most challenging. Include all the interview questions and answers in your portfolio.

What skills would you need to improve to succeed in this career?

-
-
-
-
-

Chapter Review

Vocabulary Check

Write the correct term for each clue in the crossword puzzle.

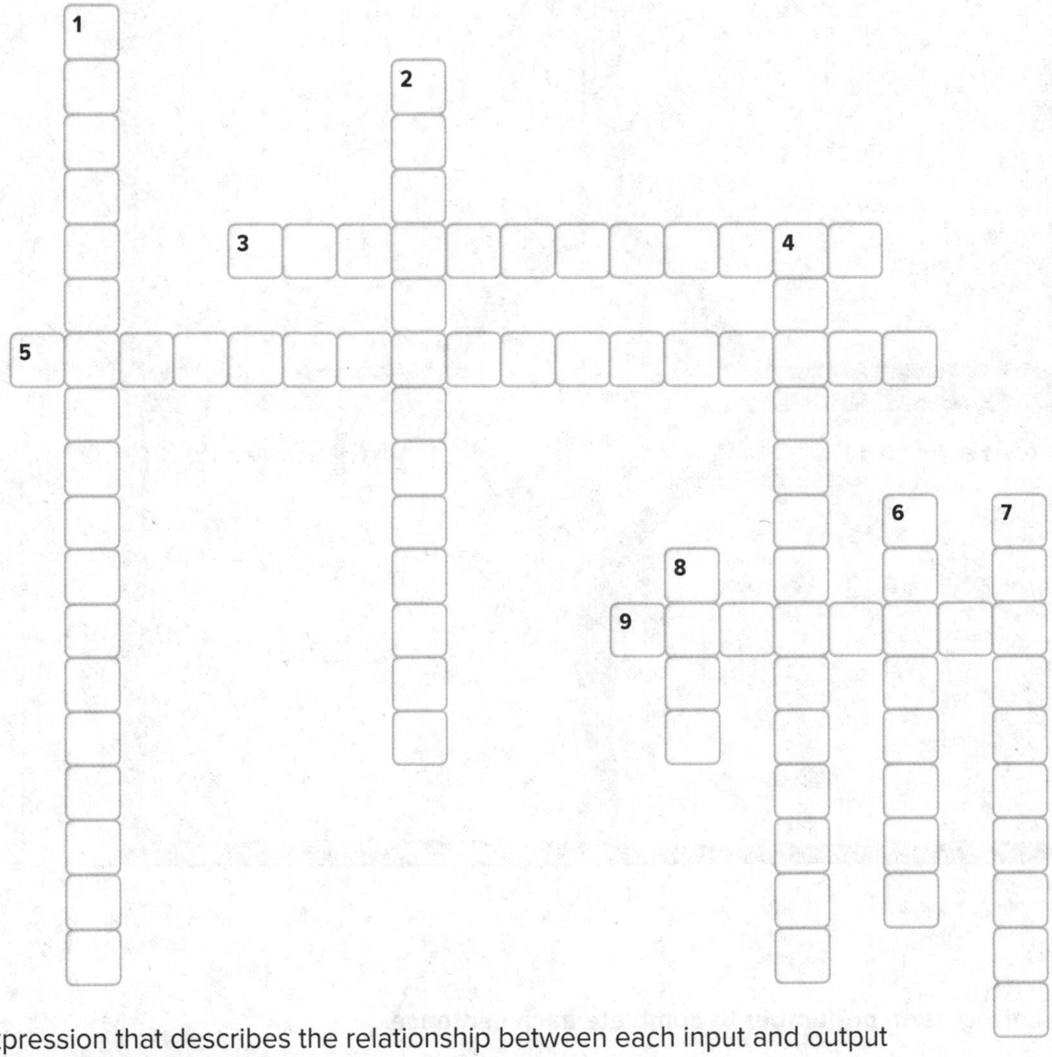

Across

3. an expression that describes the relationship between each input and output
5. found by multiplying the previous term by the same number
9. a list of numbers in a specific order

Down

1. found by adding the same number to the previous term
2. a table organizing the input, rule, and output of a function
4. a function that forms a line when graphed
6. a relationship that assigns exactly one output value to one input value
7. a mathematical sentence indicating that two quantities are not equal
8. each number in a sequence

Key Concept Check

Use Your FOLDABLES

Use your Foldable to help review the chapter.

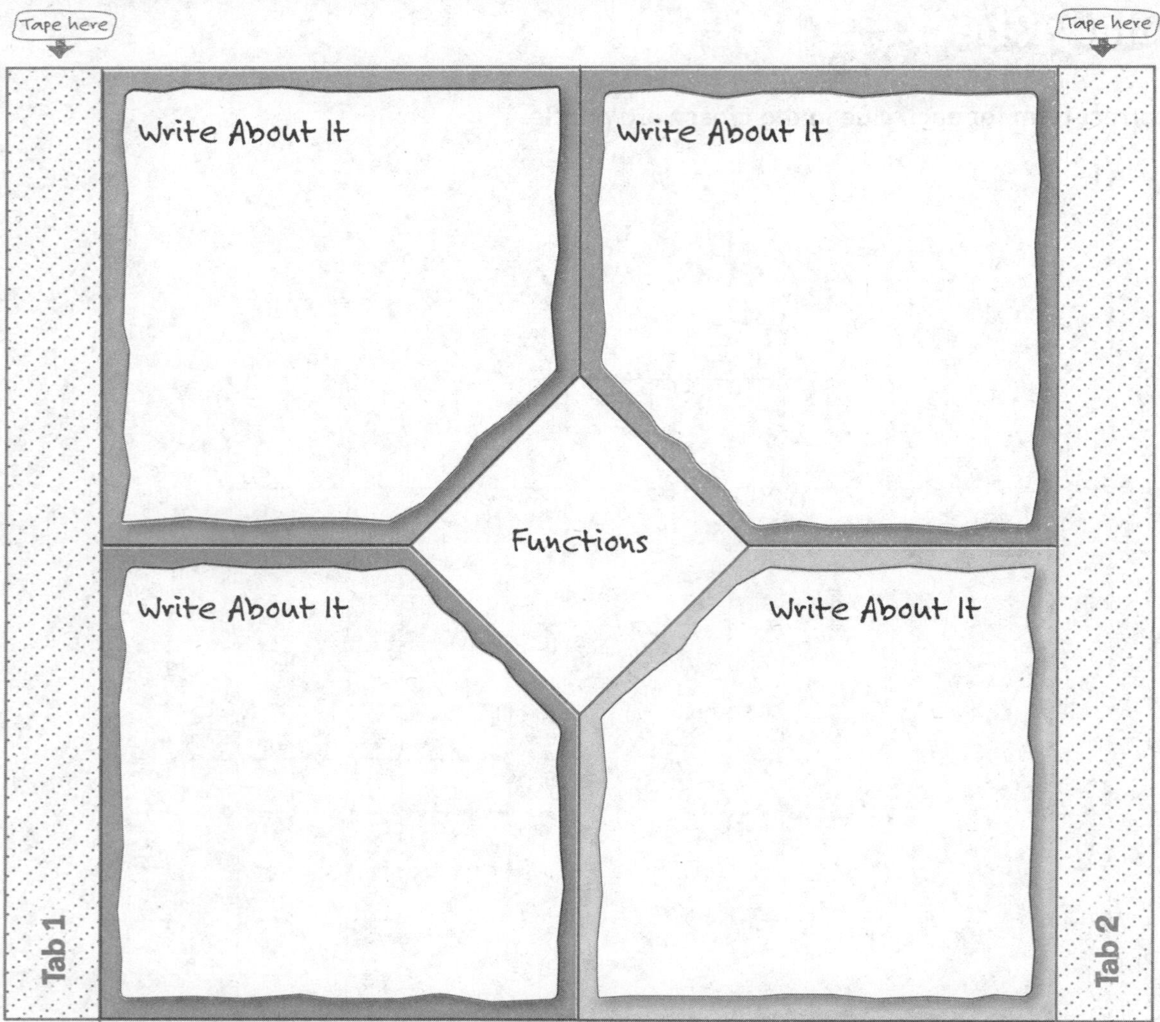

Got it?

Circle the correct term or number to complete each sentence.

1. The next number in the sequence 12, 15, 18, 21, . . . is (24, 27).
2. The output of a function is the (independent, dependent) variable.
3. A(n) (arithmetic, geometric) sequence can be found by multiplying each previous term by the same number.
4. The input of a function is the (independent, dependent) variable.
5. A(n)(inequality, function) is a relation that assigns exactly one output value to one input value.

Remodeling Project

Mr. Jacobs is installing a new kitchen floor using white and brown tiles. The relationship between the number of brown tiles and the number of white tiles is shown in the table.

White (w)	1	2	3	4	5	6
Brown (b)	4	6	8	10	?	?

Write your answers on another piece of paper. Show all of your work to receive full credit.

Part A

Fill in the missing values based on the pattern in the table. Write an equation that represents the relationship between the white tiles and the brown tiles. Let b represent brown tiles and w represent white tiles.

Part B

Each tile costs $12. Determine the cost of 60 tiles, 80 tiles, 100 tiles, and 120 tiles. Write a set of ordered pairs (number of tiles, total cost) to represent the data. Then graph the ordered pairs.

Part C

Mr. Jacobs budgeted $1,200 for tiles. His sketch of the floor requires him to use 38 white tiles. Each white tile costs $12. Write and solve an inequality to find the maximum amount he can spend on brown tiles.

Part D

The tile store has three different brown tiles Mr. Jacobs can use. The tile prices are shown in the table. Which tile(s) can he purchase to stay in his budget and meet the design for 38 white tiles? Explain your reasoning.

Tile A	Tile B	Tile C
$9.75 per tile	Tiles are $11 per tile if fewer than 50 tiles are purchased. For 50 or more tiles, tiles are $9.50 per tile.	Tiles are purchased by the box. There are 24 tiles per box. Each box is $185.

Reflect

Answering the Essential Question

Use what you learned about inequalities to complete the graphic organizer.

Essential Question

HOW are symbols, such as <, >, and =, useful?

<	>	=
What does it mean?	What does it mean?	What does it mean?
Mathematical Example	Mathematical Example	Mathematical Example
Real-World Example	Real-World Example	Real-World Example

Answer the Essential Question. HOW are symbols, such as <, >, and =, useful?

UNIT PROJECT

Watch

It's Out of This World How fast do objects in our solar system travel through space? Let's explore the orbital speed of different planets and satellites! In this project you will:

- **Collaborate** with your classmates as you investigate the orbital speed of three planets.
- **Share** the results of your research in a creative way.
- **Reflect** on how you communicate mathematical ideas effectively.

Collaborate

Go Online Work with your group to research and complete each activity. You will use your results in the Share section on the following page.

1. Choose three planets in our solar system. Use the Internet to research each planet and find its average orbital speed in miles per second or kilometers per second. Organize the information in a table.

2. Find and record the orbital distance traveled in 1, 2, and 3 seconds for each planet you chose in Exercise 1. Then describe how the orbital distance of each planet changes with time.

3. For your three planets, list the ordered pairs representing (time, distance). Graph each set of ordered pairs on a coordinate plane and connect each set of points with a line. Compare the graphs. Then write equations to represent each relationship.

4. Research artificial satellites, such as the Hubble Space Telescope, that are orbiting Earth. Use the Internet to research three different satellites and determine the purpose of those satellites. Write a summary of your findings.

5. For each satellite you found in Exercise 4, find and record its average orbital speed in miles per second or kilometers per second. Organize the information in a table. Compare the orbital speeds.

With your group, decide on a way to present what you have learned from each of the activities. Some suggestions are listed below, but you can also think of other creative ways to present your information. Remember to show how you used math to complete each of the activities of this project!

- Create a presentation using the data you collected. Your presentation should include a spreadsheet, graph, and one other visual display.
- Write an article that would be published in a magazine from the perspective of a scientist. Include any important information that you found while researching the orbital speed of each planet.

Check out the note on the right to connect this project with other subjects.

with Social Studies

Global Awareness Research the history of space exploration and write a summary of your findings. Some questions to consider are:

- What have scientists in the U.S. and other countries discovered recently about the solar system?
- Which countries have contributed the most to space exploration?

6. **Answer the Essential Question** How can you communicate mathematical ideas effectively?

a. How did you use what you learned about expressions and equations to communicate mathematical ideas effectively in this project?

b. How did you use what you learned about functions and inequalities to communicate mathematical ideas effectively in this project?

UNIT 4 Geometry

CCSS

Essential Question

HOW can you use different measurements to solve real-life problems?

Chapter 9 Area

A composite figure can be decomposed into triangles and other shapes. In this chapter, you will find the area of triangles, quadrilaterals, and composite figures.

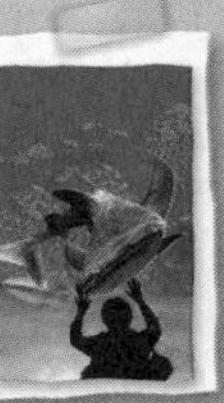

Chapter 10 Volume and Surface Area

Prisms and pyramids are examples of three-dimensional figures. In this chapter, you will find the volume and surface area of three-dimensional figures in the context of solving real-world and mathematical problems.

Unit Project Preview

A New Zoo Have you ever wondered how a zoo is designed? The space needed for each type of animal must be considered to determine the zoo's total area.

Suppose you are getting a new pet. What information will be important to gather to determine the pet's living area? Design a space. Then calculate the area. Explain your reasoning.

At the end of Chapter 10, you'll complete a project that involves designing a new zoo. Get ready to embark on your journey to the wild side!

Where My Pet Lives

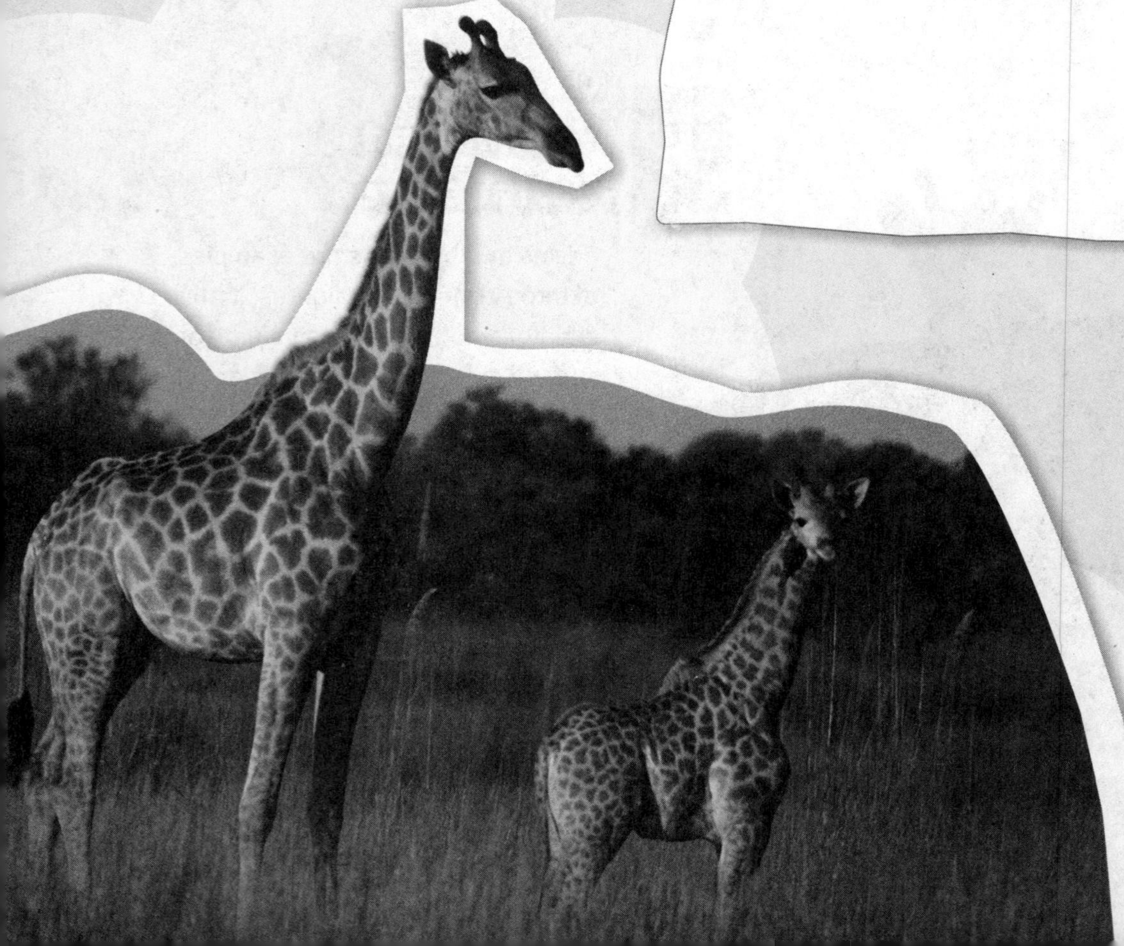

Chapter 9 Area

Essential Question

HOW does measurement help you solve problems in everyday life?

Common Core State Standards

Content Standards
6.G.1, 6.G.3, 6.NS.8

MP Mathematical Practices
1, 2, 3, 4, 5, 6, 7, 8

Math in the Real World

Gardens A garden designer plants dahlias in a 5 foot by 3 foot plot. What area of the garden do the dahlias cover? In the diagram below, shade the area covered by dahlias.

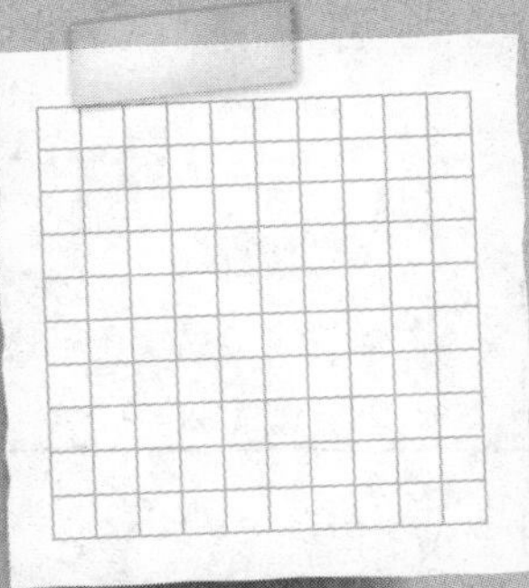

Area = ____________

FOLDABLES Study Organizer

1. Cut out the Foldable on page FL9 of this book.

2. Place your Foldable on page 728.

3. Use the Foldable throughout this chapter to help you learn about area.

What Tools Do You Need?

Vocabulary

base
composite figure
congruent
formula
height
parallelogram
polygon
rhombus

Review Vocabulary

Using a graphic organizer can help you to remember important vocabulary terms. Fill in the graphic organizer below for the word *area*.

Area

Definition

Units of Measure

Real-World Examples

What Do You Already Know?

Read each statement. Decide whether you agree (A) or disagree (D). Place a checkmark in the appropriate column and then justify your reasoning.

Area			
Statement	A	D	Why?
The area of a parallelogram is the same as the area of a rectangle.			
All parallelograms can be split into two congruent triangles.			
The bases of a trapezoid are always two horizontal sides.			
A circle is an example of a polygon.			
The formula to find the area of a triangle is $A = \frac{1}{2}bh$.			
When the dimensions of a triangle are multiplied by x, then the perimeter of the polygon changes by $x \cdot x$ or x^2.			

When Will You Use This?

Here are a few examples of how two-dimensional figures are used in the real world.

Activity 1 Work with a group of 3–4 students. Hide something in your classroom or around the school. Write a set of clues that can be used to find your hidden object. Trade clues with another group, and find each other's hidden objects.

Activity 2 Go online at **connectED.mcgraw-hill.com** to read the graphic novel ***Scavenger Hunt***. What is the third clue on the clue sheet?

Are You Ready?

Try the Quick Check below.
Or, take the online Readiness Quiz.

Quick Review

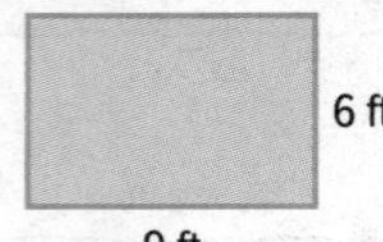

Common Core Review 4.MD.3, 5.NF.4

Example 1

Find the area of the rectangle.

(Rectangle: 9 ft by 6 ft)

$A = \ell w$ — Area of a rectangle

$A = 9 \cdot 6$ — Replace ℓ with 9 and w with 6.

$A = 54$ — Multiply.

The area of the rectangle is 54 square feet.

Example 2

Find $\frac{1}{2} \times 16$.

$\frac{1}{2} \times 16 = \frac{1}{2} \times \frac{16}{1}$ — Write 16 as $\frac{16}{1}$.

$= \frac{1 \times \overset{8}{\cancel{16}}}{\underset{1}{\cancel{2}} \times 1}$ — Divide the numerator and the denominator by 2.

$= \frac{8}{1}$ or 8 — Simplify.

Quick Check

Area **Find the area of each rectangle.**

1. (Rectangle: 8 cm by 4 cm) ______

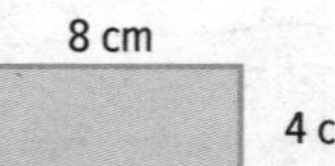

2. (Rectangle: 6 in. by 15 in.) ______

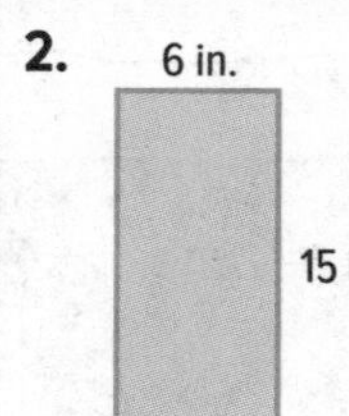

3. (Rectangle: 3 cm by 6 cm) ______

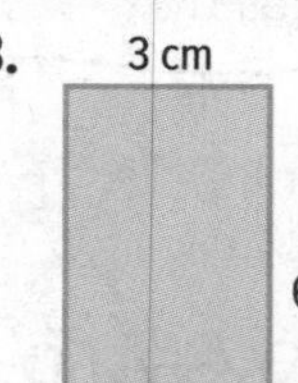

4. The playing area of a board game is a rectangle with a length of 14 inches and a width of 20 inches. What is the area of the board game? ______

Fractions **Multiply. Write in simplest form.**

5. $\frac{1}{2} \times 28 =$ ______

6. $\frac{1}{3} \times 27 =$ ______

7. $\frac{1}{7} \times 84 =$ ______

How Did You Do?

Which problems did you answer correctly in the Quick Check? Shade those exercise numbers below.

1 2 3 4 5 6 7

Inquiry Lab

Area of Parallelograms

HOW does finding the area of a parallelogram relate to finding the area of a rectangle?

Content Standards
6.G.1

Mathematical Practices
1, 2, 3, 5

Elise wants to make a banner in the shape of a parallelogram. Her parallelogram has a base of 2 feet and a height of 3 feet. What is the area of her parallelogram?

Hands-On Activity 1

Another type of quadrilateral is a *parallelogram*. A parallelogram has opposite sides parallel and congruent.

Parallelograms	Not Parallelograms

Make a parallelogram to represent Elise's banner.

Step 1 Start with a rectangle. Trace the rectangle shown at the right.

3 feet

2 feet

Step 2 Cut a triangle from one side of the rectangle you traced and move it to the other side to form a parallelogram. Tape the parallelogram to the right.

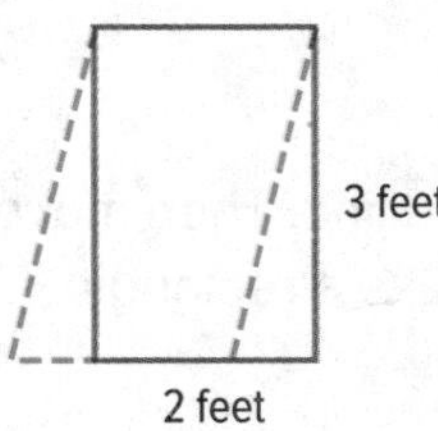

The rectangle was rearranged to form the parallelogram. Nothing was removed or added, so the parallelogram has ______ area as the rectangle.

Step 3 Multiply the base and height of the parallelogram to find the area. The base of the parallelogram is 2 feet and the height is 3 feet.

☐ feet × ☐ feet = ☐ square feet

Hands-On Activity 2

Find the area of the parallelogram below.

Step 1 Trace the parallelogram on grid paper and cut it out.

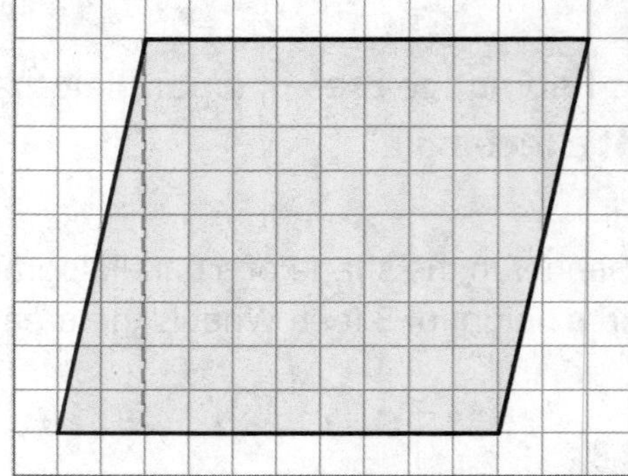

Step 2 Fold and cut along the dotted line.

Step 3 Move the triangle to the right to make a rectangle. Tape the rectangle in the space provided.

Step 4 Count the number of square units in the rectangle.

The area is ☐ square units.

Hands-On Activity 3

Find the area of the parallelogram below.

Step 1 Trace the parallelogram and cut it out.

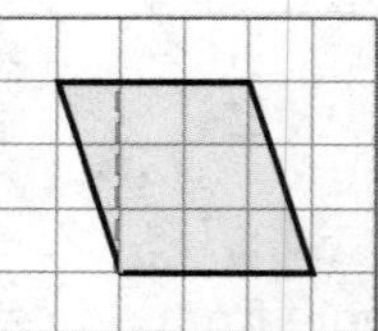

Step 2 Fold and cut along the dotted line. Then move the triangle to the right to make a rectangle. Tape it in the space provided.

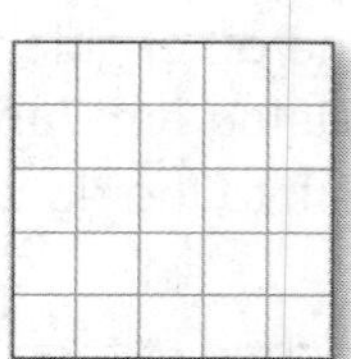

Step 3 Count the number of square units in the rectangle.

The area is ☐ square units.

Investigate

MP Use Math Tools Work with a partner. Find the area of each parallelogram.

1. $A =$ ______ square units

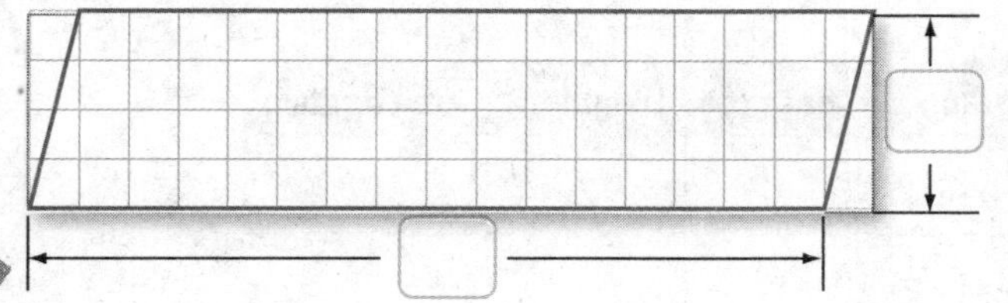

2. $A =$ ______ square units

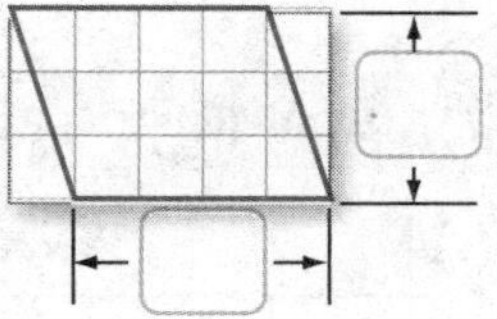

3. $A =$ ______ square units

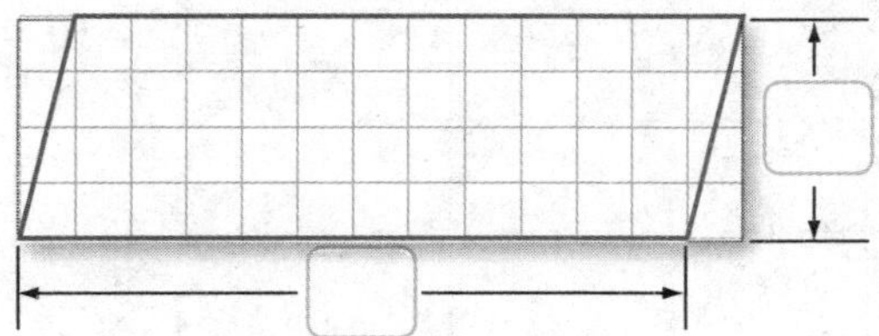

4. $A =$ ______ square units

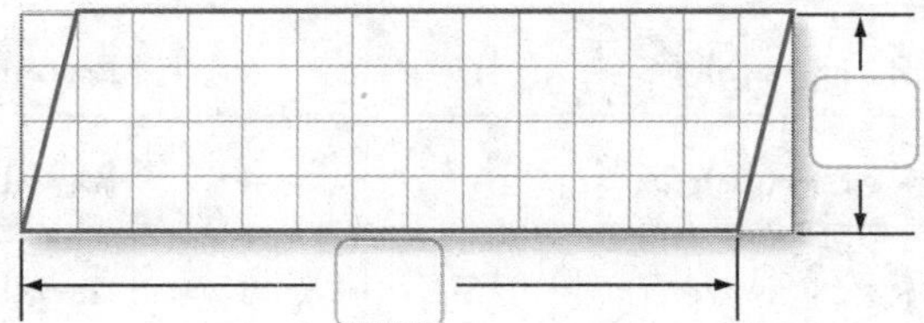

5. $A =$ ______ square units

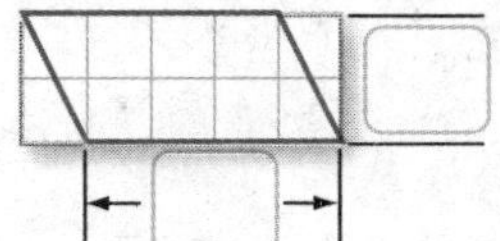

6. $A =$ ______ square units

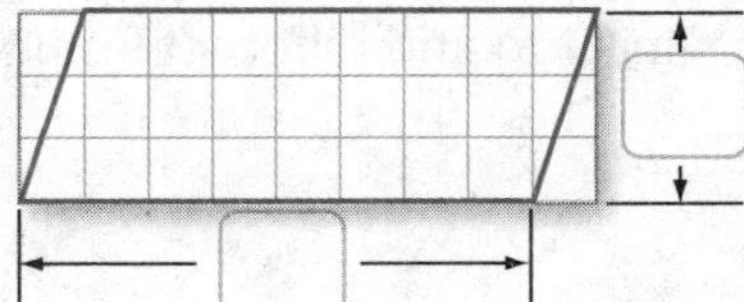

7. $A =$ ______ square units

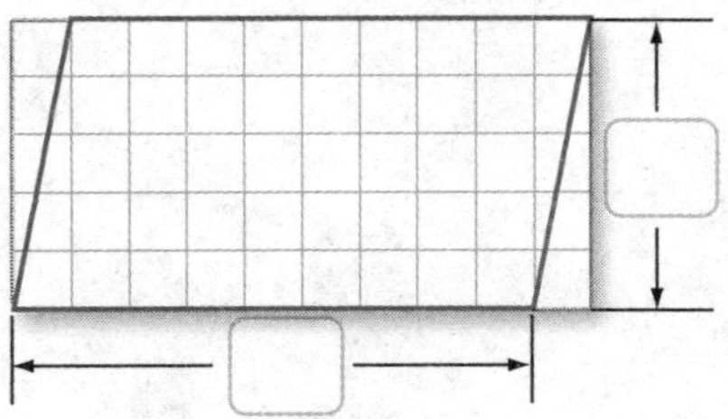

8. $A =$ ______ square units

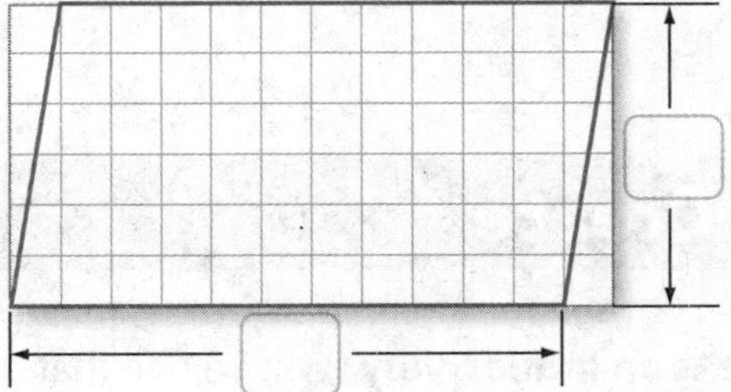

9. $A =$ ______ square units

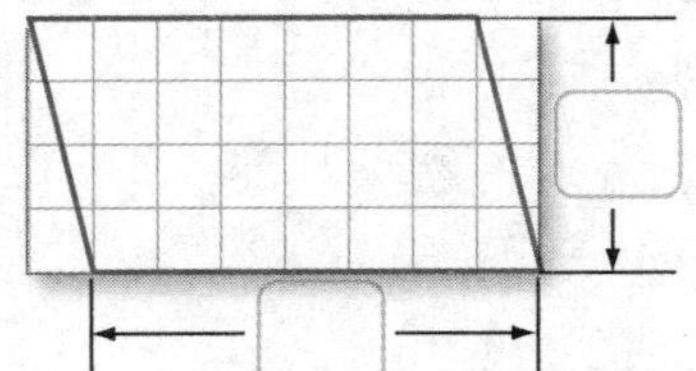

10. $A =$ ______ square units

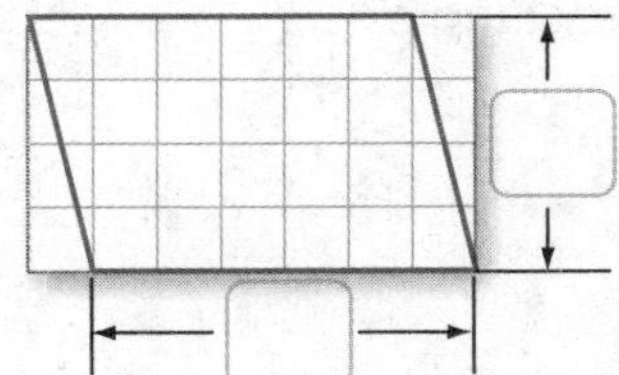

Analyze and Reflect

The table shows the dimensions of several rectangles and the corresponding dimensions of several parallelograms if each rectangle was rearranged to form a parallelogram. Work with a partner to complete the table. The first one is done for you.

	Rectangle	Length (ℓ)	Width (w)	Parallelogram	Base (b)	Height (h)	Area (units^2)
	Rectangle 1	6	2	Parallelogram 1	6	2	12
11.	Rectangle 2	12	4	Parallelogram 2			
12.	Rectangle 3	7	3	Parallelogram 3			
13.	Rectangle 4	5	4	Parallelogram 4			
14.	Rectangle 5	10	6	Parallelogram 5			
15.	Rectangle 6	6	4	Parallelogram 6			
16.	Rectangle 7	15	9	Parallelogram 7			
17.	Rectangle 8	9	3	Parallelogram 8			

18. A rectangle was rearranged to form a parallelogram. How is the height of the parallelogram similar to and different from the width of the rectangle?

19. MP **Reason Abstractly** If you were to draw three different parallelograms, each with a base of 6 units and a height of 4 units, how would the areas compare?

Create

20. MP **Reason Inductively** Write a rule that gives the area of a parallelogram.

21. Inquiry HOW does finding the area of a parallelogram relate to finding the area of a rectangle?

Lesson 1

Area of Parallelograms

Vocabulary Start-Up

A **polygon** is a closed figure formed by 3 or more straight lines. A **parallelogram** is a quadrilateral with opposite sides parallel and opposite sides the same length. A **rhombus** is a parallelogram with four equal sides. Fill in the lines in the diagram with polygon, parallelogram, or rhombus and draw an example of each.

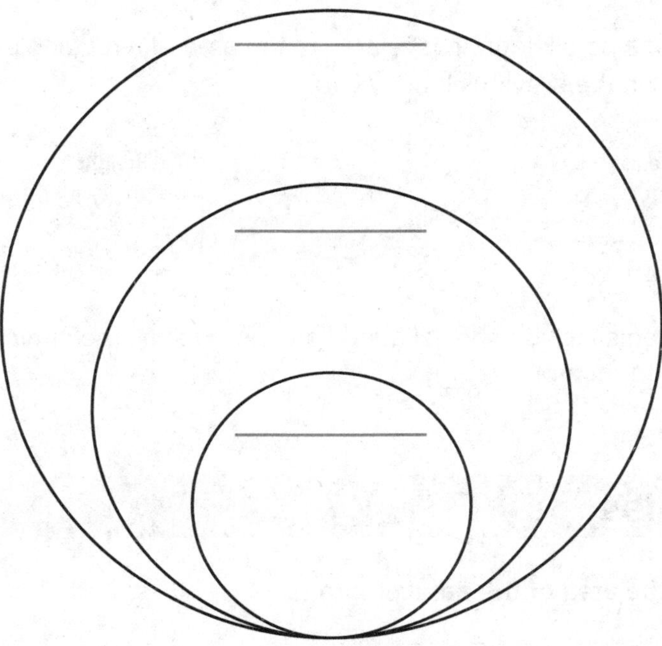

Essential Question

HOW does measurement help you solve problems in everyday life?

Vocabulary

polygon
parallelogram
rhombus
base
height
formula

Common Core State Standards

Content Standards
6.G.1

MP Mathematical Practices
1, 3, 4, 7

Real-World Link

Stairs Expert skateboarders can slide down the railings of stairs safely. A parallelogram is used to build a staircase. How many sets of parallel lines are shown in the parallelogram to the right?

Which MP Mathematical Practices did you use? Shade the circle(s) that applies.

1. Persevere with Problems
2. Reason Abstractly
3. Construct an Argument
4. Model with Mathematics
5. Use Math Tools
6. Attend to Precision
7. Make Use of Structure
8. Use Repeated Reasoning

Key Concept

Area of a Parallelogram

Watch 

Words The area A of a parallelogram is the product of its base b and its height h.

Model

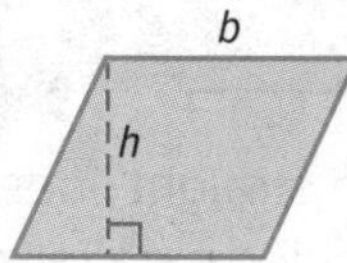

Symbols $A = bh$

Work Zone

The area of a parallelogram is related to the area of a rectangle as you discovered in the previous Inquiry Lab.

The **base** of a parallelogram can be any one of its sides.

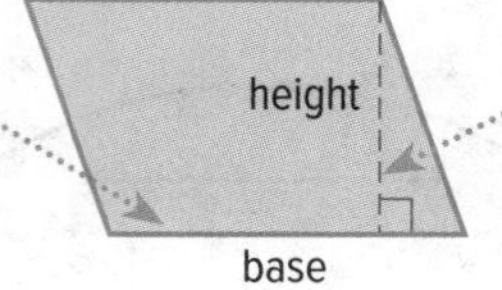

The **height** is the perpendicular distance from the base to the opposite side.

Parallelograms include special quadrilaterals, such as rectangles, squares, and rhombi.

Examples

Tutor 

1. Find the area of the parallelogram.

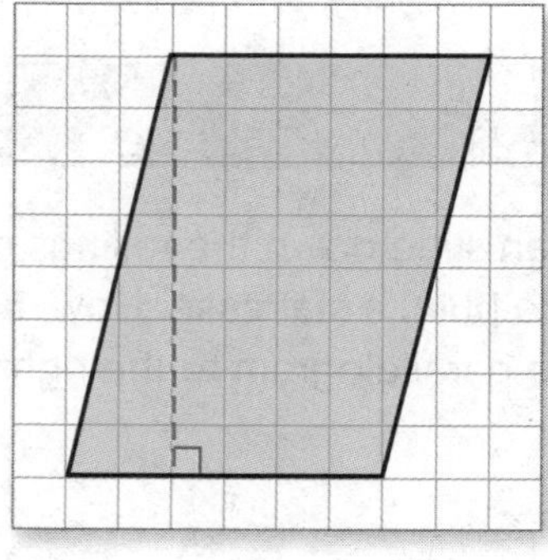

The base is 6 units, and the height is 8 units.

$A = bh$ Area of parallelogram

$A = \mathbf{6} \cdot \mathbf{8}$ Replace b with 6 and h with 8.

$A = 48$ Multiply.

The area is 48 square units or 48 units^2.

Area Measurement

An area measurement can be written using abbreviations and an exponent of 2.

For example:

square units = units^2

square inches = in^2

square feet = ft^2

square meters = m^2

2. Find the area of the parallelogram.

Estimate $A \approx 20 \cdot 10$ or 200 cm^2

$A = bh$	Area of parallelogram
$A = 20 \cdot 11$	Replace b with 20 and h with 11.
$A = 220$	**Check for Reasonableness** $220 \approx 200$ ✓

The area is 220 square centimeters or 220 cm^2.

Got it? Do these problems to find out.

a.

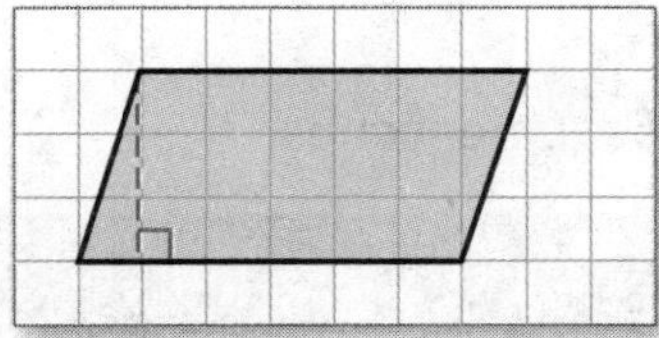

b.

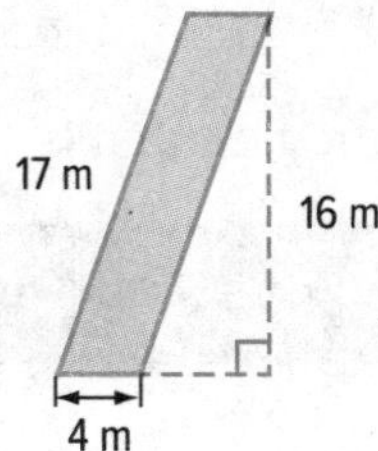

a. ______

b. ______

Find Missing Dimensions

A **formula** is an equation that shows a relationship among certain quantities. To find missing dimensions, use the formula for the area of a parallelogram. Replace the variables with the known measurements. Then solve the equation for the remaining variable.

Example

3. Find the missing dimension of the parallelogram.

$A = bh$	Area of a parallelogram
$45 = 9 \cdot h$	Replace A with 45 and b with 9.
$\frac{45}{9} = \frac{9 \cdot h}{9}$	Divide each side by 9.
$5 = h$	Simplify.

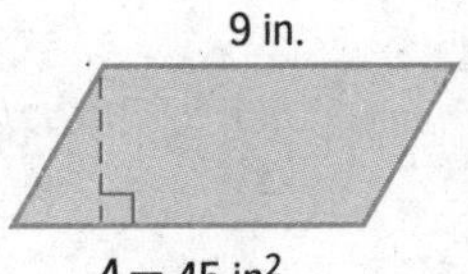

So, the height is 5 inches.

Checking Your Work

To check your work, replace b and h in the formula with 9 and 5.

$A = bh$

$A = 9 \cdot 5$

$A = 45$ ✓

Got it? Do these problems to find out.

c.

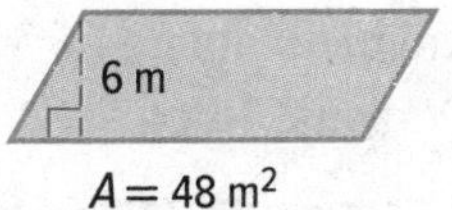

d.

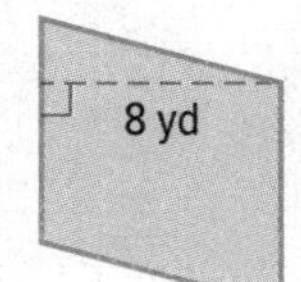

$A = 96\text{ yd}^2$

c. ______

d. ______

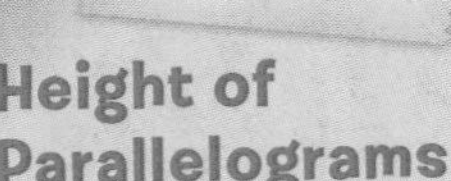

Height of Parallelograms

For the parallelogram formed by the area shaded black in Example 4, its height, 12 inches, is labeled outside the parallelogram.

Example

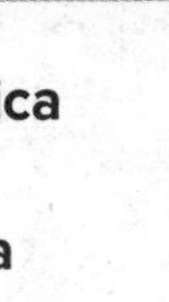

4. Romilla is painting a replica of the national flag of Trinidad and Tobago for a research project. Find the area of the black stripe.

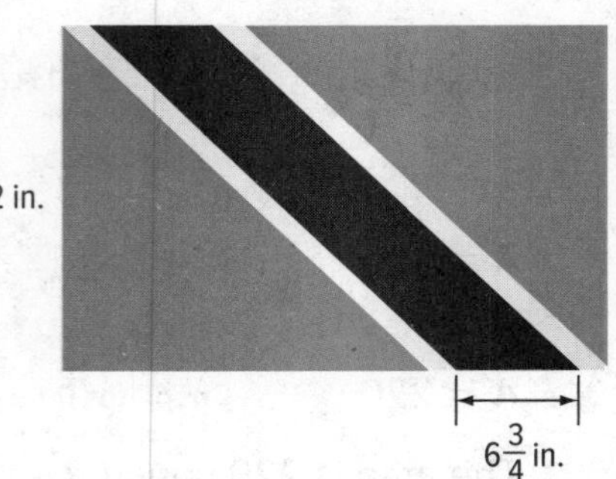

The black stripe is shaped like a parallelogram. So, use the formula $A = bh$.

$A = bh$	Area of parallelogram
$A = 6\frac{3}{4} \cdot 12$	Replace b with $6\frac{3}{4}$ and h with 12.
$A = 81$	$6\frac{3}{4} \cdot 12 = \frac{27}{4} \cdot 12$, or 81

The area of the flag that is black is 81 square inches.

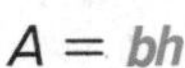

Guided Practice

Find the area of each parallelogram. (Examples 1 and 2)

1. ____________

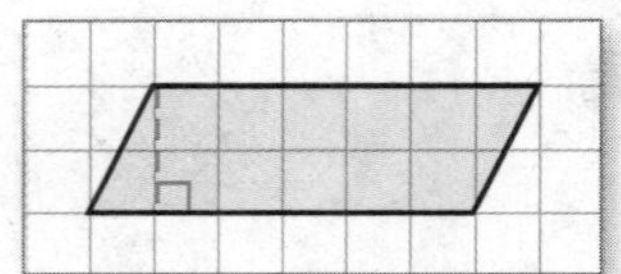

2. ____________

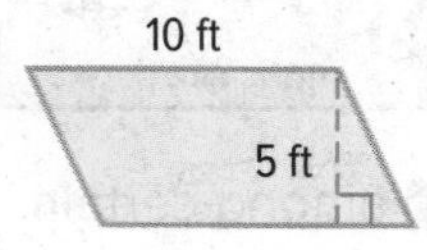

3. ____________

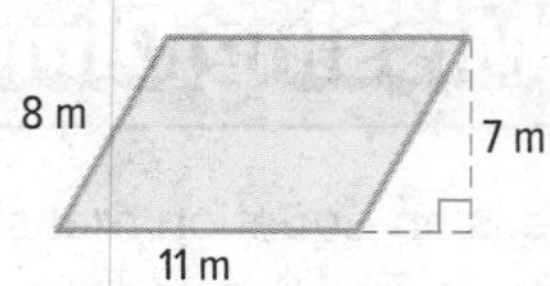

4. Find the height of a parallelogram if its base is 35 centimeters and its area is 700 square centimeters.

(Example 3) ____________

5. The size of the parallelogram piece in a set of tangrams is shown at the right. Find the area of the piece. (Example 4)

____

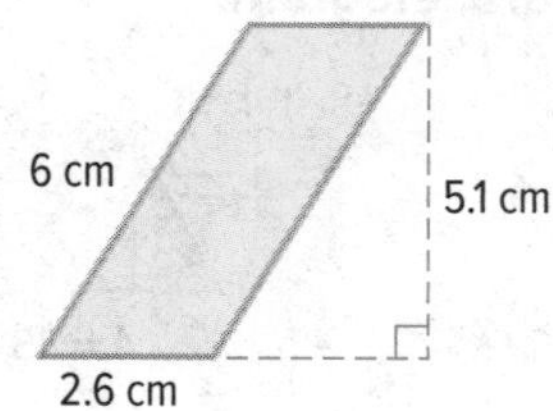

6. **Building on the Essential Question** How are parallelograms related to triangles and rectangles?

____

Rate Yourself!

How confident are you about the area of parallelograms? Shade the ring on the target.

For more help, go online to access a Personal Tutor.

FOLDABLES Time to update your Foldable!

Name ______________________ My Homework ______________________

Independent Practice

Go online for Step-by-Step Solutions

Find the area of each parallelogram. (Examples 1 and 2)

1. ______

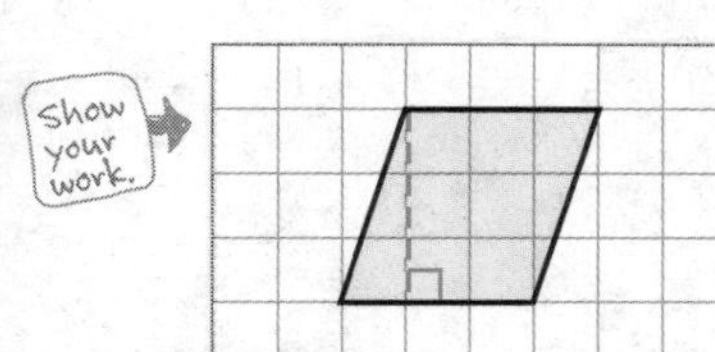

2. base, 6 millimeters; height, 4 millimeters

______

3. ______

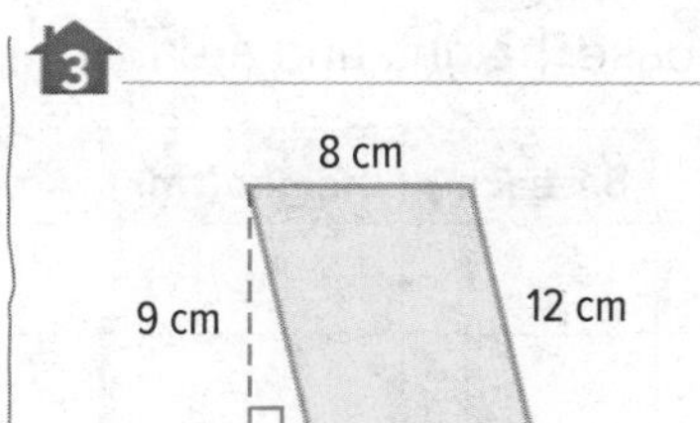

4. Find the base of a parallelogram with an area of 24 square feet and height 3 feet. (Example 3) ______

5. Find the area of the parking space shown to the right. (Example 4) ______

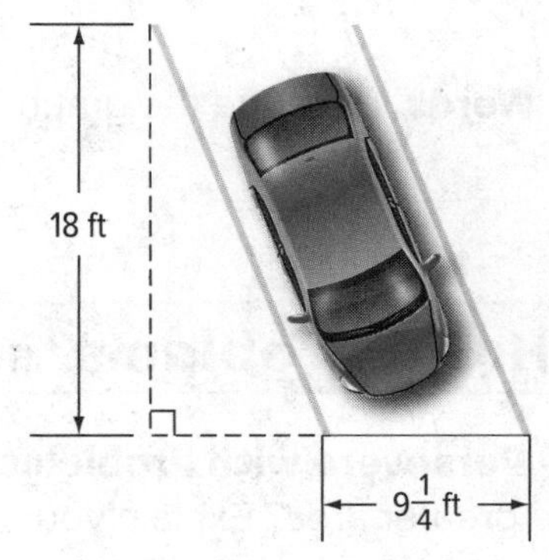

6. **STEM** An architect designed three different parallelogram-shaped brick patios. Write the missing dimensions in the table.

Patio	Base (ft)	Height (ft)	Area (ft^2)
1	$15\frac{3}{4}$		147
2		$11\frac{1}{4}$	$140\frac{5}{8}$
3	$10\frac{1}{4}$		$151\frac{3}{16}$

7. The base of a building is shaped like a parallelogram. The first floor has an area of 20,000 square feet. If the base of this parallelogram is 250 feet, can its height be 70 feet? Explain.

______

8. **MP Identify Structure** Draw and label a parallelogram with a base twice as long as the height and an area less than 60 square inches. Find the area. ______

9. MP **Multiple Representations** Draw five parallelograms that each have a height of 4 centimeters and different base measurements on centimeter grid paper.

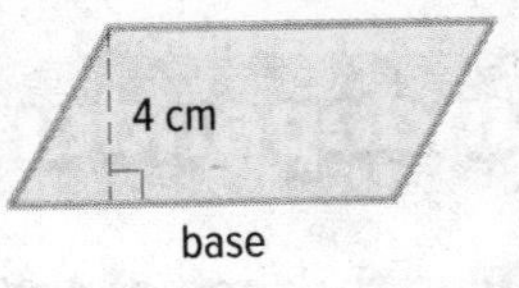

a. **Table** Make a table with a column for base, height, and area.

Base (cm)	Height (cm)	Area (cm^2)
	4	
	4	
	4	
	4	
	4	

b. **Graph** Graph the ordered pairs (base, area).

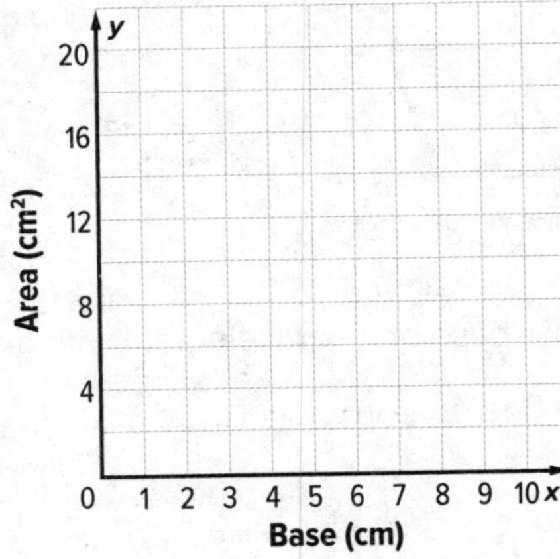

c. **Words** Describe the graph.

H.O.T. Problems Higher Order Thinking

10. MP **Persevere with Problems** If $x = 5$ and $y < x$, which figure has the greater area? Explain your reasoning.

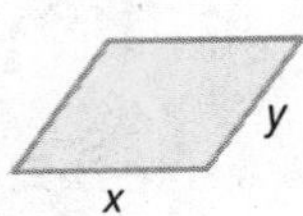

11. MP **Reason Inductively** Explain how the formula for the area of a parallelogram is related to the formula for the area of a rectangle.

12. MP **Reason Inductively** Give an example of a triangle and a parallelogram that have the same area. Describe the bases and heights of each figure. Then state the area.

Name ______________________ My Homework ______________

Extra Practice

Find the area of each parallelogram.

13. 20 units^2

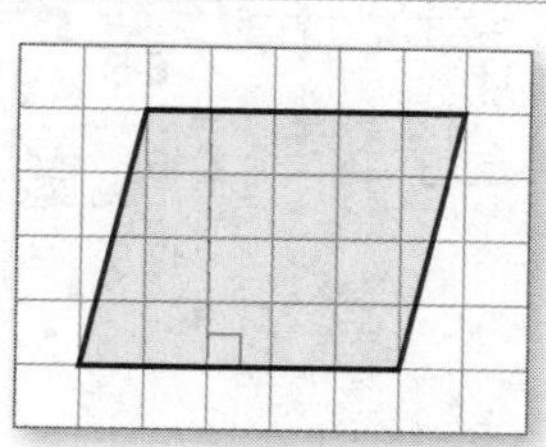

Homework Help

$A = bh$

$A = 5 \cdot 4$

$A = 20$

14. ______________

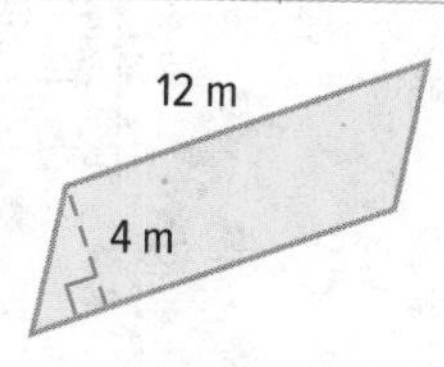

15. base, 12 inches; height, 15 inches

____

16. Find the height of a parallelogram with base 6.75 meters and an area of 218.7 square meters.

____

17. Find the area of a parallelogram with base 15 yards and height $21\frac{2}{3}$ yards.

____

18. What is the area of the region shown on the map? ______________

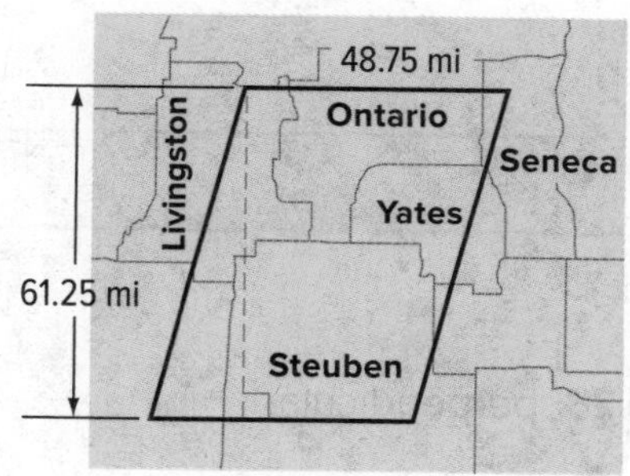

19. What is the height of the parallelogram-shaped pattern block shown below?

____

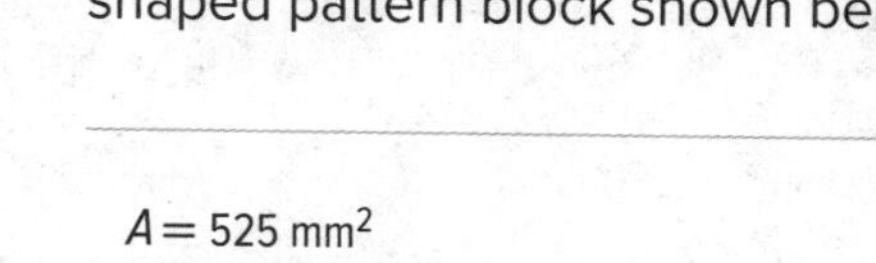

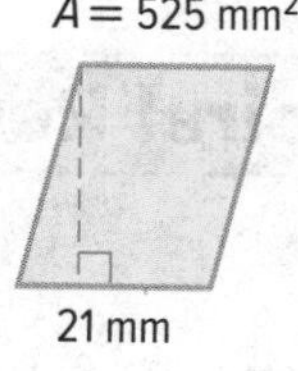

Draw and label each figure. Then find the area.

20. a parallelogram with an equal base and height and an area greater than 64 square meters

____

Show your work.

21. a parallelogram with a base four times the height and an area less than 200 square feet

____

MP Identify Structure Find the area of the shaded region in each figure.

22. ______________

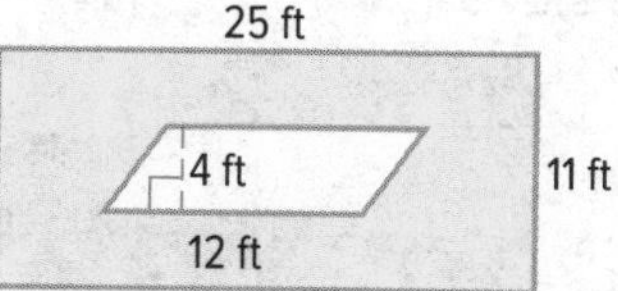

23. ______________

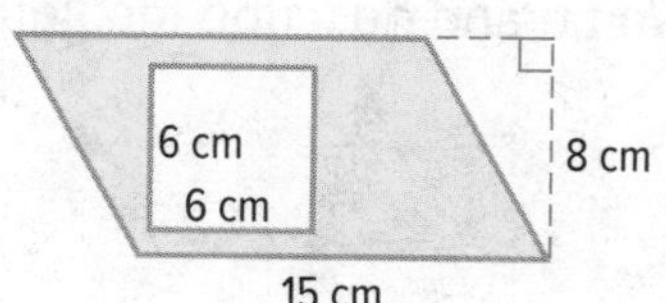

CCSS Power Up! Common Core Test Practice

24. The table shows the dimensions of 4 parallelograms. Sort the parallelograms from least to greatest area.

Parallelogram	Base (cm)	Height (cm)
A	4.75	22
B	13	6.5
C	7.25	16
D	5	13.5

	Parallelogram	Area (cm^2)
Least		
Greatest		

Which parallelogram has the greatest area? ☐

25. A family has a flower garden in the shape of a parallelogram in their backyard. They planted grass in the rest of the yard.

Fill in the boxes to complete each statement.

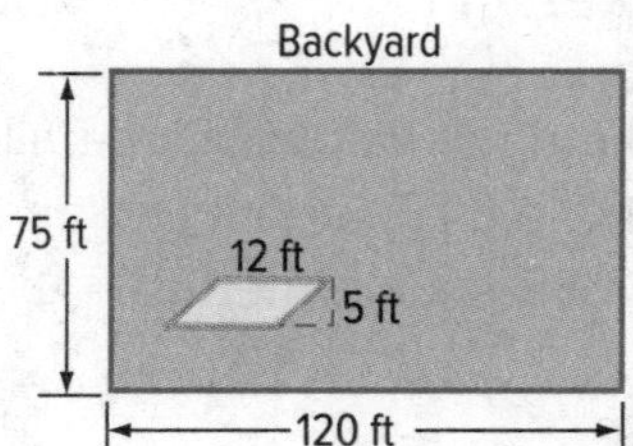

a. The total area of the backyard is ☐ square feet.

b. The area of the flower garden is ☐ square feet.

c. The area of the backyard that is planted with grass is 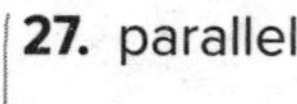 square feet.

CCSS Common Core Spiral Review

Draw each pair of lines. 4.G.1

26. intersecting

27. parallel

28. perpendicular

29. Rosa has 22 songs in her music library. Michael has half as many. How many songs does Michael have in his music library? **4.NBT.6**

30. Name and describe the figure based on the lengths of its sides. **5.G.4**

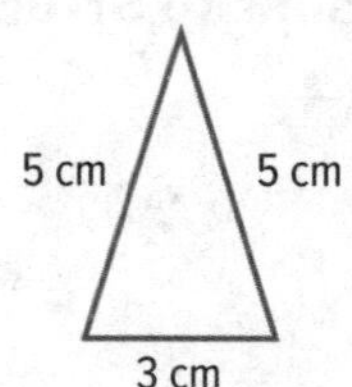

Inquiry Lab
Area of Triangles

HOW can you use the area of a parallelogram to find the area of a triangle?

Content Standards
6.G.1

Mathematical Practices
1, 3, 7, 8

Yurri is making a mosaic and is cutting rectangular tiles to make triangular tiles. He wants to find the area of the triangular tiles he is cutting.

What do you know? ______________________________

What do you need to know? ______________________________

Hands-On Activity 1

Yurri starts with a rectangular piece that is 4 inches by 6 inches, similar to the size of an index card.

Step 1 Find the area of an index card.

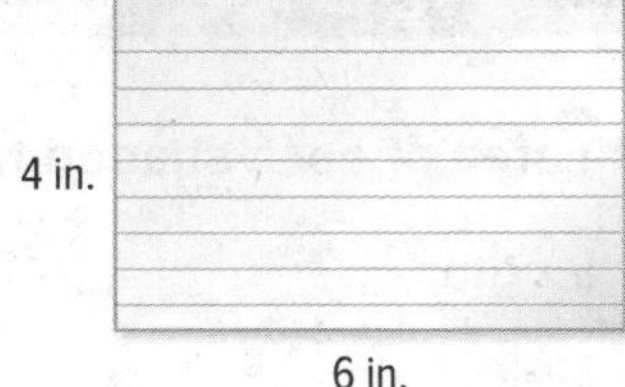

A = length × width

A = ☐ inches × ☐ inches

A = ☐ square inches

Step 2 Use an index card. Draw a diagonal line across your index card from one corner to another. Then cut across the line. Draw the resulting figures in the space below.

Step 3 Find the area of one of the remaining triangles. The triangle is exactly half the size of the related rectangle.

So, the area of the rectangle can be divided by 2 to find the area of one triangle.

The area is ☐ ÷ 2, or ☐ square inches.

Hands-On Activity 2

You can also find the area of a triangle from the area of a related parallelogram.

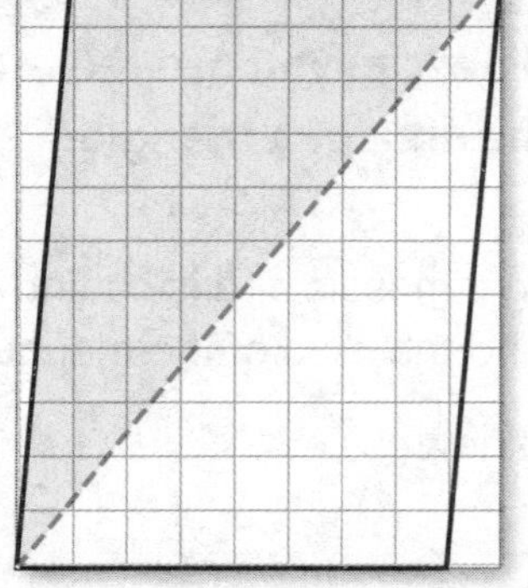

Step 1 Copy the parallelogram shown on grid paper.

Step 2 Draw a diagonal as shown by the dashed line. Cut out the parallelogram. The area of the parallelogram is ☐ square units.

Step 3 Cut along the diagonal to form two triangles. Then find the area of one triangle. The triangle is half the size of the parallelogram. So, the area of the parallelogram can be divided by 2 to find the area of one triangle.

The area of one triangle is ☐ ÷ 2 or ☐ square units.

Collaborate

Investigate

Work with a partner to find the area of each shaded triangle.

1.

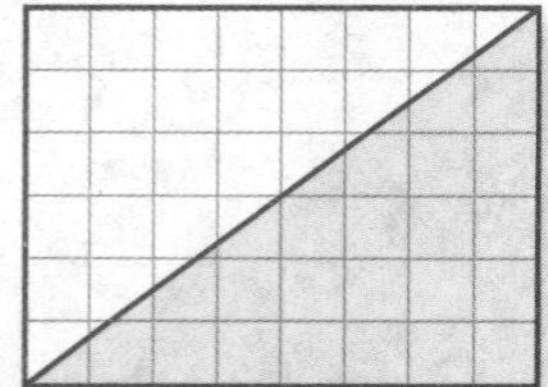

length: ________

width: ________

area: ________ × ________ = ________

area of triangle = ________ square units

2.

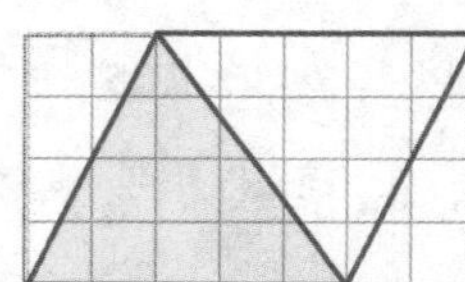

base: ________

height: ________

area: ________ × ________ = ________

area of triangle = ________ square units

3.

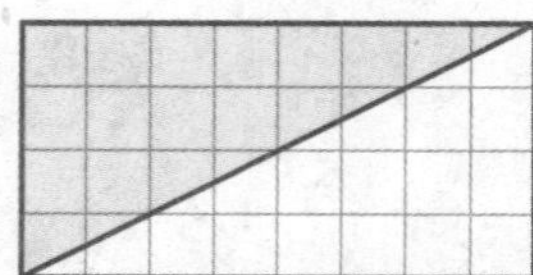

length: ________

width: ________

area: ________ × ________ = ________

area of triangle = ________ square units

4.

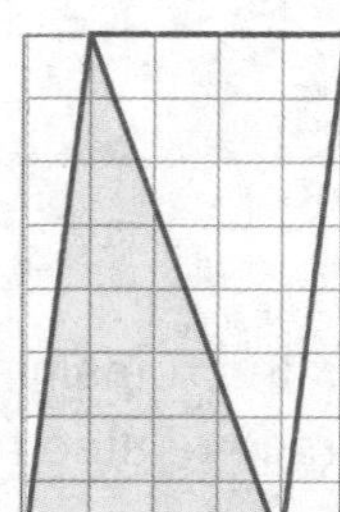

base: ________

height: ________

area: ________ × ________ = ________

area of triangle = ________ square units

Investigate

Work with a partner to find the area of each shaded triangle.

5. $A =$ ______ square feet

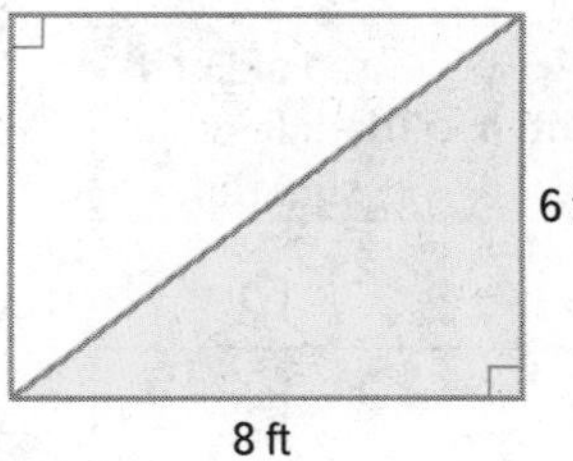

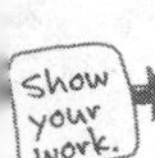

6. $A =$ ______ square meters

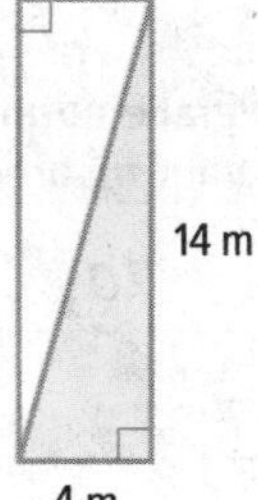

7. $A =$ ______ square centimeters

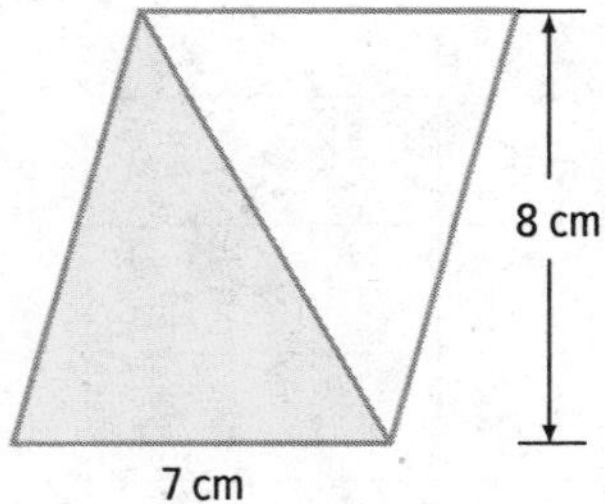

8. $A =$ ______ square feet

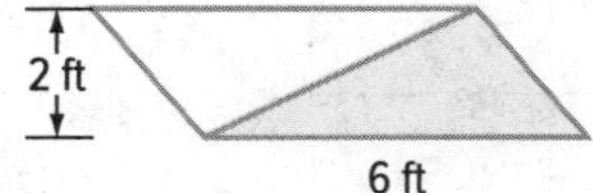

Identify Structure **Draw dotted lines to show the parallelogram or rectangle that can be used to find the area of each triangle. Then find the area of each triangle.**

9. $A =$ ______ square inches

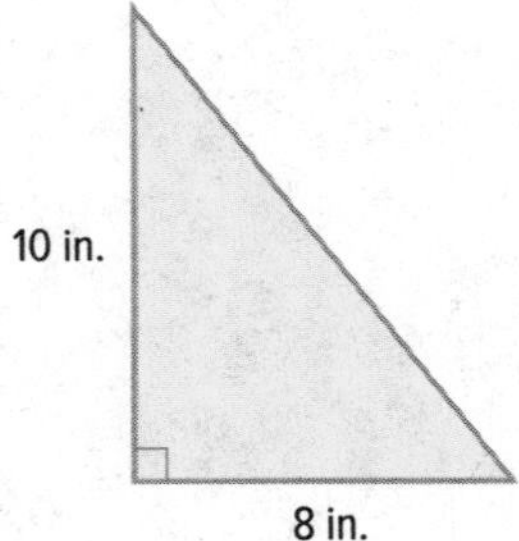

10. $A =$ ______ square yards

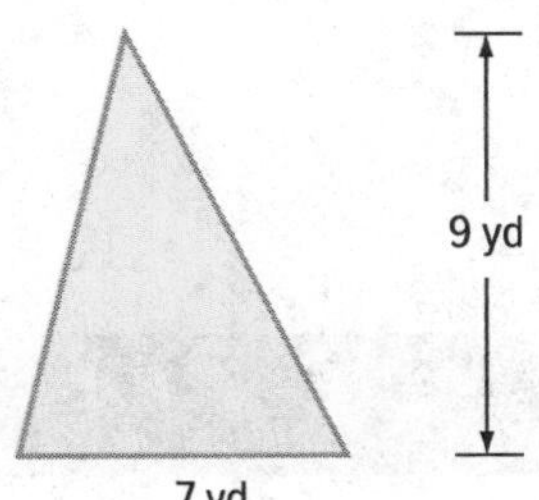

11. $A =$ ______ square centimeters

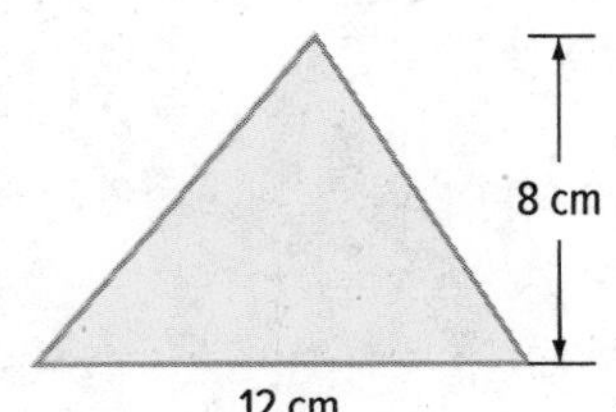

12. $A =$ ______ square feet

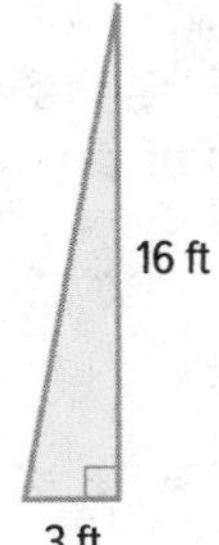

Analyze and Reflect

The table shows the dimensions of several parallelograms. Use the area of each parallelogram to find the missing information for each triangle. Work with a partner to complete the table. The first one is already done for you.

	Parallelogram	Base, b	Height, h	Area of Parallelogram (units squared)	Triangle created with diagonal	Base, b	Height, h	Area of Each Triangle (units squared)
	A	4	5	20	A	4	5	10
13.	B	4	6		B	4		12
14.	C	2	5		C	2	5	
15.	D	3	4		D	3	4	
16.	E	6	3		E		3	9
17.	F	8	5		F	8	5	
18.	G	5	7		G	5		17.5
19.	H	9	7		H	9	7	
20.	I	11	5		I	11	5	

21. **Reason Inductively** How is the area of the parallelogram related to the area of a triangle with the same base and height?

Create

22. **Identify Repeated Reasoning** Write a formula that relates the area A of a triangle to the lengths of its base b and height h.

23. Inquiry HOW can you use the area of a parallelogram to find the area of a triangle?

Lesson 2

Area of Triangles

Real-World Link

Biosphere The Biosphere 2 complex in Tucson, Arizona, researches Earth and its living systems. Sections of the building are interlocking triangles of the same size.

1. There are two triangles that are outlined in the photo.
   They have the ________ size and the ________ shape.

2. Draw the figure formed by the two triangles.

3. How many small triangles make up the outlined parallelogram? How many small triangles make up each outlined triangle? ________

4. Describe the relationship between the area of one outlined triangle and the area of the outlined parallelogram. ________

5. Draw another parallelogram like the one in the photo. Separate it into two triangles. Describe the relationship between the area of one triangle and the parallelogram. ________

Essential Question

HOW does measurement help you solve problems in everyday life?

Vocabulary

congruent

Common Core State Standards

Content Standards
6.G.1

Mathematical Practices
1, 3, 4, 8

Which MP **Mathematical Practices** did you use? Shade the circle(s) that applies.

① Persevere with Problems
② Reason Abstractly
③ Construct an Argument
④ Model with Mathematics
⑤ Use Math Tools
⑥ Attend to Precision
⑦ Make Use of Structure
⑧ Use Repeated Reasoning

Key Concept

Work Zone

Area of a Triangle

Words The area A of a triangle is one half the product of the base b and its height h.

Symbols $A = \frac{1}{2}bh$ or $A = \frac{bh}{2}$

Model

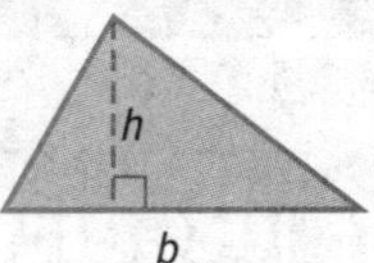

Congruent figures are figures that are the same shape and size.

A parallelogram can be formed by two congruent triangles. Since congruent triangles have the same area, the area of a triangle is one half the area of the parallelogram.

The base of a triangle can be any one of its sides. The height is the perpendicular distance from that base to the opposite vertex.

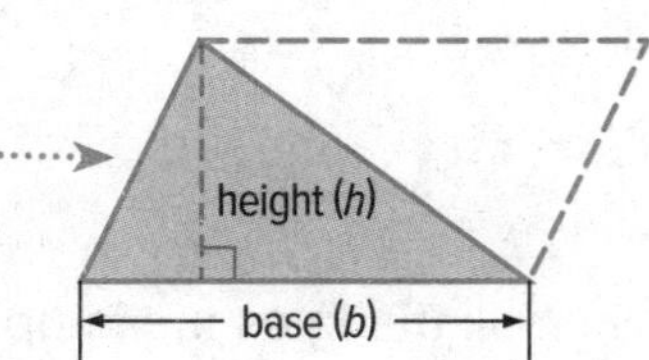

Examples

1. Find the area of the triangle.

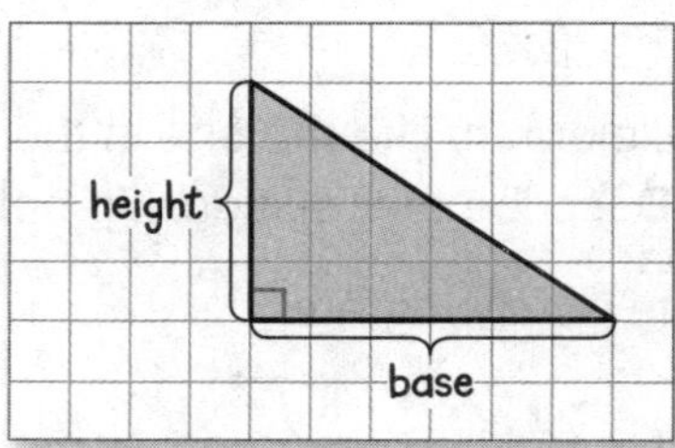

By counting, you find that the measure of the base is 6 units and the height is 4 units.

$A = \frac{1}{2}bh$	Area of a triangle
$A = \frac{1}{2}(6)(4)$	Replace b with 6 and h with 4.
$A = \frac{1}{2}(24)$	Multiply.
$A = 12$	Multiply.

The area of the triangle is 12 square units.

Mental Math

You can use mental math to multiply $\frac{1}{2}(6)(4)$. Think: Half of 6 is 3, and 3×4 is 12.

2. **Find the area of the triangle.**

12.1 m
6.4 m

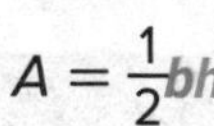

$A = \frac{1}{2}bh$	Area of a triangle
$A = \frac{1}{2}(12.1)(6.4)$	Replace b with 12.1 and h with 6.4.
$A = \frac{1}{2}(77.44)$	Multiply.
$A = 38.72$	Divide. $\frac{1}{2}(77.44) = 77.44 \div 2$, or 38.72

The area of the triangle is 38.72 square meters.

Got it? **Do these problems to find out.**

a.

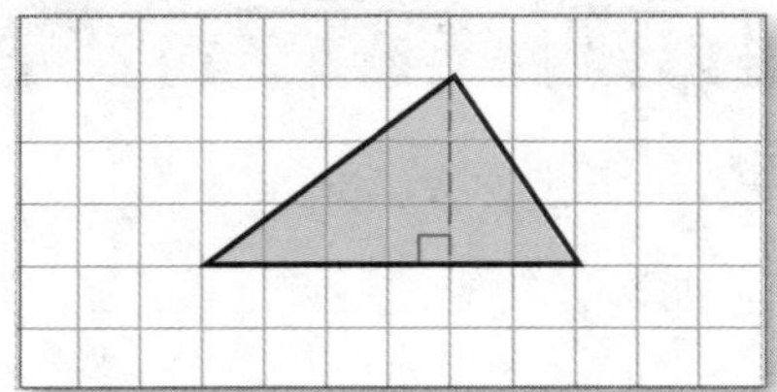

b.

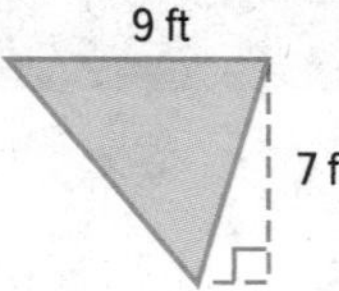

a. ____________

b. ____________

Find Missing Dimensions

Use the formula for the area of a triangle to find missing dimensions.

Example

3. **Find the missing dimension of the triangle.**

6 cm
b
$A = 24\text{ cm}^2$

$A = \frac{bh}{2}$	Area of a triangle
$24 = \frac{b \cdot 6}{2}$	Replace A with 24 and h with 6.
$24(2) = \frac{b \cdot 6}{2}(2)$	Multiply each side by 2.
$48 = b \cdot 6$	Simplify.
$\frac{48}{6} = \frac{b \cdot 6}{6}$	Divide each side by 6.
$8 = b$	Simplify.

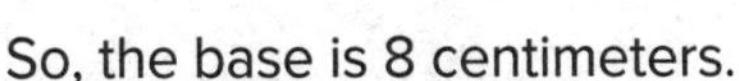

So, the base is 8 centimeters.

Check for Reasonableness

To check your work, replace b and h with the measurements and solve to find the area.

Got it? **Do these problems to find out.**

c.

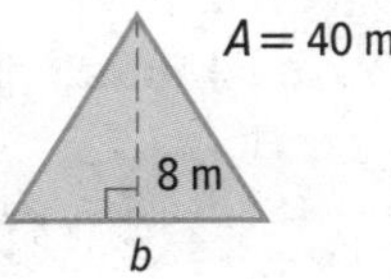

d.

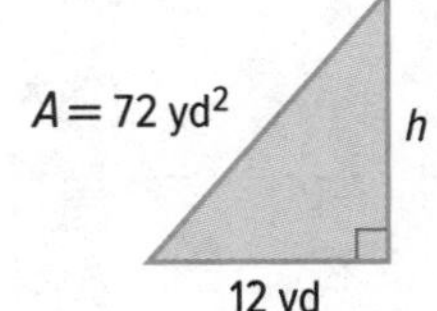

c. ____________

d. ____________

Example

4. The front of a camping tent has the dimensions shown. How much material was used to make the front of the tent?

$A = \frac{1}{2}bh$ Area of a triangle

$A = \frac{1}{2}(5)(3)$ Replace b with 5 and h with 3.

$A = \frac{1}{2}(15)$ or 7.5 Multiply.

The front of the tent has an area of 7.5 square feet.

Guided Practice

Find the area of each triangle. (Examples 1 and 2)

1. ____________

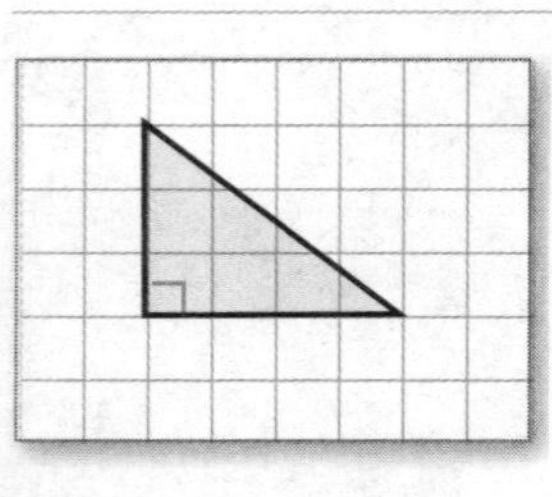

2. ____________

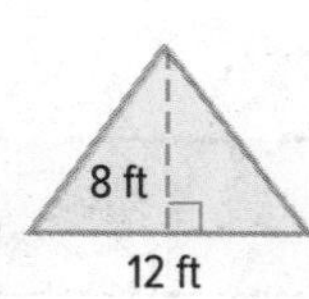

3. ____________

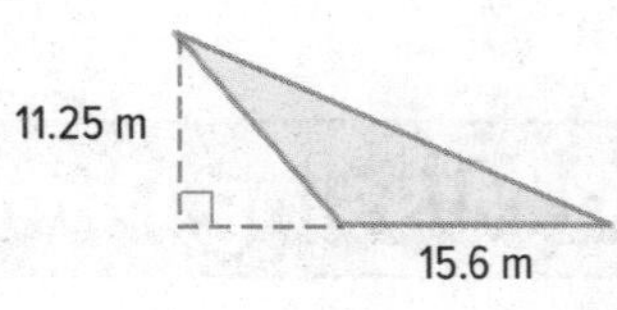

4. Tayshan designs uniquely-shaped ceramic floor tiles. What is the base of the tile shown? (Example 3)

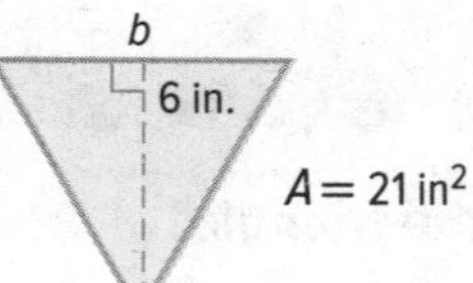

____

5. Consuela made a triangular paper box as shown. What is the area of the top of the box? (Example 4)

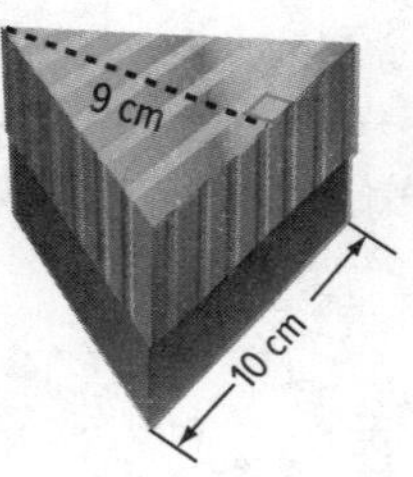

____

6. **Building on the Essential Question** How is the formula for the area of a triangle related to the formula for the area of a parallelogram?

____

____

____

____

Rate Yourself!

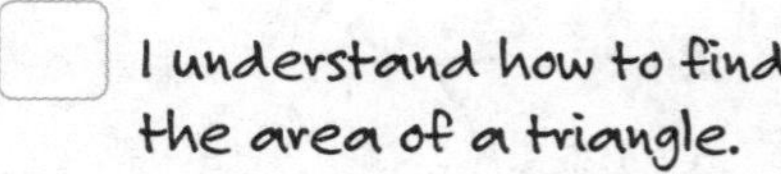

I still have some questions about the area of a triangle.

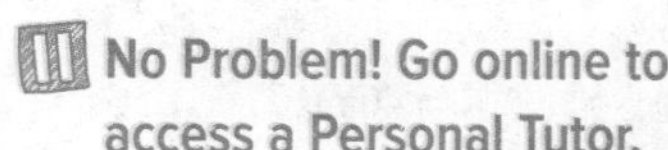

FOLDABLES Time to update your Foldable!

Name ______________________ My Homework ______________________

Independent Practice

Go online for Step-by-Step Solutions 

Find the area of each triangle. (Examples 1 and 2)

1. ______________

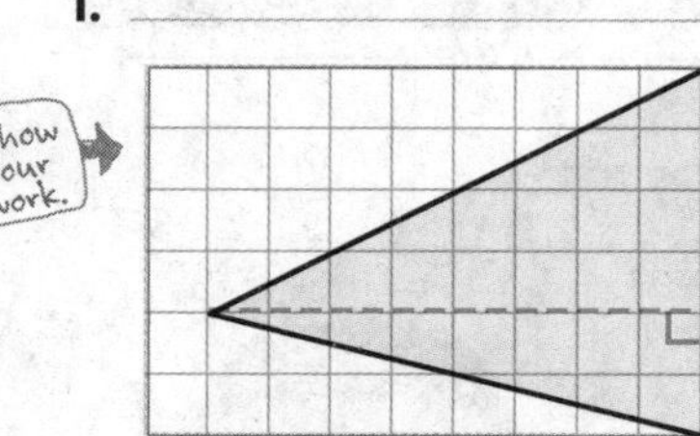

2. ______________

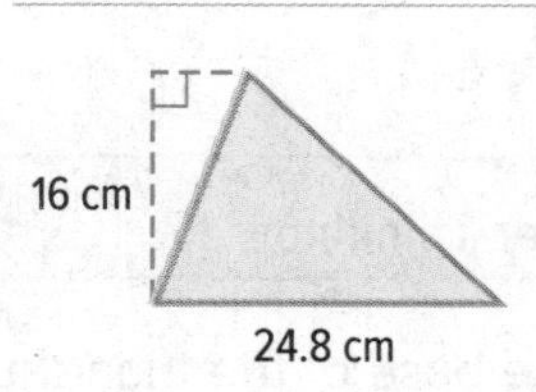

3 ______________

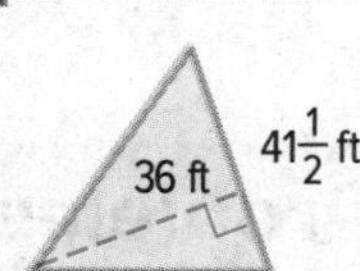

Find the missing dimension of each triangle described. (Example 3)

4. height: 14 in.
area: 245 in^2

____

5. base: 27 cm
area: 256.5 cm^2

____

6. Ansley is going to help his father shingle the roof of their house. What is the area of the triangular portion of one end of the roof? (Example 4)

____

7 **MP** **Multiple Representations** The table shows the areas of a triangle where the base of the triangle stays the same but the height changes.

Area of Triangles		
Base (units)	Height (units), x	Area (units^2), y
5	2	5
5	4	10
5	6	15
5	8	20
5	x	?

a. Algebra Write an algebraic expression that can be used to find the area of a triangle that has a base of 5 units and a height of x units. ______________

b. Graph Graph the ordered pairs (height, area).

Area (units^2): 3, 6, 9, 12, 15, 18, 21, 24, 27, 30 (y)
Height (units): O, 1, 2, 3, 4, 5, 6, 7, 8, 9, 10 (x)

c. Words Describe the graph.

____

8. What is the area of the triangle on the flag of the Philippines in inches? Explain your reasoning. ____________

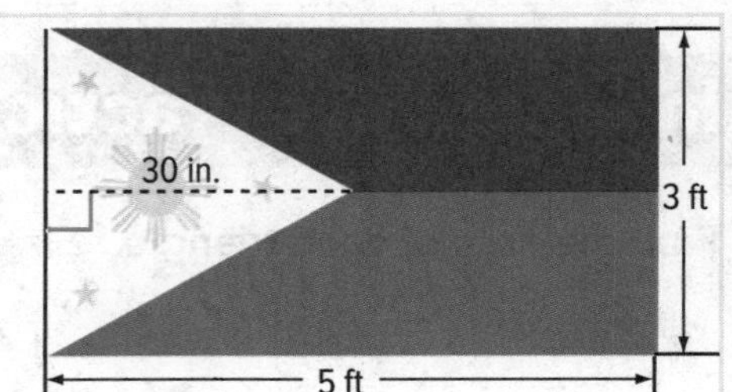

H.O.T. Problems Higher Order Thinking

9. **MP Find the Error** Dwayne is finding the base of the triangle shown. Its area is 100 square meters. Find his mistake and correct it.

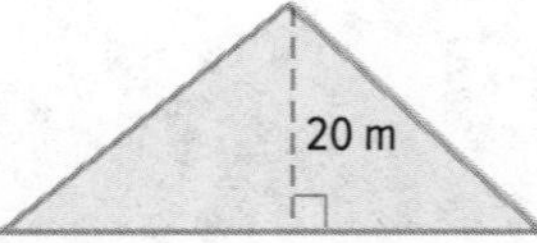

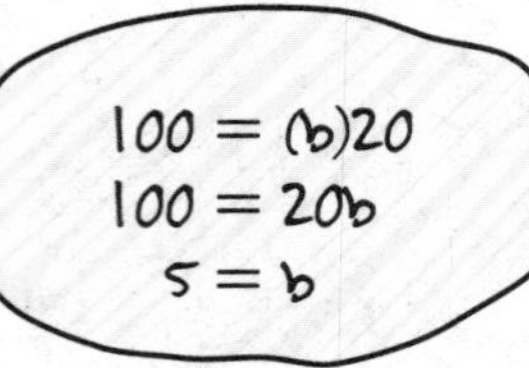

10. **MP Persevere with Problems** How can you use triangles to find the area of the hexagon shown? Draw a diagram to support your answer.

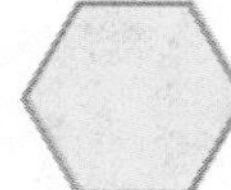

11. **MP Identify Repeated Reasoning** Draw a triangle and label its base and height. Draw another triangle that has the same base, but a height twice that of the first triangle. Find the area of each triangle. Then write a ratio that expresses the area of the first triangle to the area of the second triangle.

12. **MP Reason Inductively** The triangle shown has an area of $8\frac{15}{16}$ square feet. What is the height, in inches? ____________

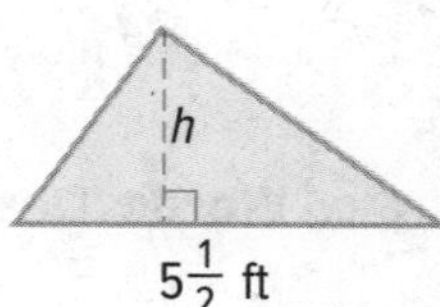

Name ______________________ My Homework ______________________

Extra Practice

Find the area of each triangle.

13. $7\frac{1}{2}$ units^2

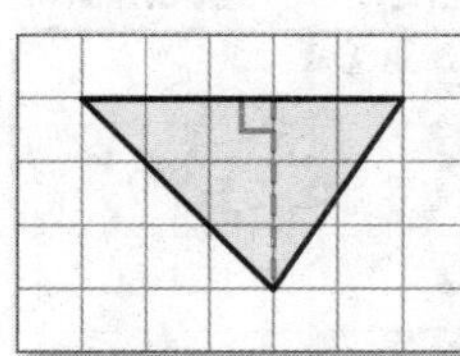

Homework Help

$A = \frac{bh}{2}$

$A = \frac{5 \cdot 3}{2}$

$A = \frac{15}{2}$ or $7\frac{1}{2}$

14. ______________________

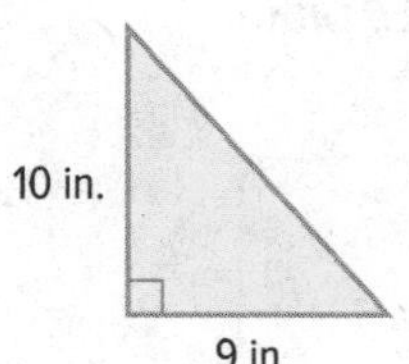

15. ______________________

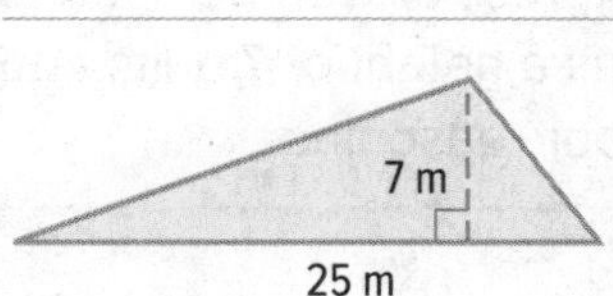

Find the missing dimension of each triangle described.

16. height: 7 in., area: 21 in^2

17. base: 11 m, area: 115.5 m^2

18. base: 14.2 yd, area: 63.9 yd^2

19. height: 11 cm, area: 260.15 cm^2

20. **STEM** An architect is designing a building on a triangular plot of land. If the base of the triangle is 100.8 feet and the height is 96.3 feet, find the available floor area of the building.

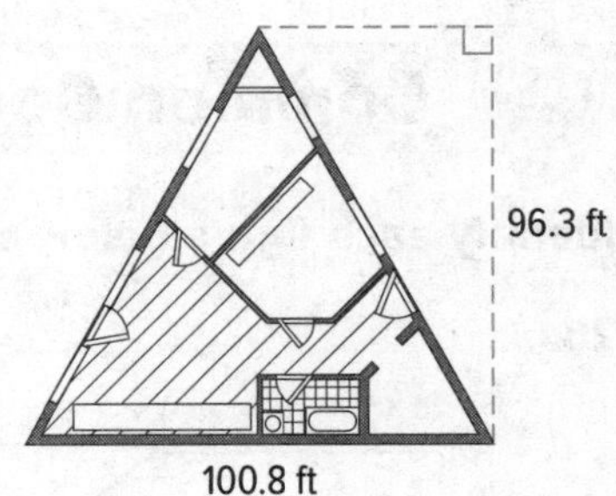

21. A flower bed in a parking lot is shaped like a triangle as shown.

a. Find the area of the flower bed in square feet.

b. If one bag of topsoil covers 10 square feet, how many bags are needed to cover this flower bed?

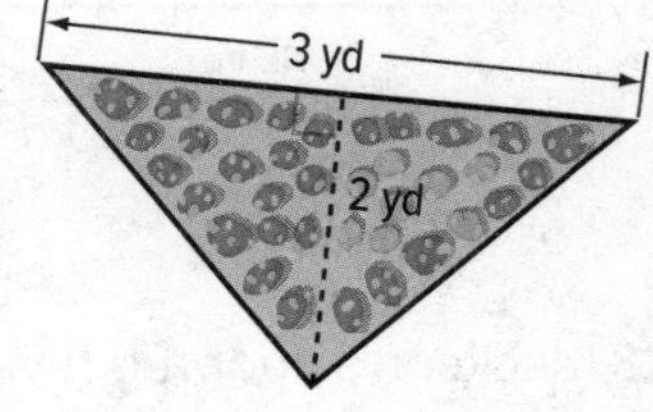

22. MP **Identify Repeated Reasoning** Refer to parallelogram *KLMN* at the right. If the area of parallelogram *KLMN* is 35 square inches, what is the area of triangle *KLN*?

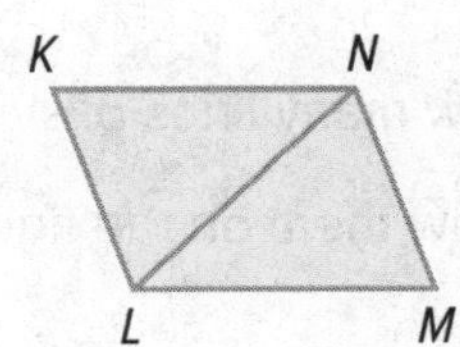

CCSS Power Up! Common Core Test Practice

23. The table shows the areas of a triangle where the height of the triangle stays the same but the base changes. What expression can be used to find the area of a triangle that has a height of 7 units and a base of x units? Explain your reasoning.

Areas of Triangles		
Height (units)	**Base (units)**	**Area (square units)**
7	2	7
7	3	$10\frac{1}{2}$
7	4	14
7	5	$17\frac{1}{2}$
7	x	?

24. Norma cut a triangle out of construction paper for an art project. The area of the triangle is 84.5 square centimeters.

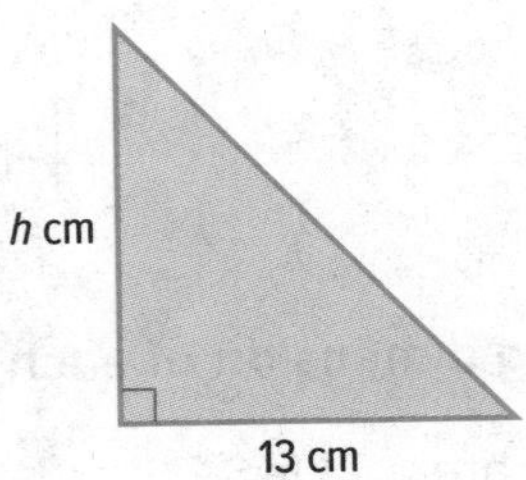

Select the correct values to complete the formula below to find the height of the triangle.

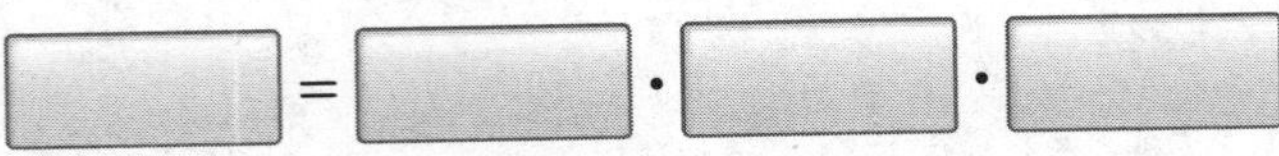

☐ = ☐ · ☐ · ☐

What is the height of the triangle? ☐

CCSS Common Core Spiral Review

Identify each figure below as a *rectangle, rhombus,* or *trapezoid.* 5.G.4

25. ______________

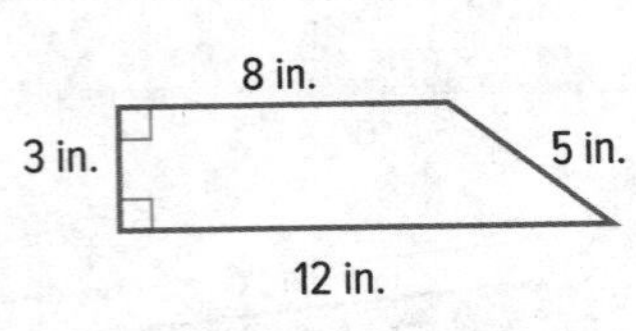

26. ______________

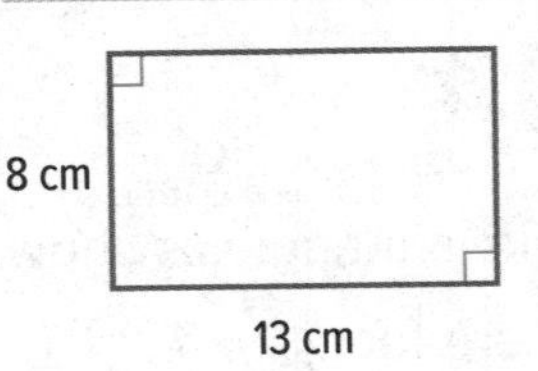

27. ______________

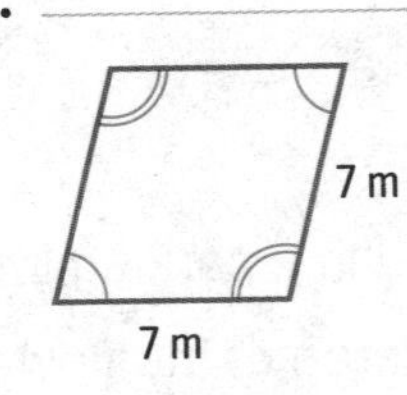

28. Jackson's floor rug has four 90° angles. All four sides are 18 inches long. The rug has two sets of parallel sides. What shape is Jackson's floor rug?

5.G.4 ______________

29. How many lines of symmetry can be drawn for the figure shown?

Draw them on the figure. 4.G.3 ______________

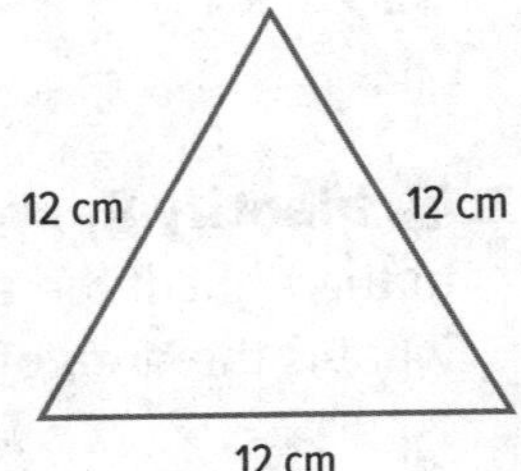

Inquiry Lab

Area of Trapezoids

HOW can you use the area of a parallelogram to find the area of a corresponding trapezoid?

Content Standards
6.G.1

Mathematical Practices
1, 3, 5, 7

Lizette is building a garden in the shape of a trapezoid. The garden is 6 feet wide in the back, 10 feet wide in the front, and 5 feet from back to front. She wants to find the area of the garden.

Hands-On Activity 1

Find the area of a trapezoid by drawing the related parallelogram.

Step 1 Trace the trapezoid below on grid paper. Label the height h and label the bases b_1 and b_2.

A trapezoid has two bases, b_1 and b_2. The height h of a trapezoid is the perpendicular distance between the bases.

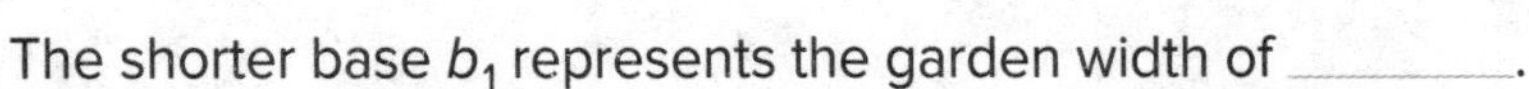

The shorter base b_1 represents the garden width of ______.

The longer base b_2 represents the garden width of ______.

The height h represents the garden dimension of ______.

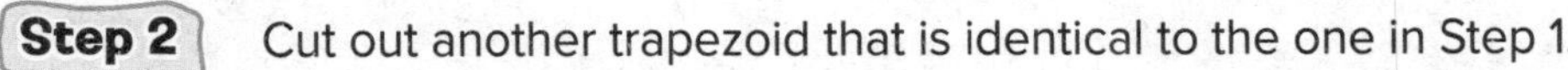

Step 2 Cut out another trapezoid that is identical to the one in Step 1.

Step 3 Tape the trapezoids together as shown.

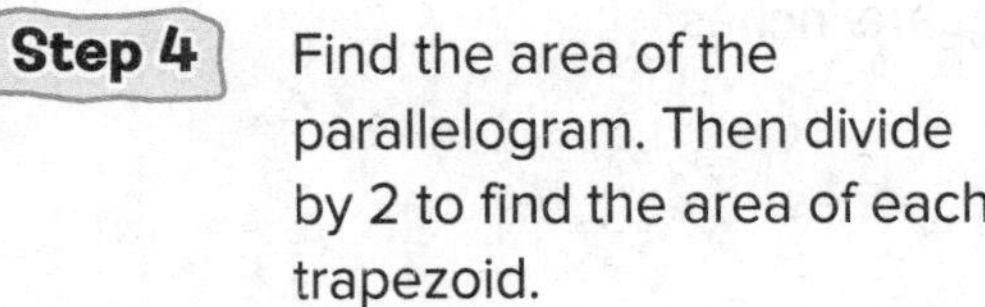

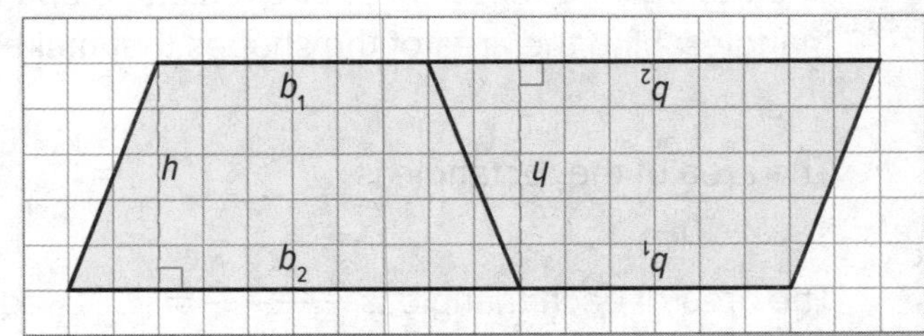

Step 4 Find the area of the parallelogram. Then divide by 2 to find the area of each trapezoid.

☐ × ☐ = ☐ ☐ ÷ 2 = ☐

So, the area of the garden is ☐ square feet.

Hands-On Activity 2

Discover the formula for the area of a trapezoid.

Step 1 What figure is formed by the two trapezoids in Activity 1?

____

Write an addition expression to represent the length of the base of the entire figure. ______________________________

Step 2 Write a formula for the area A of the parallelogram using b_1, b_2, and h. ______________________________

Step 3 How does the area of each trapezoid compare to the area of the parallelogram? ______________________________

____

Step 4 Write a formula for the area A of each trapezoid using b_1, b_2, and h.

____

Hands-On Activity 3

Another way to find the area of a trapezoid is to deconstruct it to determine which figures form the trapezoid. Find the area of the trapezoid shown below.

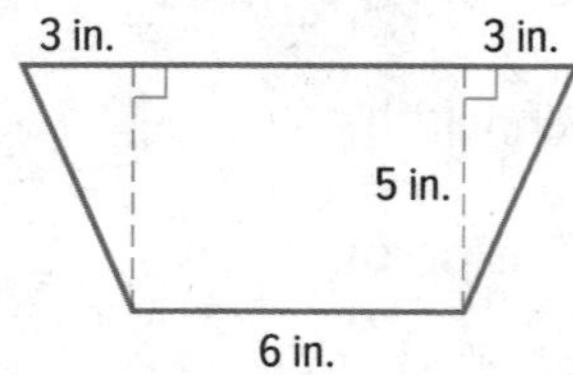

Step 1 The trapezoid is made up of one rectangle and two congruent triangles. Find the area of the shapes that make up the trapezoid.

The area of the rectangle is ☐ × ☐ = ☐ square inches.

The area of each triangle is $\frac{\square \times \square}{\square}$ = ☐ square inches.

Step 2 Add the areas.

☐ + ☐ + ☐ = ☐ square inches

MP Use Math Tools **Work with a partner. Find the area of each trapezoid by drawing the related parallelogram.**

1. $A =$ ______ square units

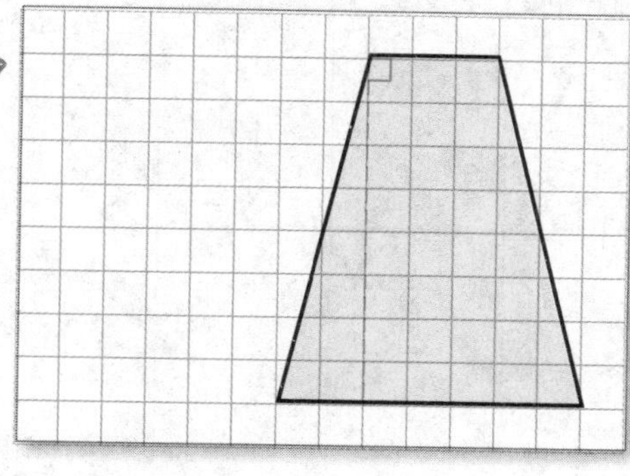

2. $A =$ ______ square units

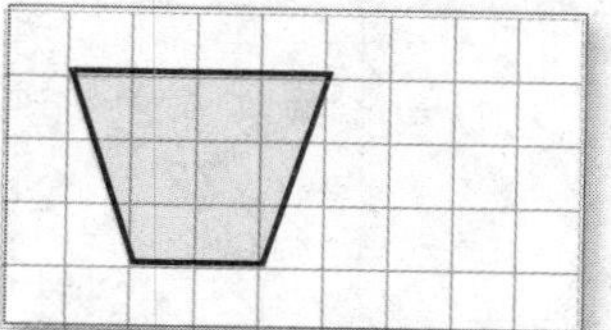

Work with a partner. Find the area of each trapezoid by using the formula.

3. $A = \frac{(\square + \square)\square}{\square}$

$A =$ ______ square units

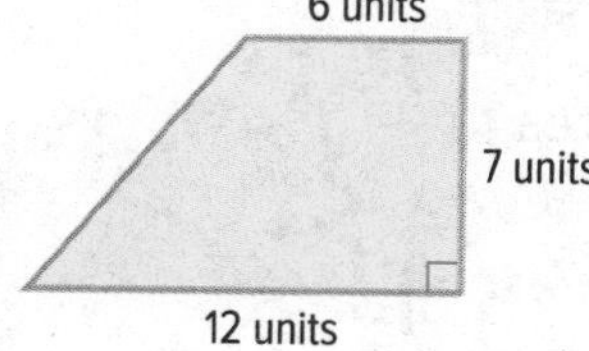

4. $A = \frac{(\square + \square)\square}{\square}$

$A =$ ______ square units

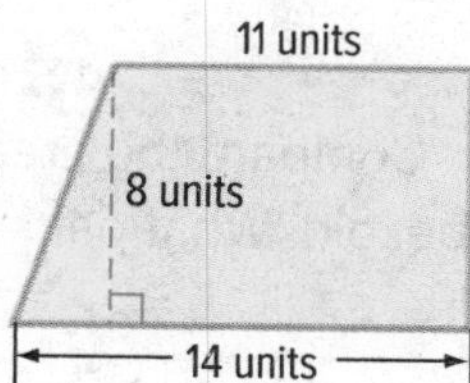

Work with a partner. Decompose each trapezoid to find the area.

5. $A =$ ______ square units

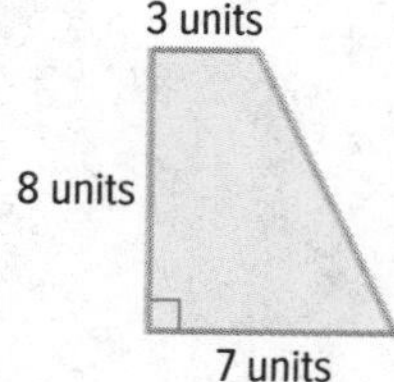

6. $A =$ ______ square units

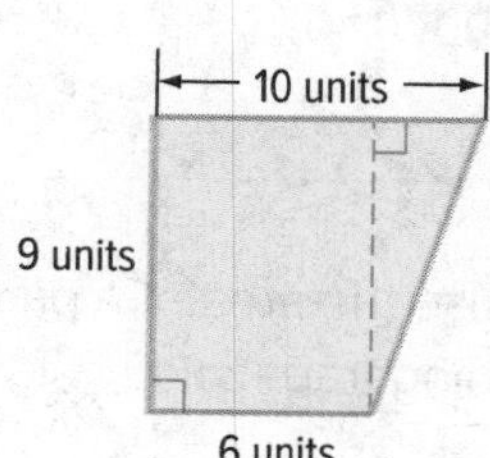

Analyze and Reflect

The table shows the dimensions of several parallelograms and corresponding trapezoids. Work with a partner to complete the table. The first one is done for you.

	Dimensions of Parallelogram	Area of Parallelogram	Length of Trapezoid b_1	Length of Trapezoid b_2	Trapezoid Height	Area of Trapezoid
	height 4, base 7	28	2	5	4	14
7.	height 6, base 11		5	6	6	
8.	height 5, base 12		8	4	5	
9.	$b = 11$ $h = 3$		7	4	3	

10. **Reason Inductively** Compare the dimensions of the parallelogram to the dimensions of the corresponding trapezoid. What pattern do you see in the table? ______

11. **Reason Inductively** Compare the area of the parallelogram to the area of the corresponding trapezoid. What pattern do you see in the table?

Create

12. **Identify Structure** Write the formula for the area A of a trapezoid with bases b_1 and b_2 and height h.

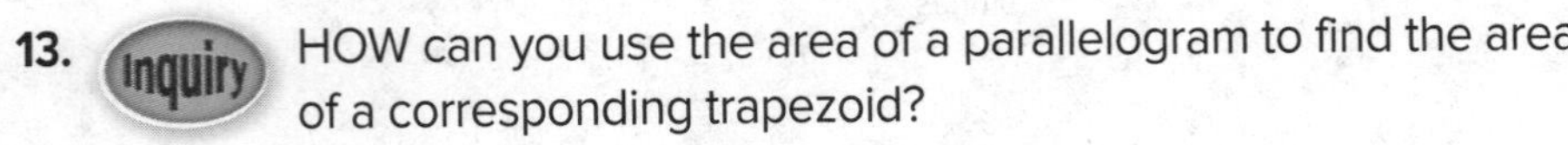

13. Inquiry HOW can you use the area of a parallelogram to find the area of a corresponding trapezoid?

Lesson 3
Area of Trapezoids

Real-World Link

Window Seat Kiana has a bay window in her room. The window seat is in the shape of a trapezoid. She needs to measure the seat in order to sew a cushion for the seat. The blue trapezoid in the diagram below represents the dimensions of the window seat.

Use the diagram below to describe the relationship between trapezoids and rectangles.

1. Find the dimensions of each figure.

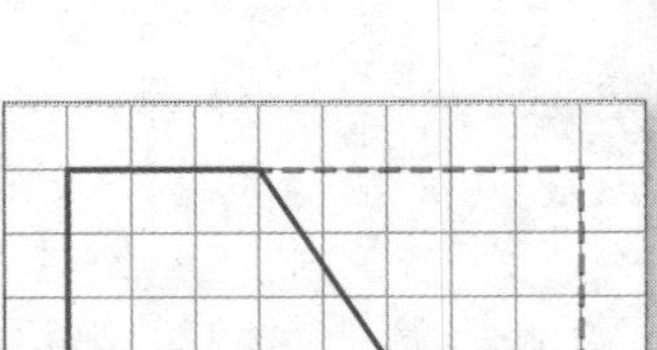

Trapezoid	**Rectangle**
base 1: ☐ units	length: ☐ units
base 2: ☐ units	height: ☐ units
height: ☐ units	

2. What is the relationship between the measures of the rectangle and the measures of the trapezoid?

____

____

____

3. MP **Make a Conjecture** How is the area of a trapezoid related to the area of a rectangle? ___________________________

____

____

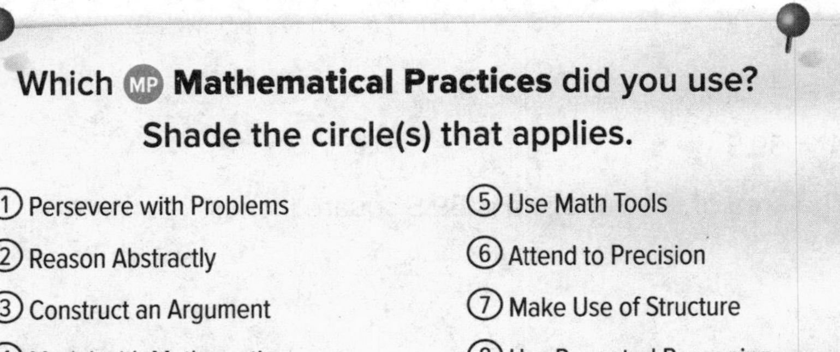

Which MP **Mathematical Practices** did you use?
Shade the circle(s) that applies.

① Persevere with Problems
② Reason Abstractly
③ Construct an Argument
④ Model with Mathematics
⑤ Use Math Tools
⑥ Attend to Precision
⑦ Make Use of Structure
⑧ Use Repeated Reasoning

Essential Question

HOW does measurement help you solve problems in everyday life?

Common Core State Standards

Content Standards
6.G.1

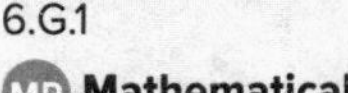

1, 2, 3, 4, 7, 8

Key Concept

Work Zone

Area of a Trapezoid

Words The area A of a trapezoid is one half the product of the height h and the sum of the bases b_1 and b_2.

Model

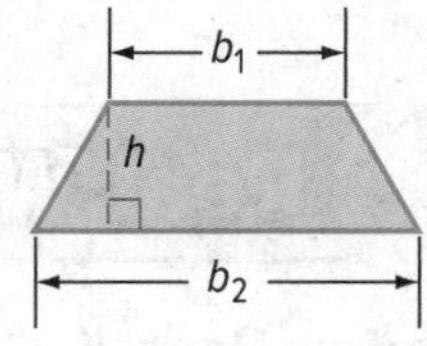

Symbols $A = \frac{1}{2}h(b_1 + b_2)$

A trapezoid has two bases, b_1 and b_2. The height of a trapezoid is the distance between the bases.

The height is the perpendicular distance between the bases.

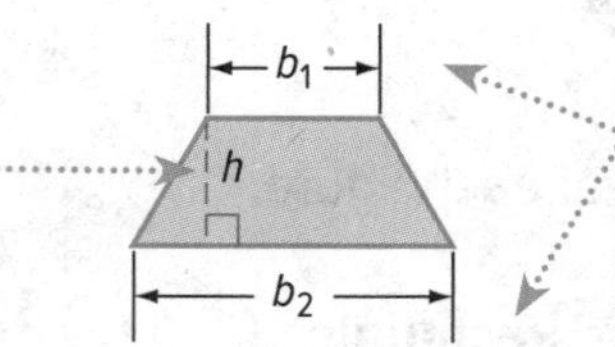

The two bases are parallel. They will always be the same distance apart.

When finding the area of a trapezoid, it is important to follow the order of operations. In the formula, the bases are to be added before multiplying by $\frac{1}{2}$ of the height h.

Examples

1. **Find the area of the trapezoid.**

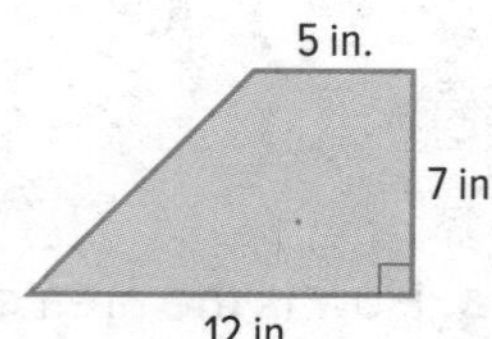

The bases are 5 inches and 12 inches.
The height is 7 inches.

$A = \frac{1}{2}h(b_1 + b_2)$ Area of a trapezoid

$A = \frac{1}{2}(7)(5 + 12)$ Replace h with 7, b_1 with 5, and b_2 with 12.

$A = \frac{1}{2}(7)(17)$ Add 5 and 12.

$A = 59.5$ Multiply.

The area of the trapezoid is 59.5 square inches.

2. Find the area of the trapezoid.

7 m, 9.8 m, 12 m

$A = \frac{1}{2}h(b_1 + b_2)$ Area of a trapezoid

$A = \frac{1}{2}(9.8)(7 + 12)$ Replace h with 9.8, b_1 with 7, and b_2 with 12.

$A = \frac{1}{2}(9.8)(19)$ Add 7 and 12.

$A = 93.1$ Multiply.

So, the area of the trapezoid is 93.1 square meters.

Got it? Do these problems to find out.

a.

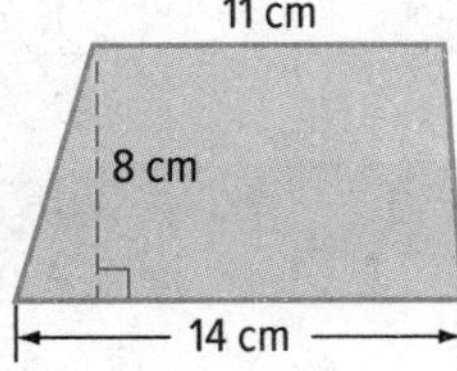

b.

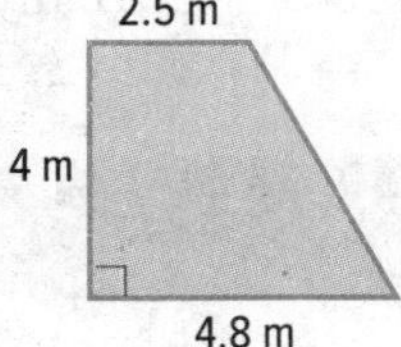

c.

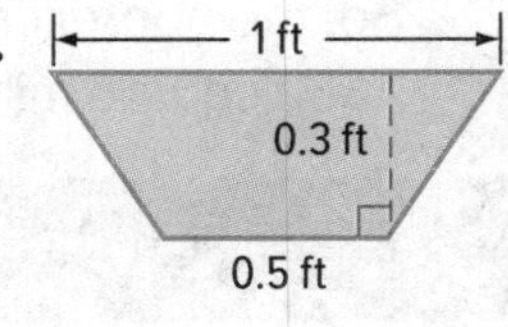

Show your work.

a. ________

b. ________

c. ________

Find the Missing Height

Use the related formula, $h = \frac{2A}{b_1 + b_2}$, to find the height of a trapezoid.

Example

3. The trapezoid has an area of 108 square feet. Find the height.

$h = \frac{2A}{b_1 + b_2}$ Height of a trapezoid

$h = \frac{2(108)}{12 + 15}$ Replace A with 108, b_1 with 12, and b_2 with 15.

$h = \frac{216}{27}$ Multiply 2 and 108. Add 12 and 15.

$h = 8$ Divide.

So, the height of the trapezoid is 8 feet.

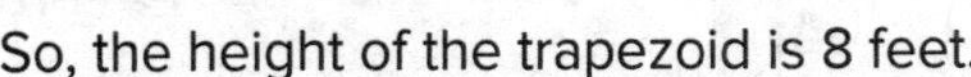

Be Precise

Check your answer by using the formula for the area of a trapezoid.

Got it? Do these problems to find out.

d. $A = 24\text{ cm}^2$
$b_1 = 4\text{ cm}$
$b_2 = 12\text{ cm}$
$h = ?$

e. $A = 21\text{ yd}^2$
$b_1 = 2\text{ yd}$
$b_2 = 5\text{ yd}$
$h = ?$

d. ________

e. ________

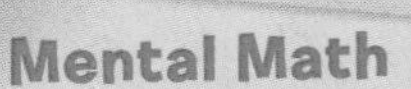

Mental Math
To multiply $\frac{1}{2}(51)(64)$, it is easier to use the Commutative Property to reorder the factors as $\frac{1}{2}(64)(51)$ and take half of 64 instead of half of 51.

Example

4. The shape of Osceola County, Florida, resembles a trapezoid. Find the approximate area of this county.

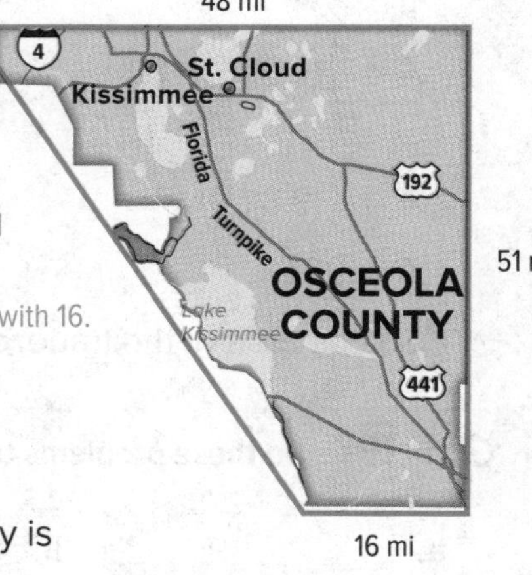

$A = \frac{1}{2}h(b_1 + b_2)$ Area of a trapezoid

$A = \frac{1}{2}(51)(48 + 16)$ Replace h with 51, b_1 with 48, and b_2 with 16.

$A = \frac{1}{2}(51)(64)$ Add 48 and 16.

$A = 1{,}632$ Multiply.

So, the approximate area of the county is 1,632 square miles.

Guided Practice

Find the area of each trapezoid. Round to the nearest tenth if necessary. (Examples 1 and 2)

1. ________

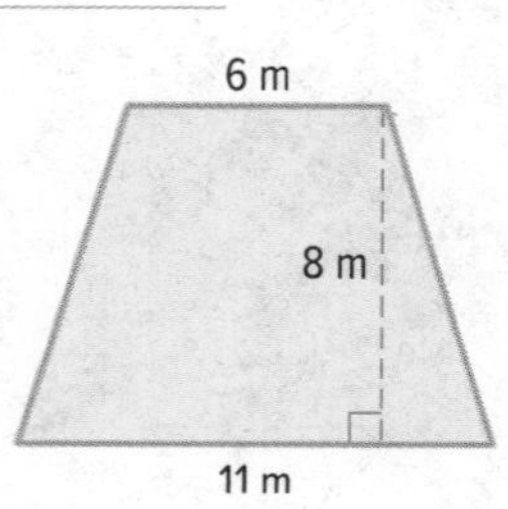

Show your work.

2. ________

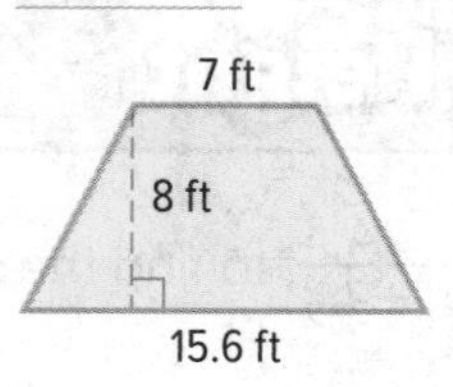

3. A trapezoid has an area of 15 square feet. If the bases are 4 feet and 6 feet, what is the height of the trapezoid? (Example 3) ________

4. In the National Hockey League, goaltenders can play the puck behind the goal line only in a trapezoid-shaped area, as shown at the right. Find the area of the trapezoid. (Example 4) ________

18 ft
11 ft
28 ft

5. **Building on the Essential Question** How is the formula for the area of a trapezoid related to the formula for the area of a parallelogram? ________

Rate Yourself!

Are you ready to move on? Shade the section that applies.

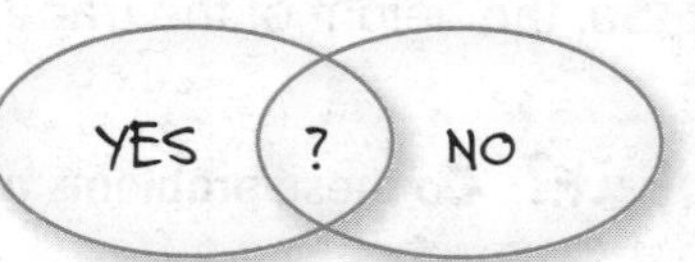

For more help, go online to access a Personal Tutor.

FOLDABLES Time to update your Foldable!

Independent Practice

Go online for Step-by-Step Solutions

Find the area of each trapezoid. Round to the nearest tenth if necessary. (Examples 1 and 2)

1. ______

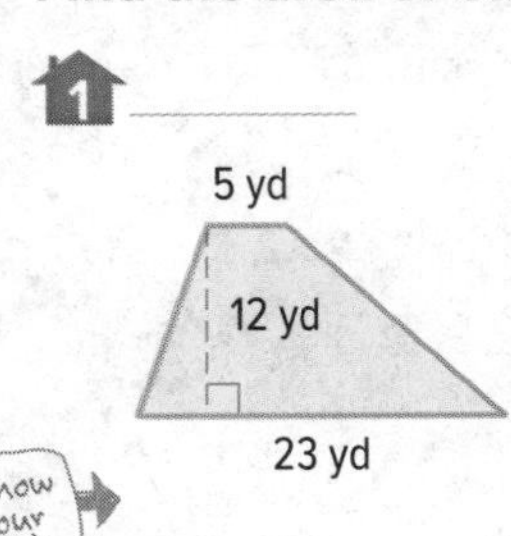

Show your work.

2. ______

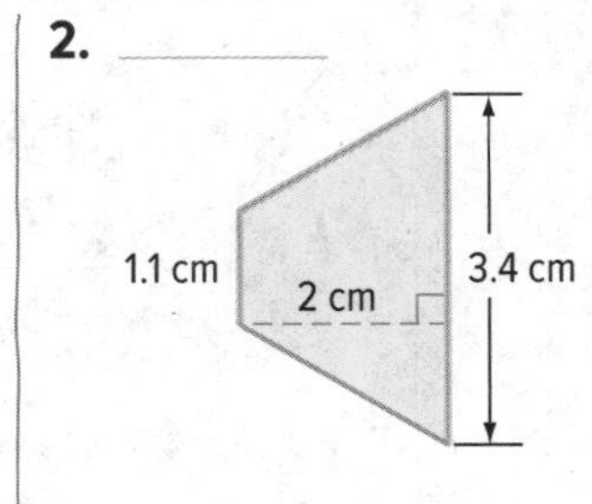

3. ______

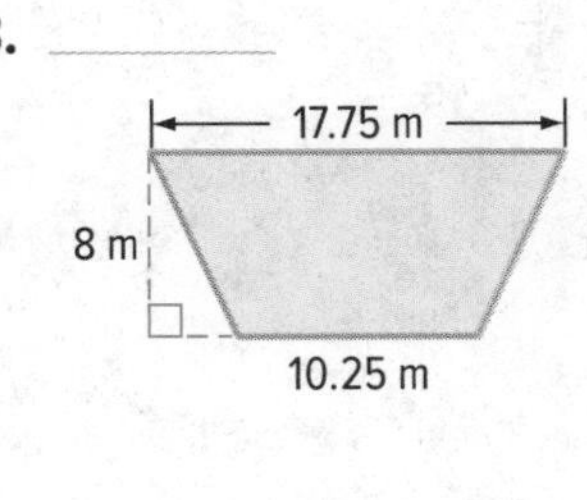

4. A trapezoid has an area of 150 square meters. If the bases are 14 meters and 16 meters, what is the height of the trapezoid? (Example 3)

5. A trapezoid has an area of 400 square millimeters. The bases are 14 millimeters and 36 millimeters. What is the height of the trapezoid? (Example 3)

6. Find the area of the patio shown. (Example 4)

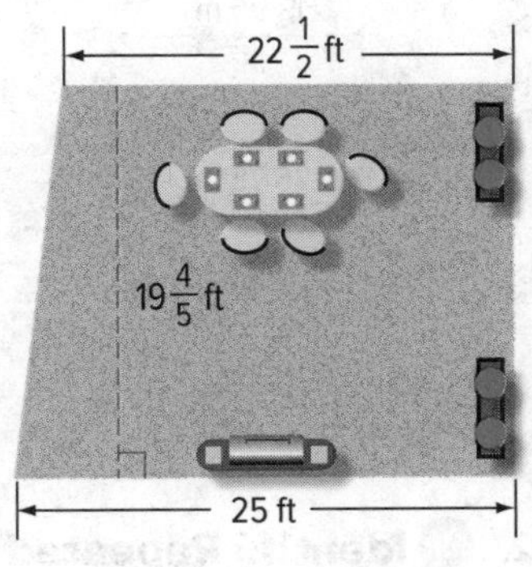

7. Use the diagram that shows the lawn that surrounds an office building.

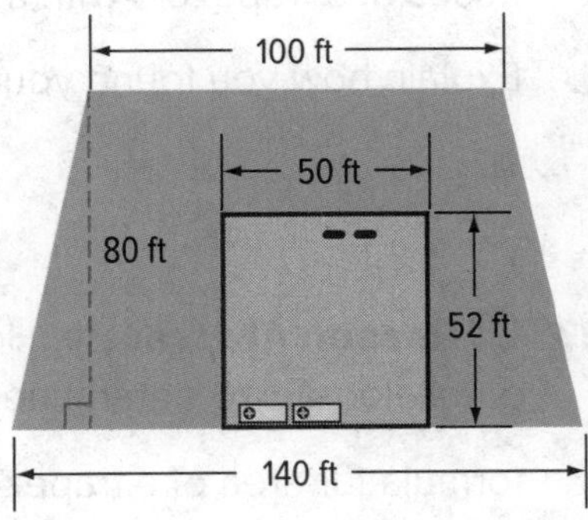

a. What is the area of the lawn? ______

b. If one bag of grass seed covers 2,000 square feet, how many bags are needed to seed the lawn?

8. MP **Reason Abstractly** Tiles are being placed in front of a fireplace to create a trapezoidal hearth. The hearth will have a height of 24 inches and bases that are 48 inches and 60 inches. If the tiles cover 16 square inches, how many tiles will be needed?

Draw and label each figure. Then find the area.

9. a trapezoid with no right angles and an area less than 12 square centimeters

10. a trapezoid with a right angle and an area greater than 40 square inches

Show your work.

H.O.T. Problems Higher Order Thinking

11. **MP Persevere with Problems** Apply what you know about rounding to explain how to estimate the height h of the trapezoid shown if the area is 235.5 m^2.

19.95 m

h

26.75 m

12. **MP Identify Repeated Reasoning** Find two possible lengths of the bases of a trapezoid with a height of 1 foot and an area of 9 square feet.

Explain how you found your answer.

13. **MP Reason Abstractly** How can you use the formula for area of a parallelogram to determine the area of a trapezoid if you forgot the formula for area of a trapezoid?

14. **MP Reason Inductively** The area of a trapezoid is 36 square inches. The height is 4 inches and one base is twice the length of the other base. What are the lengths of the bases?

Extra Practice

Find the area of each figure. Round to the nearest tenth if necessary.

15. 121 cm^2

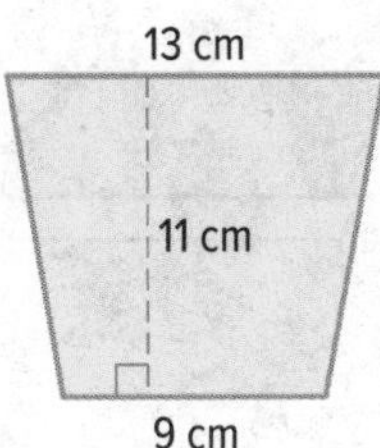

Homework Help

$A = \frac{1}{2}h(b_1 + b_2)$

$A = \frac{1}{2}(11)(13 + 9)$

$A = \frac{1}{2}(11)(22)$

$A = 121$

16. ________

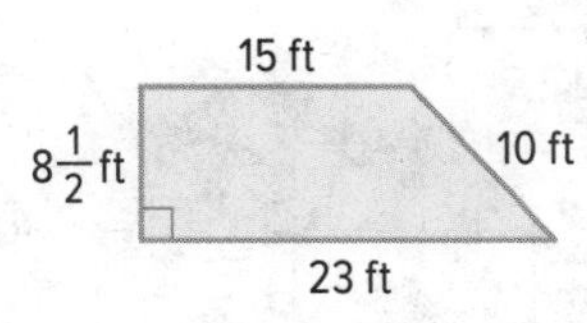

17. ________

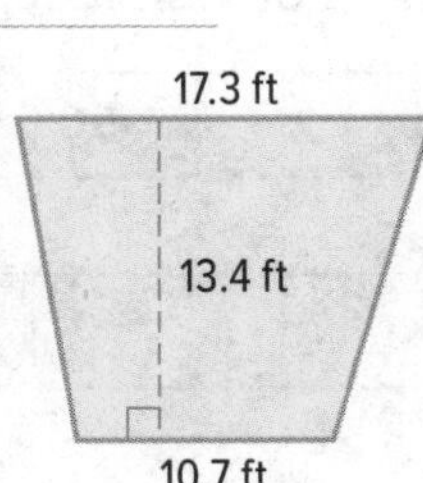

18. A trapezoid has an area of 50 square inches. The bases are 3 inches and 7 inches. What is the height of the trapezoid?

____

19. A trapezoid has an area of 18 square miles. The bases are 5 miles and 7 miles. What is the height of the trapezoid?

____

20. A county is shaped like a trapezoid. Its northern border is about 9.6 miles across, and the southern border is approximately 25 miles across. The distance from the southern border to the northern border is about 90 miles. Find the approximate area of the county.

____

21. A play tent is shown. How much fabric was used to make the front and back of the play tent?

____

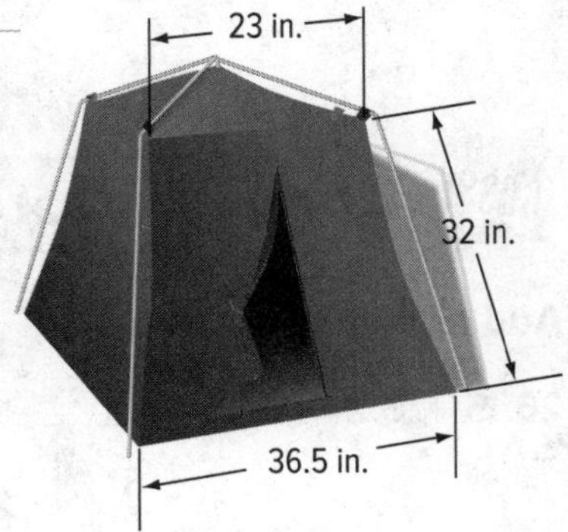

MP **Identify Structure** **Each figure below is made up of congruent trapezoids. Find the area of each figure.**

22. ________

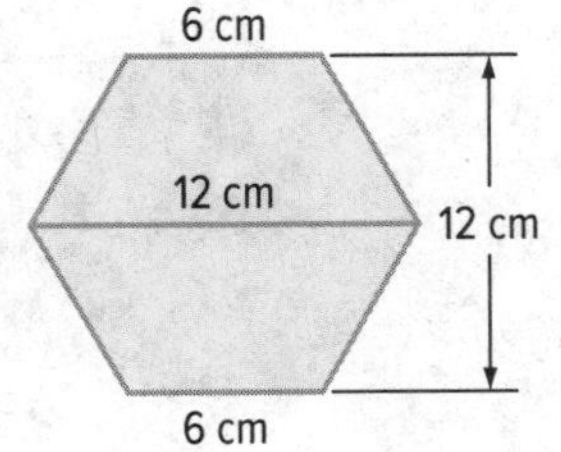

23. ________

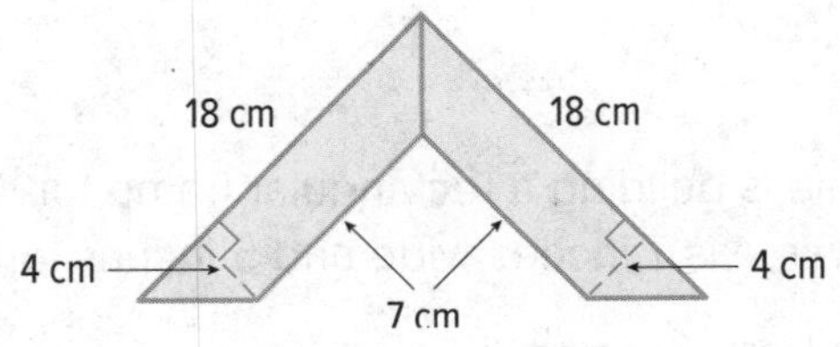

Power Up! Common Core Test Practice

24. A piece of sod is shaped like a trapezoid with the dimensions shown. Select the correct values to complete the formula to find the area of the piece of sod.

50 cm
80 cm
225 cm

$A = \square \cdot \square (\square + \square)$

What is the area of the piece of sod? ______

25. Serina designed the bags shown for her salon products.

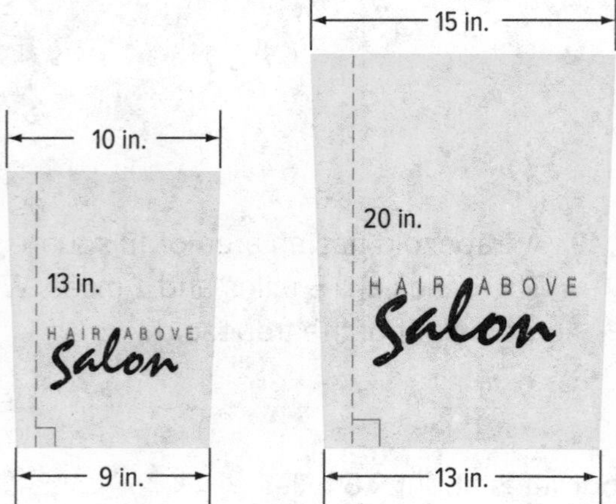

Determine if each statement is true or false.

a. It takes 123.5 in^2 of fabric to make the front of the small bag. ☐ True ☐ False

b. It takes 260 in^2 of fabric to make the front of the large bag. ☐ True ☐ False 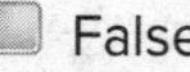

Common Core Spiral Review

Add or multiply. 5.NBT.7

26. $5 + 6.2 + 8.8 =$ ______

27. $8 \times 8 \times 4 =$ ______

28. $725 + 315 + 4 =$ ______

29. Delanie is building a rectangular frame for her favorite photograph. The frame is 7 inches wide and 5 inches long. What is the perimeter of the frame? 4.MD.3 ______

MP Problem-Solving Investigation
Draw a Diagram

Content Standards
6.G.1

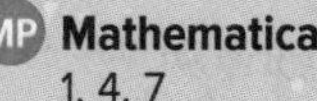
MP Mathematical Practices
1, 4, 7

Case #1 Amazing Array

A designer wants to arrange 12 mosaic tiles into a rectangular shape with the least perimeter possible.

What are the dimensions of the rectangle?

1 Understand *What are the facts?*

Twelve tiles will be arranged with the least perimeter possible.

2 Plan *What is your strategy to solve this problem?*

Use graph paper. Make diagrams of 12 squares to represent 12 tiles.

3 Solve *How can you apply the strategy?*

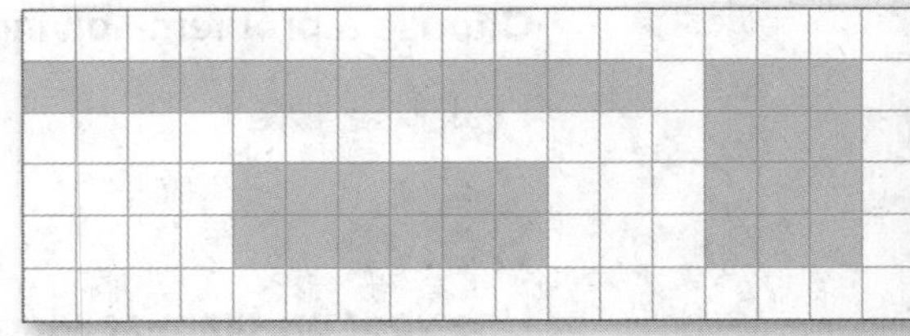

A rectangle with dimensions of 12 and 1 has a perimeter of ________.

A rectangle with dimensions of 3 and 4 has a perimeter of ________.

A rectangle with dimensions of 2 and 6 has a perimeter of ________.

So, the least perimeter has dimensions of ________.

4 Check *Does the answer make sense?*

Use addition to check your answer.

$3 + 4 + 3 + 4 = 14$ $2 + 6 + 2 + 6 = 16$  $12 + 1 + 12 + 1 = 26$

Analyze the Strategy

MP Identify Structure Describe a design with a perimeter and an area of 16.

____

____

Case #2 Dynamic Dimensions

For a school assignment, Santiago has to give three different possibilities for the dimensions of a rectangle that has a perimeter of 28 feet and an area greater than 30 square feet. One of the diagrams he drew is shown at the right.

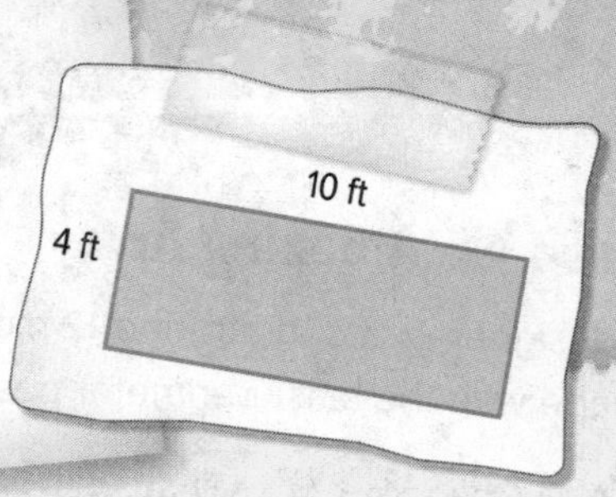

What are two other possibilities for the dimensions of the rectangle?

1 Understand

Read the problem. What are you being asked to find?

I need to find ______________________.

Underline key words and values in the problem. What information do you know?

The perimeter of the rectangle is ☐ feet, and the area is greater than ________.

2 Plan

Choose a problem-solving strategy.

I will use the ______________________ strategy.

3 Solve

Use your problem-solving strategy to solve the problem.

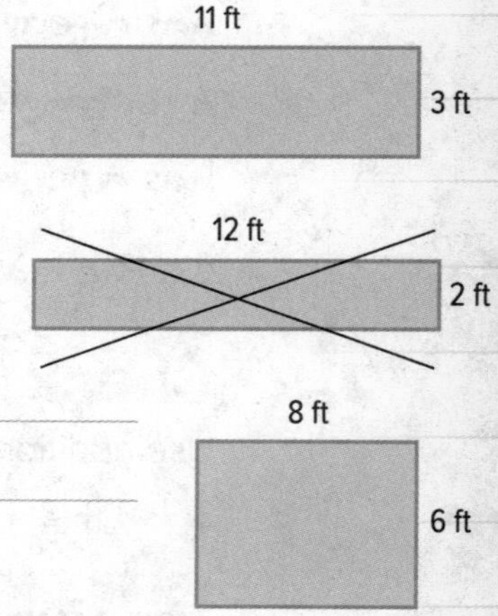

Draw rectangles with perimeters of ☐ feet.
Then, multiply length times width to find the area.

The product must be greater than ☐.

11 ft
3 ft

12 ft
2 ft

8 ft
6 ft

____

____

So, the dimensions of two possible rectangles are

______________________.

Check

Use information from the problem to check your answer.

Reread the problem. Check that both conditions have been met.

Perimeter: ☐ = 28 Area: ☐ > 30 and ☐ > 30

Work with a small group to solve the following cases. Show your work on a separate piece of paper.

Case #3 Decorations

A rectangular table that is placed lengthwise against a wall is 8 feet long and 4 feet wide. Balloons will be attached 8 inches apart along the three exposed sides, with one balloon at each of the four corners.

How many balloons are needed?

Case #4 Geography

The mall is 15 miles from your home. Your school is one-half of the way from your home to the mall. The library is two-fifths of the way from your school to the mall.

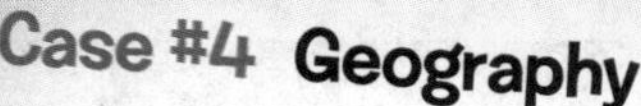

How many miles is it from your home to the library?

Case #5 Paint

Miller's Hardware store was having a sale on pints and gallons of paint. There were 107 people who bought pints of paint and 132 people who bought gallons of paint. 92 customers bought only pints. Some people bought both pints and gallons, and 48 customers did not buy any pints or gallons of paint.

How many customers shopped during the sale?

Case #6 Geometry

Make a figure that contains three triangles, a parallelogram, and a trapezoid using 7 congruent line segments. Draw your figure at the right.

Mid-Chapter Check

Vocabulary Check

1. **MP Be Precise** Define *polygon*. Give an example of a figure that is a polygon and an example of a figure that is not a polygon. (Lesson 1)

2. Fill in the blank in the sentence below with the correct term(s). (Lesson 2)

Congruent figures have the ________ size and the ________ shape.

Skills Check and Problem Solving

Find the area of each figure. (Lessons 1 and 2)

3. ________

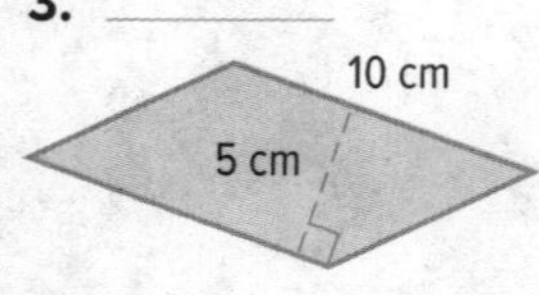

4. ________

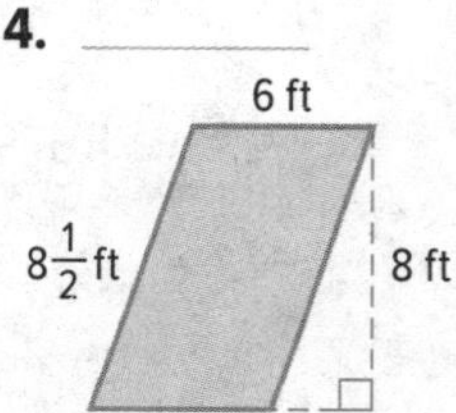

5. ________

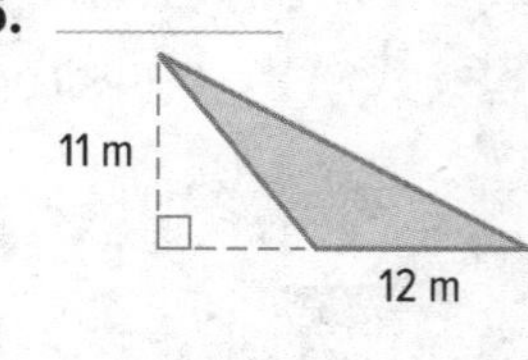

Find the missing dimension of each figure. (Lessons 1 and 3)

6. parallelogram: $h = 5\frac{1}{4}$ ft; $A = 12$ ft^2

7. trapezoid: $b_1 = 3$ m; $b_2 = 4$ m; $A = 7$ m^2

8. **MP Model with Mathematics** A corner table is in the shape of a trapezoid. Find the area of the tabletop. (Lesson 3) ________

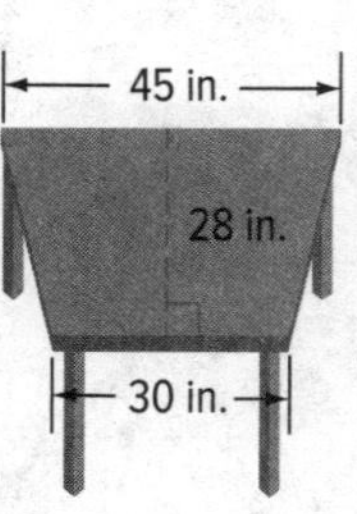

9. **MP Reason Inductively** The area of a triangle is 56 square centimeters. Give all possible sets of whole number dimensions for the base and height of the triangle. (Lesson 2) ________

Lesson 4

Changes in Dimension

Real-World Link

Construction Mr. Blackwell is building a rectangular dog house. The floor of the dog house is 4 feet long and 2 feet wide.

1. Draw the floor of the dog house on the graph paper below.

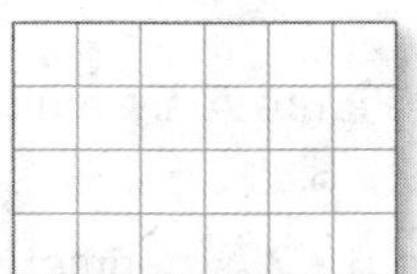

2. Add the lengths of the sides to find the perimeter.

3. Multiply the length and width to find the area.

4. Mr. Blackwell doubles the width of the dog house. Draw the new floor below.

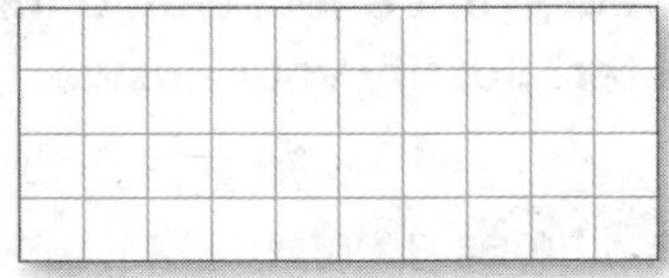

5. How did the perimeter and area of the floors change from the first to the second dog house? ______

Essential Question

HOW does measurement help you solve problems in everyday life?

Common Core State Standards

Content Standards
6.G.1

MP Mathematical Practices
1, 2, 3, 4, 7

Which **MP Mathematical Practices** did you use?
Shade the circle(s) that applies.

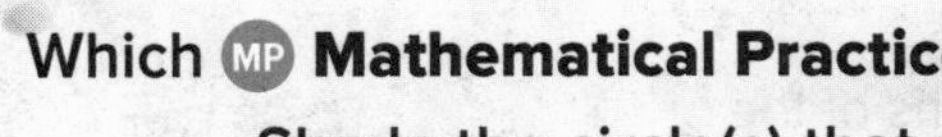

① Persevere with Problems
② Reason Abstractly
③ Construct an Argument
④ Model with Mathematics
⑤ Use Math Tools
⑥ Attend to Precision
⑦ Make Use of Structure
⑧ Use Repeated Reasoning

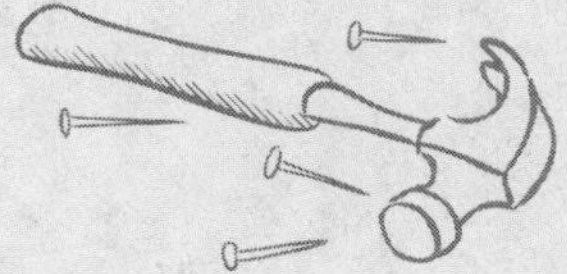

Work Zone

Key Concept: Changing Dimensions: Effect on Perimeter

Words If the dimensions of a polygon are multiplied by x, then the perimeter of the polygon changes by a factor of x.

Model

Figure A

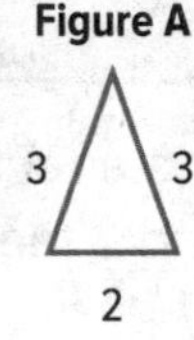

Figure B

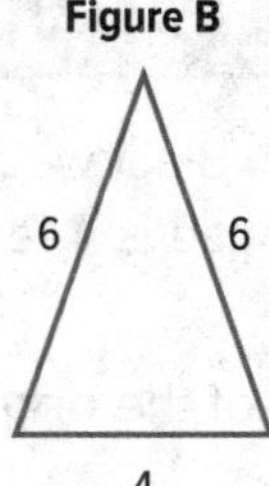

Example The dimensions of Figure A are multiplied by 2 to produce the dimensions of Figure B.

$$\underbrace{\text{perimeter of Figure A}}_{8} \cdot 2 = \underbrace{\text{perimeter of Figure B}}_{16}$$

Notice that all the dimensions of the figure must change using the same factor, x.

Example

Tutor

1. Suppose the side lengths of the parallelogram at the right are tripled. What effect would this have on the perimeter? Justify your answer.

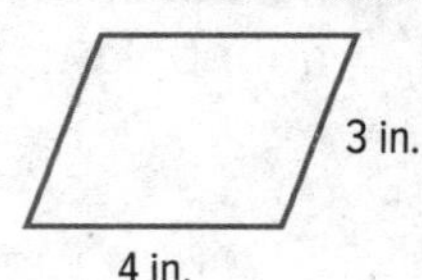

The dimensions are 3 times greater.

original perimeter: $2(4) + 2(3) = 14$ in.

new perimeter: $2(12) + 2(9) = 42$ in.

compare perimeters: 42 in. $\div$ 14 in. $= 3$

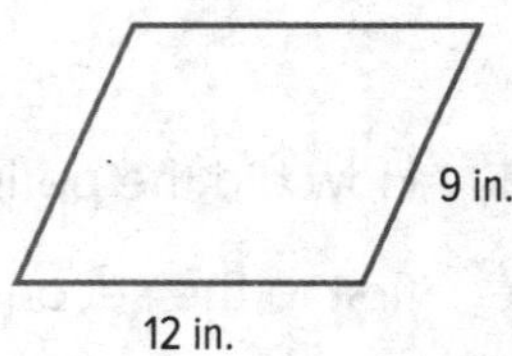

So, the perimeter is 3 times the perimeter of the original figure.

Show your work.

Got it? Do this problem to find out.

a. Suppose the side lengths of the trapezoid at the right are multiplied by $\frac{1}{2}$. What effect would this have on the perimeter? Justify your answer.

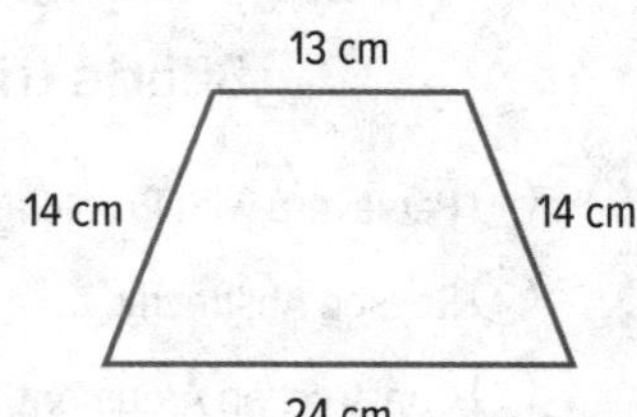

a. ________

Changing Dimensions: Effect on Area

Key Concept

Words If the dimensions of a polygon are multiplied by x, then the area of the polygon changes by $x \cdot x$ or x^2.

Model

Figure A

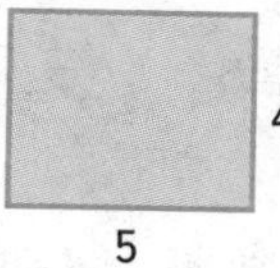

Figure B

Example The dimensions of Figure A are multiplied by 2 to produce the dimensions of Figure B.

$$\underbrace{\text{area of Figure A}}_{20} \cdot \underbrace{2^2}_{4} = \underbrace{\text{area of Figure B}}_{80}$$

Notice that all the dimensions of the figure must change using the same factor, x.

Example

Tutor

2. The side lengths of the triangle at the right are multiplied by 5. What effect would this have on the area? Justify your answer.

The dimensions are 5 times greater.

original area: $\frac{1}{2} \cdot 2 \cdot 1 = 1\text{ cm}^2$

new area: $\frac{1}{2} \cdot 10 \cdot 5 = 25\text{ cm}^2$

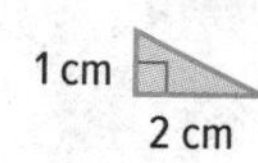

compare areas:
$25\text{ cm}^2 \div 1\text{ cm}^2 = 25$ or 5^2

So, the area is 5^2 or 25 times the area of the original figure.

Got it? **Do this problem to find out.**

b. A rectangle measures 2 feet by 4 feet. Suppose the side lengths are multiplied by 2.5. What effect would this have on the area? Justify your answer.

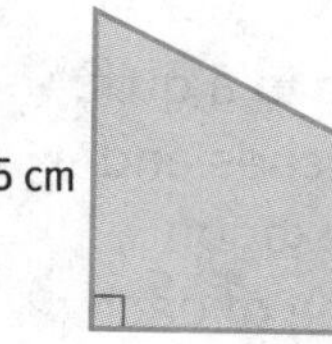

b. ____________

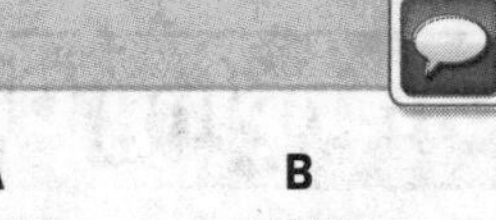

Example

3. A stop sign is in the shape of a regular octagon. Sign A shown at the right has an area of 309 square inches. What is the area of sign B?

Since $8 \times 1.5 = 12$, the area of sign B is 1.5^2 times the area of sign A.

$309 \cdot 1.5^2 = 309 \cdot 2.25$ or 695.25

So, the area of sign B is 695.25 square inches.

Guided Practice

Refer to the figure at the right for Exercises 1 and 2. Justify your answers. (Examples 1–2)

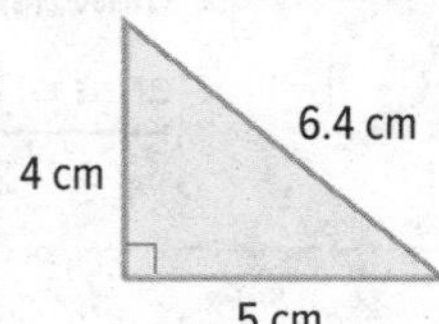

1. Each side length is doubled. Describe the change in the perimeter.

2. Each side length is tripled. Describe the change in the area.

3. Different sizes of regular hexagons are used in a quilt. Each small hexagon has side lengths of 4 inches and an area of 41.6 square inches. Each large hexagon has side lengths of 8 inches. What is the area of each large hexagon? (Example 3)

4. 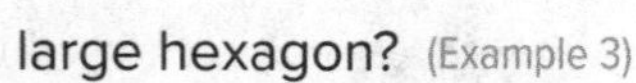 **Building on the Essential Question** How can exponents help you find the area of a rectangle if each side length is multiplied by x?

Rate Yourself!

How confident are you about changes in dimension? Check the box that applies.

For more help, go online to access a Personal Tutor.

Name ____________________ My Homework ____________________

Independent Practice

Go online for Step-by-Step Solutions 

1. Each side length of the parallelogram at the right is multiplied by 4. Describe the change in the perimeter. Justify your answer. (Example 1)

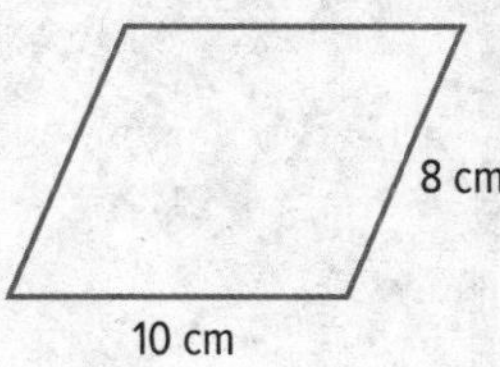

2. The base and height of the triangle at the right are multiplied by 4. Describe the change in the area. Justify your answer. (Example 2)

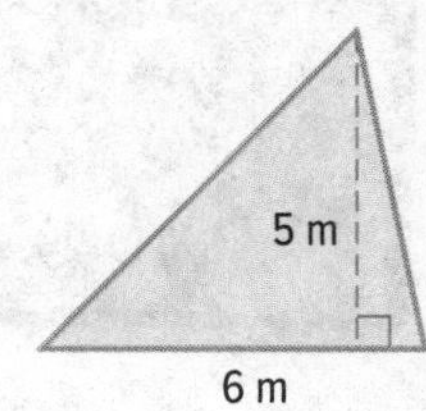

3. Each side length of the rectangle is multiplied by $\frac{1}{3}$. Describe the change in the area. Justify your answer. (Example 2)

4. Different sizes of regular pentagons are used in a stained glass window. Each small pentagon has side lengths of 4 inches and an area of 27.5 square inches. Each large pentagon has side lengths of 8 inches. What is the area of each large pentagon? (Example 3)

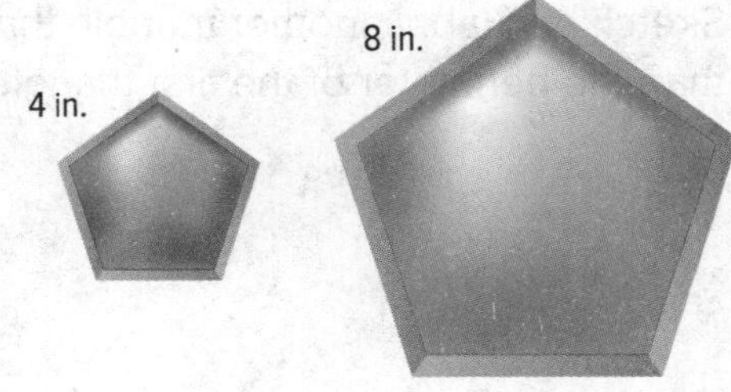

5. MP **Justify Conclusions** A dollhouse has a bed with dimensions $\frac{1}{12}$ the size of a queen-size bed. A queen-size bed has an area of 4,800 square inches, and a length of 80 inches. What are the side lengths of the dollhouse bed? Justify your answer.

6. **MP Reason Abstractly** Refer to the graphic novel frame below for Exercises a–b.

a. What is the original area of the triangle? ______

b. What is the new area if the sides are all two times longer?

______

H.O.T. Problems Higher Order Thinking

7. **MP Identify Structure** Sketch a triangle with the side lengths labeled. Sketch and label another triangle that has a perimeter two times greater than the perimeter of the first triangle.

8. **MP Persevere with Problems** The corresponding side lengths of two figures have a ratio of $\frac{a}{b}$. What is the ratio of the perimeters? the ratio of the areas?

______

9. **MP Reason Inductively** The larger square shown has a perimeter of 48 units. The smaller square inside has a perimeter that is 2 times smaller. What are the side lengths of the larger and smaller square? Explain. ______

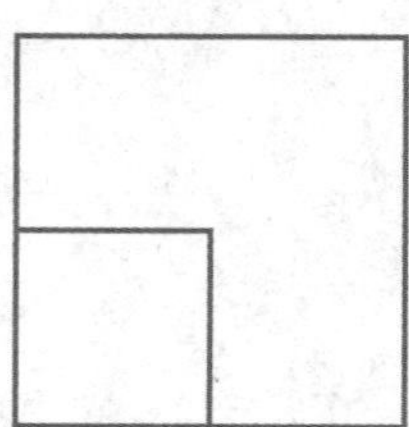

______

______

______

Extra Practice

Refer to the parallelogram at the right for Exercises 10–12. Justify your answers.

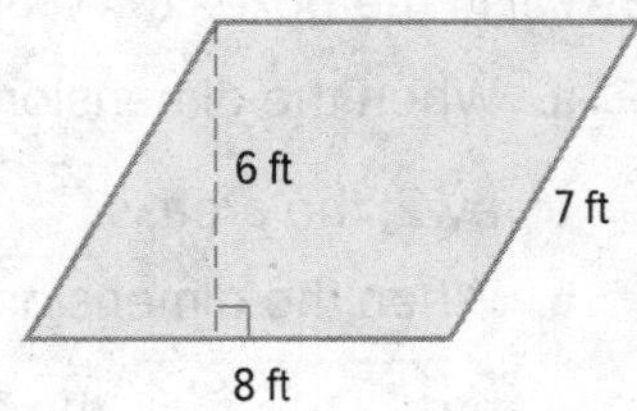

10. Suppose the base and height are each multiplied by $\frac{1}{2}$. What effect would this have on the area?

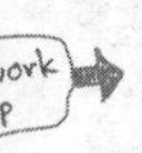

original area: $8 \cdot 6$ or 48 square feet

new dimensions: base $= 8 \cdot \frac{1}{2}$ or 4 ft, height $= 6 \cdot \frac{1}{2}$ or 3 ft

new area: $4 \cdot 3$ or 12 square feet; $12 \text{ ft}^2 \div 48 \text{ ft}^2 = \frac{1}{4}$;

So, the area is $\frac{1}{2} \cdot \frac{1}{2}$ or $\frac{1}{4}$ times the area of the original figure.

11. Suppose the side lengths are multiplied by 6. Describe the change in the perimeter. ______

12. Suppose the base and height are each multiplied by 3.5. Describe the change in the area. ______

13. Refer to the triangle at the right. Suppose the side lengths and height of the triangle were divided by 4. What effect would this have on the perimeter? the area? Justify your answer.

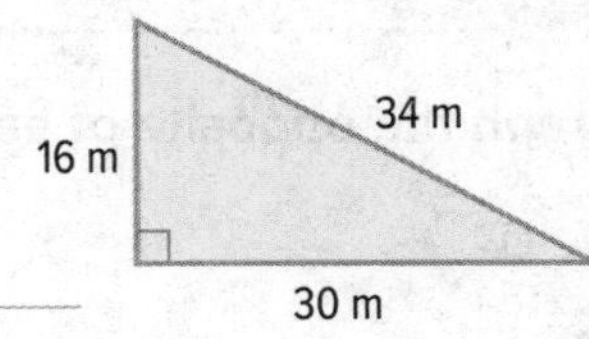

14. MP **Justify Conclusions** A model car has a windshield with dimensions $\frac{1}{18}$ the size of a real car windshield. The rectangular windshield of the real car has an area of about 2,318 square inches, with a width of 61 inches. What are the side lengths of the model car's windshield? Round to the nearest hundredth. Justify your answer.

Power Up! Common Core Test Practice

15. Fill in the boxes to complete each statement about the trapezoid at the right.

a. When the dimensions of the trapezoid are multiplied by 2, the area is ☐ times greater.

b. When the dimensions of the trapezoid are multiplied by ☐, the area is 16 times greater.

c. When the dimensions of the trapezoid are multiplied by 5, the area is ☐ times greater.

3 mm
6 mm
10.5 mm

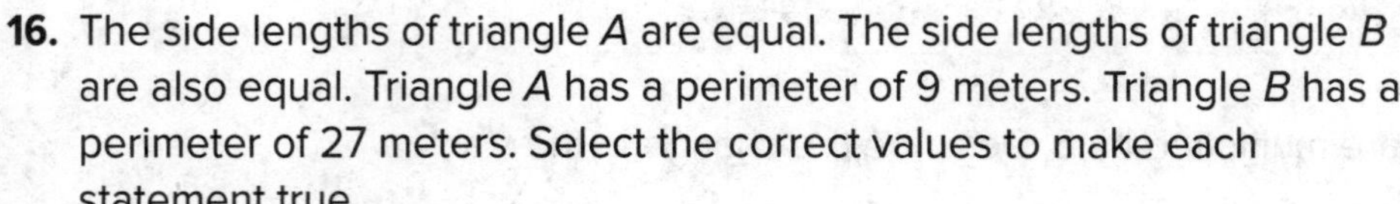

16. The side lengths of triangle *A* are equal. The side lengths of triangle *B* are also equal. Triangle *A* has a perimeter of 9 meters. Triangle *B* has a perimeter of 27 meters. Select the correct values to make each statement true.

3	23.4
6	27
9	35.1
11.7	

a. The length of each side in triangle *A* is ☐ meters.

b. The length of each side in triangle *B* is ☐ meters.

c. The area of triangle *A* is about 3.9 square meters. The area of triangle *B* is about ☐ square meters.

Common Core Spiral Review

Graph the opposite of each number on a number line. 6.NS.6a

17. 0

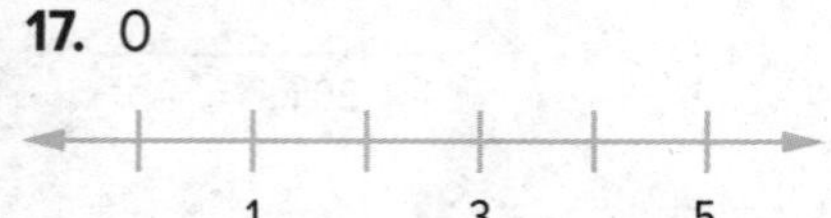

18. −7

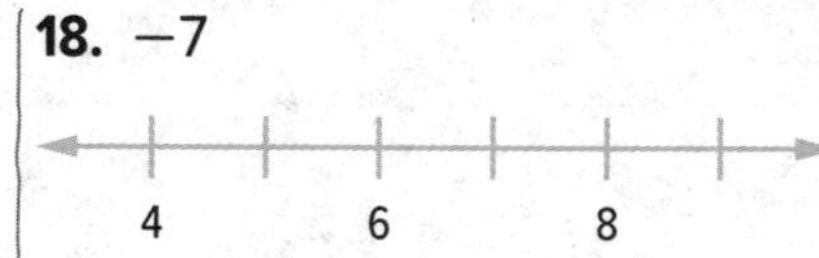

19. 5

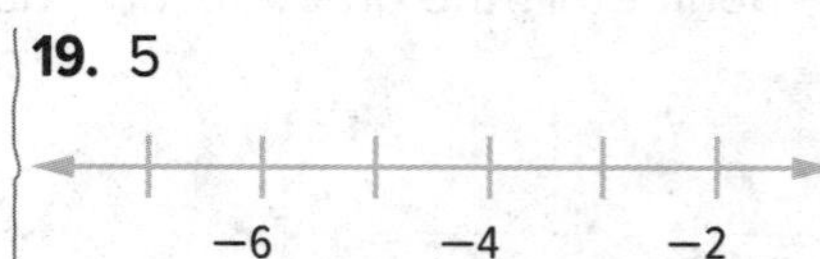

20. Graph 2 and 9. Then use the number line to find the distance between 9 and 2. 6.NS.8, 6.NS.5 ______

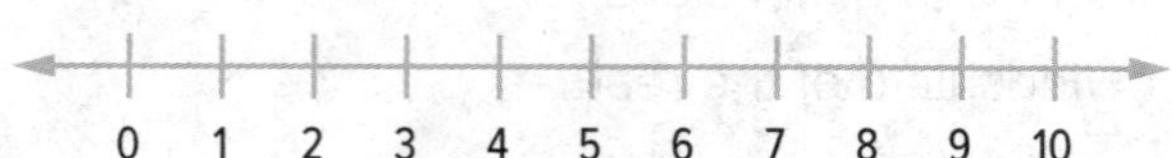

21. John and his dad are playing catch on the football field. John is standing on the 10-yard line. His dad is standing on the 25-yard line. How far is John from his dad? If his dad moves to the 20-yard line, what is the distance between them now? 4.OA.3

______

Lesson 5

Polygons on the Coordinate Plane

Real-World Link

Maps Graph points on a coordinate plane to draw a map of an outdoor stadium. Complete the table to identify each shape.

Location	Vertices	Shape
Stage	(2, 6), (2, 9), (6, 9), (6, 6), (5, 5), (3, 5)	
Bleachers	(7, 5), (7, 9), (9, 9), (9, 5)	
Concession Stand	(5, 2), (5, 4), (7, 4), (7, 2)	

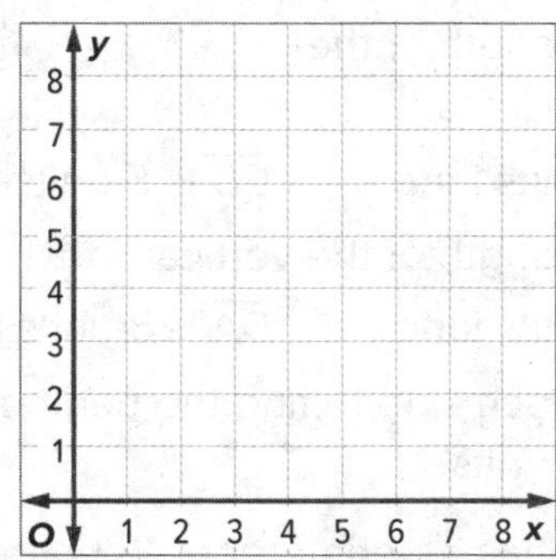

1. Find the dimensions of the bleachers.

   Length: ________ Height: ________

2. The length of the line from point (2, 6) to point (2, 9) is 3 units long. How can you use the y-coordinates to find the length of the line?

   ____

Essential Question

HOW does measurement help you solve problems in everyday life?

Common Core State Standards

Content Standards
6.G.1, 6.G.3, 6.NS.8

MP Mathematical Practices
1, 2, 3, 4, 5, 7

Which **MP Mathematical Practices** did you use?
Shade the circle(s) that applies.

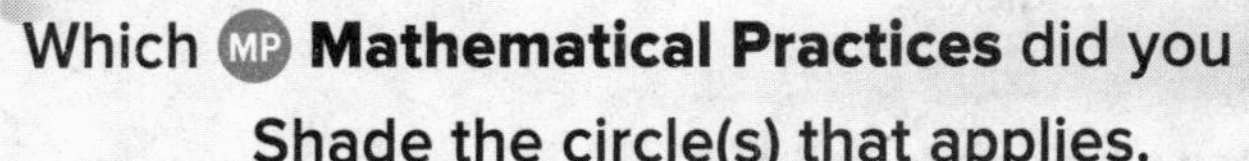

① Persevere with Problems
② Reason Abstractly
③ Construct an Argument
④ Model with Mathematics
⑤ Use Math Tools
⑥ Attend to Precision
⑦ Make Use of Structure
⑧ Use Repeated Reasoning

Work Zone

Find Perimeter

You can use the coordinates of a figure to find its dimensions by finding the distance between two points. To find the distance between two points with the same *x*-coordinates, subtract their *y*-coordinates. To find the distance between two points with the same *y*-coordinates, subtract their *x*-coordinates.

Examples

1. A rectangle has vertices *A*(2, 8), *B*(7, 8), *C*(7, 5), and *D*(2, 5). Use the coordinates to find the length of each side. Then find the perimeter of the rectangle.

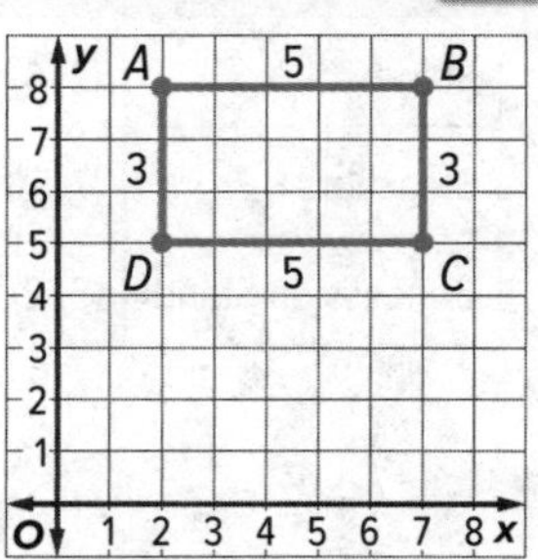

Width: Find the length of the horizontal lines.

$\overline{AB}$ is 5 units long. $\overline{CD}$ is 5 units long.

Length: Find the length of the vertical lines.

$\overline{BC}$ is 3 units long. $\overline{DA}$ is 3 units long.

Add the lengths of each side to find the perimeter.
$5 + 5 + 3 + 3 = 16$ units

So, rectangle *ABCD* has a perimeter of 16 units.

Perimeter and Area
Remember that perimeter is the distance around a closed figure. Area is the number of square units needed to cover the surface enclosed by a geometric figure.

2. Rectangle *ABCD* has vertices *A*(2, 1), *B*(2, 5), *C*(4, 5), and *D*(4, 1). Use the coordinates to find the length of each side. Then find the perimeter of the rectangle.

Width: Subtract *y*-coordinates.

AB: $5 - 1 = 4$ units *CD*: $5 - 1 = 4$ units

Length: Subtract *x*-coordinates.

AD: $4 - 2 = 2$ units *BC*: $4 - 2 = 2$ units

Add the lengths of each side to find the perimeter.
$4 + 2 + 4 + 2 = 12$ units

Got it? Do these problems to find out.

Use the coordinates to find the length of each side. Then find the perimeter of the rectangle.

a. *E*(3, 6), *F*(3, 8), *G*(7, 8), *H*(7, 6)

b. *I*(1, 4), *J*(1, 9), *K*(8, 9), *L*(8, 4)

a. ________

b. ________

Example

3. Each grid square on the zoo map has a length of 200 feet. Find the total distance, in feet, around the zoo.

When *x*-coordinates are the same, subtract the *y*-coordinates. When *y*-coordinates are the same, subtract the *x*-coordinates.

$10 + 7 + 3 + 4 + 4 + 4 + 3 + 7 = 42$ units

Multiply by 200 feet to find the total distance.

$42 \times 200 = 8{,}400$ feet. The total distance is 8,400 feet.

Got it? Do this problem to find out.

c. The coordinates of the vertices of a garden are (0, 1), (0, 4), (8, 4), and (8, 1). If each unit represents 12 inches, find the perimeter in inches of the garden.

c. ____________

Find Area

You can find the area of a figure that has been drawn grid paper or graphed on the coordinate plane.

Example

4. Find the area of the figure in square units.

The figure can be separated into a rectangle and a trapezoid.

Area of rectangle

$A = \ell \times w$

$A = 5 \times 2$ or 10

Area of trapezoid

$A = \frac{1}{2}h(b_1 + b_2)$

$A = \frac{1}{2}(2)(3 + 4)$ or 7

So, the area of the figure is $10 + 7$ or 17 square units.

Got it? Do this problem to find out.

d. Find the area, in square units, of the figure at the right.

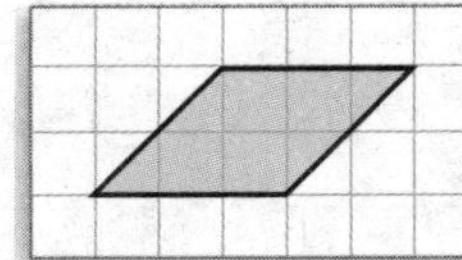

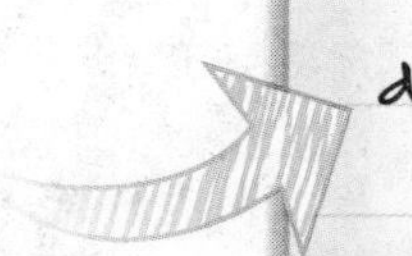

d. ____________

Example

5. A figure has vertices $A(2, 5)$, $B(2, 8)$, and $C(5, 8)$. Graph the figure and classify it. Then find the area.

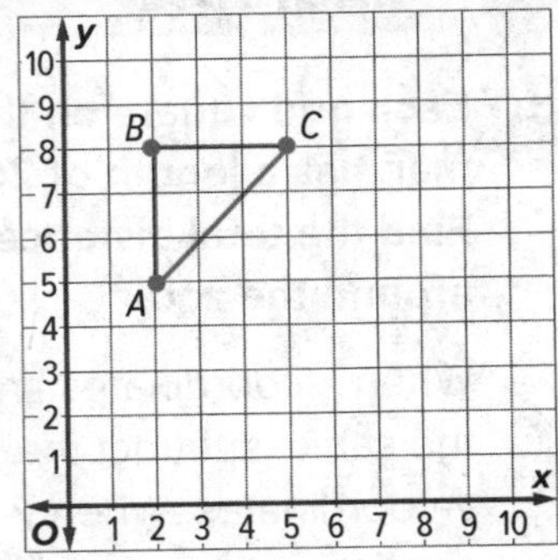

Plot the points. Connect the vertices. The figure is a right triangle.

The height from point A to point B is 3 units. The base from point B to point C is 3 units.

$A = \frac{1}{2}bh$	Area formula of a triangle
$A = \frac{1}{2}(3)(3)$	Replace b with 3 and h with 3.
$A = 4.5$	Multiply.

Triangle ABC has an area of 4.5 square units.

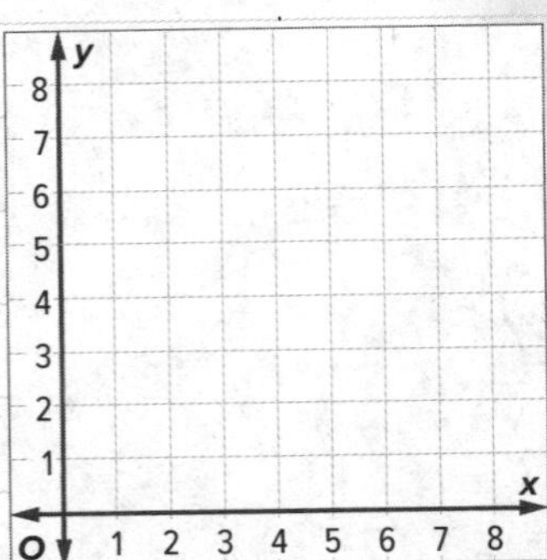

Got it? **Do this problem to find out.**

Graph the figure and classify it. Then find the area.

e. $A(3, 3)$, $B(3, 6)$, $C(5, 6)$, $D(8, 3)$

e. ________

Guided Practice

Use the coordinates to find the length of each side of the rectangle. Then find the perimeter. (Examples 1 and 2)

1. $L(3, 3)$, $M(3, 5)$, $N(7, 5)$, $P(7, 3)$

2. $P(3, 0)$, $Q(6, 0)$, $R(6, 7)$, $S(3, 7)$

3. Mrs. Piel is building a fence around the perimeter of her yard for her dog. The coordinates of the vertices of the yard are (0, 0), (0, 10), (5, 10), and (5, 0). If each grid square has a length of 100 feet, find the amount of wire, in feet, needed for the fence. What is the shape of her yard? (Example 3) ________

4. **Building on the Essential Question** How can coordinates help you to find the area of figures on the coordinate plane? ________

Rate Yourself!

How well do you understand polygons on the coordinate plane? Circle the image that applies.

Clear | Somewhat Clear | Not So Clear

For more help, go online to access a Personal Tutor.

Name ______________________ My Homework ______________________

Independent Practice

Go online for Step-by-Step Solutions

Use the coordinates to find the length of each side of the rectangle. Then find the perimeter. (Examples 1 and 2)

1. $D(1, 2)$, $E(1, 7)$, $F(4, 7)$, $G(4, 2)$

2. $Q(0, 0)$, $R(4, 0)$, $S(4, 4)$, $T(0, 4)$

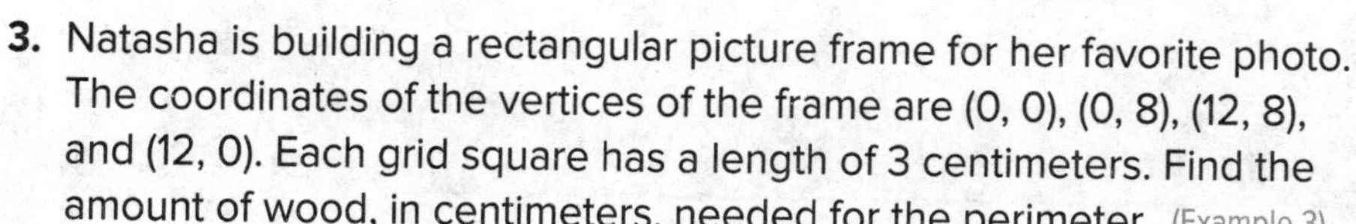

3. Natasha is building a rectangular picture frame for her favorite photo. The coordinates of the vertices of the frame are (0, 0), (0, 8), (12, 8), and (12, 0). Each grid square has a length of 3 centimeters. Find the amount of wood, in centimeters, needed for the perimeter. (Example 3)

Find the area of each figure in square units. (Example 4)

4. ______

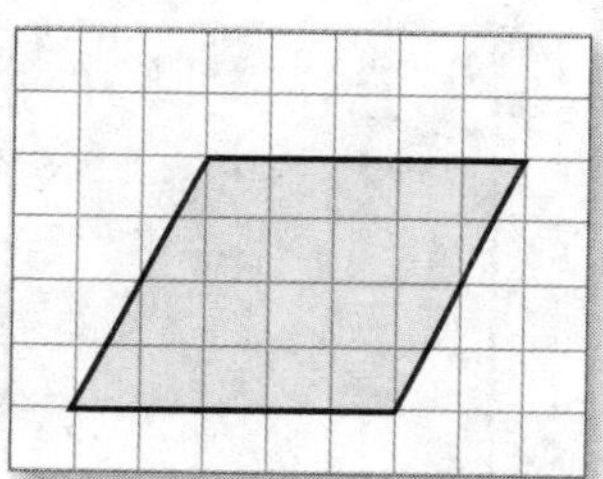

5. ______

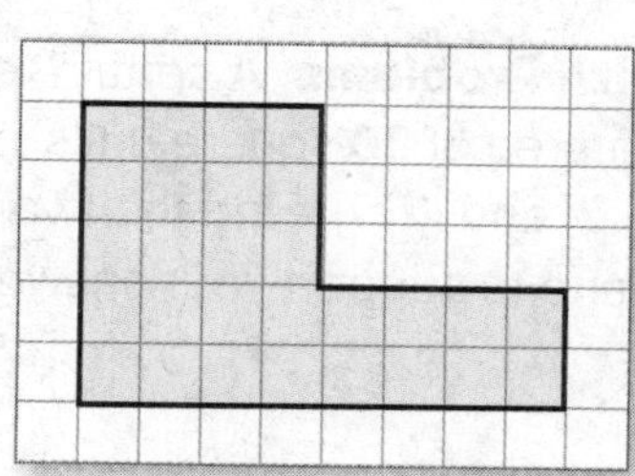

Graph each figure and classify it. Then find the area. (Example 5)

6. $R(3, -2)$, $S(7, -2)$, $T(8, -6)$, $V(1, -6)$

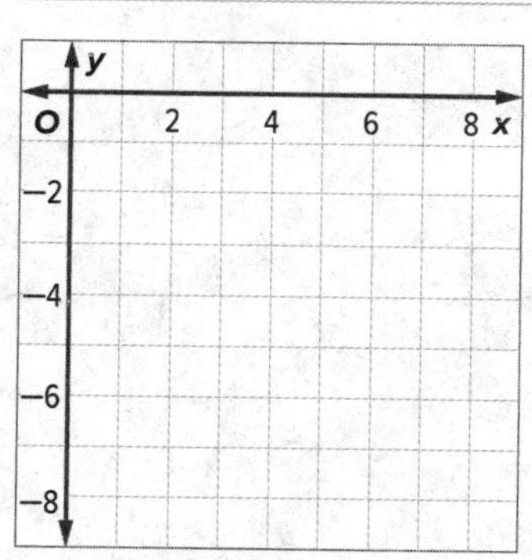

7. $A(-3, -4)$, $B(-3, 5)$, $C(2, 5)$, $D(2, -4)$

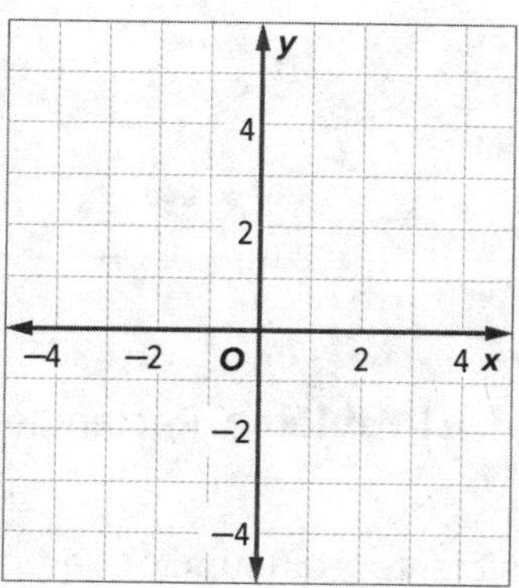

8. MP **Use Math Tools** A rectangle has a perimeter of 20 units. The coordinates of three of the vertices are (0, 0), (6, 0), and (6, 4) as shown on the graph.

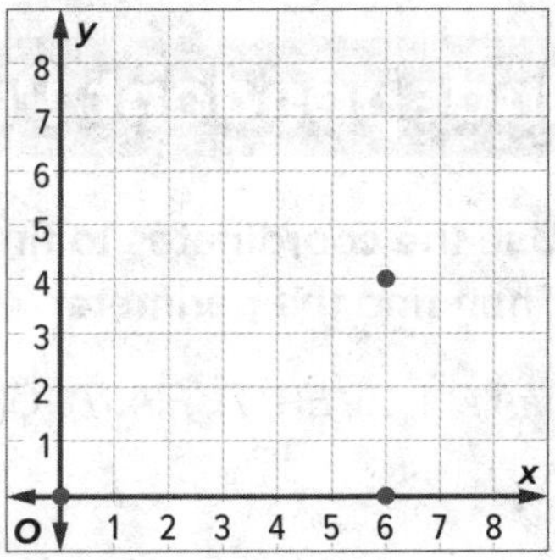

a. What is the coordinate of the missing vertex?

____

b. Plot points (6, 6) and (2, 4). Connect these points to create a composite figure.

c. What is the area of the composite figure? ______________

H.O.T. Problems Higher Order Thinking

9. MP **Use Math Tools** Draw a rectangle on a coordinate plane that has a perimeter of 16 units. Label all of the vertices with the coordinates. Then find the area of the rectangle. ______________

10. MP **Persevere with Problems** A certain rectangle has a perimeter of 22 units and an area of 30 square units. Two of the vertices have coordinates at (2, 2) and (2, 7). Find the two missing coordinates. Use the coordinate plane to support your answer.

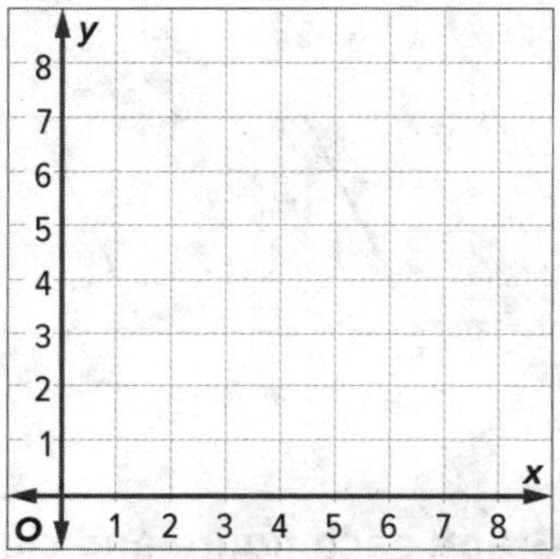

11. MP **Identify Structure** Explain the steps you would use to find the perimeter of a rectangle using the coordinates of the vertices.

12. MP **Persevere with Problems** Rectangle *QRST* has vertices *Q*(3, 2) and *S*(7, 8).

a. Give two possible coordinates for vertices *R* and *T*.

____

b. Find the perimeter and area of the rectangle.

____

Name ______________________ My Homework ______________

Extra Practice

Use the coordinates to find the length of each side of the rectangle. Then find the perimeter.

13. *A*(5, 2), *B*(5, 4), *C*(2, 4), *D*(2, 2)

Homework Help

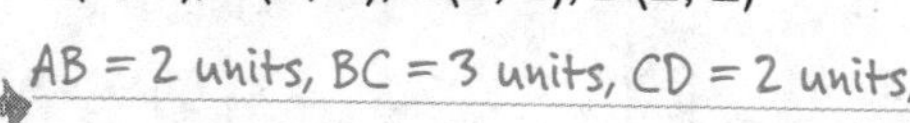

DA = 3 units; 10 units

14. *M*(1, 1), *N*(1, 9), *P*(7, 9), *Q*(7, 1)

____

____

15. MP **Reason Abstractly** Andre is creating a border around his rectangular patio with paver bricks. The coordinates of the vertices of the patio are (1, 5), (6, 5), (6, 1), and (1, 1). Each grid square has a length of 3 feet. Find the amount of brick, in feet, needed for the perimeter. ______________

Find the area of each figure in square units.

16. ______________________

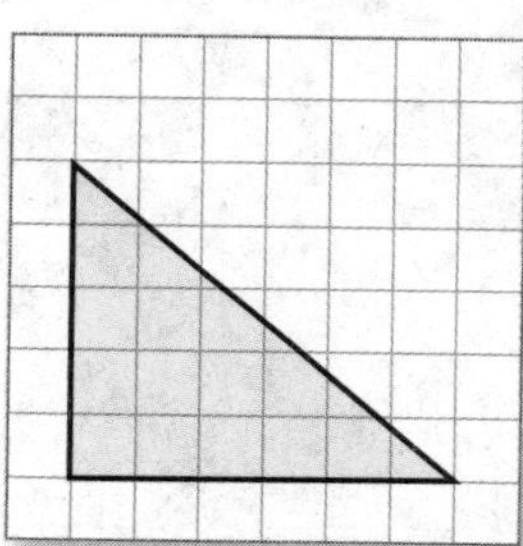

17. ______________________

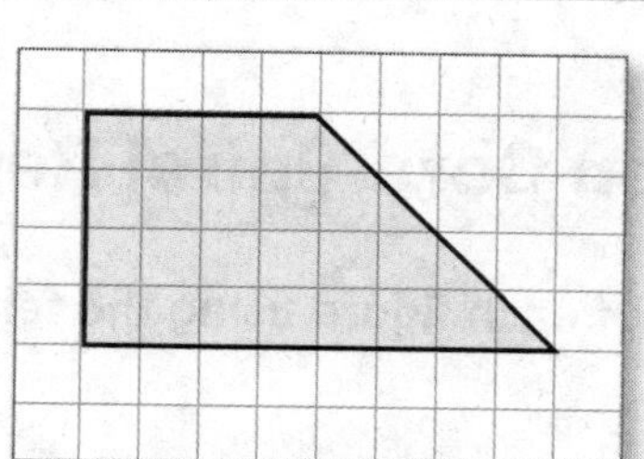

Graph each figure and classify it. Then find the area.

18. *G*(−4, 1), *H*(4, 1), *I*(3, −3), *J*(−1, −3)

____

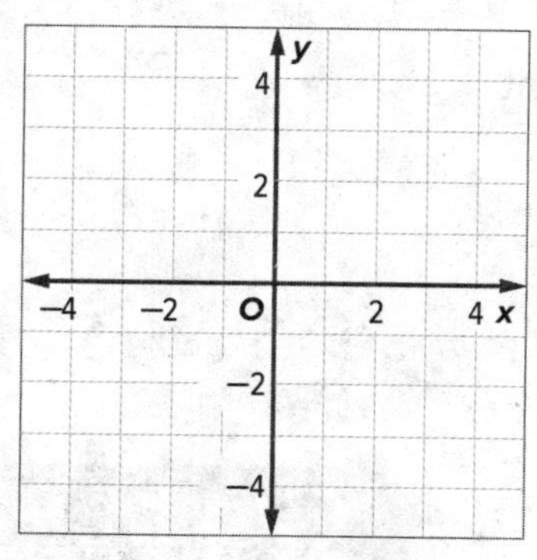

19. *X*(−7, 2), *Y*(−7, 6), *Z*(−4, 2)

____

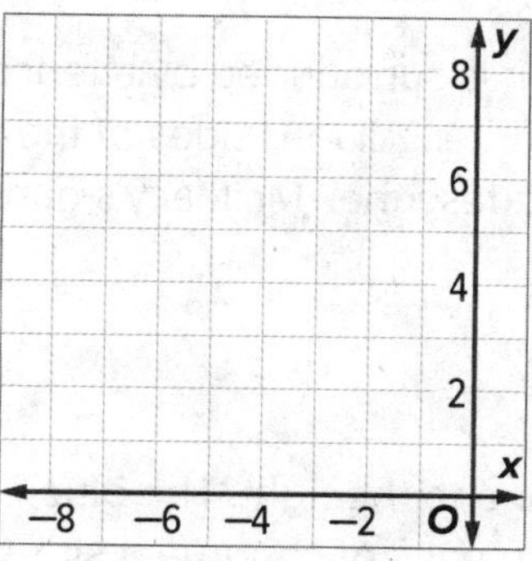

Copy and Solve **Graph each figure and classify it. Then find the area.**

20. *K*(−2, 2), *L*(3, 2), *M*(2, −2), *N*(−3, −2)

21. *Q*(−2, 4), *R*(0, −2), *S*(−4, −2)

Power Up! Common Core Test Practice

22. Figure *BCDEFG* has vertices located at *B*(1, 3), *C*(1, 7), *D*(4, 7), *E*(4, 5), *F*(8, 5), and *G*(8, 3). Sketch the figure on the coordinate plane and connect the vertices.

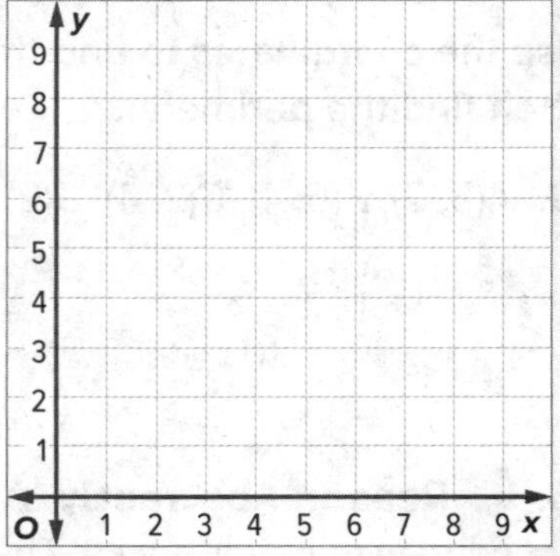

What is the area of the figure?

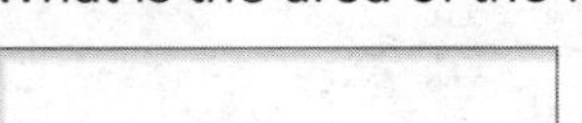

23. A quadrilateral has vertices with coordinates of *A*(8, 5), *B*(7, 2), *C*(4, 2), and *D*(2, 5). Which of the following are characteristics of the quadrilateral? Select all that apply.

- ☐ one set of parallel sides
- ☐ two sets of parallel sides
- ☐ four vertices
- ☐ two acute angles

Common Core Spiral Review

Describe the sides of each figure using the terms *parallel*, *perpendicular*, and *congruent*. 5.G.4

24. parallelogram ______________________

____

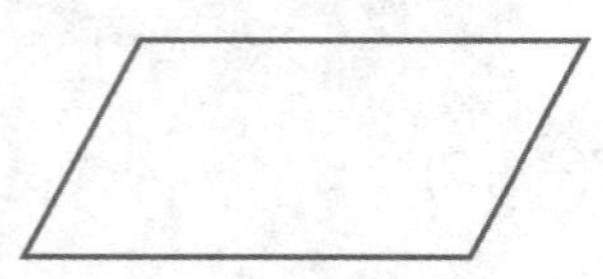

25. trapezoid ______________________

____

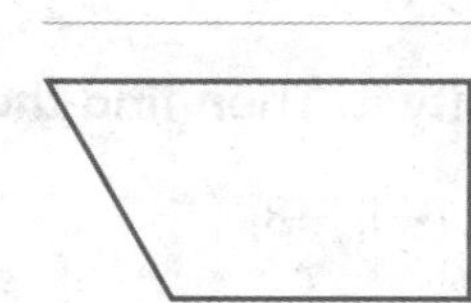

26. Mr. Macy's garden is surrounded by a fence. The fence makes four right angles at each corner. All four sides of the fence are 14 meters long. What shape best describes Mr. Macy's garden? 5.G.4

____

____

27. Gary drew the logo to the right. The blue figure has two pairs of parallel sides, two pairs of congruent sides, and four right angles. What is the shape of the blue figure? 5.G.4

GO LEFT
PRODUCTIONS

____

____

Inquiry Lab

Area of Irregular Figures

 **HOW can you estimate the area of an irregular figure?**

 **Content Standards**
6.G.1

 **Mathematical Practices**
1, 3, 4, 5

The Ramirez family is putting a koi pond in their backyard. They need to estimate the area of the pond to know how many fish they can put in the pond. A scale drawing of the pond is shown below. In the drawing, each square represents one square foot.

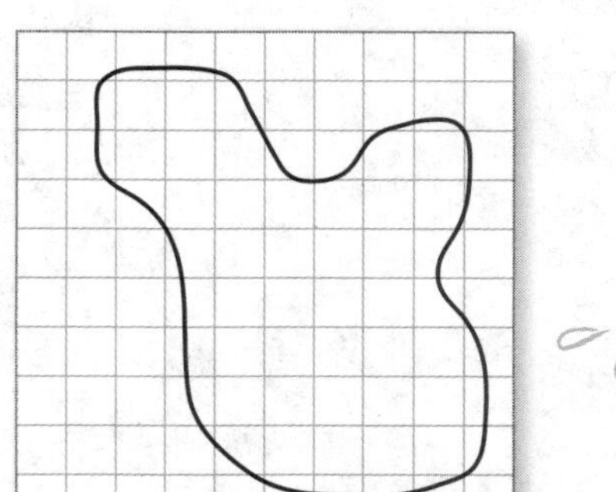

What do you know? ______________________________

____

What do you need to know? ______________________________

____

Hands-On Activity 1

Step 1 Shade and count the number of whole squares the pond covers. ☐

Step 2 Estimate the number of whole squares covered by the partial squares altogether. ☐

Step 3 Add your answers from Steps 1 and 2.

☐ + ☐ = ☐

So, the area of the pond is about ☐ square feet.

Hands-On Activity 2

Another way to estimate the area of an irregular figure is to separate the figure into simpler shapes. Then find the sum of these areas.

Step 1 First, separate the figure into a triangle and a rectangle.

triangle
481 mi
100 mi
170 mi
IDAHO
300 mi
rectangle

Step 2 Find the area of each figure.

Area of a triangle

$A = \frac{1}{2}bh$

$= \frac{1}{2} \cdot 200 \cdot 311$ $b = 300 - 100$ or 200; $h = 481 - 170$ or 311

$= 31{,}100$ Simplify.

Area of rectangle

$A = \ell w$

$= 300 \cdot 170$ or $51{,}000$ $\ell = 300$ and $w = 170$

Step 3 Add to find the total area.

☐ + ☐ = ☐

The area of Idaho is about ☐ square miles.

Investigate

MP Use Math Tools Work with a partner to estimate the area of each irregular figure.

1. $A \approx$ ____________

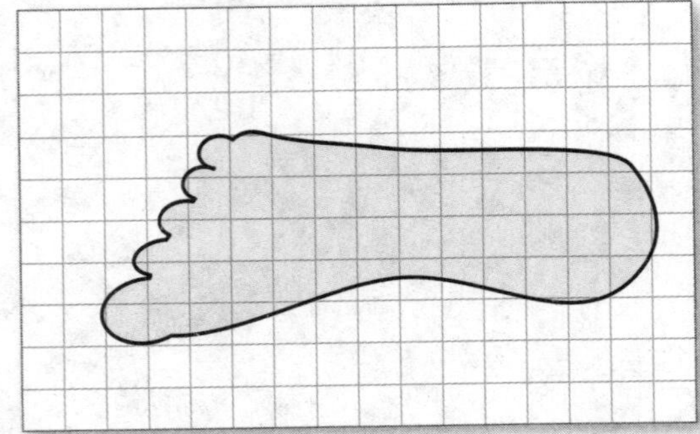

2. $A \approx$ ____________

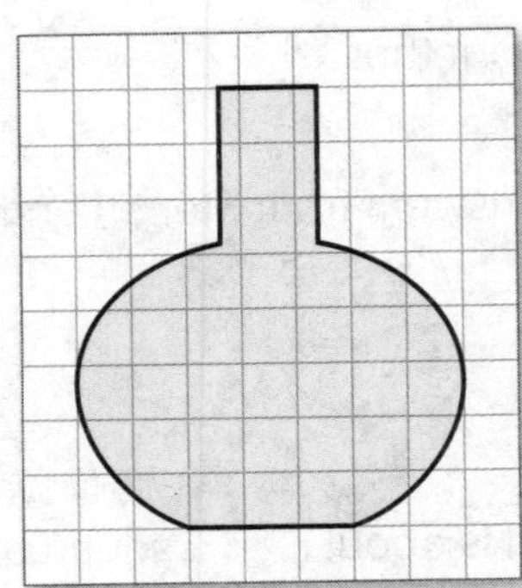

Work with a partner to estimate the area of each irregular figure.

3. $A \approx$ ____________

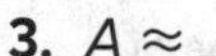

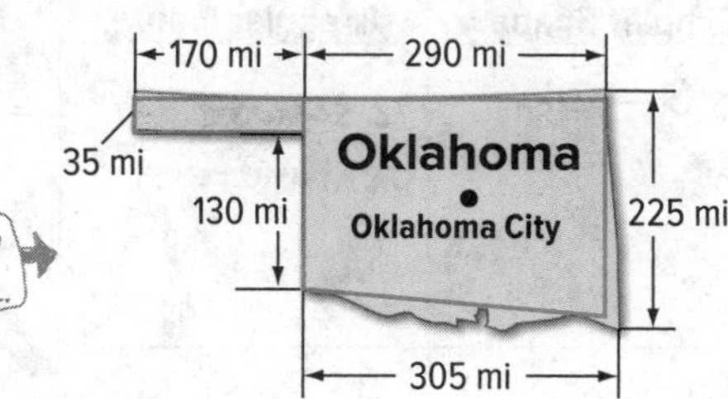

4. $A \approx$ ____________

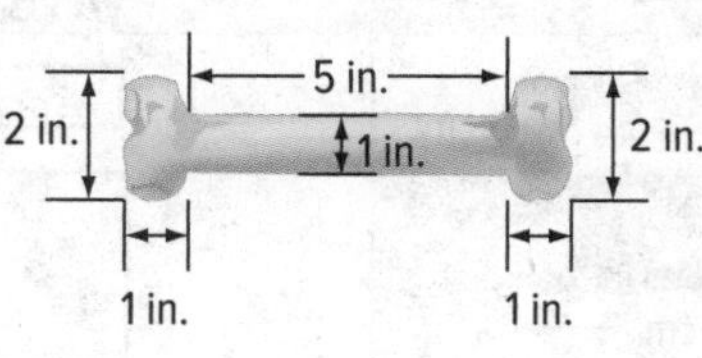

5. $A \approx$ ____________

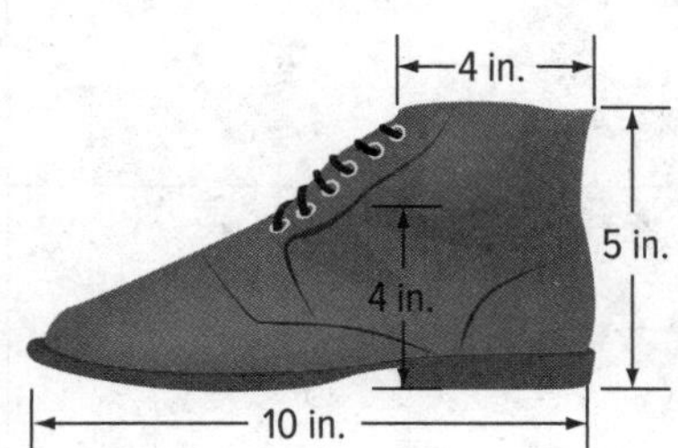

6. $A \approx$ ____________

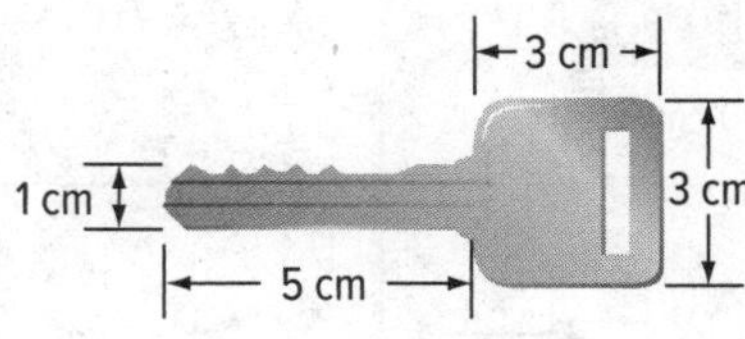

7. $A \approx$ ____________

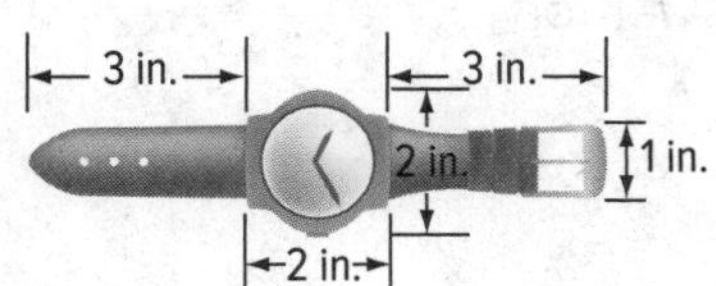

8. $A \approx$ ____________

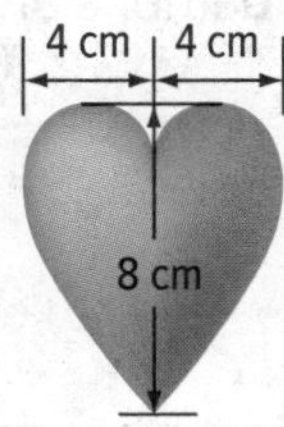

9. $A \approx$ ____________

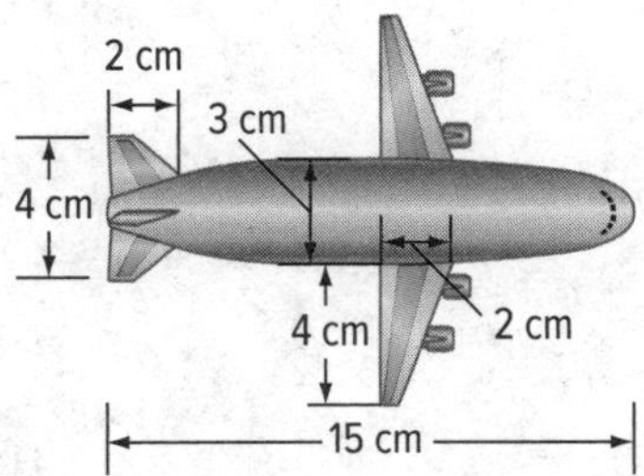

10. $A \approx$ ____________

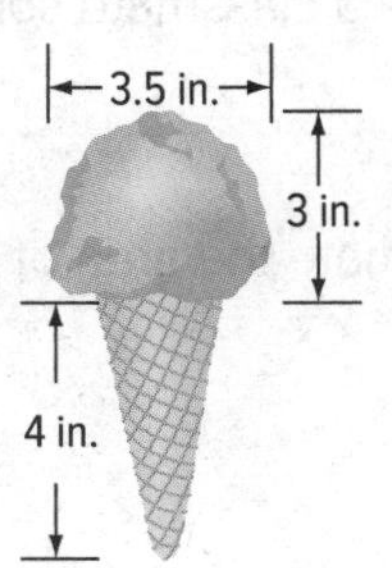

Analyze and Reflect

Collaborate

Work with a partner to complete the table. The first one is done for you.

	Irregular Figure	Draw the simpler shapes you can make.	Area of Each Simpler Shape	Estimated Area of Irregular Figure
	8 cm, 3 cm, 4 cm, 12 cm		$8 \times 3 = 24$ $12 \times 4 = 48$	72 square centimeters
11.	15 in., 6 in., USA, 20 in.			
12.	4 cm, 4 cm, 7 cm, 5 cm, 9 cm			
13.	1 in., 2 in., 1 in., 2 in., 3 in., 6 in.			

14. MP **Reason Inductively** Heather solves Exercise 11 by subtracting the area of two triangles from the area of a large rectangle and finds the answer 105 square inches. How does Heather's answer compare to your answer for Exercise 11?

Create

On Your Own

15. MP **Model with Mathematics** Draw an irregular figure. Write a problem about your figure. Then have a classmate solve the problem.

16. Inquiry HOW can you estimate the area of an irregular figure?

Lesson 6

Area of Composite Figures

Vocabulary Start-Up

A **composite figure** is a figure made of two or more two-dimensional figures. The composite figure shown to the right is made of two rectangles.

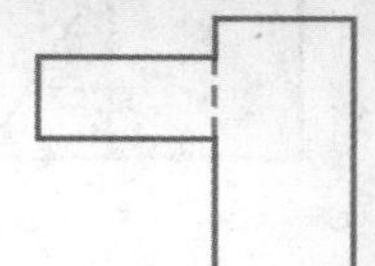

Draw a composite figure made of a rectangle and a right triangle on the graph paper below.

Essential Question

HOW does measurement help you solve problems in everyday life?

Vocabulary

composite figure

Common Core State Standards

Content Standards
6.G.1

Mathematical Practices
1, 2, 3, 4, 6, 7

Real-World Link

Pools The dimensions of the city pool are shown.

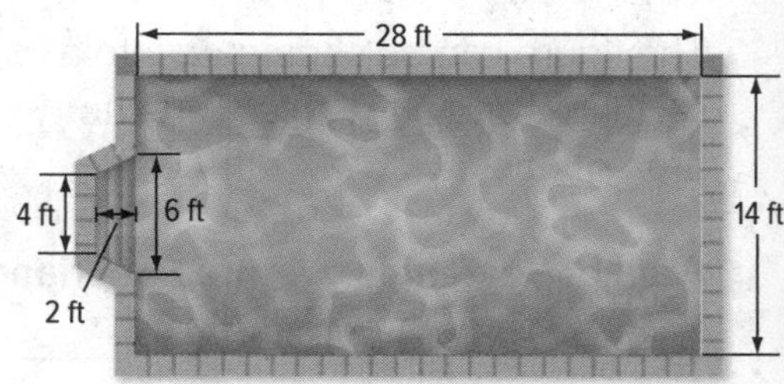

1. What two-dimensional figures are used to make the shape of the pool?

2. How could you determine the area of the pool floor?

Which **Mathematical Practices** did you use? Shade the circle(s) that applies.

① Persevere with Problems
② Reason Abstractly
③ Construct an Argument
④ Model with Mathematics
⑤ Use Math Tools
⑥ Attend to Precision
⑦ Make Use of Structure
⑧ Use Repeated Reasoning

Work Zone

Find the Area of a Composite Figure

You can decompose some trapezoids into a square and a triangle to find the area.

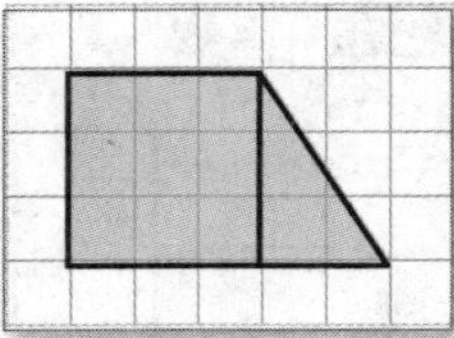

Area of Square

$A = \ell \cdot w$

$A = 3 \cdot 3$, or 9

Area of Triangle

$A = \frac{1}{2}bh$

$A = \frac{1}{2}(2)(3)$, or 3

Then add the area of the square and the area of the triangle to find the area of the trapezoid. The area of the trapezoid is 9 + 3 or 12 square units.

You can find the area of a composite figure using the same strategy. To find the area of a composite figure, separate it into figures with areas you know how to find. Then add those areas.

Example

1. Find the area of the figure at the right.

The figure can be separated into a rectangle and a triangle. Find the area of each.

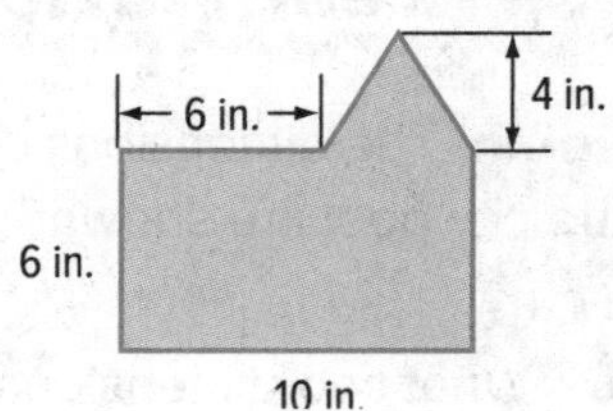

Area of Rectangle

6 in.

10 in.

$A = \ell w$

$A = 10 \cdot 6$ or 60

Area of Triangle

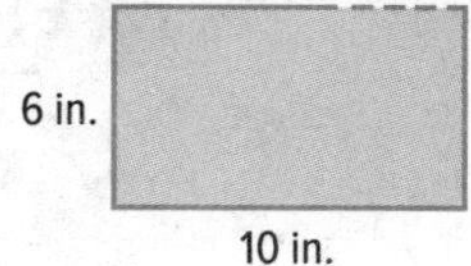

$A = \frac{1}{2}bh$

$A = \frac{1}{2}(4)(4)$ or 8

The base of the triangle is 10 − 6 or 4 inches.

The area is 60 + 8 or 68 square inches.

Got it? **Do these problems to find out.**

Find the area of each figure.

a.

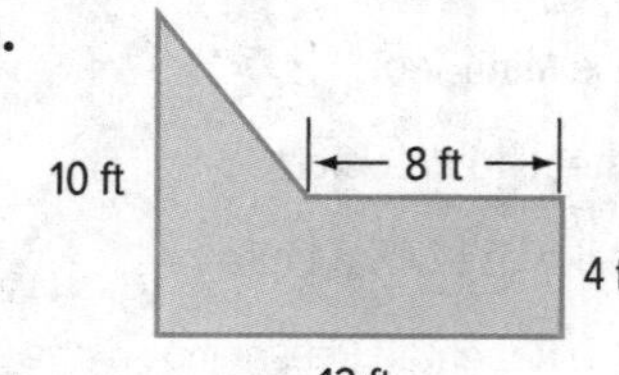

b.

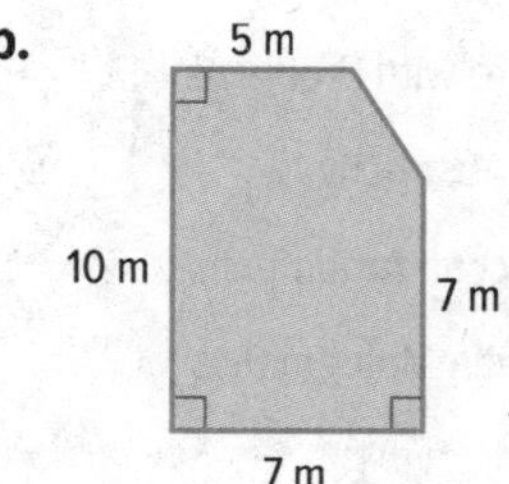

Show your work.

a. ________

b. ________

Example

2. Find the area of the pool's floor.

Separate the figure into a rectangle and a trapezoid.

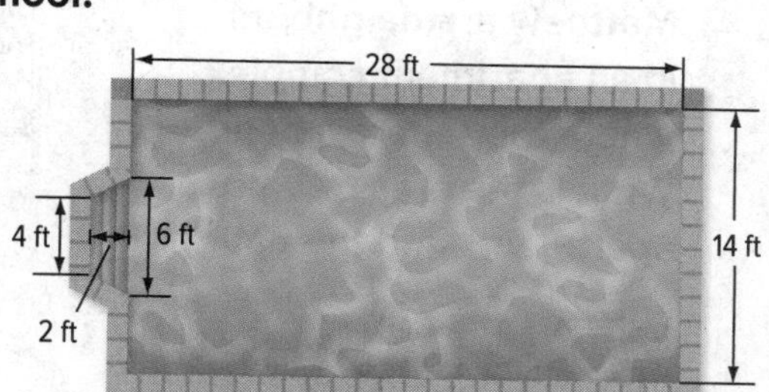

Rectangle: 28×14 or 392

Trapezoid: $\frac{1}{2}(2)(4 + 6)$ or 10

So, the area of the pool's floor is 392 + 10 or 402 square feet.

Got it? Do this problem to find out.

c.

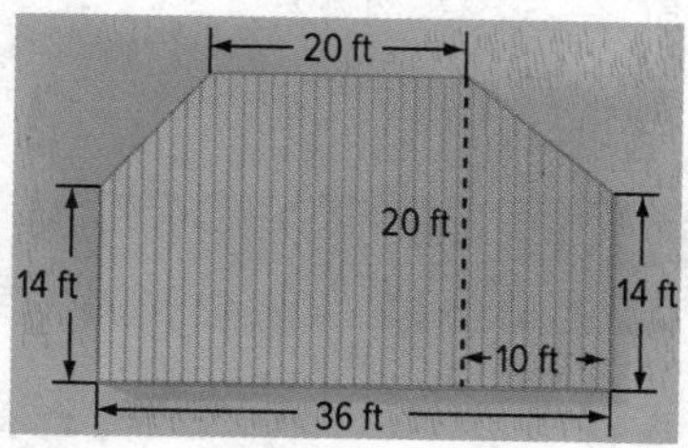

c. ____________

Find the Area of Overlapping Figures

To find the area of overlapping figures, decompose the figures.

Example

3. Find the area of the figure at the right.

Square: 12×12 or 144

Rectangle: 15×12 or 180

The sum of the areas: 144 + 180 or 324

Overlapping area: 6×7 or 42

Subtract the overlapping area. $324 - 42 = 282$

So, the area of the figure is 282 square centimeters.

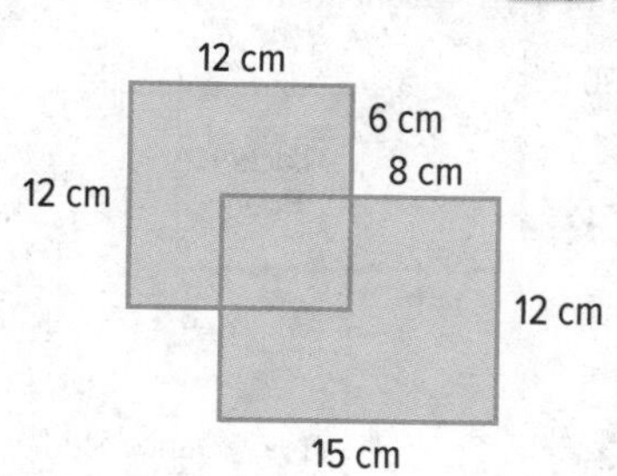

Be Precise

It is important not to count the area of the overlapping portion twice when finding the area of overlapping figures.

Got it? Do this problem to find out.

d.

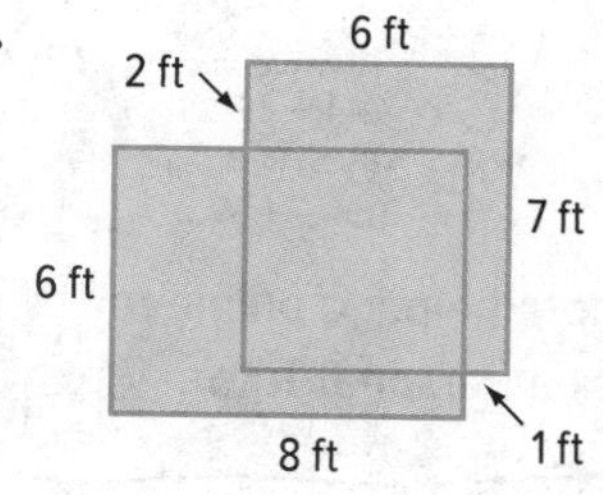

d. ____________

Example

4. Charlie and his brother Matthew are neighbors in an apartment complex where they share a patio. What is the area of both apartments and the patio?

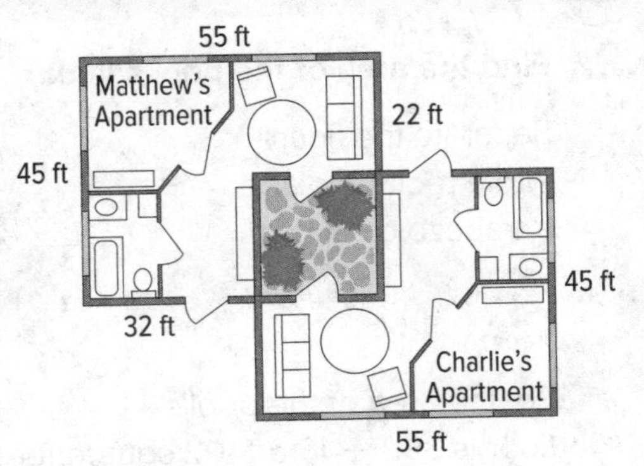

Each apartment:
55×45 or 2,475

The sum of the areas:
$2{,}475 + 2{,}475$ or 4,950

Patio: 23×23 or 529

Subtract the overlapping area. $4{,}950 - 529 = 4{,}421$

So, the total area is 4,421 square feet.

Guided Practice

1. The manager of an apartment complex will install new carpeting in a studio apartment. The floor plan is shown at the right. What is the total area that needs to be carpeted? (Examples 1 and 2)

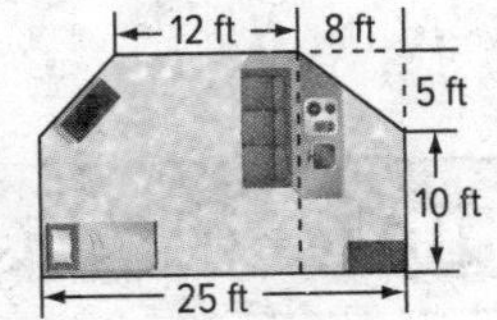

Show your work.

2. Finn Fitness has an entrance to the locker room from both the dance studio and the weight room. What is the total area of Finn Fitness? (Examples 3 and 4)

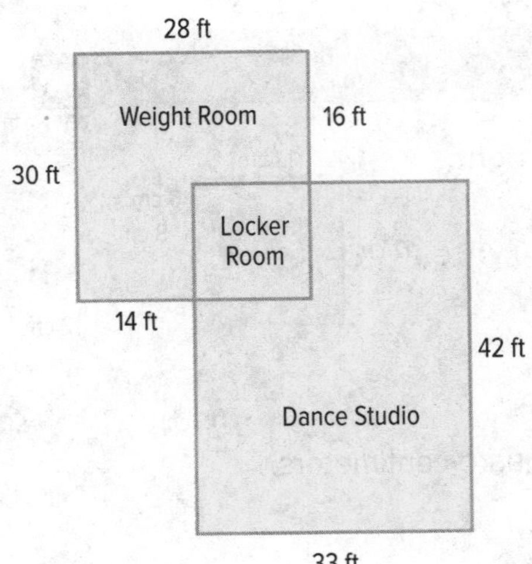

3. **Building on the Essential Question** How can you decompose figures to find area?

Rate Yourself!

Are you ready to move on? Shade the section that applies.

I have a few questions. | I'm ready to move on. | I have a lot of questions.

For more help, go online to access a Personal Tutor.

Tutor

Name ______ My Homework ______

Independent Practice

Go online for Step-by-Step Solutions

Find the area of each figure. Round to the nearest tenth if necessary. (Example 1)

1. ______

Show your work.

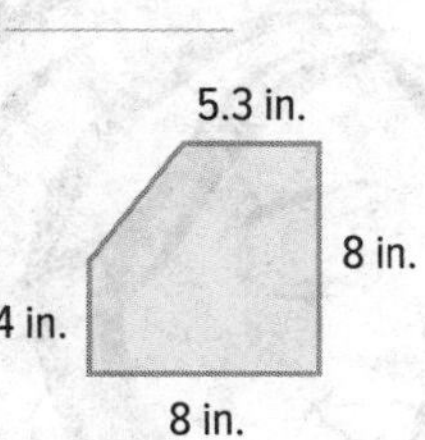

2. ______

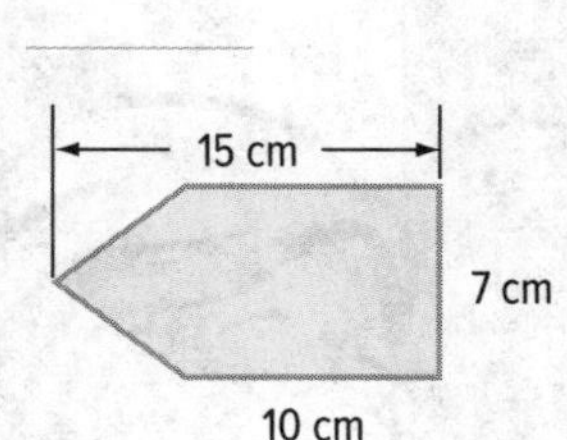

3. The floor plan of a kitchen is shown at the right. If the entire kitchen floor is to be tiled, how many square feet of tile are needed? (Example 2)

______

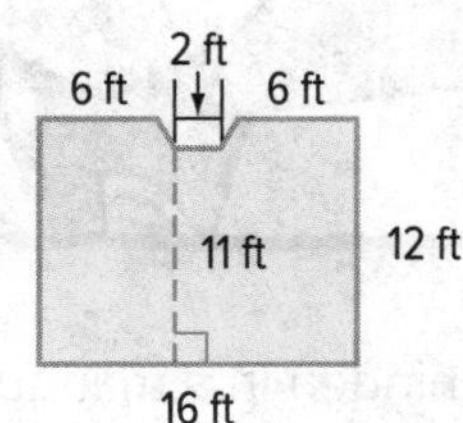

4. Ms. Friedman and Mrs. Elliot both teach sixth grade math. They share a storage closet. What is the total area of both rooms and the storage closet? (Examples 3 and 4)

______

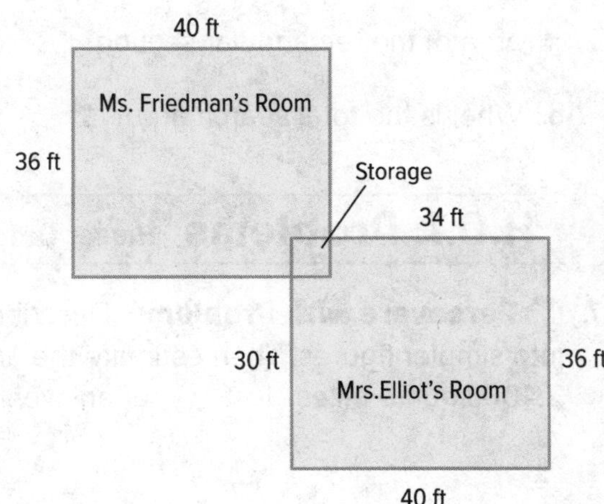

5. The diagram shows one side of a storage barn.

a. This side needs to be painted. Find the total area to be painted. ______

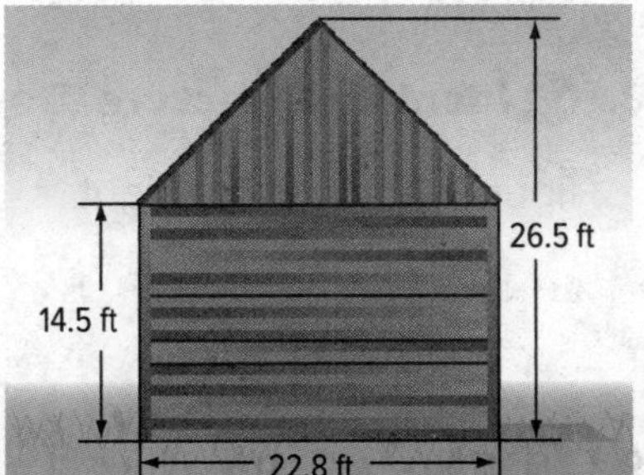

b. Each gallon of paint costs $20 and covers 350 square feet. Find the total cost to paint this side once. Justify your answer.

______

______

______

6. **MP Reason Abstractly** Refer to the graphic novel frame below for Exercises a–b.

a. The first clue was hidden in a triangular section of the park with an area of 600 square feet. The second clue was hidden in a rectangular section with a height of 30 feet and a width of 24 feet. What was the area of the rectangular section? ______

b. What is the total search area? ______

H.O.T. Problems Higher Order Thinking

7. **MP Persevere with Problems** Describe how to separate the figure into simpler figures. Then estimate the area. One square unit equals 2,400 square miles. Justify your answer.

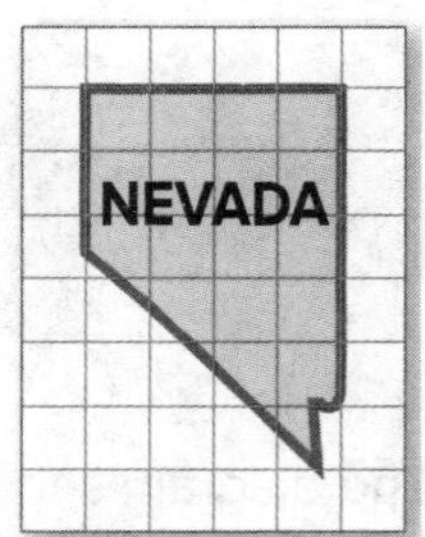

8. **MP Identify Structure** Describe how you would find the area of the figure shown at the right. ______

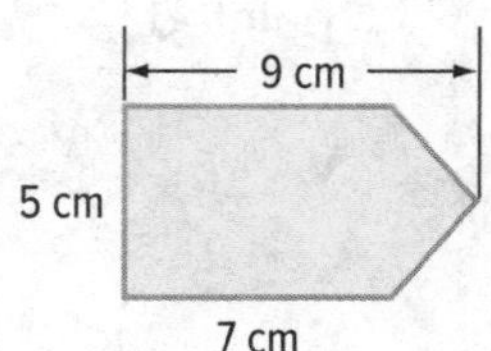

9. **MP Make a Conjecture** Refer to the composite figure at the right. Make a conjecture about how the area of the composite figure changes if each dimension given is doubled. Then test your conjecture by doubling the dimensions and finding the area.

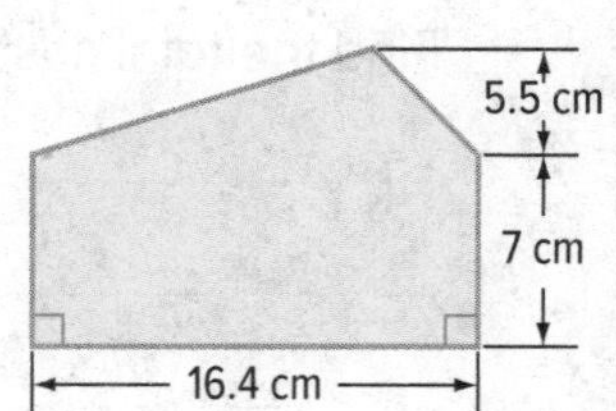

Extra Practice

Find the area of each figure. Round to the nearest tenth if necessary.

10. 69.5 ft^2

11.3 ft
8 ft
4.3 ft

$A = \frac{1}{2}(8)(11.3) = 45.2$

$A = \frac{1}{2}(4.3)(11.3) \approx 24.3$

$45.2 + 24.3 = 69.5$

11. ______

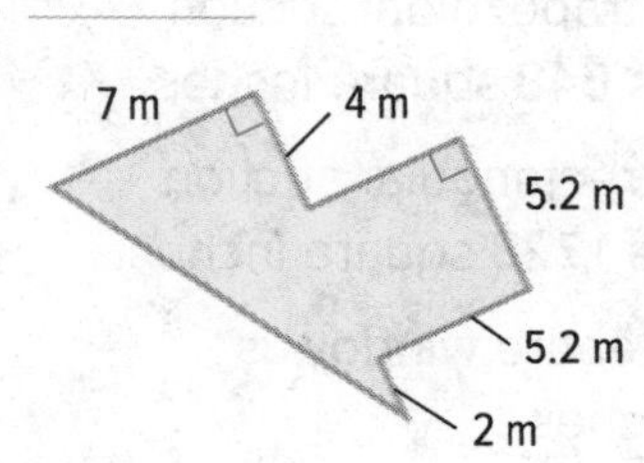

12. The diagram gives the dimensions of a swimming pool. If a cover is needed for the pool, what will be the approximate area of the cover? ______

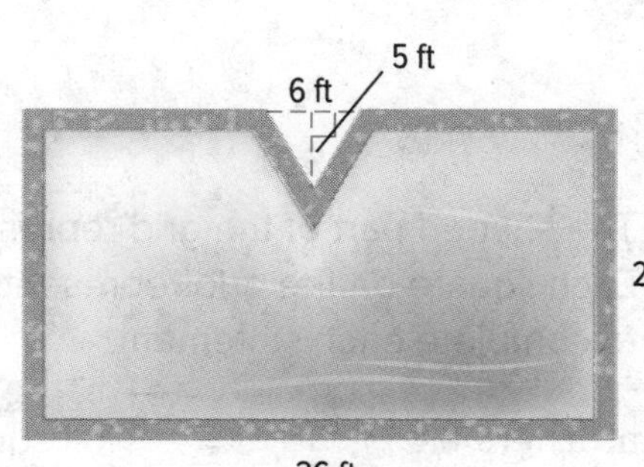

13. At the local zoo, the aquarium can be seen from the Reptile Room and the Amphibian Arena. What is the total area of both rooms and the aquarium?

______

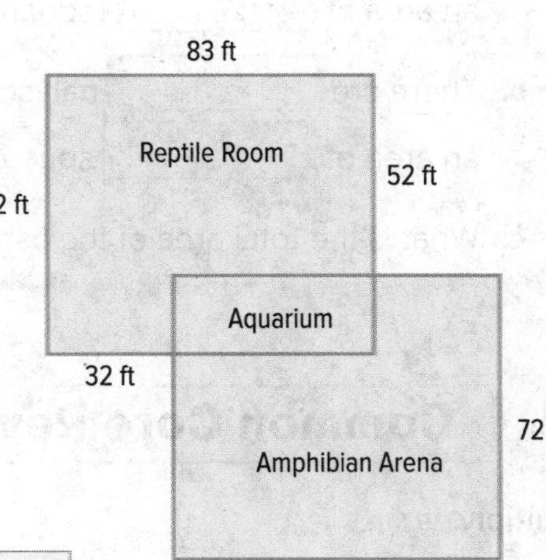

14. MP **Persevere with Problems** The diagram shows one wall of Sadie's living room.

a. This wall needs to be painted. Find the total area to be painted.

______

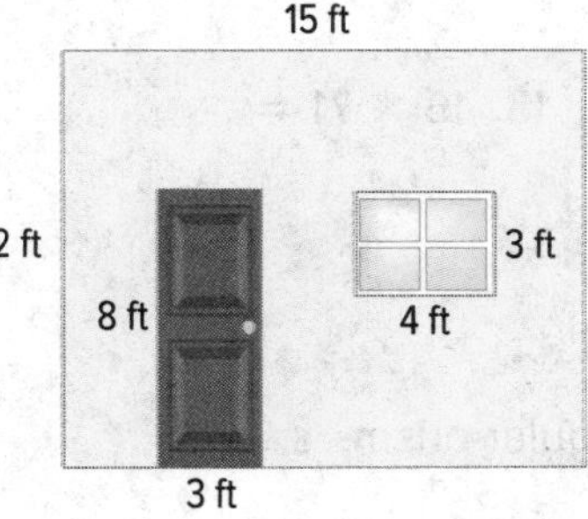

b. Each quart of paint costs $8 and covers 90 square feet. Find the total cost to paint this wall once. Justify your answer.

______

______

______

Power Up! Common Core Test Practice

15. A window has the dimensions shown. Determine if each statement is true or false.

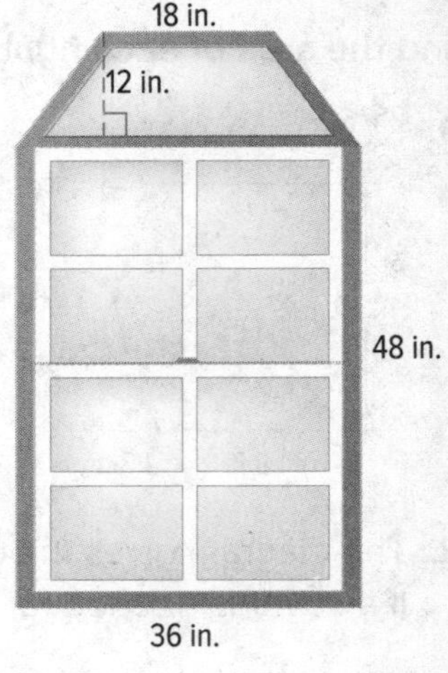

a. The area of the trapezoidal section of the window is 648 square inches. ☐ True ☐ False

b. The area of the rectangular section of the window is 1,728 square inches. ☐ True ☐ False

c. The area of the entire window is 2,376 square inches. ☐ True ☐ False

16. The shaded part of the grid represents the plans for a fish pond. Each square on the grid represents 5 square feet. Fill in the boxes to complete each statement.

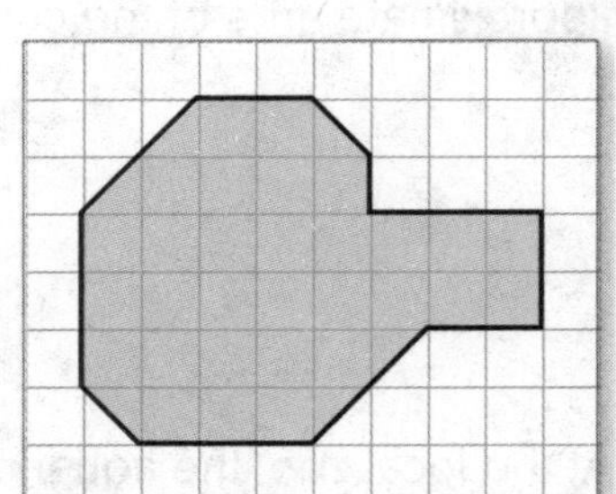

a. There are ☐ full squares in the pond. This represents an area of ☐ square feet.

b. There are ☐ half squares in the pond. This represents an area of ☐ square feet.

c. What is the total area of the fish pond? ☐

Common Core Review

Multiply. 5.NBT.5

17. 36 × 12 = ________

18. 15 × 71 = ________

19. 72 × 200 = ________

20. Find the volume of the rectangular prism. 5.MD.5b

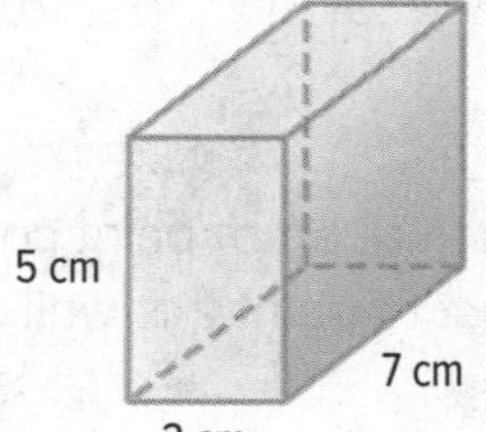

21. Hiking burns about 144 Calories each half hour. About how many Calories can be burned if someone hikes 3 days a week for an hour?

4.OA.3 ________________________________

21ST CENTURY CAREER in Community Planning

Parks and Recreation Planner

Do you enjoy thinking about how your community might look 10 years in the future? If so, a career in parks and recreation planning might be a perfect fit for you. Most planners are employed by local governments. They assess the best use for the land and create short and long term plans for various parks and recreation areas. They make recommendations based on the location of roads, schools, and residential areas. A parks and recreation planner uses mathematics, science, and computer software to complete their work.

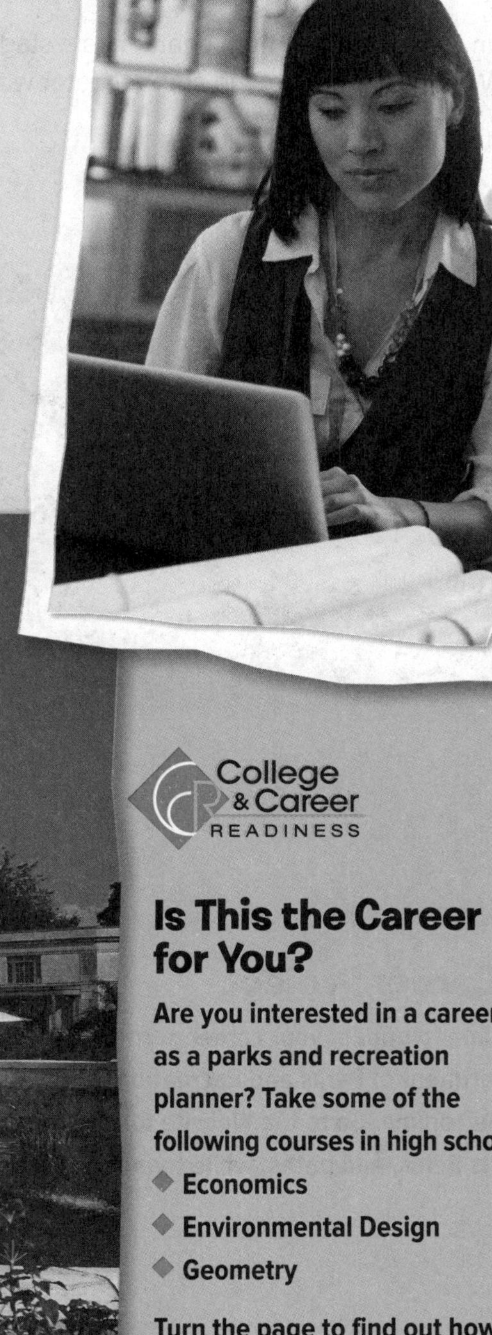

College & Career READINESS

Is This the Career for You?

Are you interested in a career as a parks and recreation planner? Take some of the following courses in high school.

- Economics
- Environmental Design
- Geometry

Turn the page to find out how math relates to a career in Community Planning.

You be the Parks and Recreation Planner!

For each problem, use the information in the designs.

1. What is the area of the playground in Design 2? ______

2. In Design 2, how much larger is the area of the soccer field than the area of the playground? ______

3. In Design 1, the amphitheater has a stage. What is the area of the amphitheater without the stage? ______

4. The cost of building the amphitheater including the stage is $225 a square yard. The budget provided to build the amphitheater is $65,000. Are they within budget? Explain.

______

______

______

______

Design 1

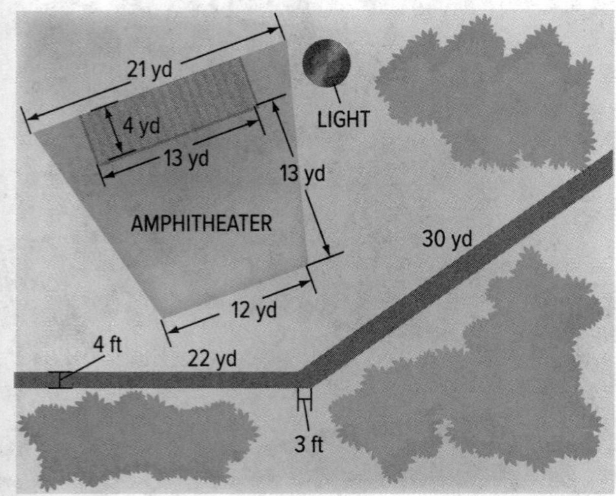

Design 2

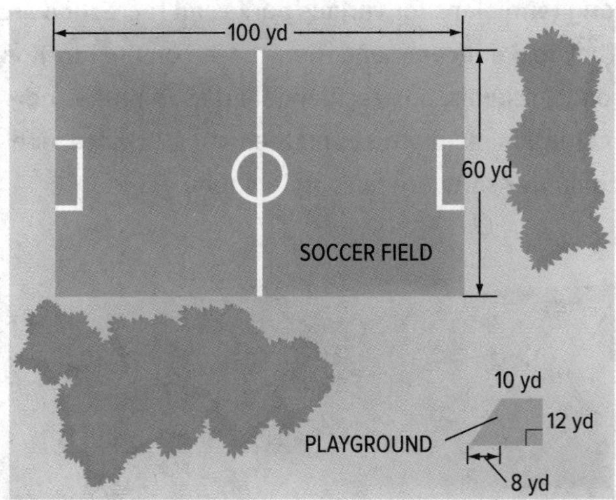

Career Project

It's time to update your career portfolio! The New York City Department of Parks and Recreation has a free "Park Planner Game" online. Go to the Website to create your own park with trees, sports fields, and paths, while trying to stay under budget.

______

______

______

______

______

What is something you really want to do in the next ten years?

- ______
- ______
- ______
- ______
- ______

Chapter Review

Vocabulary Check

Unscramble each of the clue words.

SEBA ☐☐☐☐

HGEHTI ☐☐☐☐☐☐

LYNPOOG ☐☐☐☐☐☐☐

LAEGARLAPLORM ☐☐☐☐☐☐☐☐☐☐☐☐☐

MHOBRUS ☐☐☐☐☐☐☐

NETRUGNOC ☐☐☐☐☐☐☐☐☐

POMECSOTI ERFUGI ☐☐☐☐☐☐☐☐☐ ☐☐☐☐☐☐

AROMLUF ☐☐☐☐☐☐☐

Complete each sentence using one of the unscrambled words above.

1. A ________________ is a simple closed figure formed by three or more straight line segments.
2. The shortest distance from the base of a parallelogram to its opposite side is the ________________.
3. A ________________ is a quadrilateral with opposite sides parallel and opposite sides congruent.
4. Any side of a parallelogram is a ________________.
5. A parallelogram with four congruent sides is a ________________.
6. If two shapes have the same measure they are ________________.
7. A figure made of triangles, quadrilaterals, and other two dimensional figures is a ________________.
8. A ________________ is an equation that shows a relationship among certain quantities.

Key Concept Check

Use Your FOLDABLES

Use your Foldable to help review the chapter.

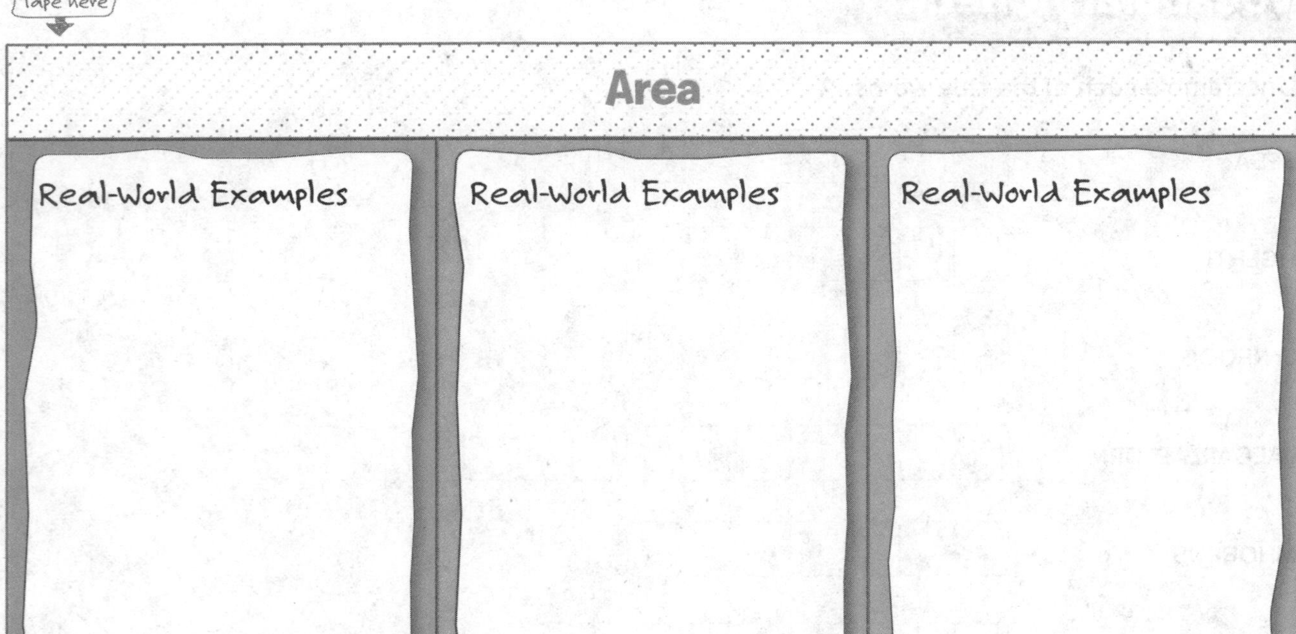

Got it?

Match each expression with correct steps used to find the area of the trapezoid.

7 m
9.8 m
12 m

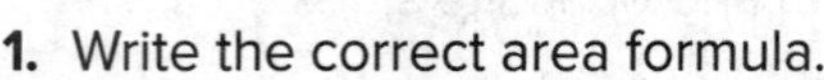

1. Write the correct area formula.
2. Replace h with 9.8.
3. Replace b_1 with 7 and replace b_2 with 12.
4. Add.
5. Multiply.

a. $A = \frac{1}{2}(9.8)(b_1 + b_2)$

b. $A = \frac{1}{2}bh$

c. $A = \frac{1}{2}(9.8)(19)$

d. $A = \frac{1}{2}h(b_1 + b_2)$

e. $A = 93.1$

f. $A = \frac{1}{2}(9.8)(7 + 12)$

Family Land

The Hernandez family owns a plot of land as shown.

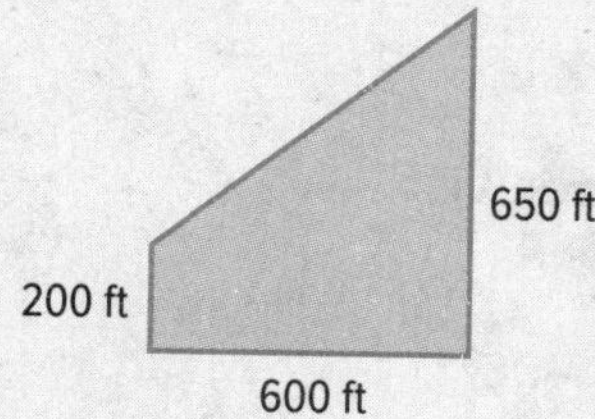

Write your answers on another piece of paper. Show all of your work to receive full credit.

Part A

A house on the land measures 45 feet by 38 feet, a wooded area covers 118 feet by 60 feet, and the front yard is 78 feet by 40 feet. The remainder of the land is farmed. How many acres of the land are farmed? Round to the nearest tenth. Explain your answer. (*Hint:* 1 acre = 43,560 square feet)

Part B

It costs \$0.05 per square foot to seed crops on 4 acres. The remainder of the farm land is seeded with grass for grazing animals. It costs \$0.03 a square foot to seed the grass. What is the total cost to seed the farm land?

Part C

Graph the vertices of the plot of land on a coordinate plane. The vertices are: (4, 3), (9, 3), (9, 8.4), and (4, 5). There is also a road that extends from (4, 3) westward, where it intersects with the main highway at (0, 3). Determine the length of the road from the main highway to the western edge of the plot of land. Also, determine the length of the road from the main highway to the eastern edge of the plot of land. Explain how you got your answers.

Reflect

Answering the Essential Question

Use what you learned about area to complete the graphic organizer. List several real-world examples for each figure.

Essential Question

HOW does measurement help you solve problems in everyday life?

area

parallelograms

triangles

irregular figures

trapezoids

composite figures

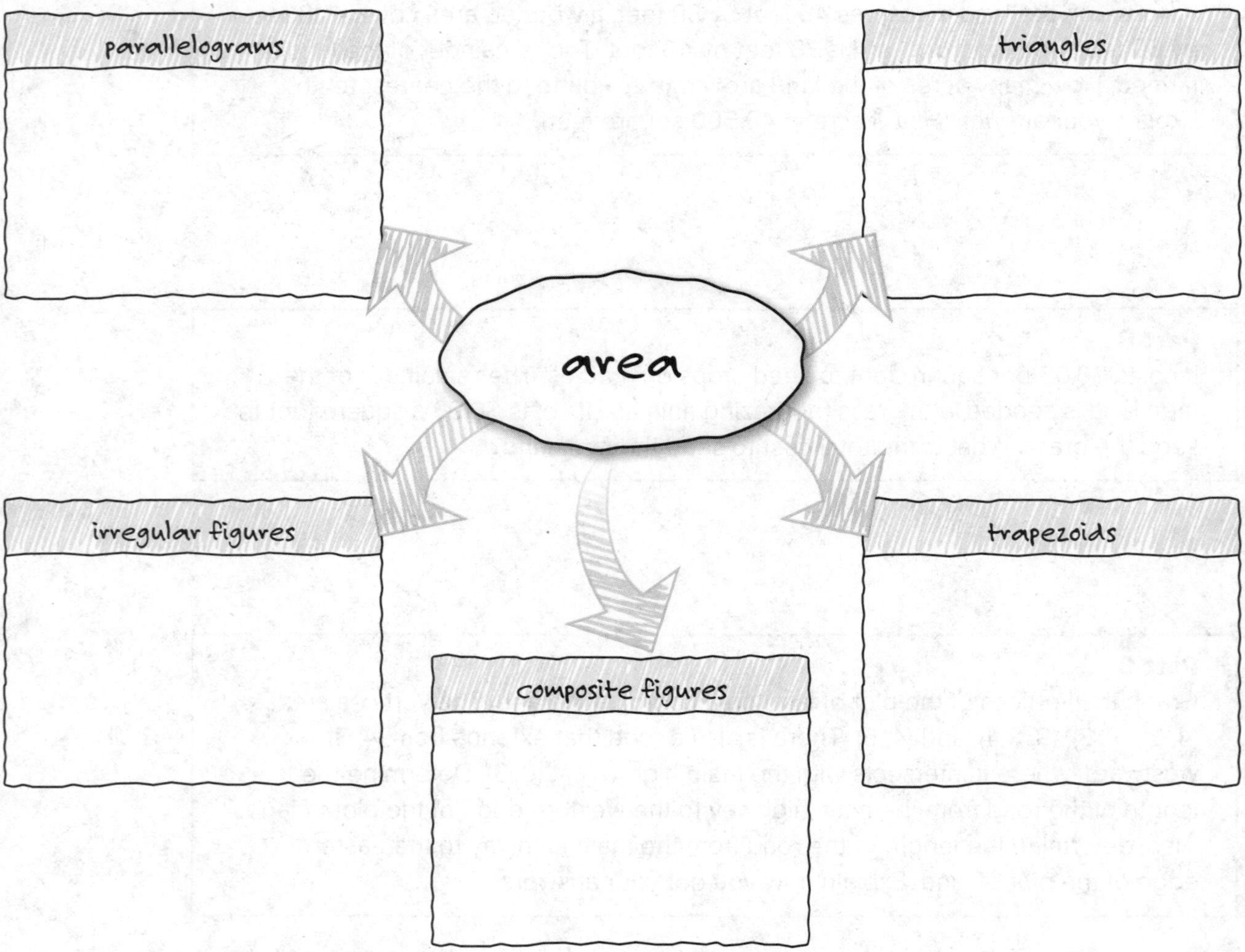

Answer the Essential Question. HOW does measurement help you solve problems in everyday life?

Chapter 10

Volume and Surface Area

Essential Question

HOW is shape important when measuring a figure?

Common Core State Standards

Content Standards
6.G.2, 6.G.4

MP Mathematical Practices
1, 2, 3, 4, 5, 6, 7, 8

Math in the Real World

Aquariums Two-dimensional figures have area, while three-dimensional figures have volume and surface area.

A 20-gallon aquarium can measure 24 inches wide, 12 inches deep, and 16 inches high. What is the area of the bottom of the aquarium?

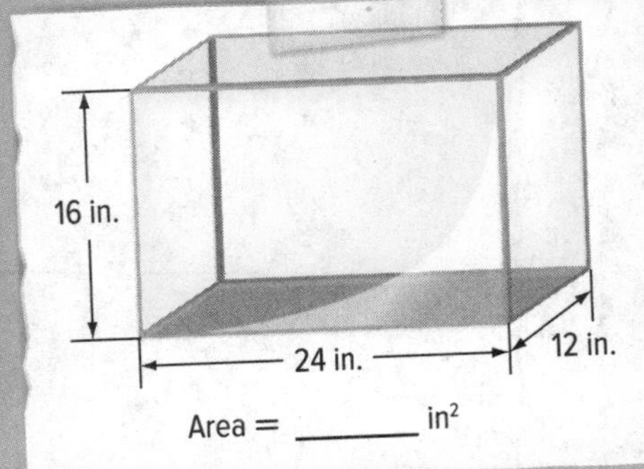

FOLDABLES® Study Organizer

Cut out the Foldable on page FL11 of this book.

Place your Foldable on page 794.

Use the Foldable throughout this chapter to help you learn about volume and surface area.

What Tools Do You Need?

Vocabulary

base
cubic units
lateral face
prism
pyramid
rectangular prism
slant height
surface area
three-dimensional figure
triangular prism
vertex
volume

Review Vocabulary

Using a graphic organizer can help you to remember important vocabulary terms. Fill in the graphic organizer below for the phrase *two-dimensional figure*.

two-dimensional figure

Definition

Real-World Examples

Drawings

The number of square units needed to cover the surface of a closed figure is the ______________.

What Do You Already Know?

List three things you already know about volume and surface area in the first section. Then list three things you would like to learn about volume and surface area in the second section.

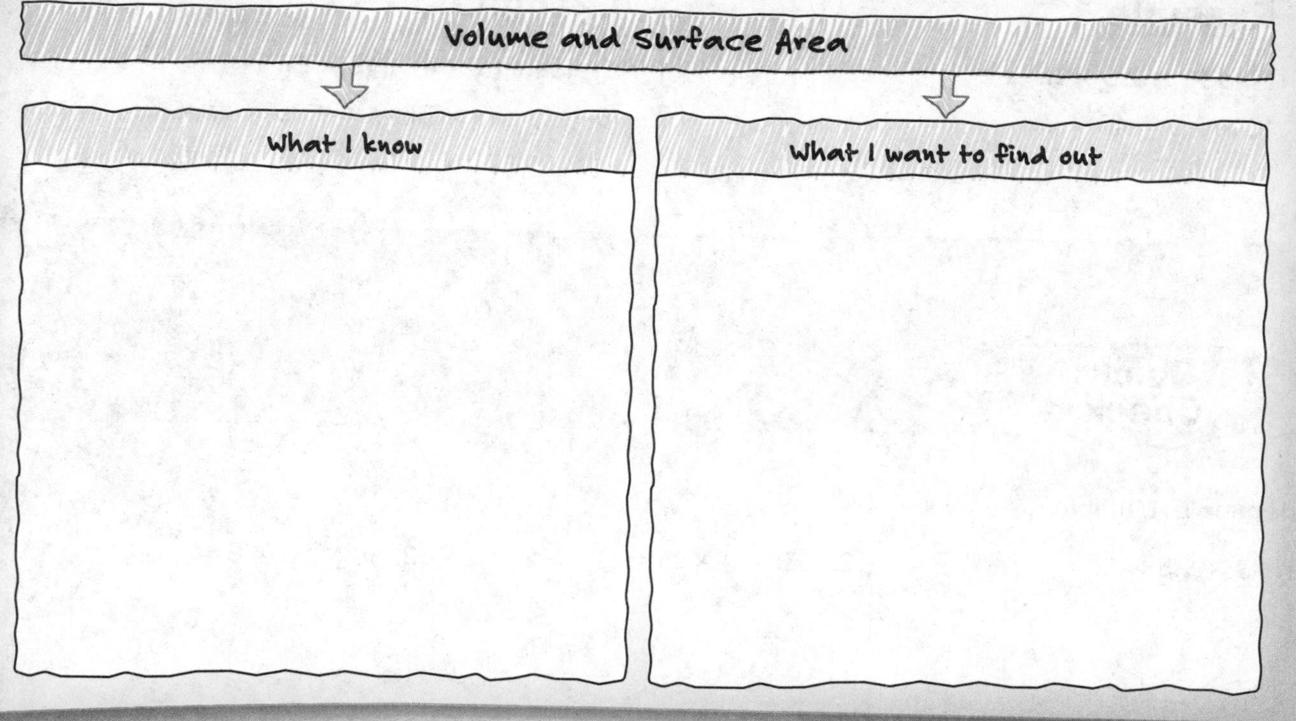

When Will You Use This?

Here are a few examples of how three-dimensional figures are used in the real world.

Activity 1 When you go to see a movie, do you buy popcorn? If so, do you base your purchase on the cost of the popcorn or the size of the container?

Activity 2 Go online at **connectED.mcgraw-hill.com** to read the graphic novel ***Popcorn Problem***. What are the dimensions of each popcorn container?

Try the Quick Check below.
Or, take the Online Readiness Quiz.

Quick Review

Common Core Review 5.OA.1, 5.NBT.5, 5.NBT.7

Example 1

Find $16 \times 2.5 \times 8$.

$16 \times 2.5 = 40$ Multiply 16 and 2.5.
$40 \times 8 = 320$ Multiply the product by 8.

Example 2

Evaluate $(6 \times 4) + (3 \times 5)$.

$(6 \times 4) + (3 \times 5) = 24 + 15$ Multiply.
$= 39$ Add.

Quick Check

Decimals **Multiply.**

1. $3 \times 5.5 \times 13 =$ ______
2. $9.8 \times 4 \times 15 =$ ______
3. $18 \times 1.6 \times 6 =$ ______

4. Dante earned $7.25 for each hour he worked. If he worked 8 hours a week for 4 weeks, how much did he earn?

______

Numerical Expressions **Evaluate each expression.**

5. $(3 \times 12) + (4 \times 2) =$ ______
6. $(9 \times 7) + (6 \times 4) =$ ______
7. $(15 \times 3) + (8 \times 7) =$ ______

Which problems did you answer correctly in the Quick Check? Shade those exercise numbers below.

      7

Inquiry Lab

Volume of Rectangular Prisms

HOW can you use models to find volume?

Content Standards 6.G.2

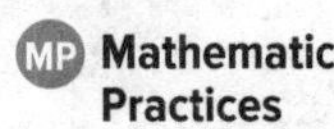

Mathematical Practices 1, 3, 4

Desmond is purchasing a storage cabinet. The cabinet is 2 feet wide, 3 feet long, and 6 feet tall. What is the volume of the cabinet?

Hands-On Activity 1

You can use centimeter cubes to find the *volume* of the cabinet. Volume is the amount of space inside a three-dimensional figure. Volume is measured in *cubic units*. Each cube of your model represents 1 cubic foot.

Step 1 Build a model that is 2 cubes wide, 3 cubes long, and 6 cubes tall.

Step 2 Count the number of cubes used to build the model. The model uses ☐ cubes.

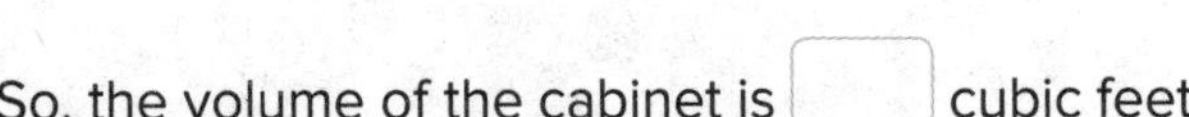

So, the volume of the cabinet is ☐ cubic feet.

Find the product of the dimensions of the cabinet.

☐ × ☐ × ☐ = ☐

The product is ______________ as the volume.

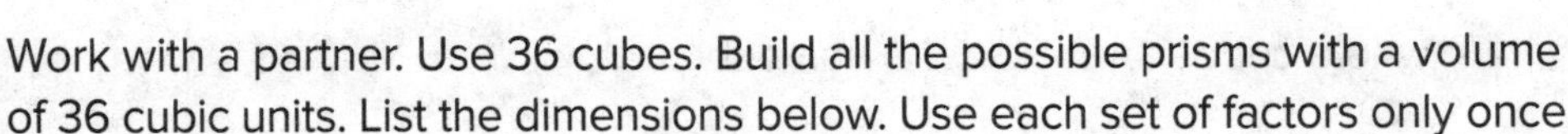

Work with a partner. Use 36 cubes. Build all the possible prisms with a volume of 36 cubic units. List the dimensions below. Use each set of factors only once.

☐ × ☐ × ☐ = 36 ☐ × ☐ × ☐ = 36

☐ × ☐ × ☐ = 36 ☐ × ☐ × ☐ = 36

☐ × ☐ × ☐ = 36 ☐ × ☐ × ☐ = 36

☐ × ☐ × ☐ = 36 ☐ × ☐ × ☐ = 36

Hands-On Activity 2

You can find the volume of rectangular prisms with fractional side lengths.

Step 1 The model to the right is ________ cubes long, ________ cube wide, and ________ cube tall.

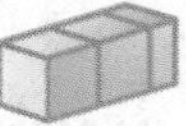

Step 2 Count the number of cubes used to build the model.

The model uses ________ cubes.

So, the volume of the model is ________ cubic feet.

Compare the product of the dimensions of the prism with its volume.

________ × ________ × ________ = ________

They are ________.

Hands-On Activity 3

You can use cubes of candy to find the volume of rectangular prisms with fractional sides.

Step 1 Cut one piece of candy into two halves.

Step 2 Make a model that is $2\frac{1}{2}$ cubes long, 2 cubes wide, and 1 cube tall. Draw a picture of your model.

Step 3 Count the number of cubes used to build the model. The model uses ________ whole cubes and ________ half-cubes.

Two halves equal one whole. So, a total of ________ cubes were used.

So, the volume of the prism is ________ cubic units.

Compare the product of the dimensions of the prism with its volume.

________ × ________ × ________ = ________

They are ________.

Investigate

MP **Model with Mathematics** **Work with a partner. Use models to determine the volume of each prism. Draw a diagram of each model in the space provided.**

Show your work.

1. length: 1
height: 1
width: 1
volume: ______

2. length: 2
height: 4
width: 1
volume: ______

3. length: 3
height: 4
width: 2
volume: ______

4. length: $\frac{1}{2}$
height: 1
width: 1
volume: ______

5. length: $2\frac{1}{2}$
height: 4
width: 1
volume: ______

6. length: $3\frac{1}{2}$
height: 2
width: 2
volume: ______

Analyze and Reflect

Work with a partner to complete the table. Use models, if needed. The first one is done for you.

	Prism	Height (units)	Length (units)	Width (units)	Volume (units^3)
	A	6	3	2	36
7.	B	$2\frac{1}{2}$	$1\frac{1}{2}$	2	
8.	C	5	$1\frac{1}{2}$	2	
9.	D	2	5	$1\frac{1}{2}$	
10.	E	5	3	4	

11. Compare the dimensions for prism *C* to the dimensions of prism *D*. Compare the volume of the two prisms. What do you notice?

12. The length and width of prisms *B* and *C* are equal. Compare the height of the two prisms. How does the change in height affect the change in volume?

13. Compare the dimensions for prism *B* to the dimensions of prism *E*. Compare the volume of the two prisms. What do you notice?

14. MP **Reason Inductively** Describe the relationship between the number of cubes needed and the dimensions of the prism.

Create

15. MP **Model with Mathematics** Write a real-world problem that involves volume of rectangular prisms. Include the dimensions and the volume of the rectangular prism in your response.

16. Inquiry HOW can you use models to find volume?

Lesson 1

Volume of Rectangular Prisms

Vocabulary Start-Up

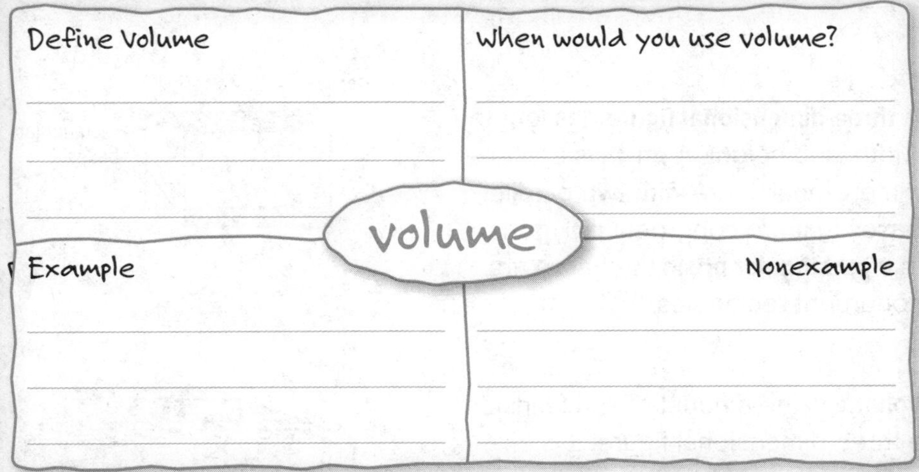

Essential Question

HOW is shape important when measuring a figure?

Vocabulary

three-dimensional figure
prism
rectangular prism
volume
cubic units

Common Core State Standards

Content Standards
6.G.2

MP Mathematical Practices
1, 3, 4, 5, 6, 7

Real-World Link

Aquarium The dimensions of an aquarium are shown.

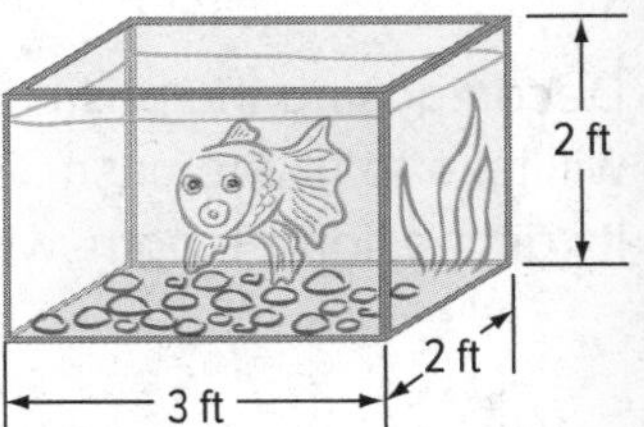

1. What is the area of the base of the aquarium? ________

2. What is the height of the aquarium? ________

3. Fill in the blanks to find the volume.

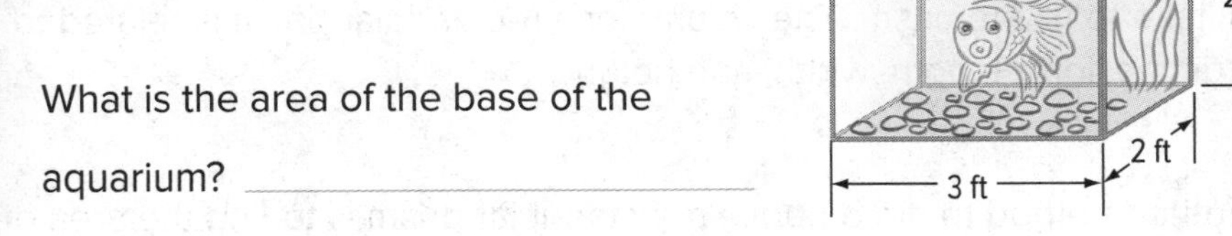

______ × ______ × ______ = 12 ft^3
length width height

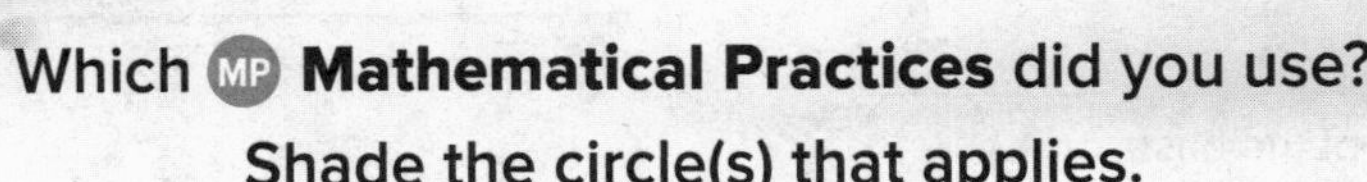

Which MP **Mathematical Practices** did you use?
Shade the circle(s) that applies.

① Persevere with Problems
② Reason Abstractly
③ Construct an Argument
④ Model with Mathematics
⑤ Use Math Tools
⑥ Attend to Precision
⑦ Make Use of Structure
⑧ Use Repeated Reasoning

Key Concept

Volume of a Rectangular Prism

Words The volume V of a rectangular prism is the product of its length ℓ, width w, and height h.

Symbols $V = \ell wh$ or $V = Bh$

Model

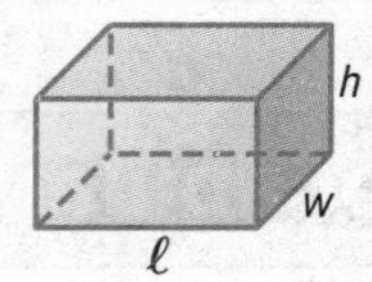

Work Zone

A **three-dimensional figure** has length, width, and height. A **prism** is a three-dimensional figure with two parallel bases that are congruent polygons. In a **rectangular prism**,the bases are congruent rectangles.

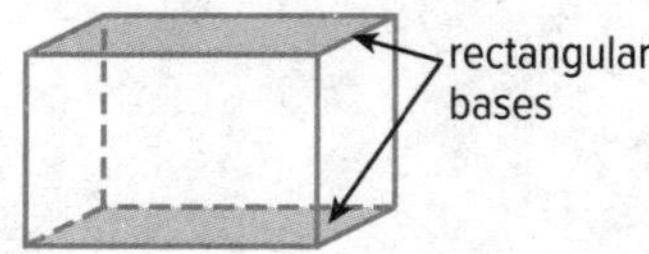

Volume is the amount of space inside a three-dimensional figure. It is measured in **cubic units**, which can be written using abbreviations and an exponent of 3, such as units^3 or in^3.

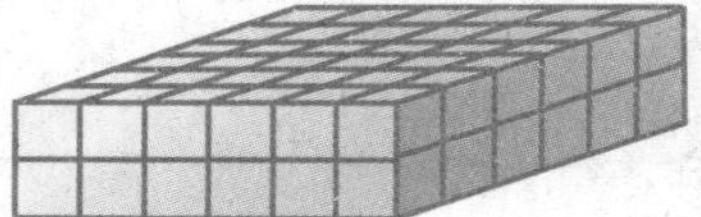

Cubes

Cubes are special rectangular prisms. All three side lengths are equal. So, the volume of a cube can be written using the formula $V = s^3$.

Decomposing the prism tells you the number of cubes of a given size it will take to fill the prism. The volume of a rectangular prism is related to its dimensions, length, width, and height.

Another method to decompose a rectangular prism is to find the area of the base (B) and multiply it by the height (h).

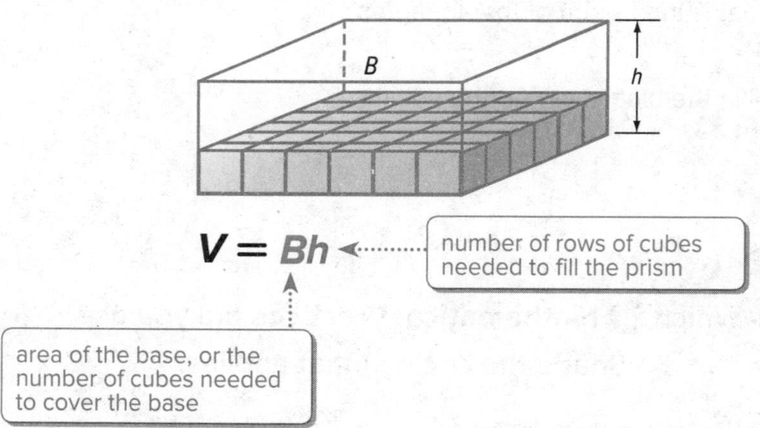

$V = Bh$

number of rows of cubes needed to fill the prism

area of the base, or the number of cubes needed to cover the base

Example

1. Find the volume of the rectangular prism.

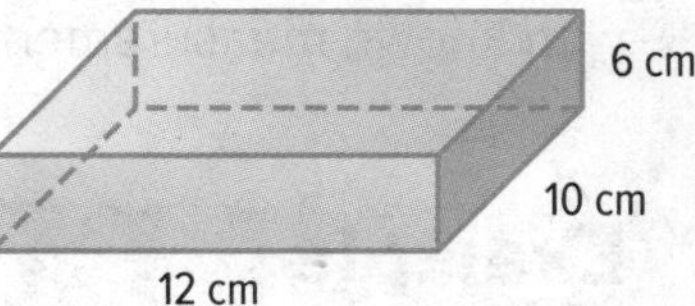

B, or the area of the base, is 10×12 or 120 square centimeters. The height of the prism is 6 centimeters.

$V = Bh$	Volume of rectangular prism
$V = 120 \times 6$	Replace B with 120 and h with 6.
$V = 720$	Multiply.

The volume is 720 cubic centimeters.

Decomposing Figures

You can think of the volume of the prism as consisting of six congruent slices. Each slice contains the area of the base, 120 cm², multiplied by a height of 1 cm.

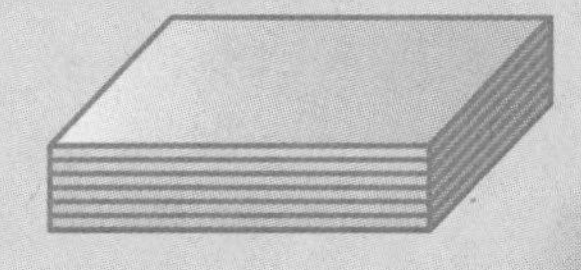

Got it? Do these problems to find out.

a.

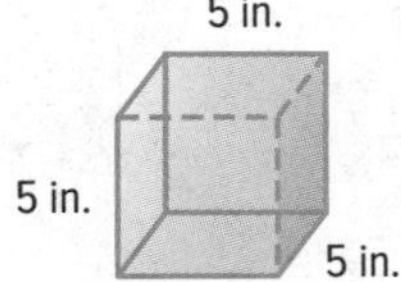

b.

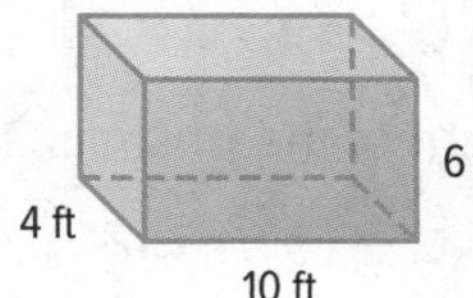

a. ______

b. ______

Example

2. A cereal box has the dimensions shown. What is the volume of the cereal box?

8 in.
Cereal
$12\frac{1}{2}$ in.
$3\frac{1}{4}$ in.

Estimate $10 \times 3 \times 10 = 300$

$V = \ell wh$	Volume of a rectangular prism.
$V = 8 \times 3\frac{1}{4} \times 12\frac{1}{2}$	Replace ℓ with 8, w with $3\frac{1}{4}$, and h with $12\frac{1}{2}$.
$V = \frac{\overset{1}{\cancel{8}}}{1} \times \frac{13}{\underset{1}{\cancel{4}}} \times \frac{25}{\underset{1}{\cancel{2}}}$	Write as improper fractions. Then divide out common factors.
$V = \frac{325}{1}$ or 325	Multiply.

The volume of the cereal box is 325 cubic inches.

Check for Reasonableness $325 \approx 300$ ✓

Got it? Do this problem to find out.

c. Find the volume of a container that measures 4 inches long, 5 inches high, and $8\frac{1}{2}$ inches wide.

c. ______

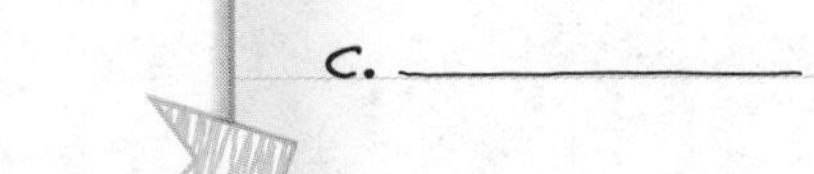

Find Missing Dimensions

To find missing dimensions of a rectangular prism, replace the variables with known measurements. Then solve for the unknown measurement.

Example

3. Find the missing dimension of the prism.

4 m, 6 m, h

$V = 84 \text{ m}^3$

$V = \ell wh$	Volume of rectangular prism
$84 = 6 \times 4 \times h$	Replace V with 84, ℓ with 6, and w with 4.
$84 = 24h$	Multiply.
$\frac{84}{24} = \frac{24h}{24}$	Divide each side by 24.
$3.5 = h$	Simplify.

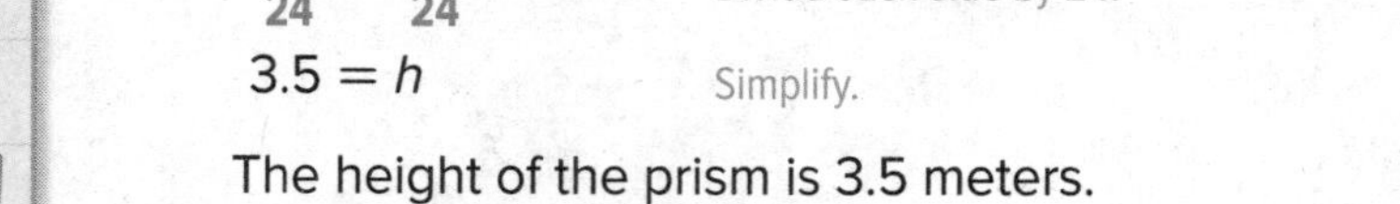

The height of the prism is 3.5 meters.

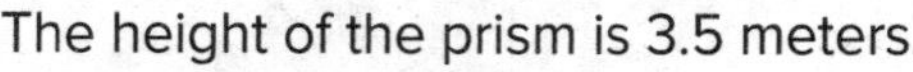

Check $6 \times 4 \times 3.5 = 84$ ✓

Got it? **Do this problem to find out.**

d. ____________

d. $V = 94.5 \text{ km}^3$, $\ell = 7$ km, $h = 3$ km, $w = ?$

Guided Practice

1. A rectangular kitchen sink is 25.25 inches long, 19.75 inches wide, and 10 inches deep. Find the amount of water that can be contained in the sink. (Examples 1 and 2) ____________

2. Find the missing dimension of a rectangular prism with a volume of 126 cubic centimeters, a width of $7\frac{7}{8}$ centimeters, and a height of 2 centimeters. (Example 3) ____________

3. **Building on the Essential Question** Why can you use either the formula $V = \ell wh$ or $V = Bh$ to find the volume of a rectangular prism?

____

Independent Practice

Go online for Step-by-Step Solutions 

Find the volume of each prism. (Example 1)

1. ________

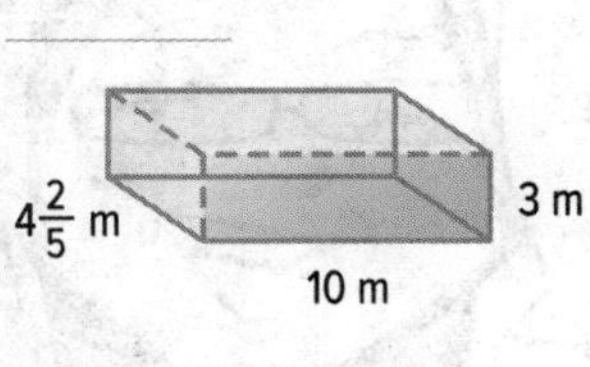

Show your work.

2. ________

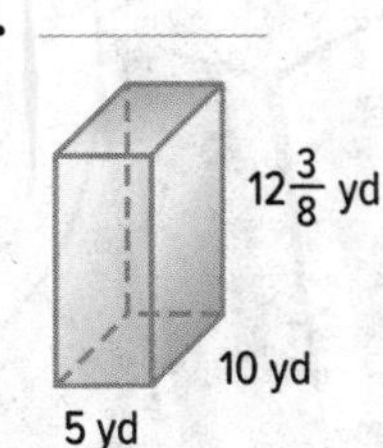

3. ________

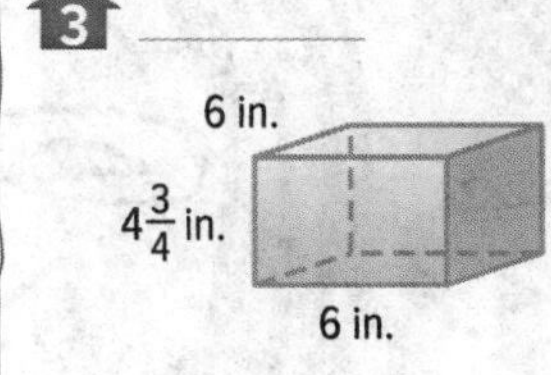

4. A fishing tackle box is 13 inches long, 6 inches wide, and $2\frac{1}{2}$ inches high. What is the volume of the tackle box? (Example 2)

____

5. Find the length of a rectangular prism having a volume of 2,830.5 cubic meters, width of 18.5 meters, and height of 9 meters. (Example 3)

____

Find the missing dimension of each prism. (Example 3)

6. ________

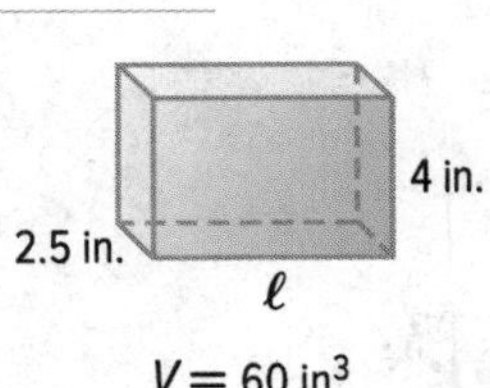

$V = 60$ in^3

7. ________

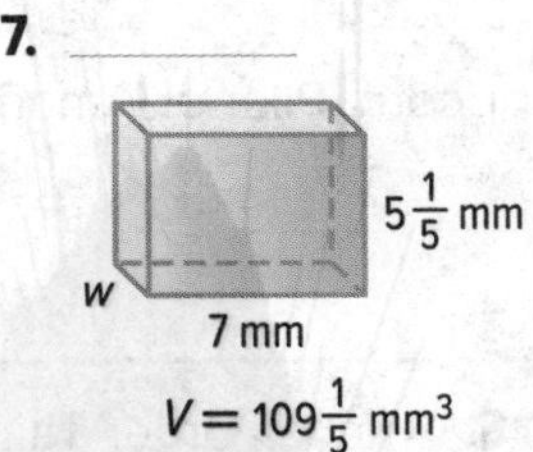

$V = 109\frac{1}{5}$ mm^3

8. MP **Be Precise** In Japan, farmers have created watermelons in the shape of rectangular prisms. Find the volume of a prism-shaped watermelon in cubic inches if its length is 10 inches, its width is $\frac{2}{3}$ foot, and its height is 9 inches.

____

9. The glass container shown is filled to a height of 2.25 inches.

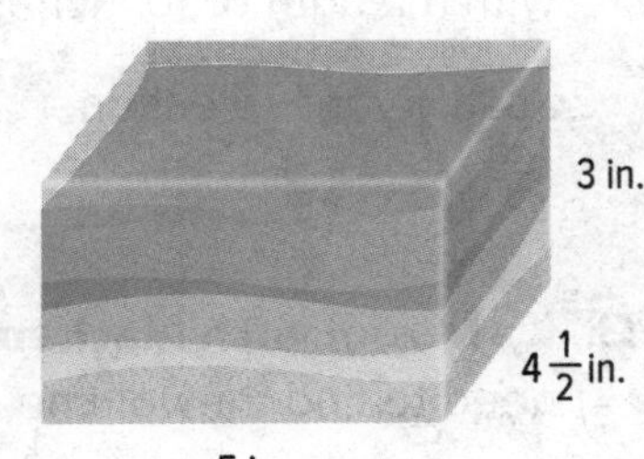

a. How much sand is currently in the container?

____

b. How much more sand could the container hold before it overflows? ________________

c. What percent of the container is filled with sand? ________________

10. **MP Identify Structure** Refer to the graphic novel frame below for Exercises a–c.

a. Pilar chose the box on the left. If it is 8 inches long, 8 inches wide, and 8 inches tall, what is the volume of Pilar's box?

______

b. Amanda chose the box on the right. If it is 8 inches long, 6 inches wide, and 10 inches tall, what is the volume of Amanda's box?

______

c. Who received more popcorn, Pilar or Amanda? How much more?

______

H.O.T. Problems Higher Order Thinking

11. **MP Persevere with Problems** Refer to the prism at the right. If all the dimensions of the prism doubled, would the volume double? Explain your reasoning.

______

______

12. **MP Justify Conclusions** Which has the greater volume: a prism with a length of 5 inches, a width of 4 inches, and a height of 10 inches, or a prism with a length of 10 inches, a width of 5 inches, and a height of 4 inches?

Justify your selection. ______

______

13. **MP Model with Mathematics** Write a real-world problem in which you need to find the volume of a right rectangular prism. Solve your problem.

______

______

Name ______________________ My Homework ______________________

Extra Practice

Find the volume of each prism.

14. 105.84 cm³

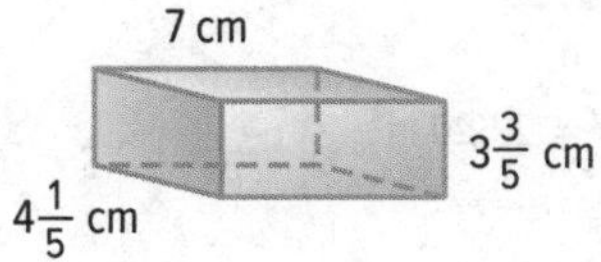

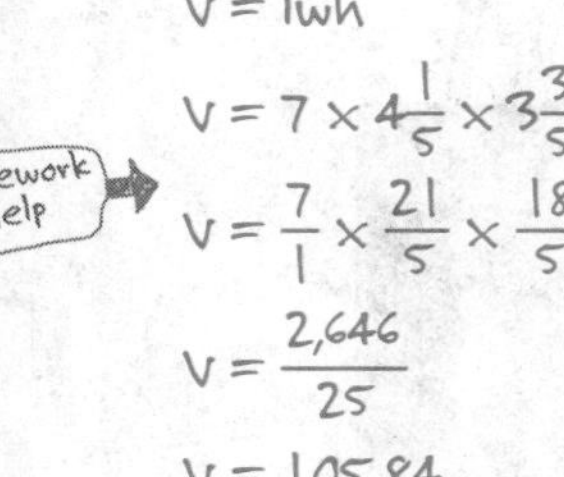

15. ________

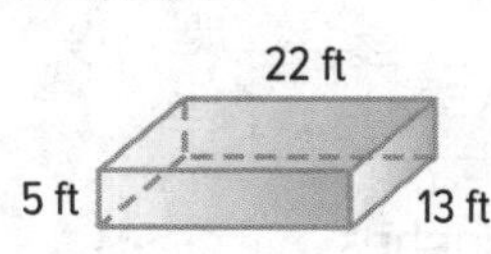

16. ________

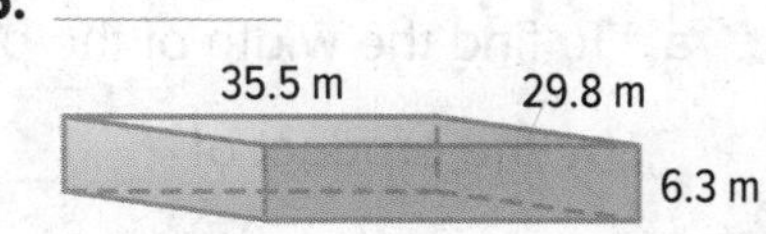

17. Find the volume of the pet carrier shown at the right.

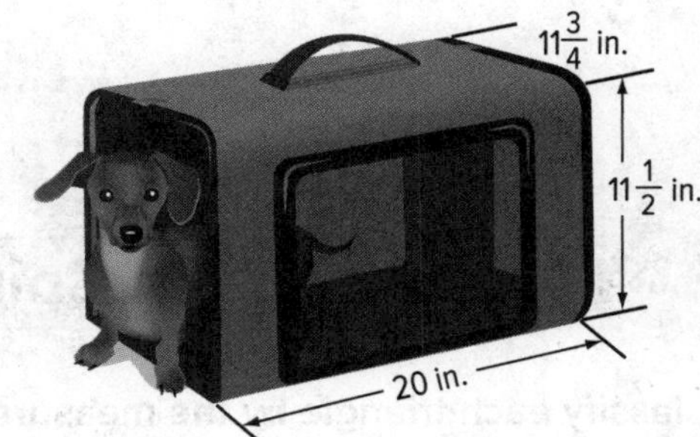

____

18. What is the width of a rectangular prism with a length of 13 feet, volume of 11,232 cubic feet, and height of 36 feet?

____

19. The Palo Duro Canyon is 120 miles long, as much as 20 miles wide, and has a maximum depth of more than 0.15 mile. What is the approximate volume of this canyon?

____

20. MP **Use Math Tools** Use the table at the right.

a. What is the approximate volume of the small truck?

____

b. The Davis family is moving, and they estimate that they will need a truck with about 1,250 cubic feet. Which truck would be best for them to rent?

____

c. About how many cubic feet greater is the volume of the Mega Moving Truck than the 2-bedroom moving truck?

____

Inside Dimensions of Moving Trucks

Truck	Length (ft)	Width (ft)	Height (ft)
Van	10	$6\frac{1}{2}$	6
Small Truck	$11\frac{1}{13}$	$7\frac{5}{12}$	$6\frac{3}{4}$
2-Bedroom Moving Truck	$14\frac{1}{2}$	$7\frac{7}{12}$	$7\frac{1}{6}$
3-Bedroom Moving Truck	$20\frac{5}{6}$	$7\frac{1}{2}$	$8\frac{1}{12}$
Mega Moving Truck	$22\frac{1}{4}$	$7\frac{7}{12}$	$8\frac{5}{12}$

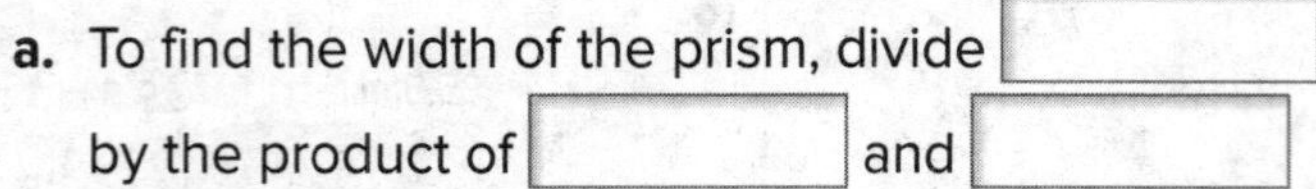

Power Up! Common Core Test Practice

21. The volume of the rectangular prism shown is 2,520 cubic inches. Fill in the boxes to complete each statement.

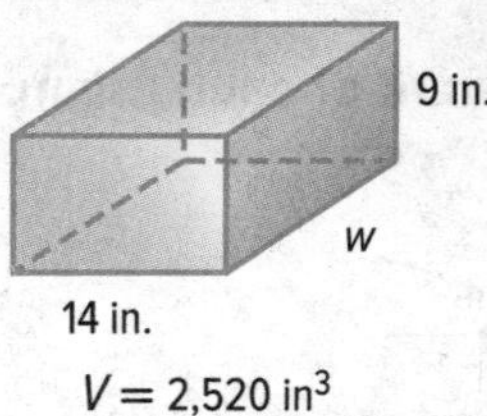

a. To find the width of the prism, divide ______ by the product of ______ and ______.

b. The width of the prism is ______ inches.

22. A pet carrier company is creating a new size carrier. It has a length of 27 centimeters, a width of 7 centimeters, and a volume of 6,426 cubic centimeters. Select values to complete the formula below to find the height h of the carrier.

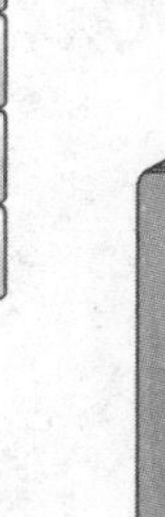

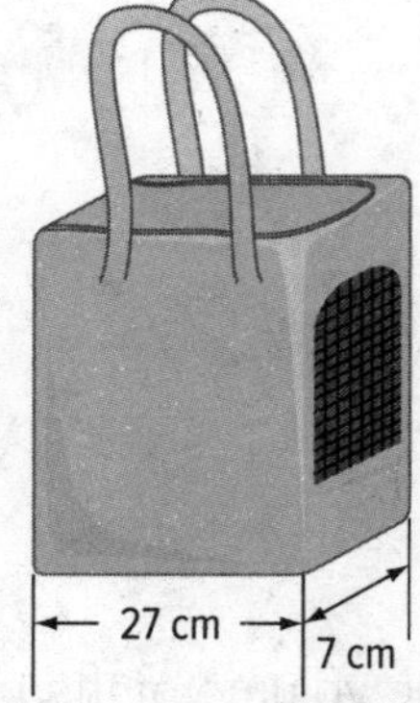

______ = ______ × ______ × ______

What is the height of the pet carrier? ______

Common Core Spiral Review

Classify each triangle by the measure of the angles. 5.G.4

23. ______

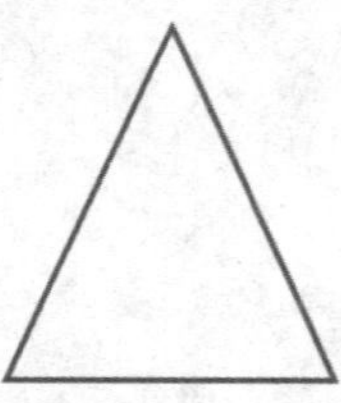

24. ______

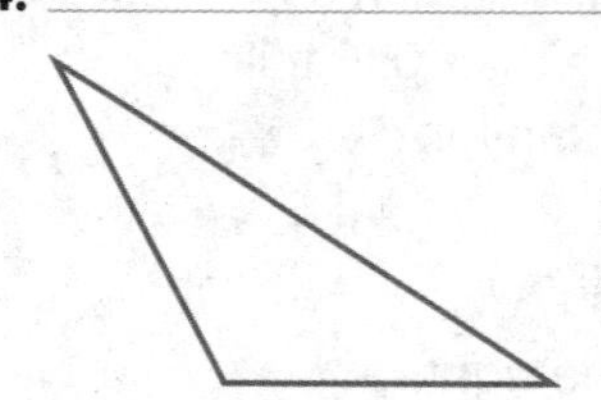

25. ______

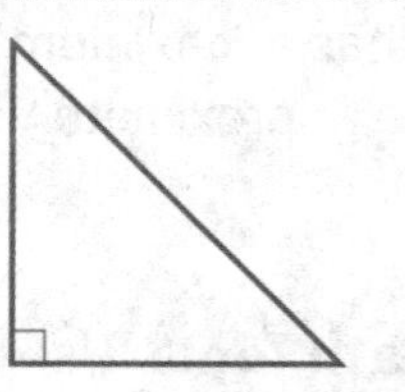

26. Draw the next figure in the pattern below. 4.OA.5, 4.G.2

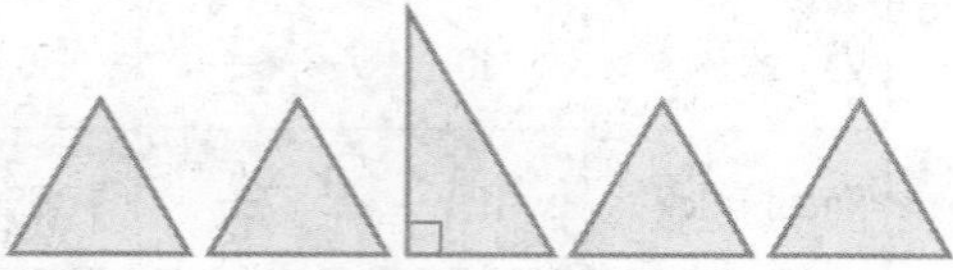

27. Triangles are often used in designing bridges. Classify the triangle shown by the measure of its sides. Explain. 5.G.4

______

______

Lesson 2

Volume of Triangular Prisms

Real-World Link

Camping Ari has a pup tent like the one shown. The opening of the tent has a base and a height of 6 feet. The length of the tent is 8 feet.

What is the area of the front triangular face? ________

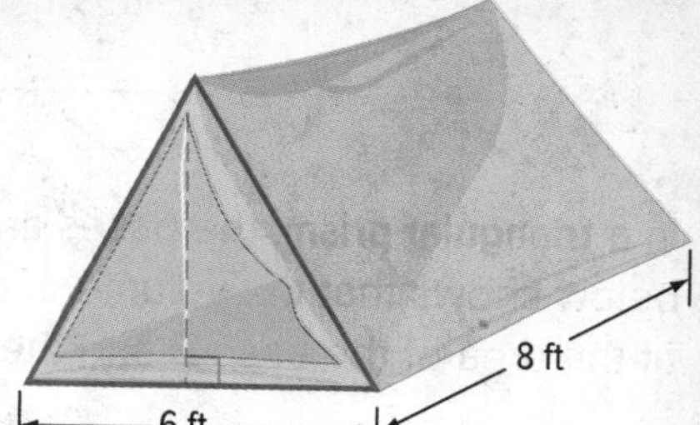

On a piece of grid paper, draw a right triangle with a base and height of 4 units as shown.

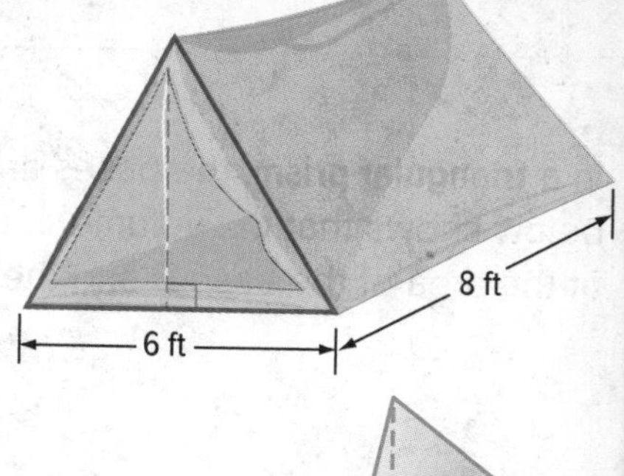

1. What is the area of the triangle?

____

2. Suppose you cover the triangle with cubes the size of one square on the grid paper. How many cubes would you use? (*Hint*: You can cut and reassemble the cubes.) ________

3. How many cubes would you use if you had 4 layers? ________

4. MP **Make a Conjecture** Write a formula to find the volume of a triangular prism. ________

Essential Question

HOW is shape important when measuring a figure?

Vocabulary

triangular prism

Common Core State Standards

Content Standards
Extension of 6.G.2

MP **Mathematical Practices**
1, 3, 4, 6, 8

Which MP **Mathematical Practices** did you use? Shade the circle(s) that applies.

① Persevere with Problems
② Reason Abstractly
③ Construct an Argument
④ Model with Mathematics
⑤ Use Math Tools
⑥ Attend to Precision
⑦ Make Use of Structure
⑧ Use Repeated Reasoning

Key Concept — Volume of a Triangular Prism

Work Zone

Words	The volume V of a triangular prism is the area of the base B times the height h.	**Model**
Symbols	$V = Bh$, where B is the area of the base	

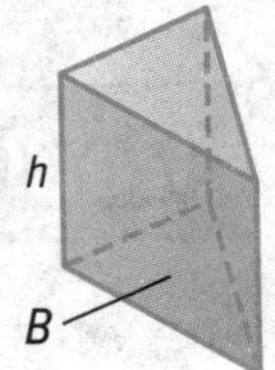

In a **triangular prism**, the bases are congruent triangles. The diagram below shows that the volume of a triangular prism is also the product of the area of the base B and the height h of the prism.

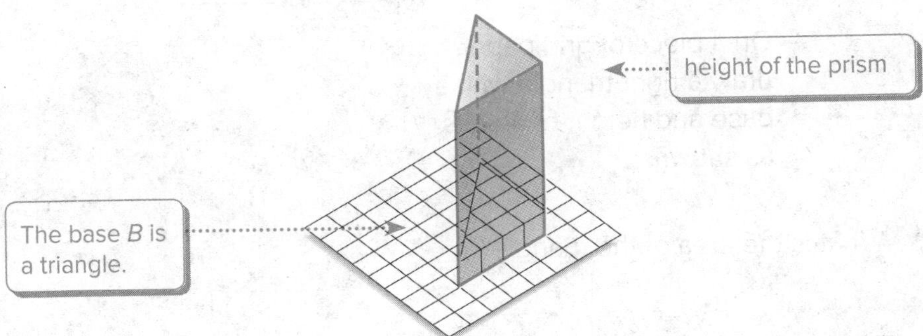

Example

1. Find the volume of the triangular prism.

8 m
13 m
10 m
12.8 m

The area of the triangle is $\frac{1}{2} \cdot 8 \cdot 10$, so B is $\frac{1}{2} \cdot 8 \cdot 10$.

$V = \mathbf{Bh}$	Volume of a prism
$V = \left(\frac{1}{2} \cdot 8 \cdot 10\right)h$	Replace B with $\frac{1}{2} \cdot 8 \cdot 10$.
$V = \left(\frac{1}{2} \cdot 8 \cdot 10\right)13$	Replace h with 13, the height of the prism.
$V = 520$	Multiply.

The volume is 520 cubic meters or 520 m^3.

Base

Before finding the volume of a triangular prism, identify the base. In Exercise b, the base is not on the "bottom." The base is one of the parallel faces.

Show your work.

Got it? Do these problems to find out.

a.

4 m
3.5 m
7 m
4 m
4 m

b.

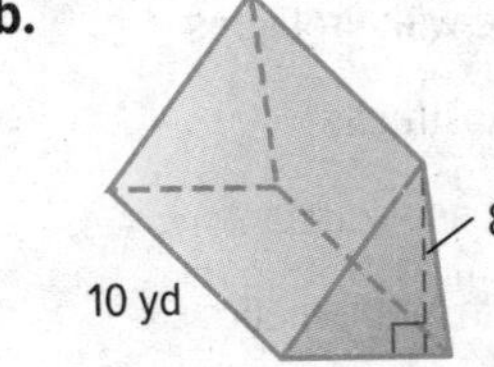

a. ________

b. ________

Example

Tutor

2. A large skateboard ramp is shown. Find the volume of the triangular prism.

The base is a triangle with a base length of 10 feet and a height of 7 feet. The height of the prism is 4 feet.

$V = Bh$ — Volume of a prism

$V = \left(\frac{1}{2} \cdot 10 \cdot 7\right)h$ — Replace B with $\frac{1}{2} \cdot 10 \cdot 7$.

$V = \left(\frac{1}{2} \cdot 10 \cdot 7\right)4$ — Replace h with 4, the height of the prism.

$V = 140$ — Multiply.

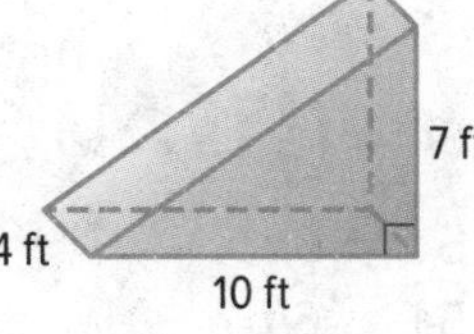

The volume is 140 cubic feet or 140 ft^3.

Got it? **Do this problem to find out.**

c. Find the volume of a triangular prism-shaped model with a base of 32 square centimeters and a height of 6 centimeters.

c. ____________

Find Missing Dimensions

To find missing dimensions of a triangular prism, replace the variables with known measurements. Then solve for the unknown measurement.

Example

Tutor

3. Find the height of the triangular prism.

$V = Bh$ — Volume of a triangular prism

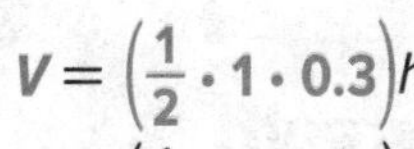

$V = \left(\frac{1}{2} \cdot 1 \cdot 0.3\right)h$ — Replace B with $\frac{1}{2} \cdot 1 \cdot 0.3$.

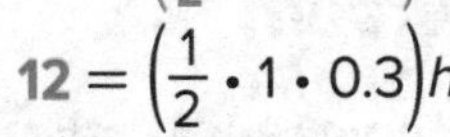

$12 = \left(\frac{1}{2} \cdot 1 \cdot 0.3\right)h$ — Replace V with 12.

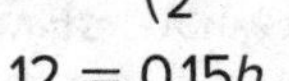

$12 = 0.15h$ — Multiply.

$\frac{12}{0.15} = \frac{0.15h}{0.15}$ — Divide each side by 0.15.

$80 = h$ — Simplify.

1 cm
0.3 cm
h
$V = 12\ cm^3$

So, the height of the prism is 80 cm.

Got it? **Do this problem to find out.**

Find the missing dimension of the triangular prism.

d. $V = 55\ km^3$, base length = 2 km, base height = 5 km, $h = ?$

d. ____________

Example

4. Dwane bought a cheese wedge for his March Madness party. The cheese wedge has the dimensions shown. The volume of the cheese wedge is 54 cubic inches. What is the height of the cheese wedge?

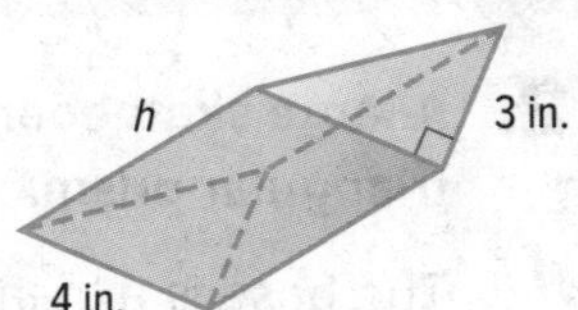

$V = Bh$	Volume of a triangular prism
$54 = \left(\frac{1}{2} \cdot 3 \cdot 4\right)h$	Replace V with 54, and B with $\frac{1}{2} \cdot 3 \cdot 4$.
$54 = 6h$	Multiply.
$\frac{54}{6} = \frac{6h}{6}$	Divide each side by 6.
$9 = h$	Simplify.

So, the height of the cheese wedge is 9 inches.

Guided Practice

Find the volume of each prism. Round to the nearest tenth if necessary. (Example 1)

1. ______

Show your work.

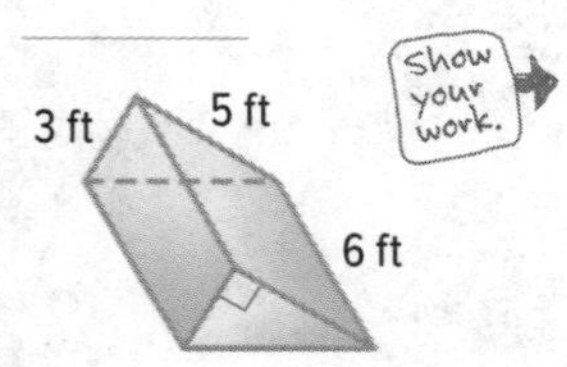

2. ______

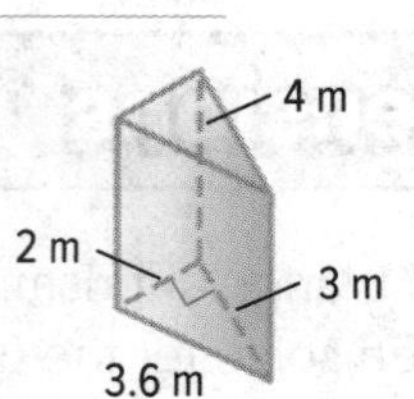

3. Dirk has a triangular-shaped piece of cheesecake in his lunch. Find the volume of the piece of cheesecake. (Example 2)

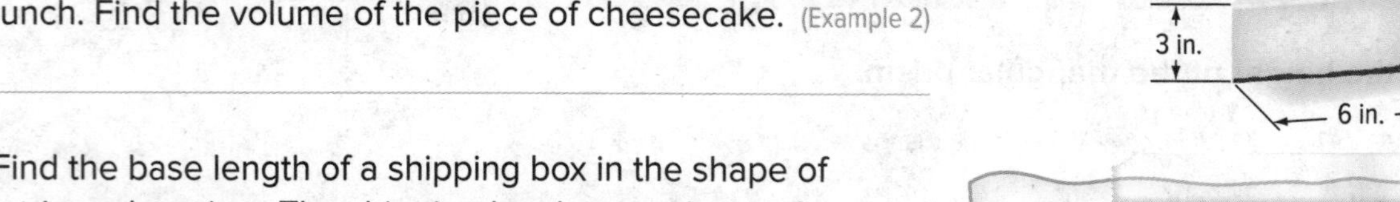

4. Find the base length of a shipping box in the shape of a triangular prism. The shipping box has a volume of 276 cubic feet, a base height of 6.9 feet, and a height of 10 feet. (Examples 3 and 4)

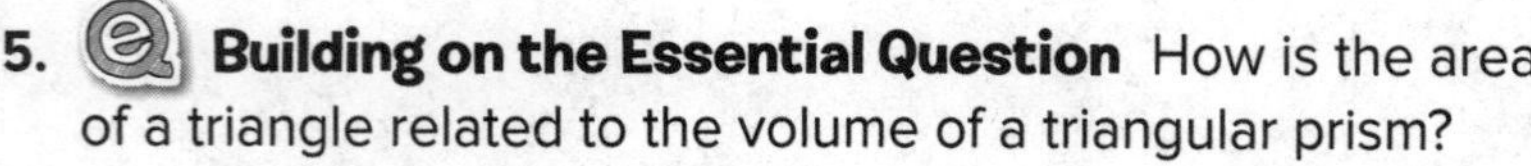

5. Building on the Essential Question How is the area of a triangle related to the volume of a triangular prism?

Rate Yourself!

How well do you understand volume of triangular prisms? Circle the image that applies.

Clear

Somewhat Clear

Not So Clear

For more help, go online to access a Personal Tutor.

FOLDABLES Time to update your Foldable!

Name ______________________ My Homework ______________________

Independent Practice

Go online for Step-by-Step Solutions

Find the volume of each prism. Round to the nearest tenth if necessary. (Example 1)

1. ______

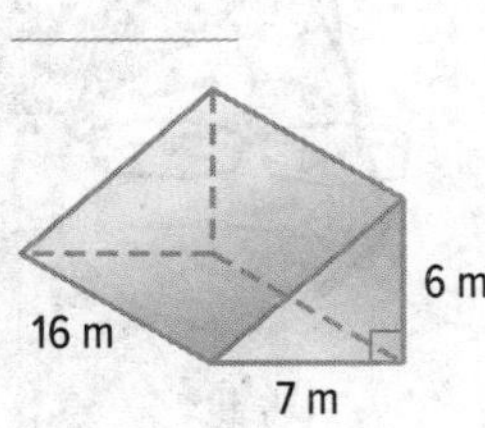

2. ______

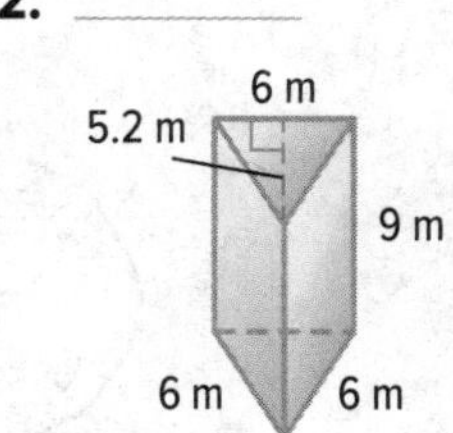

3. ______

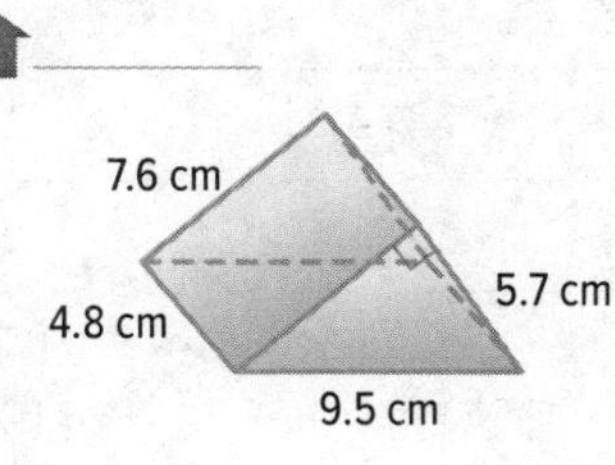

4. A wheelchair ramp is in the shape of a triangular prism. It has a base area of 37.4 square yards and a height of 5 yards. Find the volume of the ramp. (Example 2)

______

5. A triangular prism has a height of 9 inches. The triangular base has a base of 3 inches and a height of 8 inches. Find the volume of the prism. (Example 2)

______

Find the missing dimension of each triangular prism. (Example 3)

6. $x =$ ______

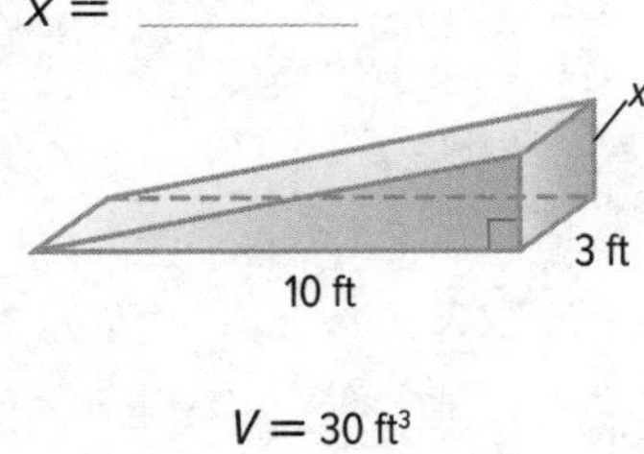

$V = 30$ ft^3

7. $x =$ ______

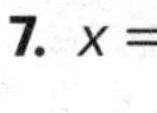

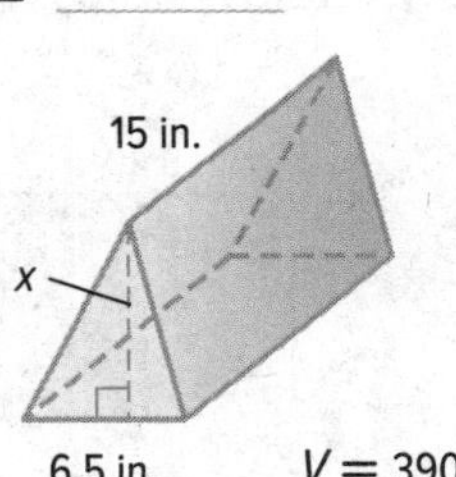

$V = 390$ in^3

8. $x =$ ______

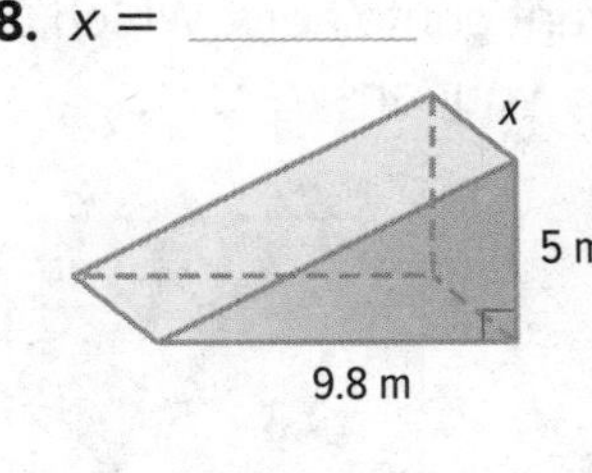

$V = 98$ m^3

9. Mr. Standford's greenhouse has the dimensions shown. The volume of the greenhouse is 90 cubic yards. Find the missing dimension of the greenhouse. (Example 4)

______

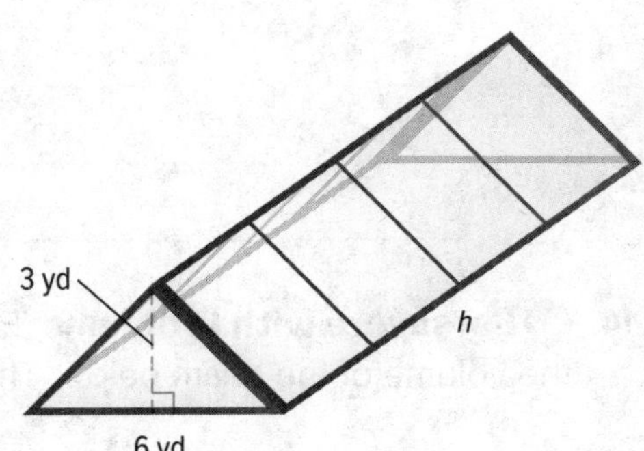

10. **Be Precise** Darcy built the dollhouse shown.

a. What is the volume of the first floor?

______

b. What is the volume of the attic space?

______

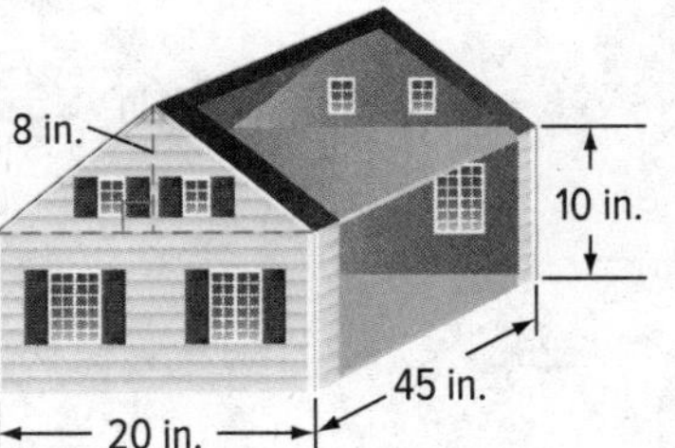

H.O.T. Problems Higher Order Thinking

11. **Find the Error** Amanda is finding the volume of the triangular prism. Find her mistake and correct it.

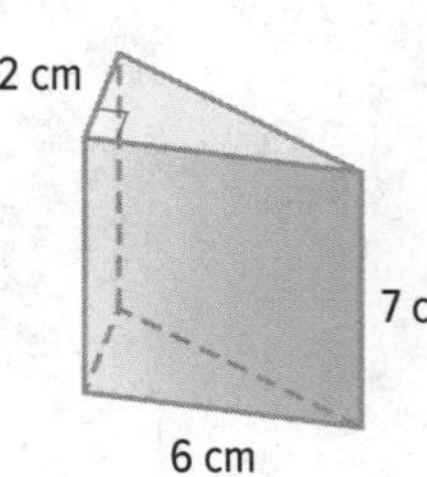

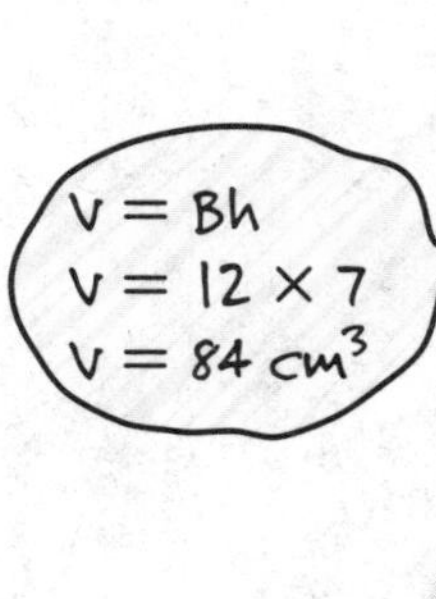

12. **Identify Repeated Reasoning** A rectangular prism and a triangular prism each have a volume of 210 cubic meters. Find possible sets of dimensions for each prism.

13. **Persevere with Problems** A candy company sells mints in two different containers. Which container shown below holds more mints? Justify your answer.

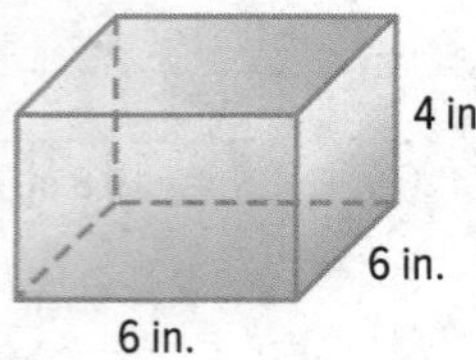

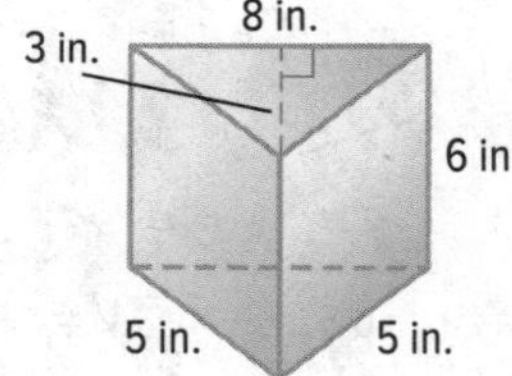

14. **Persevere with Problems** Explain a method you could use to find the volume of the prism below. Then find the volume of the prism.

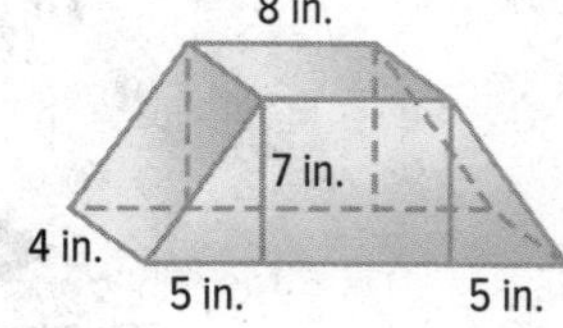

Extra Practice

Find the volume of each prism. Round to the nearest tenth if necessary.

15. 346.5 ft³

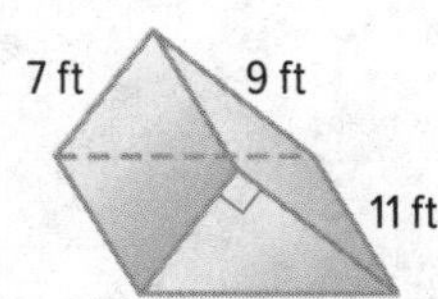

mework Help

$V = Bh$

$V = \left(\frac{1}{2} \cdot 7 \cdot 9\right)(11)$

$V = 346.5$

16. ______

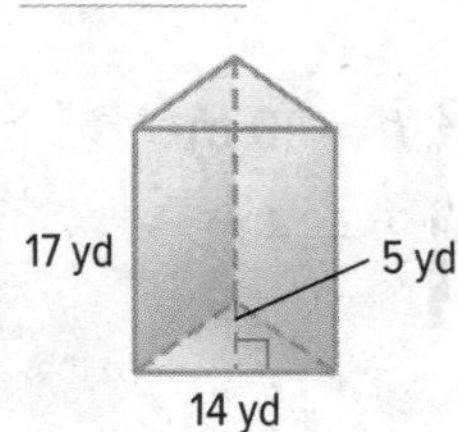

17. ______

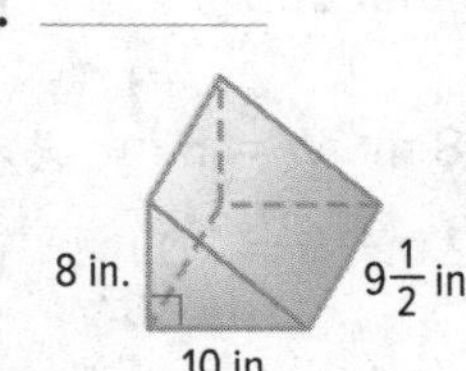

18. A candle is in the shape of a triangular prism. The base has an area of 30 square inches. The candle has a height of 6 inches. Find the volume of the candle.

______

19. A cabinet is in the shape of a triangular prism. The triangular base has a base length of 14 inches and a base height of 22 inches. The cabinet is 67.5 inches tall. What is the volume of the cabinet?

______

Find the missing dimension of each triangular prism.

20. $x =$ ______

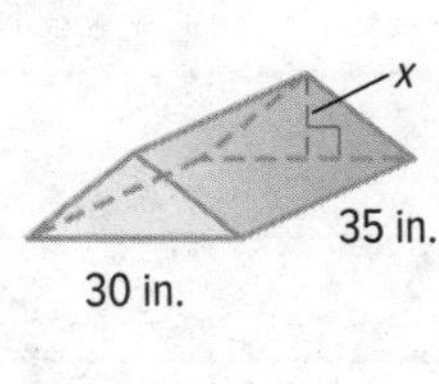

$V = 6{,}300 \text{ in}^3$

21. $x =$ ______

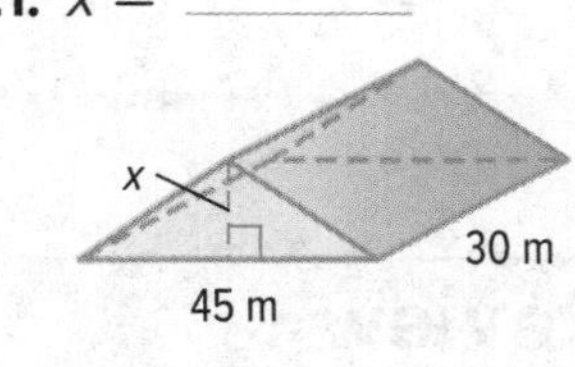

$V = 10{,}125 \text{ m}^3$

22. $x =$ ______

x, 2.5 cm, 1.4 cm

$V = 3.5 \text{ cm}^3$

23. What is the volume of the A-frame tent shown?

______

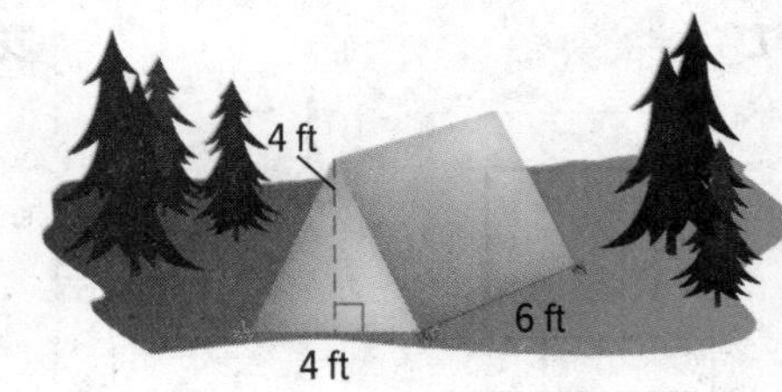

24. MP **Be Precise** A covered bridge in Vermont has the dimensions shown.

a. What is the volume of the bottom section rounded to the nearest tenth? ______

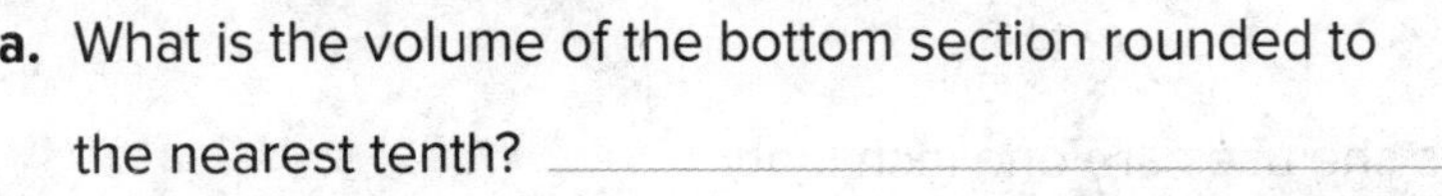

b. What is the volume of the top section, rounded to the nearest tenth? ______

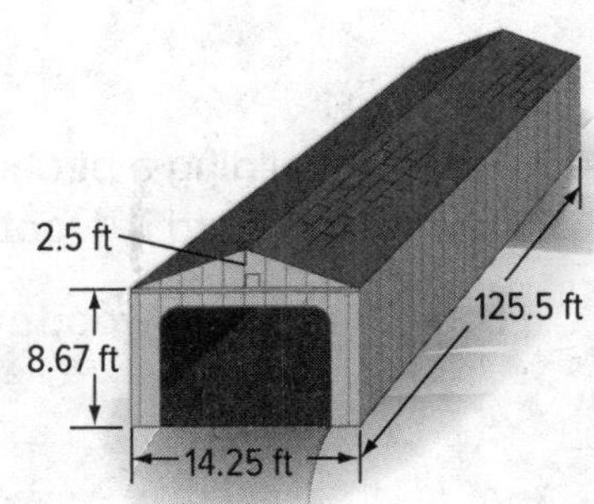

Power Up! Common Core Test Practice

25. A triangular prism has a volume of 240 cubic meters. Which of the following are possible dimensions for the area of the base and the height of the prism? Select all that apply.

- ☐ $B = 48\text{ m}^2, h = 5\text{ m}$
- ☐ $B = 24\text{ m}^2, h = 10\text{ m}$
- ☐ $B = 12\text{ m}^2, h = 20\text{ m}$
- ☐ $B = 50\text{ m}^2, h = 4\text{ m}$

26. A kitchen cabinet manufacturer offers three different size corner cabinets with the dimension shown below. Sort the volume of the cabinets from least to greatest.

	Cabinet	Volume (in^3)
Least		
Greatest		

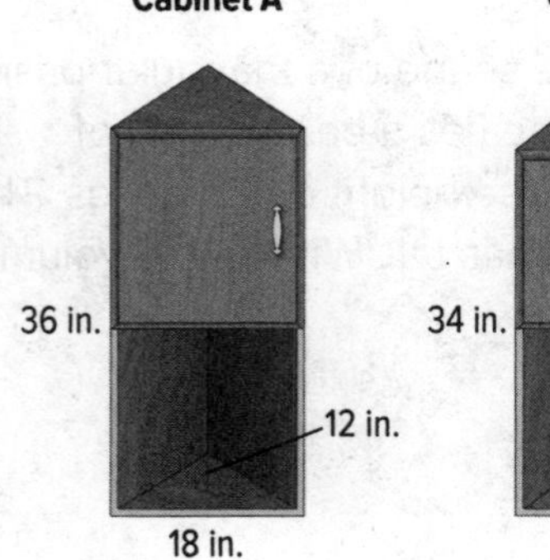

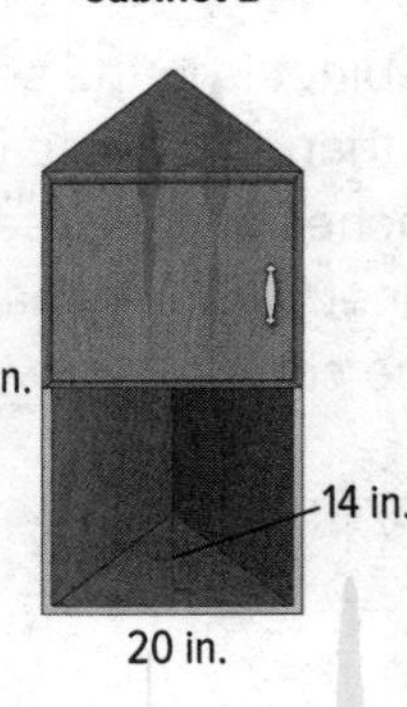

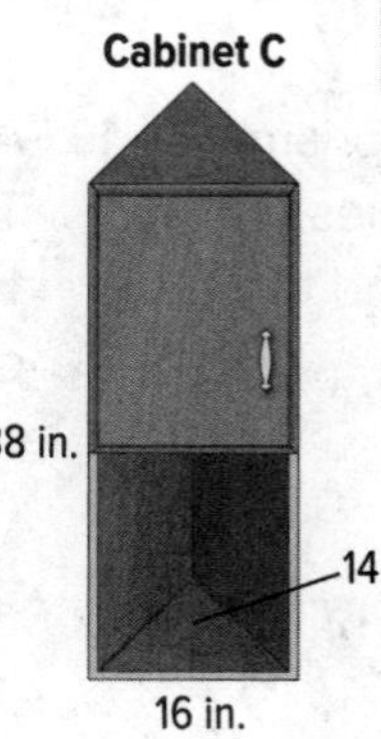

Which cabinet has the greatest volume? ☐

Common Core Spiral Review

Find the area of each figure. 4.MD.3

27. ______

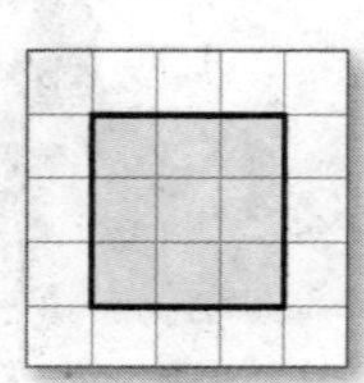

28. ______

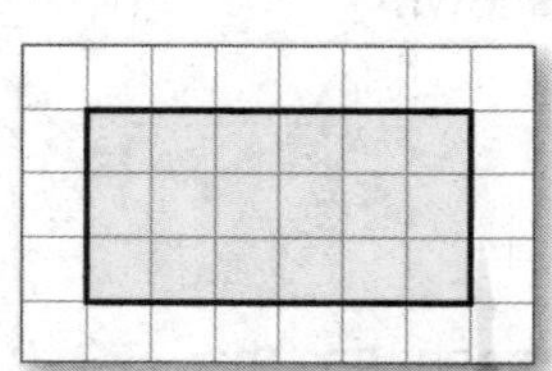

29. ______

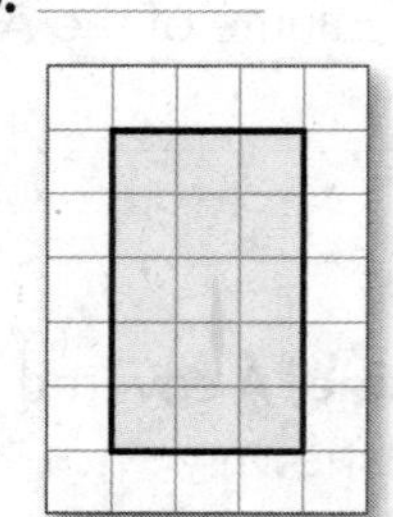

30. Sarah is building a birdhouse. The nails she uses are one inch long. The wood board is 1 foot long. How many times smaller are the nails compared to the wood? 4.MD.1 ______

Problem-Solving Investigation
Make a Model

Content Standards
6.G.2, 6.G.4

Mathematical Practices
1, 3, 4

Case #1 Scooter Storage

Nick works for a sporting goods store. He is stacking boxes of scooters in the storage space at the back of the store. The first layer has 9 boxes.

If the storage area will hold 6 layers of boxes, how many boxes will the storage space hold?

Understand *What are the facts?*

- The first layer has 9 boxes.
- The storage space will hold 6 layers.

Plan *What is your strategy to solve this problem?*

Make a model using centimeter cubes.

Solve *How can you apply the strategy?*

Make a model of one layer of the box by arranging 9 cubes in a 3 × 3 array.

Continue stacking the cubes until there are 6 layers.

So, the storage space will hold 54 boxes.

Check *Does the answer make sense?*

Use the volume formula to check your answer.
$V = 3 \times 3 \times 6$ or 54

So, the storage space will hold a total of 54 boxes.

Justify Conclusions Suppose the boxes are a different size and the first layer has 6 boxes instead. How many boxes can be stored if the storage space will hold 5 layers? Explain.

Case #2 Contain Your Fun

A storage container is made from plastic that measures $1\frac{1}{2}$ feet long, 2 feet wide, and $2\frac{1}{2}$ feet high.

Find the surface area of the plastic container, including the lid.

Understand

Read the problem. What are you being asked to find?

I need to find ______________________________.

Underline key words and values in the problem. What information do you know?

The storage container measures __________ long, __________ wide, and __________ high.

Plan

Choose a problem-solving strategy.

I will use the ______________________________ strategy.

Solve

Use your problem-solving strategy to solve the problem.

Make a model of the container using a net. Then find the area of each rectangle to find the total surface area.

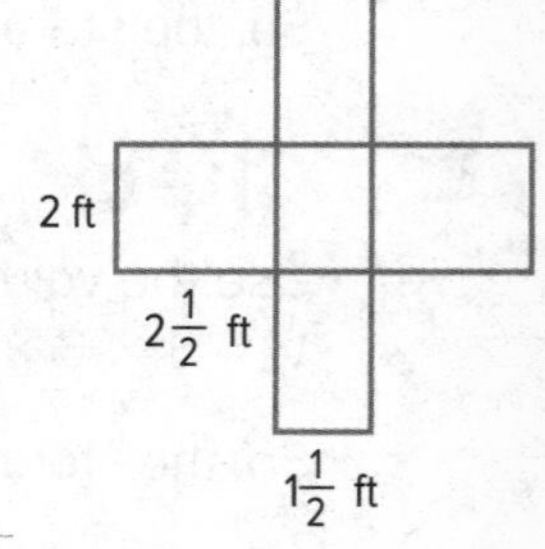

front and back: 2(________ × ________) = ________

left and right: 2(________ × ________) = ________

top and bottom: 2(________ × ________) = ________

Sum of 6 sides: ________ + ________ + ________ = ________

So, the surface area of the container is ________ square feet.

Check

Use information from the problem to check your answer.

Substitute known values into the surface area formula to check your answer.

S.A. = (____________) + (____________) + (____________) = ________ ft^2

Work with a small group to solve the following cases. Show your work on a separate piece of paper.

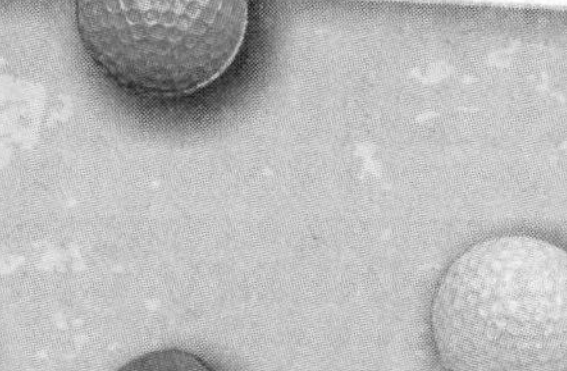

Case #3 Assembly

DJ is helping set up 7 rows of chairs for a school assembly. There are eight chairs in the first row. Each row after that has two more chairs than the previous row.

If he has 100 chairs, can he set up enough rows? Explain.

Case #4 Paper

Timothy took a piece of notebook paper and cut it in half. Then he placed the 2 pieces on top of each other and cut them in half again to have 4 pieces of paper.

If he could keep cutting the paper in this manner, how many pieces of paper would he have after 6 cuts?

Case #5 Sports

Rosario is packing a crate with boxes of miniature golf putters. Each box has a height of 1 foot, a width of 1 foot, and a length of 3 feet.

How many boxes can Rosario fit in a crate that is 4 feet high, 4 feet wide, and 3 feet long?

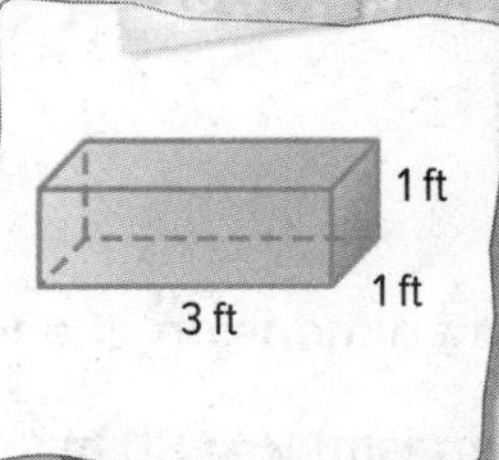

Case #6 Patterns

Draw the seventeenth figure in the pattern.

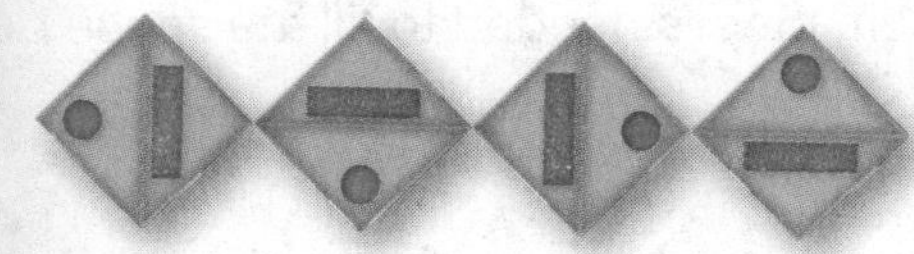

Mid-Chapter Check

Vocabulary Check

1. **MP Be Precise** Define *three-dimensional figure*. Give an example of a figure that is a three-dimensional figure and an example of a figure that is not a three-dimensional figure. (Lesson 1)

____

____

Fill in the blanks in the sentences below with the correct terms. (Lesson 1)

2. Volume is the amount of __________ inside a three-dimensional figure.

3. Volume is measured in __________ units.

Skills Check and Problem Solving

Find the volume of each prism. Round to the nearest tenth if necessary. (Lessons 1 and 2)

4. __________

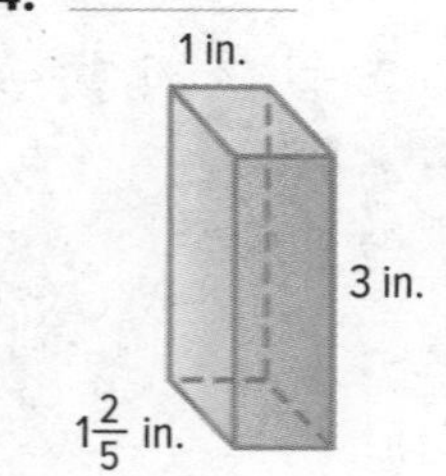

5. __________

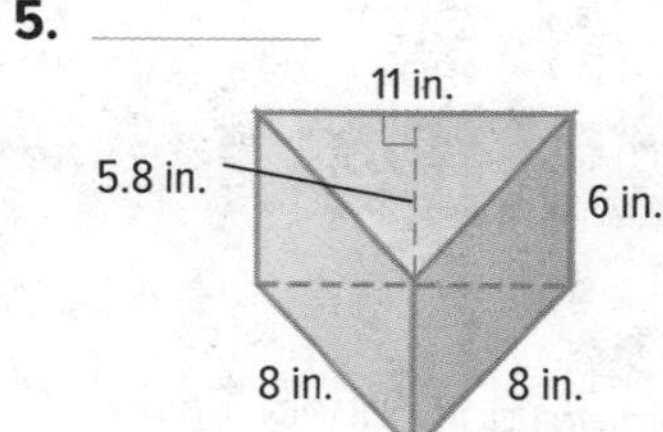

6. __________

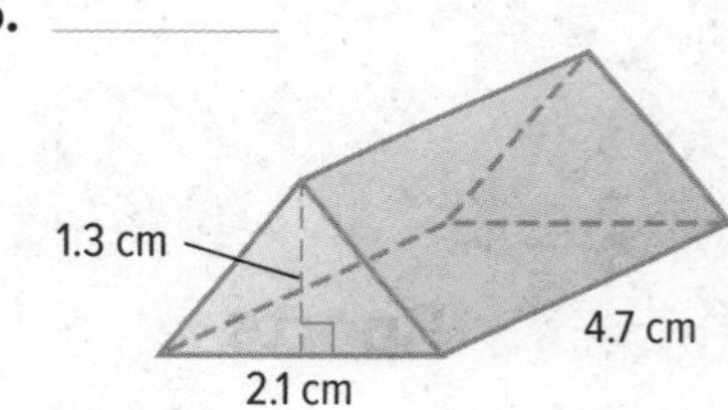

Find the missing dimension of each figure. (Lessons 1 and 2)

7. rectangular prism: $V = 80$ m^3; length = 5 m; width = 4 m

   $h =$ __________

8. triangular prism: $V = 42$ cm^3; base length = 2 cm; base height = 6 cm

   $h =$ __________

9. **MP Persevere with Problems** Janet is mailing a candle that is in the shape of a triangular prism as shown. She put the candle in a rectangular box that measures 3 inches by 5 inches by 7 inches and places foam pieces around the candle. Find the volume of the foam pieces needed to fill the space between the candle and the box. (Lesson 2) ______________________

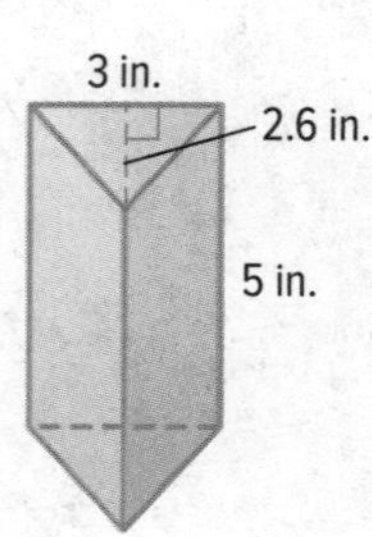

Inquiry Lab

Surface Area of Rectangular Prisms

HOW can you use nets to find surface area?

Content Standards 6.G.4

Mathematical Practices 1, 3, 4

If you want to know the amount of cereal that can fit in the box, you would find the volume. But if you want to know how much cardboard is needed to make the box, you would find the *surface area*.

Hands-On Activity 1

One way to find the surface area is to use a *net*. Nets are two-dimensional patterns of three-dimensional figures. When you construct a net, you are decomposing the three-dimensional figure into separate shapes.

Step 1 Use a cereal box in the shape of a rectangular prism. Measure and record the length, width, and height of the box on the lines below.

Length: ____________

Width: ____________

Height: ____________

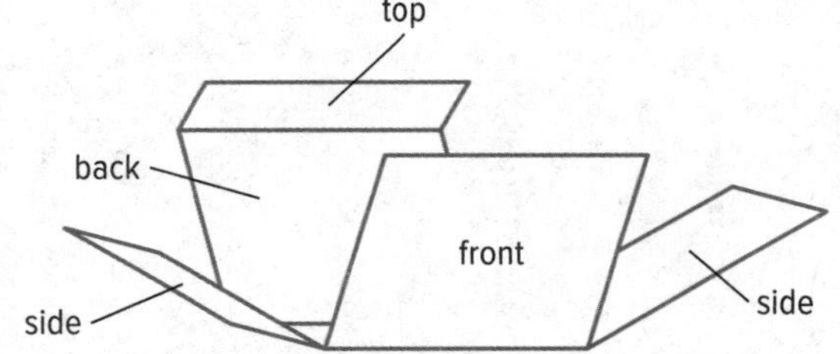

Step 2 Using a marker, label the top, bottom, front, back, and side faces of the box.

Step 3 Using scissors, carefully cut along three edges of the top face and then cut down each vertical edge.

Step 4 Measure and record the area of each face, using the dimensions of the box shown in the table.

Step 5 Add the areas of each face to find the surface area of the box.

Face	Length	Width	Area of Face
Front			
Back			
Side 1			
Side 2			
Top			
Bottom			

☐ + ☐ + ☐ + ☐ + ☐ + ☐ = ☐

So, the surface area of the box is ☐ square inches.

Hands-On Activity 2

Orthogonal drawings consist of separate views of an object taken from different angles. You can make a net from orthogonal drawings.

Step 1 Find the dimensions of each side of a rectangular prism from the orthogonal drawing.

Orthogonal Drawing		
View	**Drawing**	**Dimensions**
Front and Back		×
Sides		×
Top and Bottom		×

Step 2 Use grid paper to draw a net from the orthogonal drawing. Trace and cut out your drawing and tape it in the space below. Check the dimensions of each face using the information in the table.

Step 3 Fold the net into a three-dimensional figure. Draw the resulting figure in the space provided.

So, the figure is a ______________.

It has a surface area of [ ] square units.

Investigate

MP Model with Mathematics **Work with a partner. Use a net to determine the surface area of each prism. Draw a net of each prism on the provided grid.**

1. ________ mm^2

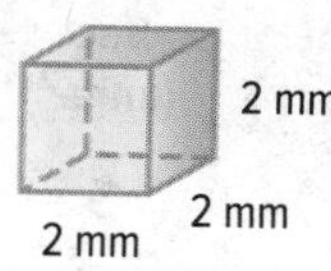

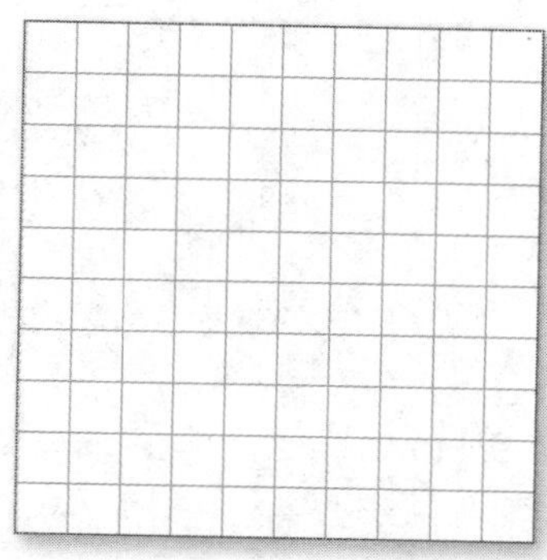

2. ________ in^2

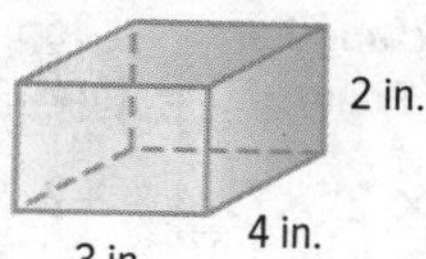

3. ________ ft^2

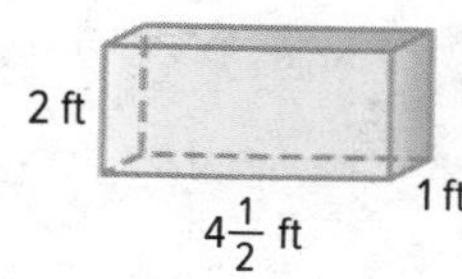

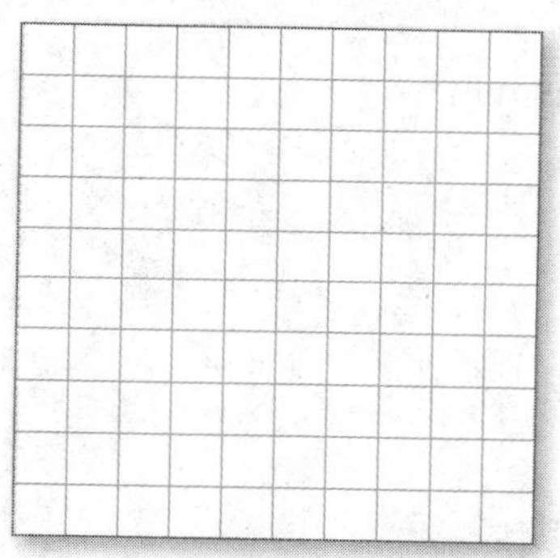

4. ________

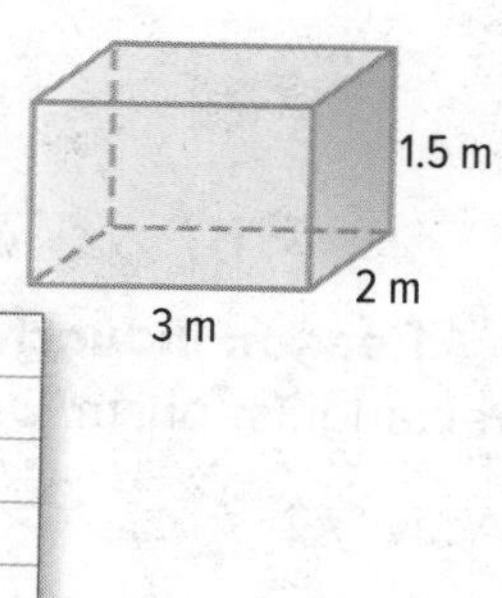

Draw a net on the grid from the orthogonal drawing. Then find the surface area of the prism.

5. ________ square units

Orthogonal Drawing	
View	**Drawing**
Front and Back	
Sides	
Top and Bottom	

Analyze and Reflect

Collaborate

Work with a partner to complete the table. The first one is done for you.

	Dimensions of Rectangular Prism	Area of Top (units^2)	Area of Bottom (units^2)	Area of Side 1 (units^2)	Area of Side 2 (units^2)	Area of Front (units^2)	Area of Back (units^2)	Surface Area (units^2)
	1 × 2 × 3	2	2	6	6	3	3	22
6.	2 × 2 × 3							
7.	3 × 3 × 3							
8.	3 × 2 × 8							
9.	6 × 6 × 6							

10. Compare the surface area for Exercise 7 to the surface area for Exercise 9. How does doubling each dimension affect the surface area?

11. MP **Reason Inductively** Write a formula to find the surface area of a rectangular prism. Use your formula to find the surface area of the prism in Activity 2.

Create

On Your Own

12. MP **Model with Mathematics** Write a real-world problem that involves the surface area of rectangular prisms. Provide the dimensions and the surface area.

13. Will the surface area of a cube ever have the same numerical value as the volume of the cube?

14. Inquiry HOW can you use nets to find surface area?

Lesson 3

Surface Area of Rectangular Prisms

Vocabulary Start-Up

Define Surface	Define Area
What is surface area?	Example:

Essential Question

HOW is shape important when measuring a figure?

Vocabulary

surface area

Common Core State Standards

Content Standards
6.G.4

Mathematical Practices
1, 3, 4, 8

Real-World Link

Gifts Roberta is wrapping a gift for her sister's quinceañera. She places it in a box with the measurements shown.

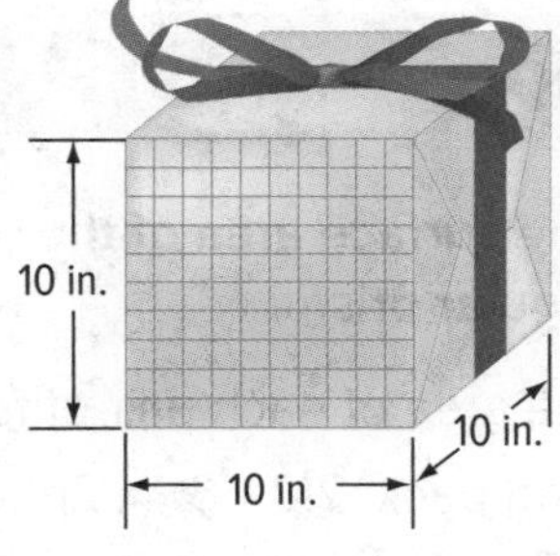

1. What is the area of one face of the box?

2. How many faces does the box have?

3. What operations would you use to find the surface area of the box?

Which  **Mathematical Practices** did you use? Shade the circle(s) that applies.

① Persevere with Problems
② Reason Abstractly
③ Construct an Argument
④ Model with Mathematics
⑤ Use Math Tools
⑥ Attend to Precision
⑦ Make Use of Structure
⑧ Use Repeated Reasoning

Key Concept

Surface Area of a Rectangular Prism

Work Zone

Words The surface area *S.A.* of a rectangular prism with length ℓ, width w, and height h is the sum of the areas of the faces.

Model

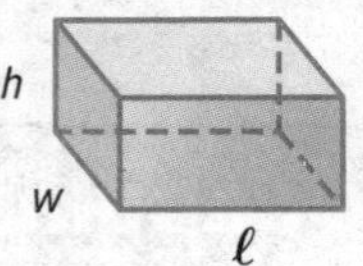

Symbols $S.A. = 2\ell h + 2\ell w + 2hw$

The surface area of a prism is the sum of the areas of its faces.

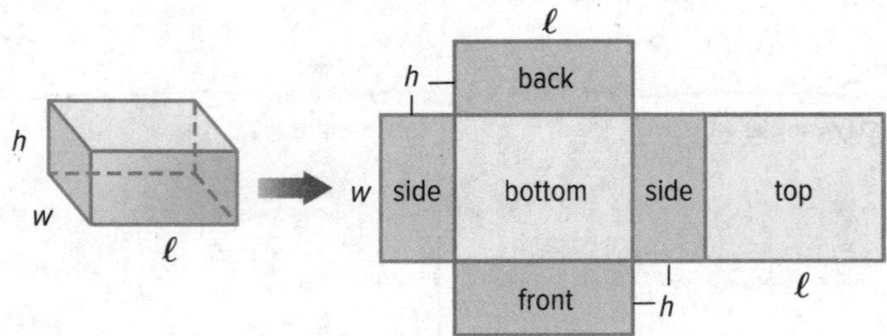

$$\left.\begin{array}{ll}\text{front and back:} & \ell h + \ell h = 2\ell h \\ \text{top and bottom:} & \ell w + \ell w = 2\ell w \\ \text{two sides:} & hw + hw = 2hw\end{array}\right\} 2\ell h + 2\ell w + 2hw$$

Example

Watch Tutor

1. Find the surface area of the rectangular prism.

Find the area of each pair of faces.

front and back: $2(8 \cdot 6) = 2(48)$

top and bottom: $2(7 \cdot 8) = 2(56)$

sides: $2(7 \cdot 6) = 2(42)$

$48 + 48 + 56 + 56 + 42 + 42 = 292$ Add the area of each face.

So, the surface area is 292 square meters.

	8 m		
	48 m²	6 m	
6 m		6 m	8 m
7 m 42 m²	56 m²	42 m²	56 m²
	48 m²	6 m	

Nets

The net shows that a rectangular prism has six faces. The faces can be grouped as three pairs of congruent sides. The colors indicate which faces are congruent.

Got it? Do this problem to find out.

a. Find the surface area of the rectangular prism.

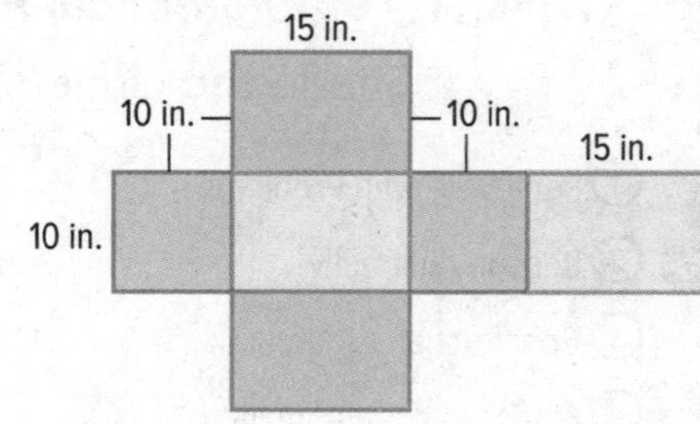

a. ____________

Find Surface Area Using a Formula

You can use nets or models to find the surface area of a rectangular prism. You can also use the surface area formula, $S.A. = 2\ell h + 2\ell w + 2hw$.

Examples

2. **Find the surface area of the rectangular prism.**

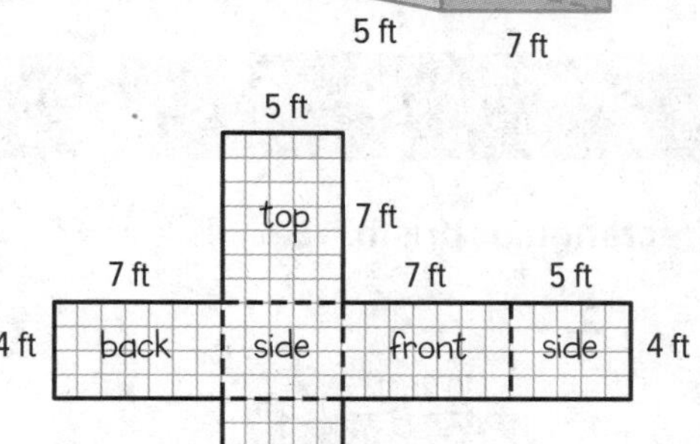

Find the area of each face.

front and back:
$2\ell h = 2(7)(4)$ or 56

top and bottom:
$2\ell w = 2(7)(5)$ or 70

left and right sides:
$2hw = 2(4)(5)$ or 40

Add to find the surface area.

The surface area is 56 + 70 + 40 or 166 square feet.

3. **Find the surface area of the rectangular prism.**

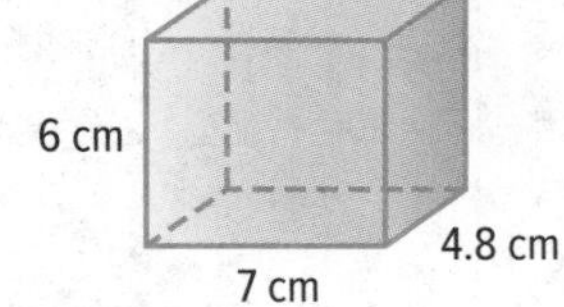

To find the area of each face, determine the dimensions.

$\ell = 7, w = 4.8, h = 6$

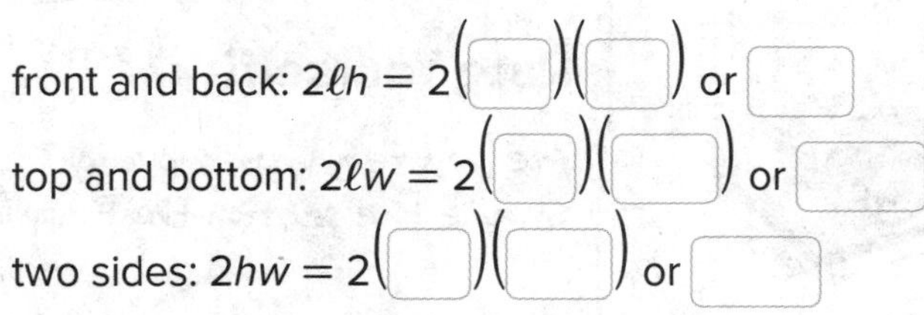

Add to find the surface area.

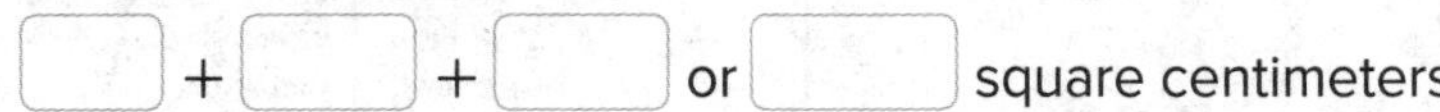

Got it? **Do this problem to find out.**

b. Find the surface area of the rectangular prism.

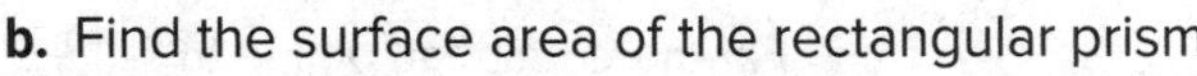

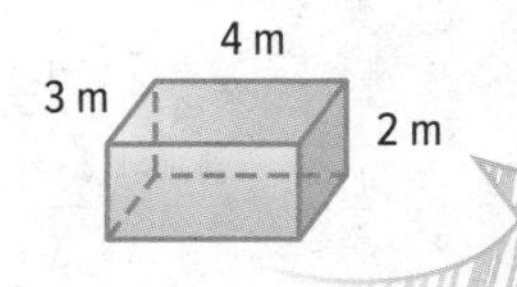

b. ____________

Example

4. **STEM** **A geode is being sent as a gift. It is packed in a box that measures 7 inches long, 3 inches wide, and 16 inches tall. What is the surface area of the box?**

$S.A. = 2\ell h + 2\ell w + 2hw$ — Surface area of a prism

$S.A. = 2(7)(16) + 2(7)(3) + 2(16)(3)$ — $\ell = 7, w = 3, h = 16$

$S.A. = 14(16) + 14(3) + 32(3)$ — Multiply.

$S.A. = 224 + 42 + 96$ — Multiply.

$S.A. = 362$ — Add.

The surface area of the box is 362 square inches.

Guided Practice

Find the surface area of each rectangular prism. (Examples 1–3)

1. ______

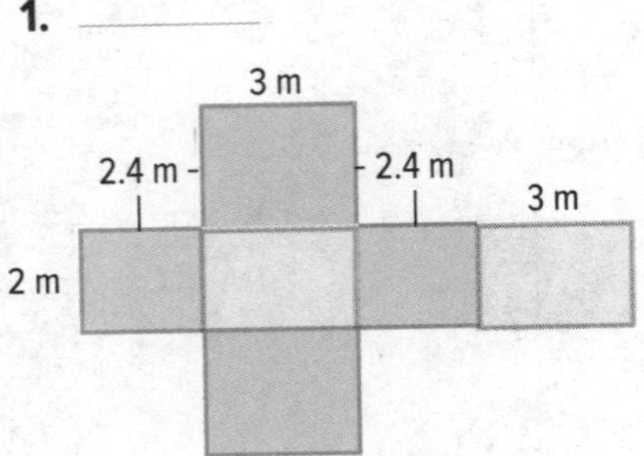

2. ______

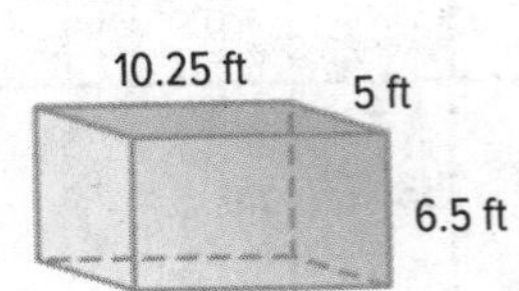

3. ______

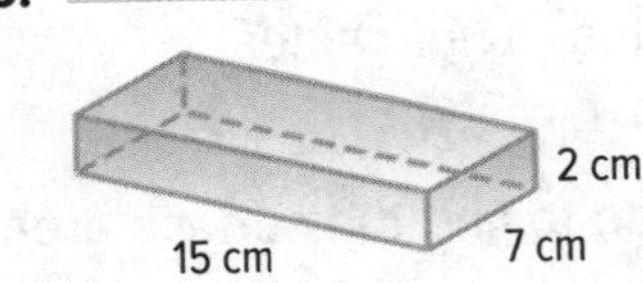

4. Tomás keeps his diecast car in a glass display case as shown. What is the surface area of the glass, including the bottom? (Example 4)

______

5. **Building on the Essential Question** What is the relationship between area and surface area?

______

______

______

______

Rate Yourself!

Are you ready to move on? Shade the section that applies.

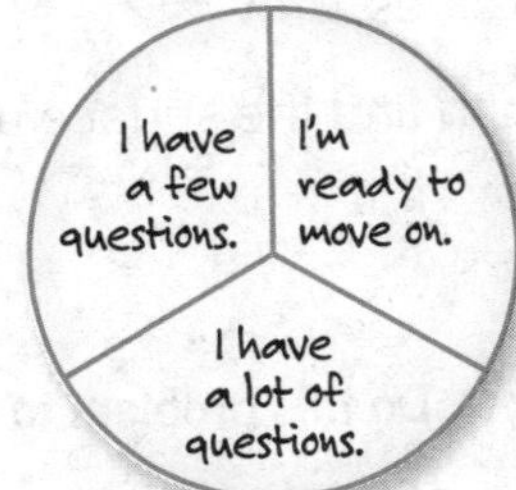

For more help, go online to access a Personal Tutor.

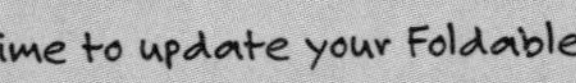

Name ________________ My Homework ________________

Independent Practice

Go online for Step-by-Step Solutions

Find the surface area of each rectangular prism. (Examples 1–3)

1. ______

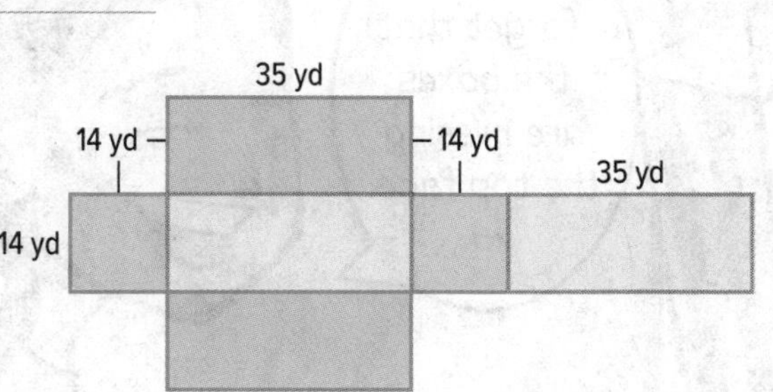

2. ______

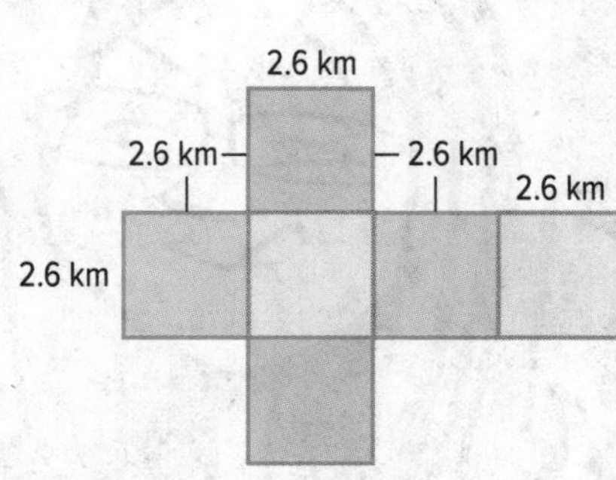

3. ______

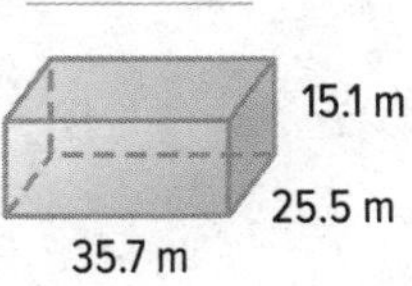

4. ______

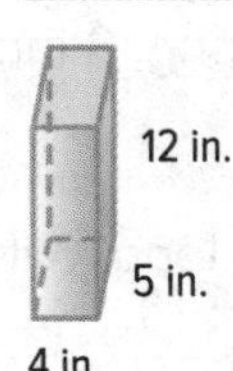

5. **STEM** A game box for video games is shaped like a rectangular prism. What is the surface area of the game box? (Example 4)

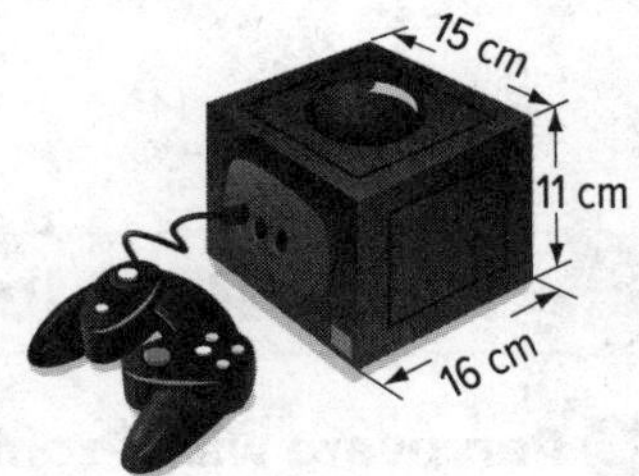

6. **MP Justify Conclusions** Martina estimates that the surface area of a rectangular prism with a length of 13.2 feet, a width of 6 feet, and a height of 8 feet is about 460 square feet. Is her estimate reasonable? Explain your reasoning.

7. **MP Justify Conclusions** Find the surface area of each shipping package. Which package has the greater surface area? Does the same package have a greater volume? Explain your reasoning to a classmate.

Package A

Package B

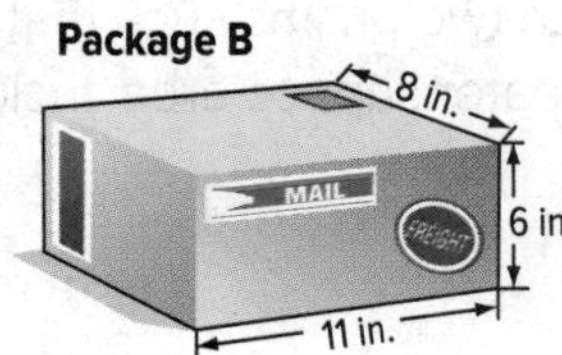

8. **Model with Mathematics** Refer to the graphic novel frame below for Exercises a–c.

a. The box on the left is 8 inches long, 8 inches wide, and 8 inches tall. What is the surface area of the box? ______

b. The box on the right is 8 inches long, 6 inches wide, and 10 inches tall. What is the surface area of the box? ______

c. How much more surface area does the larger container have?

______

H.O.T. Problems Higher Order Thinking

Persevere with Problems All of the triangular faces of the figure are congruent.

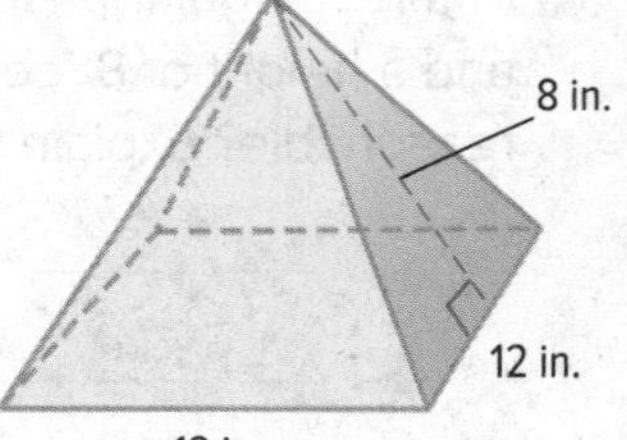

9. What is the area of one of the triangular faces? the square face?

______

10. Use what you know about finding the surface area of a rectangular prism to find the surface area of the square pyramid.

______

11. **Model with Mathematics** Sketch two prisms such that one has a greater volume and the other has a greater surface area. Include real-world units.

Name ______________________ My Homework ______________________

Extra Practice

Find the surface area of each rectangular prism.

12. 150 ft²

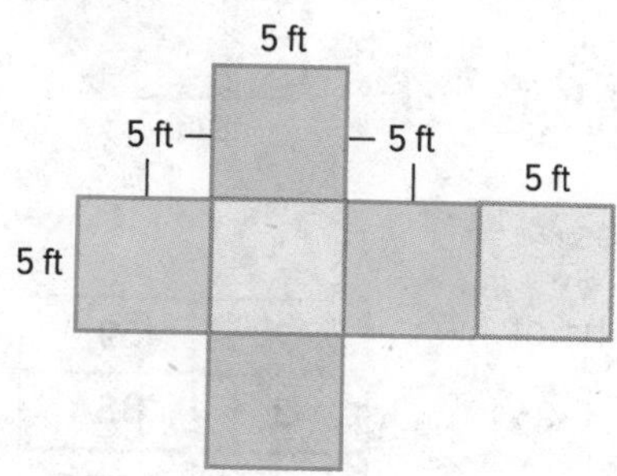

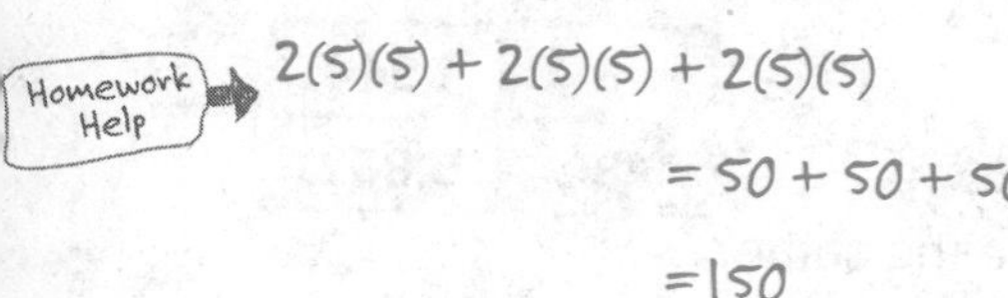

13. ______

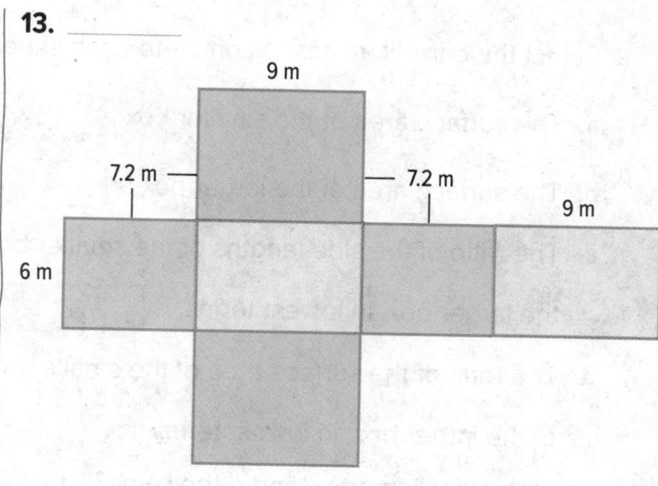

14. ______

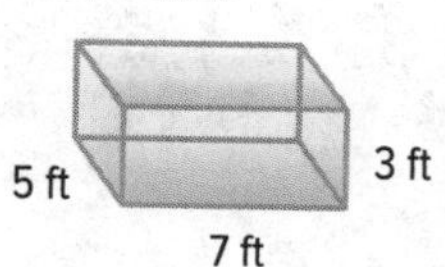

15. ______

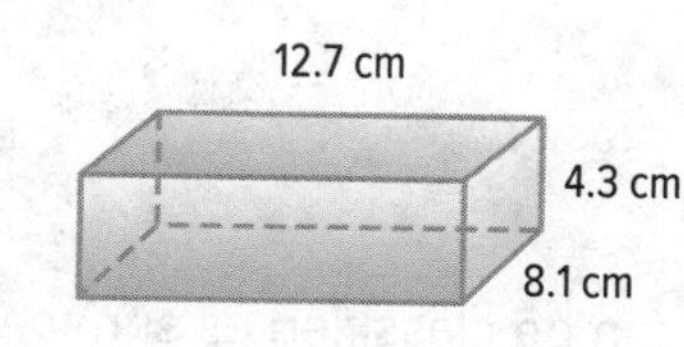

16. Nadine is going to paint her younger sister's toy chest, including the bottom. What is the approximate surface area that she will paint? ______

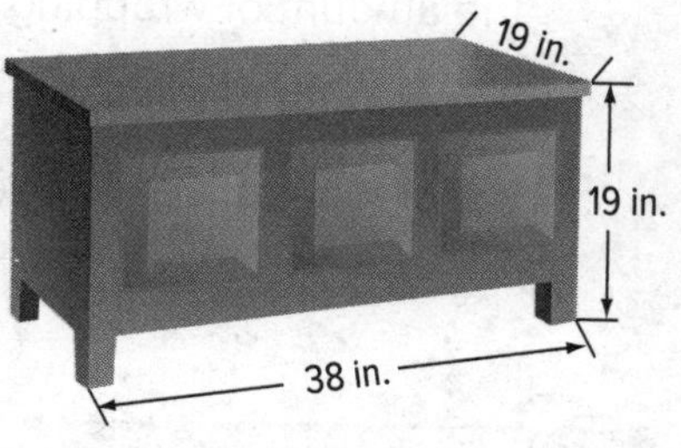

17. MP **Identify Repeated Reasoning** Chrissy is making a bird nesting box for her backyard.

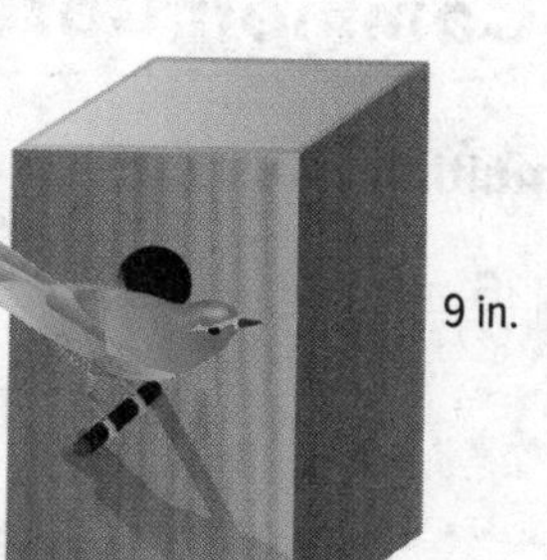

a. What is the surface area of the nesting box, including the hole? ______

b. What is the surface area if the width of 7.5 inches is doubled?

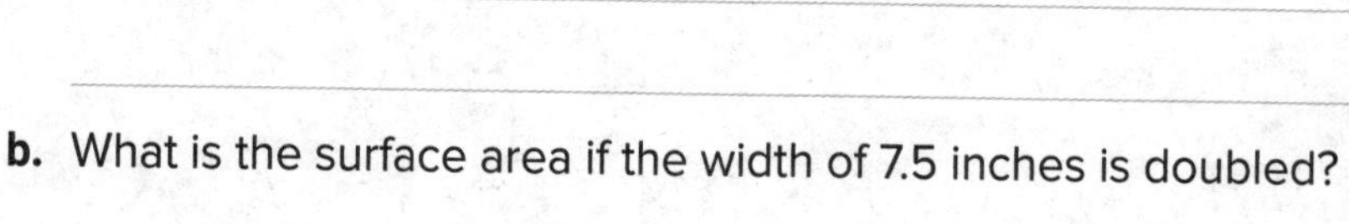

c. What is the surface area if the width of 7.5 inches is half as great?

CCSS Power Up! Common Core Test Practice

18. A company is experimenting with two new boxes for packaging merchandise. Each box is a cube with the side lengths shown.

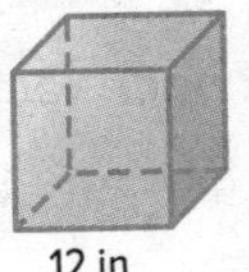

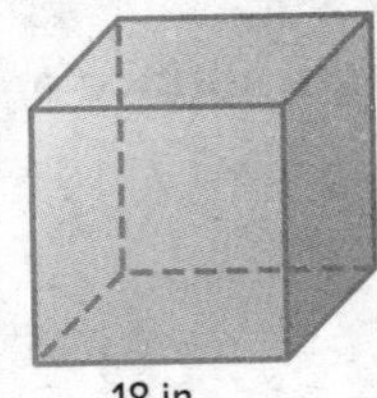

Select the correct values to complete each statement.

a. The surface area of the smaller box is ______ square inches.

b. The surface area of the larger box is ______ square inches.

c. The ratio of the side lengths of the smaller box to the side lengths of the larger box, in lowest terms, is ______ to ______.

d. The ratio of the surface area of the smaller box to the surface area of the larger box, in lowest terms, is ______ to ______.

2	9
3	864
4	1,728
6	1,944
8	5,832

Are the ratios in parts c and d the same? Did you expect them to be the same? Explain your reasoning.

19. Which measure(s) can be classified as surface area? Select all that apply.

- ☐ the amount of water in a lake
- ☐ the amount of wrapping paper needed to cover a box
- ☐ the amount of paint needed to cover a statue
- ☐ the amount of space needed to build a playset

CCSS Common Core Spiral Review

Add or multiply. 5.NBT.5, 4.NBT.4

20. $14 \times 16 =$ ______

21. $72 + 62 + 84 =$ ______

22. $27 \times 63 =$ ______

23. Classify the triangle by the measure of its sides. Explain. 5.G.4

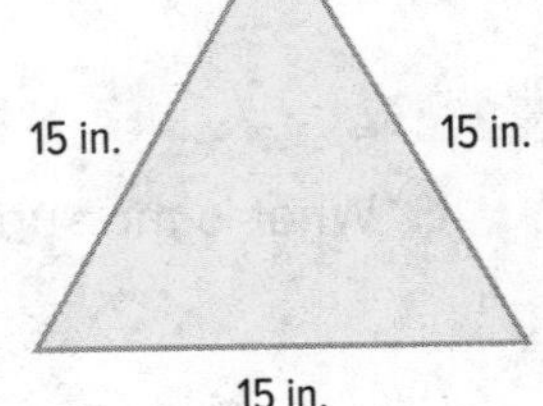

 Need more practice? Download more Extra Practice at **connectED.mcgraw-hill.com.**

Inquiry Lab

Nets of Triangular Prisms

HOW is the area of a triangle related to the surface area of a triangular prism?

Content Standards
6.G.4

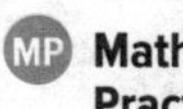

Mathematical Practices
1, 3, 4, 7

A computer hardware company packages batteries and cords in boxes shaped like triangular prisms. You can use nets and drawings to determine the surface area of the box.

Hands-On Activity

Use orthogonal drawings to find the surface area of a triangular prism. A *triangular prism* is a prism that has triangular bases.

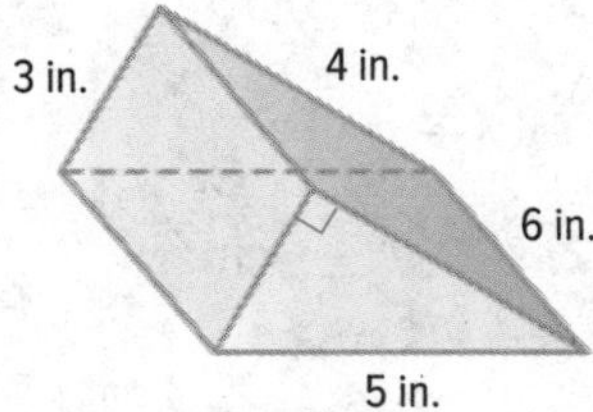

Step 1 Find the dimensions of each side of the triangular prism from the orthogonal drawing.

Orthogonal Drawing							
View	**Drawing**	**Dimensions (in.)**	**Area of Face (in^2)**	**View**	**Drawing**	**Dimensions (in.)**	**Area of Face (in^2)**
Bases		base = 3 height = 4	$\frac{1}{2}(3 \times 4) = 6$	Bottom		length = 6 width = 5	$6 \times 5 = 30$
Left		length = 6 width = 3	$6 \times 3 = 18$	Right		length = 6 width = 4	$6 \times 4 = 24$

Step 2 Use grid paper to draw a net. Check the dimensions of each face using the information in the table.

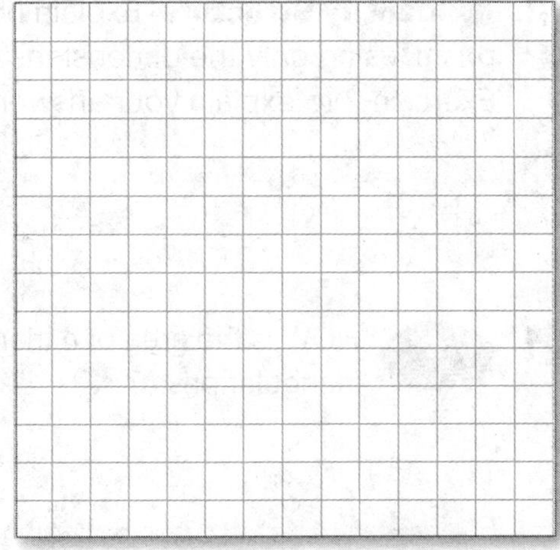

Step 3 Add the area of each face to find the surface area of the figure. Remember, there are two bases.

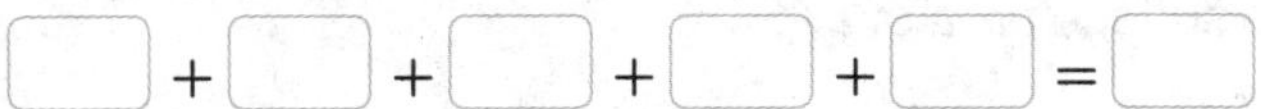

$\square + \square + \square + \square + \square = \square$

So, the surface area is $\square$ square units.

 **Model with Mathematics** Work with a partner. Use nets to determine the surface area of each prism. Draw a net of each prism on the provided grid paper.

1. ______ m^2

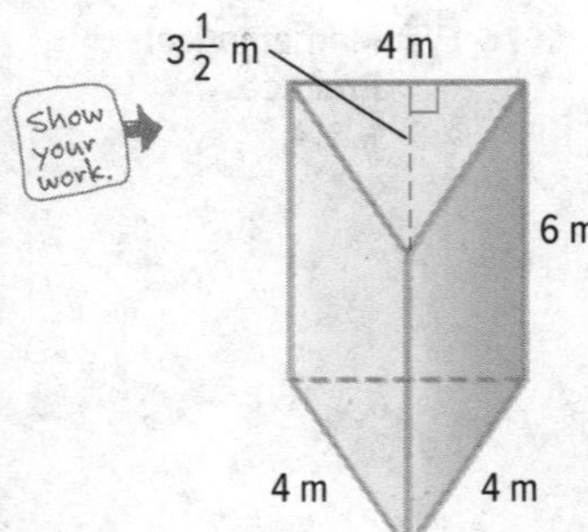

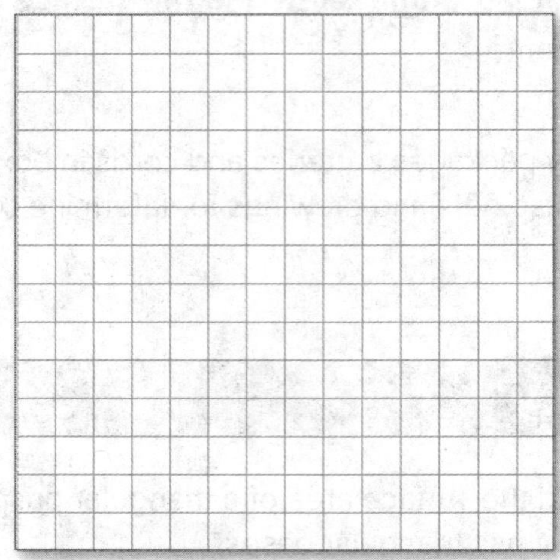

2. ______ cm^2

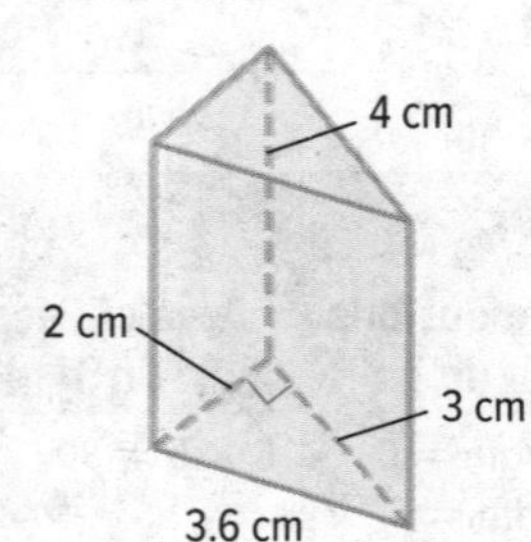

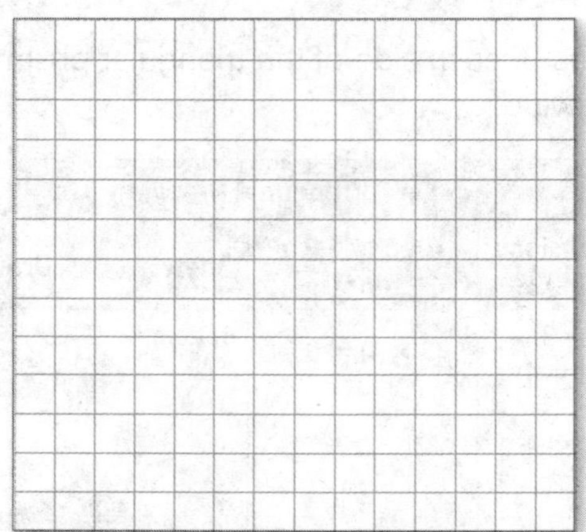

Create

3. **MP Identify Structure** Explain how to find the surface area of a triangular prism, using only the dimensions of the figure. Use the dimensions in Exercise 2 to explain your answer.

____

____

4. **Inquiry** HOW is the area of a triangle related to the surface area of a triangular prism?

____

____

Lesson 4

Surface Area of Triangular Prisms

Real-World Link

Ramp Raj and his dad are building a ramp to move his dirt bike onto a trailer.

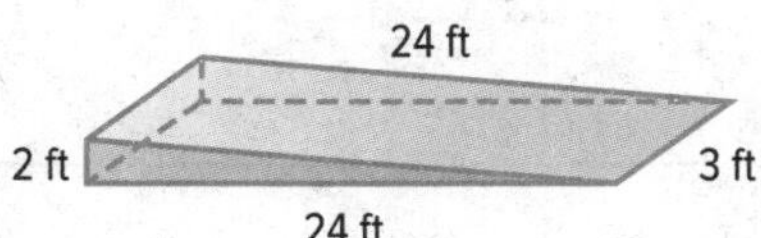

Essential Question

HOW is shape important when measuring a figure?

Common Core State Standards

Content Standards
6.G.4

MP Mathematical Practices
1, 2, 3, 4, 6

Fill in the table by drawing the sides of the ramp and naming the shape of each face.

	Face	Draw the Face	Shape of the Face
1.	Front		
2.	Back		
3.	Top		
4.	Bottom		
5.	Side		

Which MP **Mathematical Practices** did you use?
Shade the circle(s) that applies.

① Persevere with Problems
② Reason Abstractly
③ Construct an Argument
④ Model with Mathematics
⑤ Use Math Tools
⑥ Attend to Precision
⑦ Make Use of Structure
⑧ Use Repeated Reasoning

Key Concept

Surface Area of a Triangular Prism

Work Zone

Words

The surface area of a triangular prism is the sum of the areas of the two triangular bases and the three rectangular faces.

Model

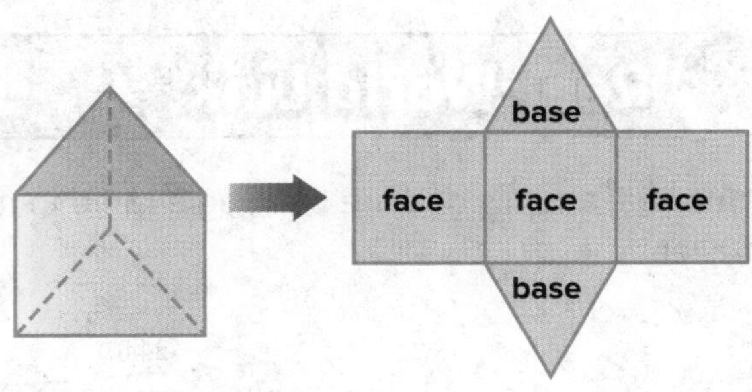

A triangular prism is a prism that has triangular bases. When the bases are equilateral triangles, the areas of the three rectangular faces are equal. You can use a net to find the surface area of a triangular prism.

Example

1. Find the surface area of the triangular prism.

To find the surface area of the triangular prism, find the area of each face and add.

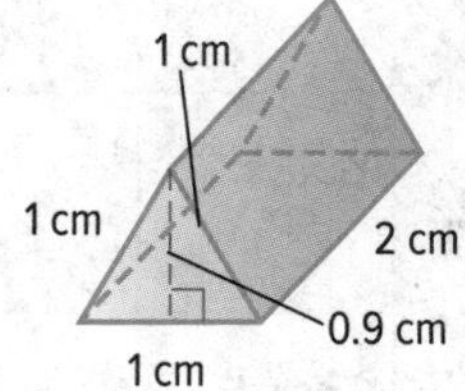

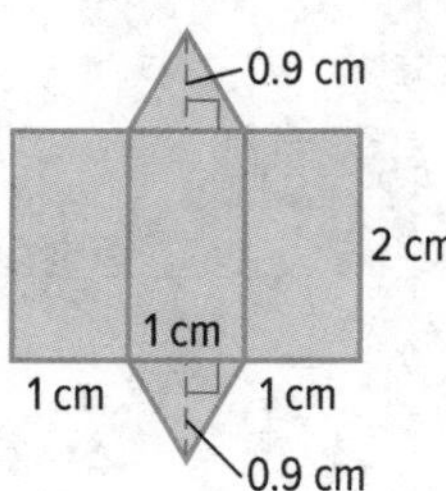

area of each triangular base: $\frac{1}{2}(1)(0.9) = 0.45$

area of each rectangular face: $1(2) = 2$

Add to find the surface area.

$0.45 + 0.45 + 2 + 2 + 2 = 6.9$ square centimeters

Got it? **Do this problem to find out.**

a. Find the surface area of the triangular prism.

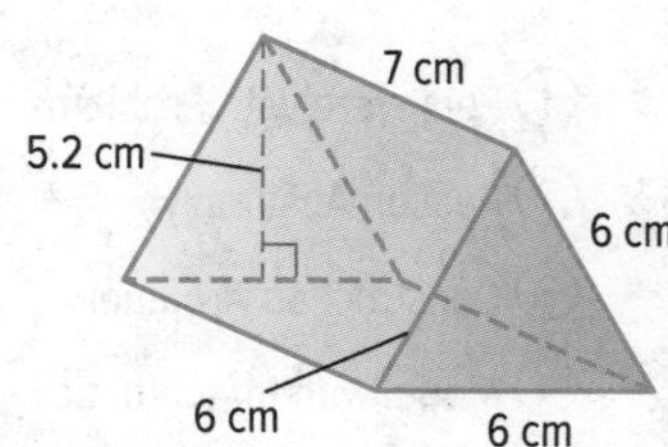

a. ____________

Surface Area of Other Triangular Prisms

You can also find the surface area of any triangular prism by adding the areas of all the sides of the prism using an orthogonal drawing.

Example

2. Find the surface area of the triangular prism.

Find the area of each face and add. For this prism, each rectangular face has a different area.

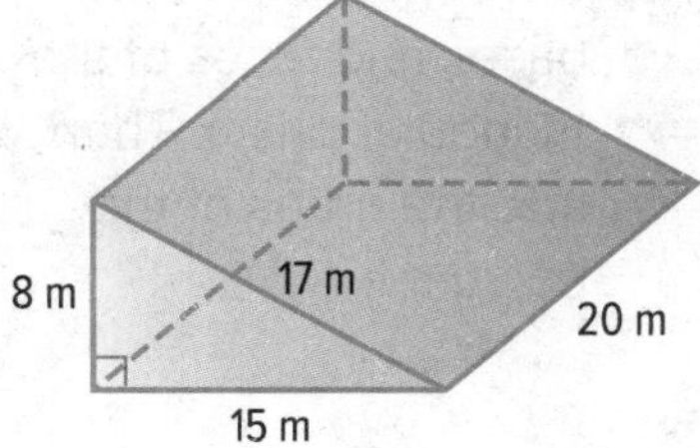

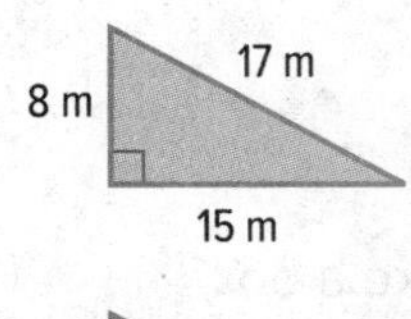

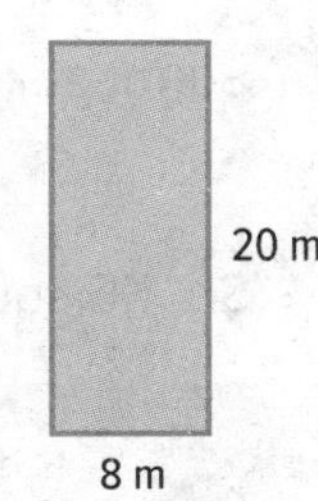

area of each triangular base: $\frac{1}{2}(15)(8) = 60$

area of the rectangular faces: $15(20) = 300$

$17(20) = 340$

$8(20) = 160$

Add to find the surface area.

$60 + 60 + 300 + 340 + 160 = 920$ square meters

Got it? Do this problem to find out.

Find the surface area of each triangular prism.

b.

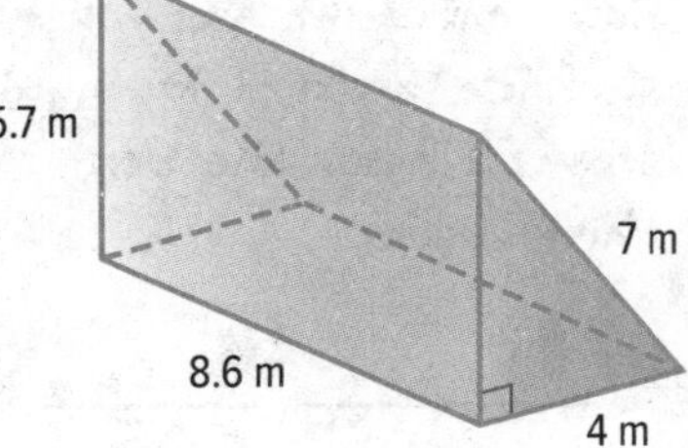

c.

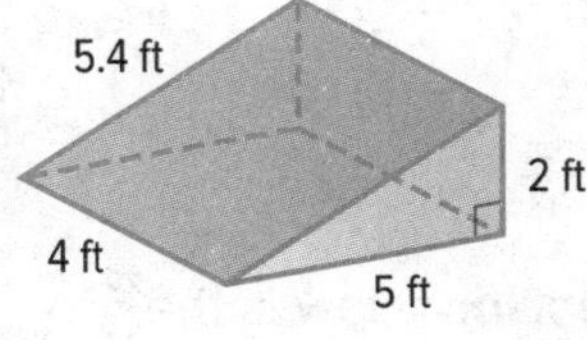

b. ____________

c. ____________

Example

3. A bakery boxes pastries in a triangular prism box, as shown. Find the amount of cardboard used to make a pastry box.

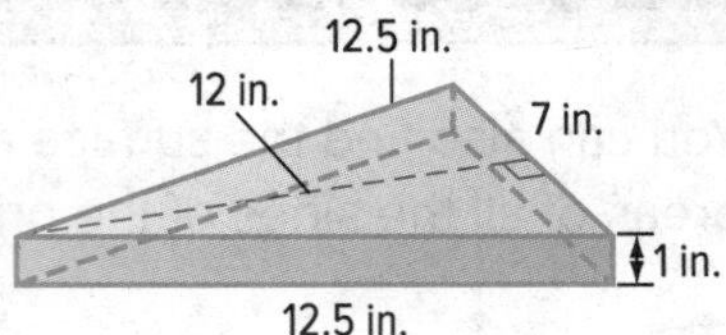

Sketch and label the bases and faces of the triangular prism. Then add the areas of the polygons.

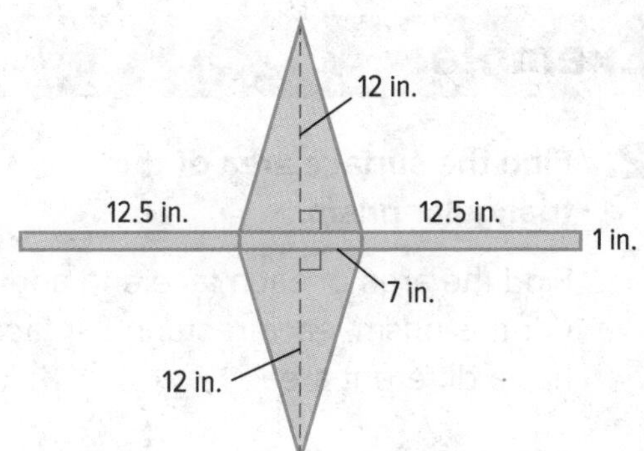

Surface area $= 2\left(\frac{1}{2} \cdot 7 \cdot 12\right) + 2(1 \cdot 12.5) + (1 \cdot 7)$

$= 84 + 25 + 7$ or 116

So, 116 square inches of cardboard is needed to make a box.

Guided Practice

1. Find the surface area of the triangular prism. (Examples 1–2) ________

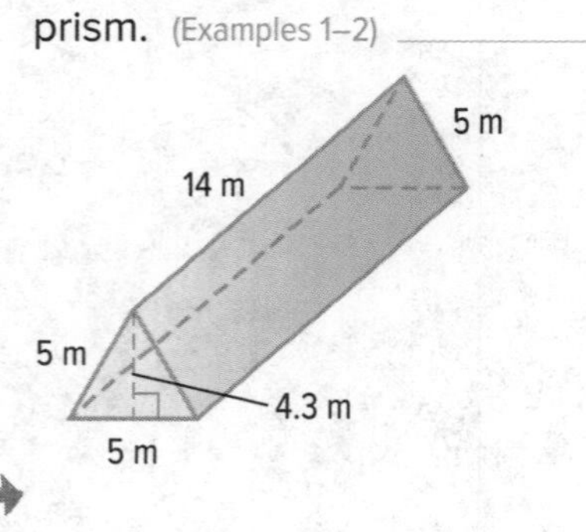

Show your work.

2. A skateboarding ramp is in the shape of a triangular prism. If the entire ramp is to be painted, what is the surface area to be painted? (Example 3) ________

50 in.
14 in.
36 in.
48 in.

3. **Building on the Essential Question** How is the area of a rectangle related to the surface area of a triangular prism? ________

Rate Yourself!

How confident are you about surface area of triangular prisms? Check the box that applies.

For more help, go online to access a Personal Tutor.

Tutor

FOLDABLES Time to update your Foldable!

Name ______________________ My Homework ______________________

Independent Practice

Go online for Step-by-Step Solutions

Find the surface area of each triangular prism. (Examples 1–2)

1. ________

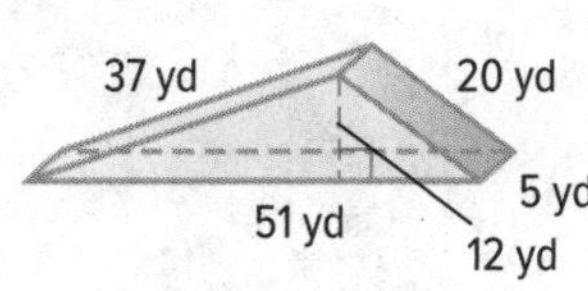

2. ________

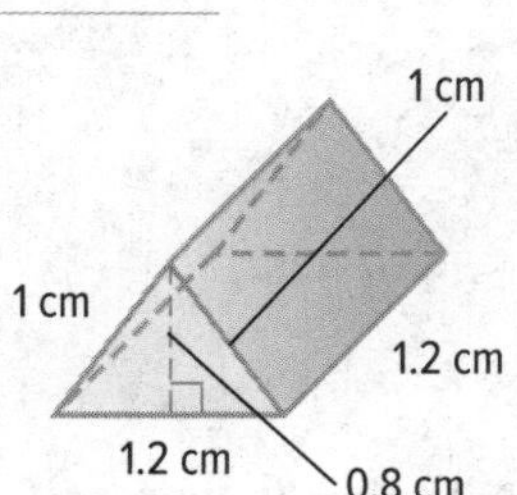

3 ________

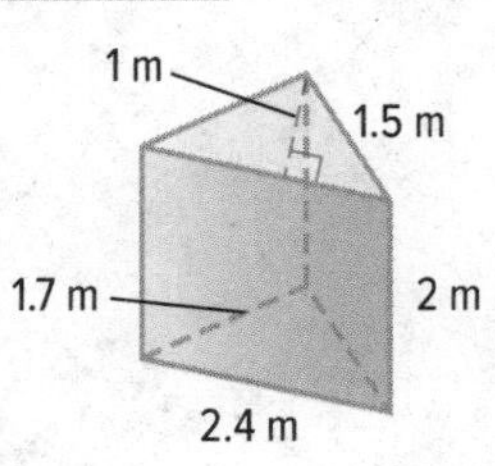

4. ________

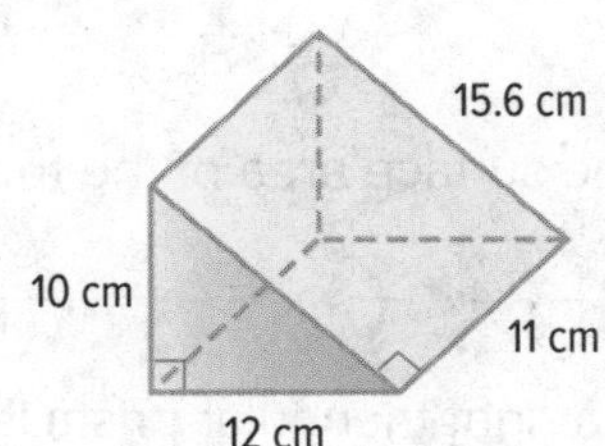

5 A tent is in the shape of a triangular prism. About how much canvas, including the floor, is used to make the tent? (Example 3)

____

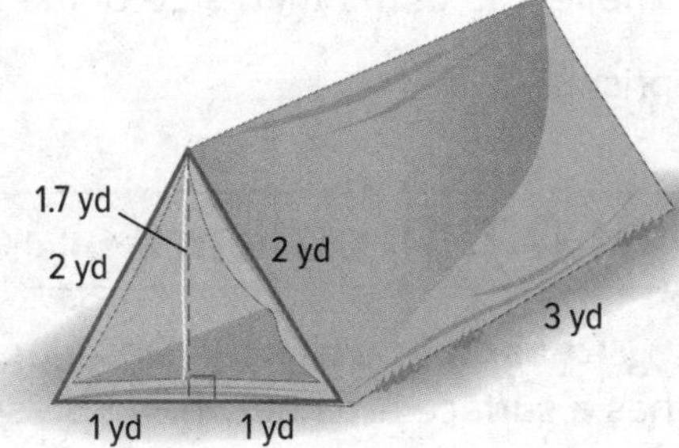

6. A decorative gift box is in the shape of a triangular prism as shown. What is the surface area of the box? (Example 3)

____

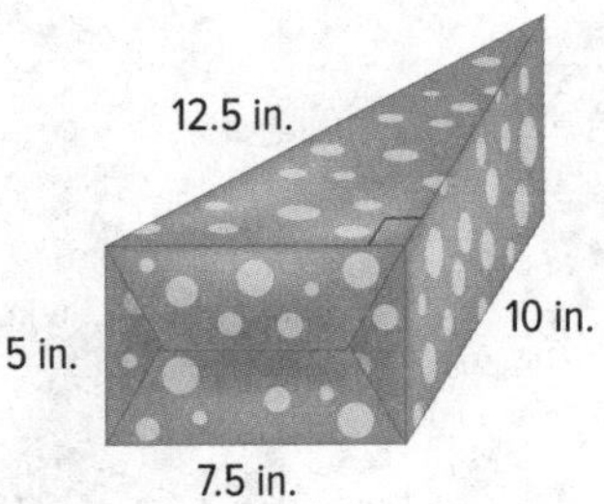

7. A mailer for posters is a triangular prism as shown. Find the surface area of the mailer. (Example 3)

____

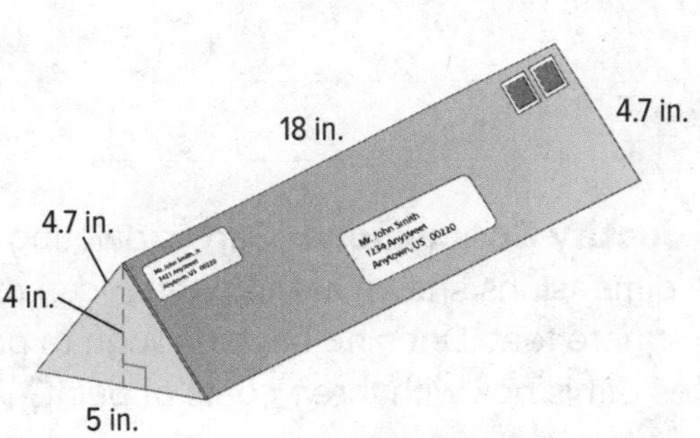

8. **MP Multiple Representations** The figure shows the dimensions of a triangular prism.

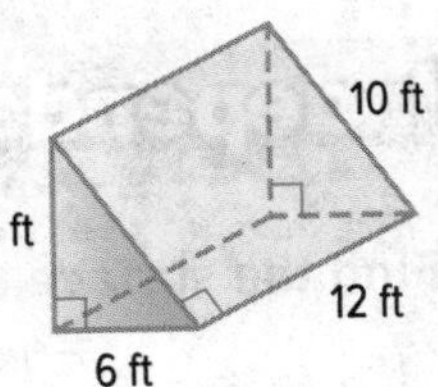

a. **Models** Draw a model of the faces and bases of the triangular prism.

b. **Words** Describe the triangular prism. ______

c. **Numbers** Find the surface area of the triangular prism using addition.

9. The surface area of a right triangular prism is 228 square inches. The base is a right triangle with a base height of 6 inches and a base length of 8 inches. The length of the third side of the base is 10 inches. Find the height of the prism. ______

H.O.T. Problems Higher Order Thinking

10. **MP Reason Abstractly** Describe the dimensions of a triangular prism that has a surface area between 550 square inches and 700 square inches.

11. **MP Persevere with Problems** Sketch and label two triangular prisms such that one has a greater volume and the other has a greater surface area.

12. **MP Justify Conclusions** Gary is painting a decorative box with the dimensions shown at the right. A can of paint covers about 25 square feet. Does he have enough to paint the rectangular faces of his box with three coats of paint? Justify your answer.

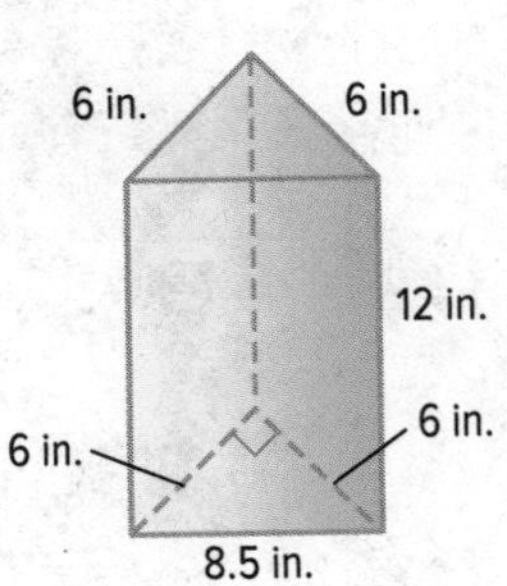

Extra Practice

MP Be Precise Find the surface area of each triangular prism. Round to the nearest tenth if necessary.

13. 537 ft²

Homework Help

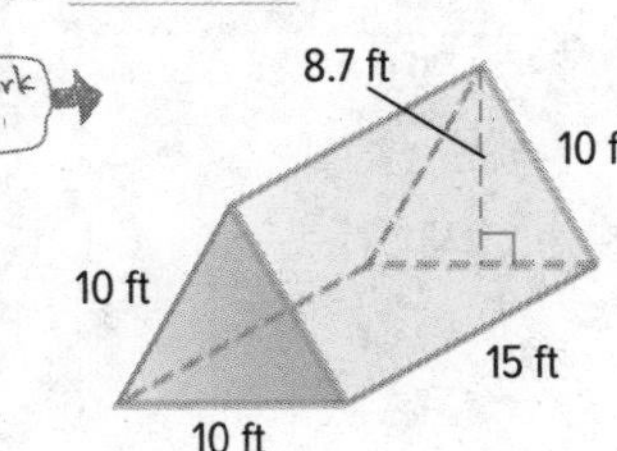

area of each base: $\frac{1}{2} \cdot 10 \cdot 8.7 = 43.5$ ft²

area of each face: $15 \cdot 10 = 150$ ft²

surface area $= 2(43.5) + 3(150)$
$= 537$ ft²

14. 331.9 m²

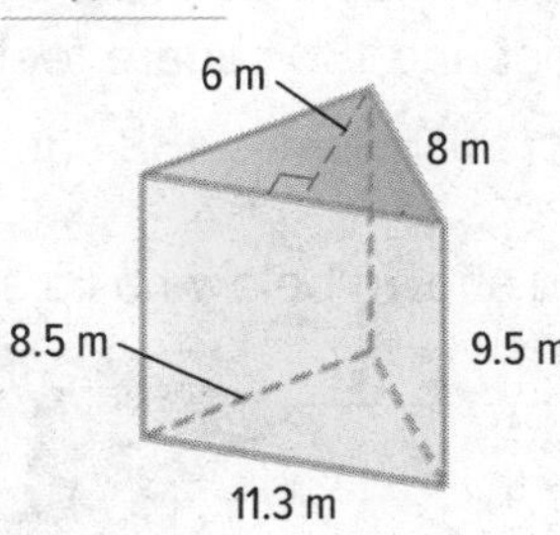

area of each base: $\frac{1}{2} \cdot 11.3 \cdot 6 = 33.9$ m²

areas of faces: $11.3 \cdot 9.5 = 107.35$ m²
$8.5 \cdot 9.5 = 80.75$ m²
$8 \cdot 9.5 = 76$ m²

surface area $= 33.9 + 33.9 + 107.35 + 80.75 + 76$ or 331.9 m²

15. ______

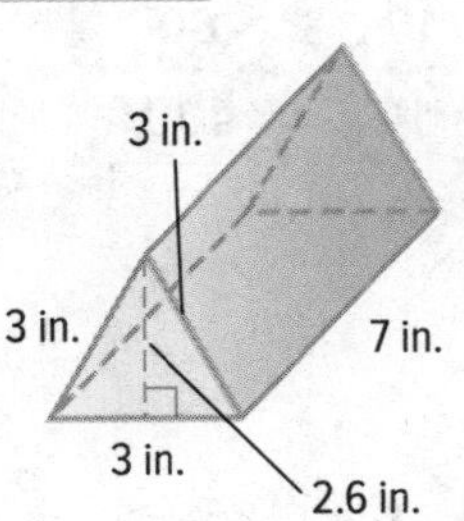

16. ______

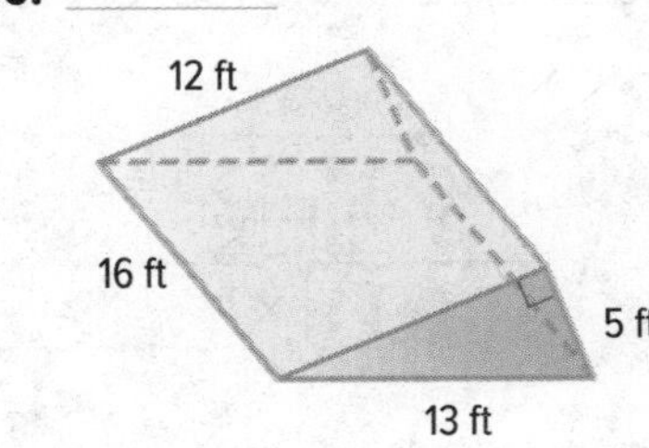

17. ______

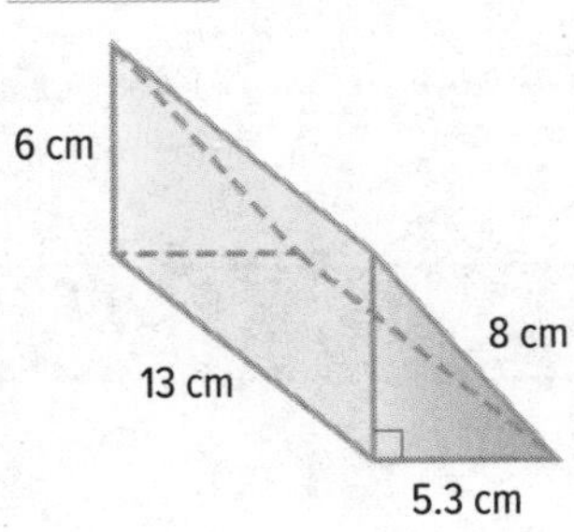

18. ______

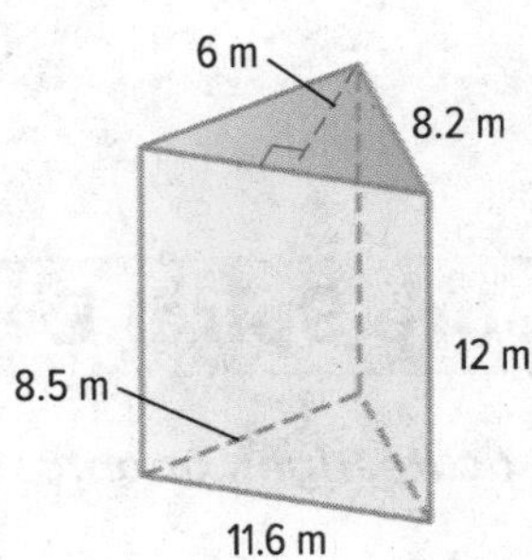

Copy and Solve Find the surface area of each triangular prism using the base triangles shown. Show your work on a separate piece of paper.

19.

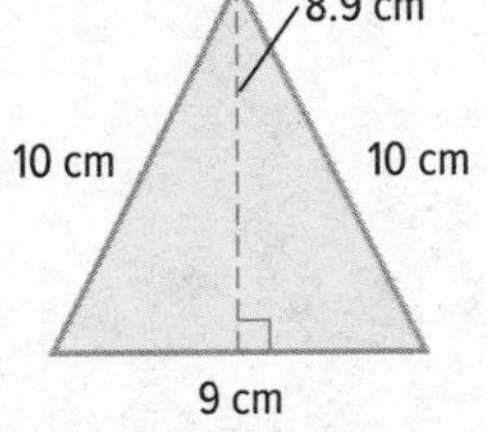

height of prism: 12 cm

20.

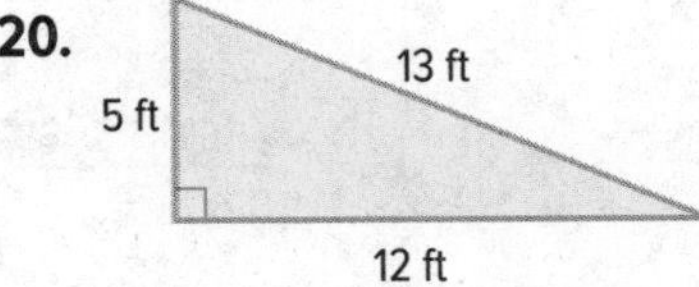

height of prism: 15 ft

Power Up! Common Core Test Practice

21. A triangular prism has the dimensions shown. Determine if each statement is true or false.

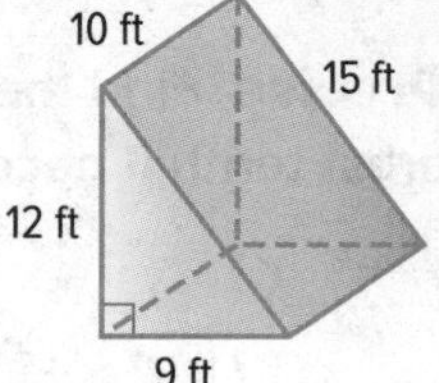

a. The combined areas of the bases is 54 ft^2. ☐ True ☐ False

b. The areas of the rectangular faces are 90 square feet, 120 square feet and 180 square feet. ☐ True ☐ False

c. The surface area of the prism is 468 square feet. ☐ True ☐ False

22. The attic of the house shown below has a solid wood floor.

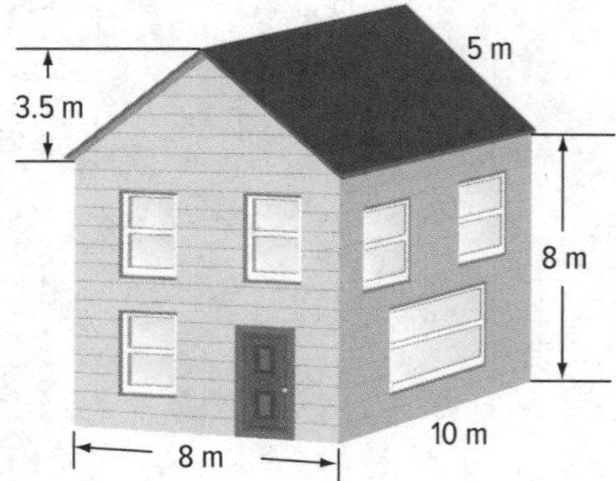

Select values to complete the model below to find how much wood is needed to make the roof of the house and the floor of the attic.

2	8	50
3.5	10	80
5	14	100

Attic Floor: ☐ × ☐ = ☐ m^2

Roof: ☐ × ☐ × ☐ = ☐ m^2

How many square meters of wood are needed to build the roof and the floor of the attic? ☐

Common Core Spiral Review

Identify each triangle as *acute, right,* or *obtuse.* 5.G.4

23. ______

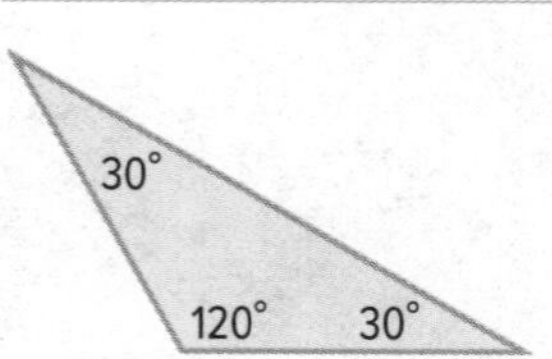

24. ______

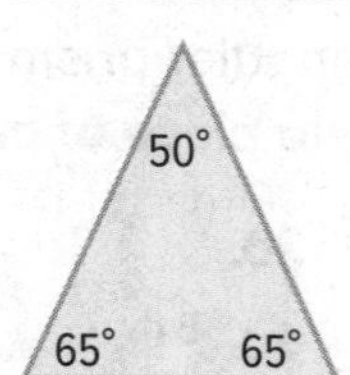

25. ______

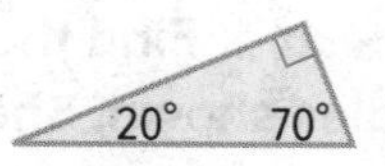

26. A certain two-dimensional figure has two pairs of parallel lines, four right angles, and four congruent sides. What is the figure? 4.G.2 ______

Inquiry Lab

Nets of Pyramids

HOW is the area of a triangle related to the surface area of a square pyramid?

Content Standards
6.G.4

Mathematical Practices
1, 3, 4

Anderson Art is designing a paper weight that is shaped like a square pyramid.

Hands-On Activity

Use orthogonal drawings to find the surface area of a square pyramid. A *square pyramid* is a three-dimensional figure with a square base and four triangular faces.

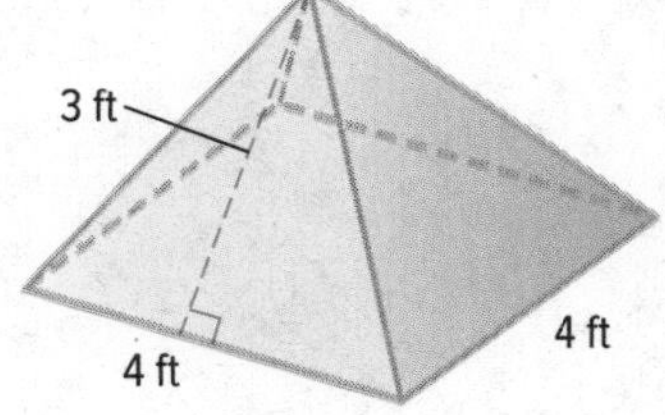

Step 1 Find the dimensions of each side of the square pyramid from the orthogonal drawing.

Orthogonal Drawing			
View	**Drawing**	**Dimensions (ft)**	**Area of Face (ft^2)**
Base	4 ft, 4 ft	length = 4 width = 4	$4 \times 4 = 16$
Triangular Faces	3 ft, 4 ft	height = 3 base = 4	$\frac{1}{2}(3 \times 4) = 6$

Step 2 Use grid paper to draw a net. Let 1 unit on the grid represent 1 foot. Check the dimensions of each face using the information in the table.

Step 3 Add the area of each face to find the surface area of the figure. Remember, there are four triangular faces.

☐ + ☐ × ☐ = ☐

So, the surface area is ☐ square feet.

Investigate

Collaborate

MP Model with Mathematics **Work with a partner. Use nets to determine the surface area of each pyramid. Draw a net of each pyramid on the provided grid paper.**

1. ______ cm^2

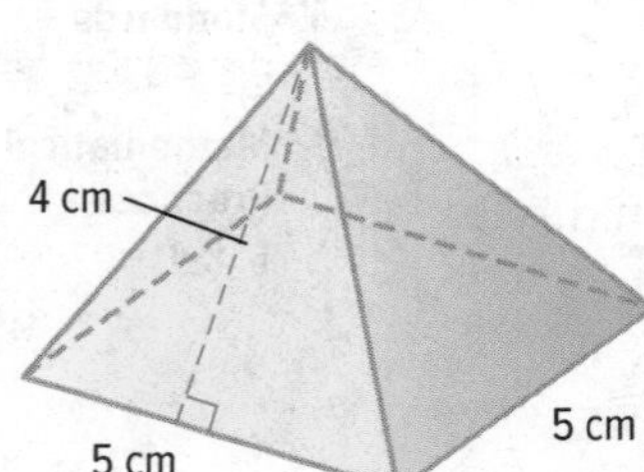

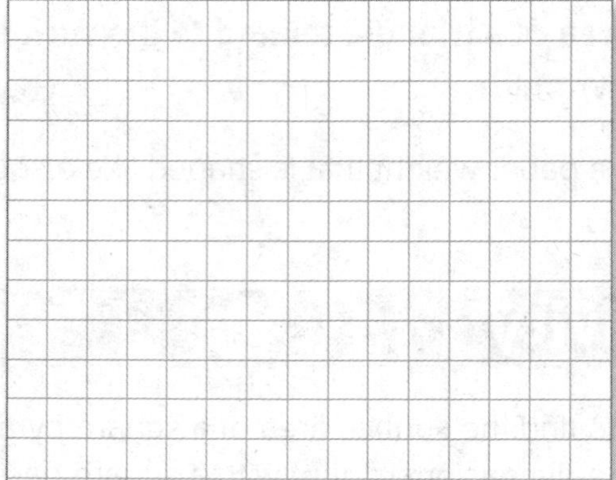

2. ______ m^2

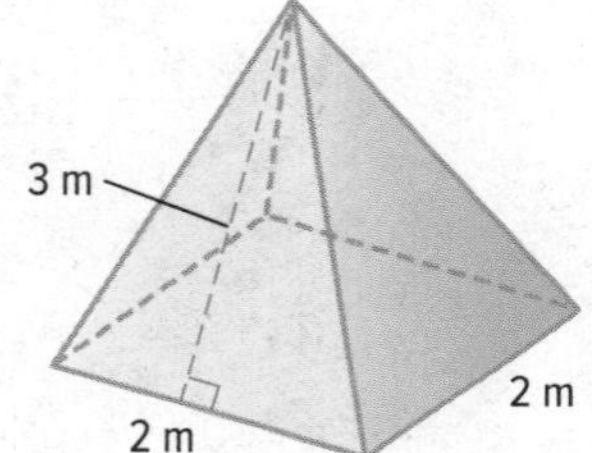

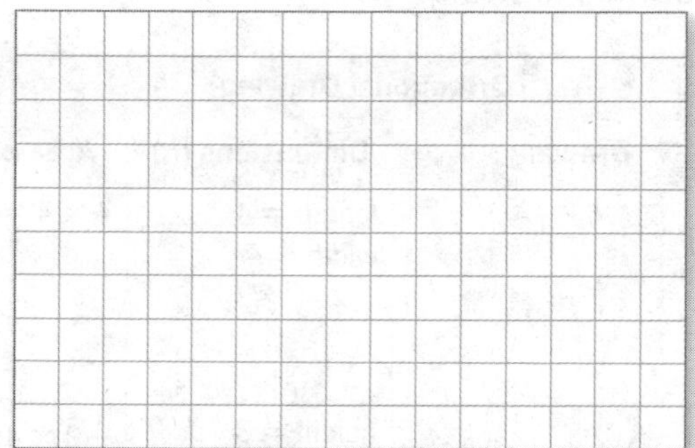

Create

On Your Own

3. **MP Construct an Argument** Explain how to find the surface area of a square pyramid, without creating a net. Use the dimensions in Exercise 1 to explain your answer.

____

____

____

____

4. **Inquiry** HOW is the area of a triangle related to the surface area of a square pyramid? ______________________________

____

____

Lesson 5

Surface Area of Pyramids

Vocabulary Start-Up

A **pyramid** is a three-dimensional figure with at least three triangular sides that meet at a common **vertex** and only one **base** that is a polygon. The triangular sides of a square pyramid are called the **lateral faces**. The **slant height** is the height of each lateral face.

Fill in the blanks on the diagram below with vocabulary words.

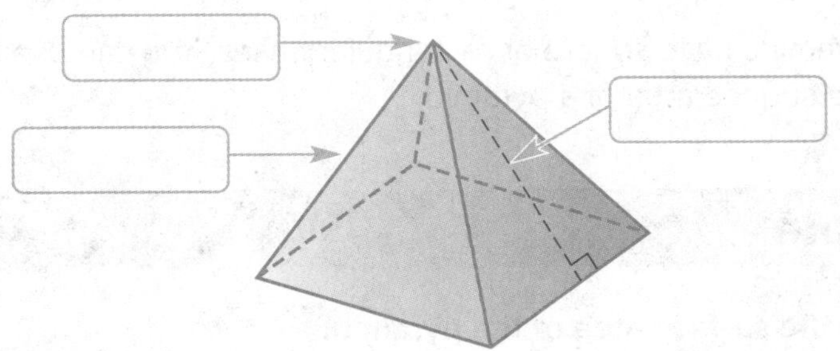

Essential Question

HOW is shape important when measuring a figure?

Vocabulary

pyramid
vertex
base
lateral face
slant height

Common Core State Standards

Content Standards
6.G.4

1, 3, 4, 6, 7

Real-World Link

Museum Claude made a model of the large pyramid in front of the Louvre museum. His model is shown.

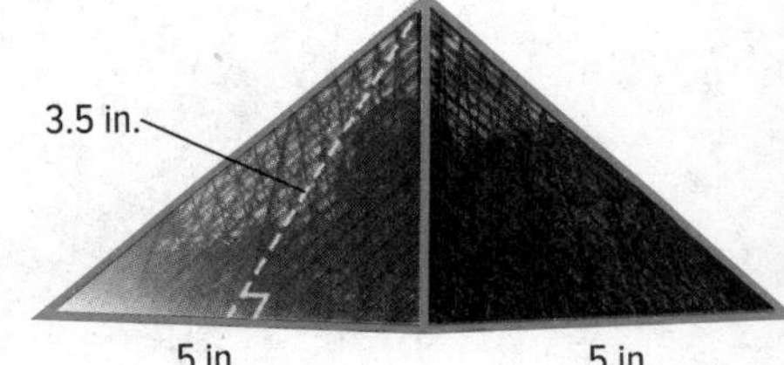

1. Draw the faces of the pyramid.

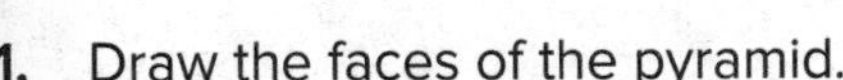

base lateral face lateral face lateral face lateral face

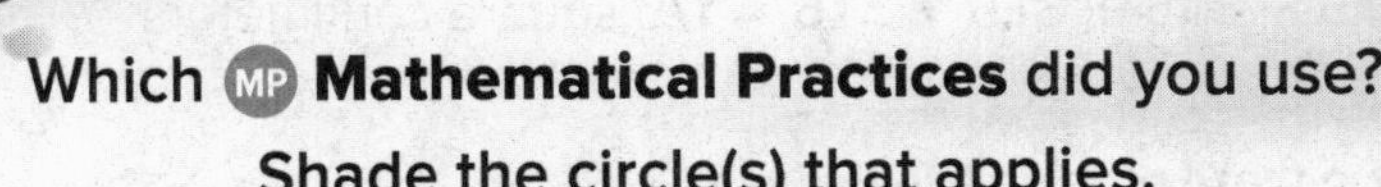

Which MP **Mathematical Practices** did you use? Shade the circle(s) that applies.

① Persevere with Problems
② Reason Abstractly
③ Construct an Argument
④ Model with Mathematics
⑤ Use Math Tools
⑥ Attend to Precision
⑦ Make Use of Structure
⑧ Use Repeated Reasoning

Key Concept

Surface Area of a Pyramid

Work Zone

Words The surface area of a pyramid is the sum of the area of the base and the areas of the lateral faces.

Model

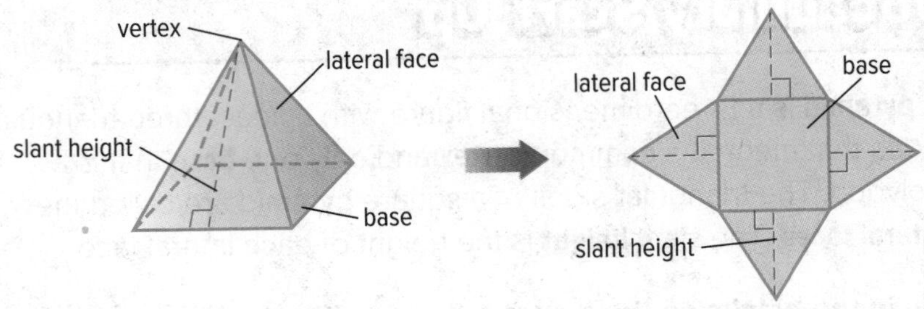

Some pyramids have square or rectangular bases. You can use a net to find the surface area of a pyramid.

Example

1. Find the surface area of the pyramid.

Use a net to find the area of each face and then add.

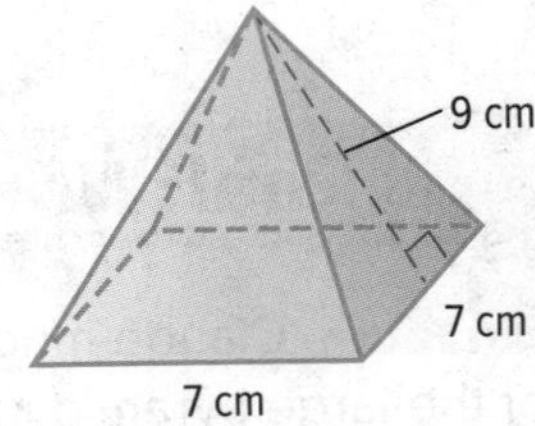

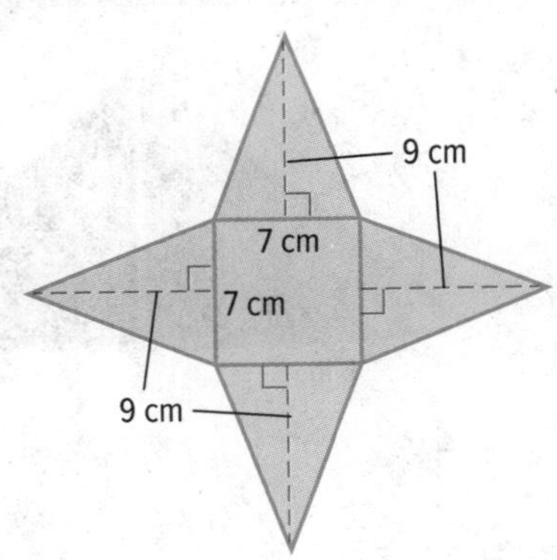

area of base: 7(7) = 49

area of each triangular side: $\frac{1}{2}(7)(9) = 31.5$

Add to find the surface area.

49 + 31.5 + 31.5 + 31.5 + 31.5 = 175 square centimeters

Got it? Do these problems to find out.

a. ______

b. ______

a.

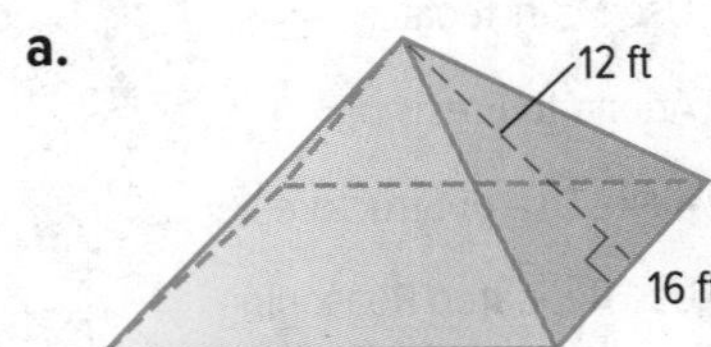

b.

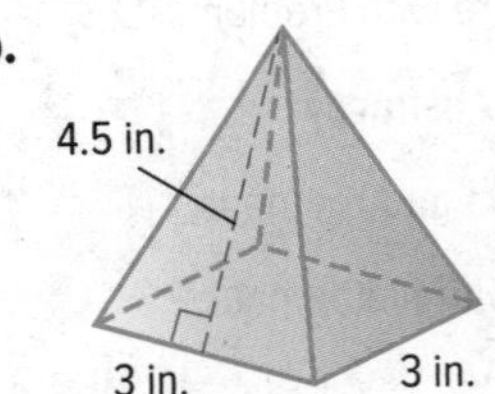

Surface Area of Pyramids with Triangular Bases

A triangular pyramid has one triangular base, and three triangular faces. If the base is an equilateral triangle, all three lateral faces are congruent. If the sides of the base triangle are different lengths, the areas of the lateral faces will also vary.

Example

2. Find the surface area of the pyramid.

Find the area of each face and add. The triangular base is an equilateral triangle because all three sides are 4 feet long.

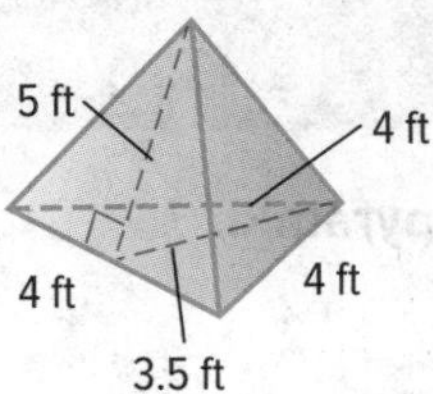

base **lateral faces**

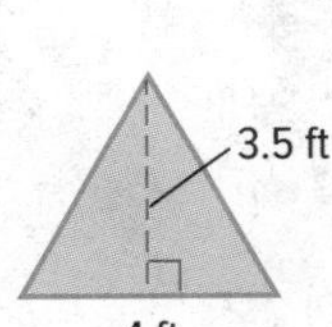

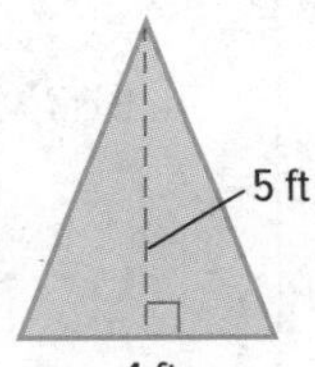

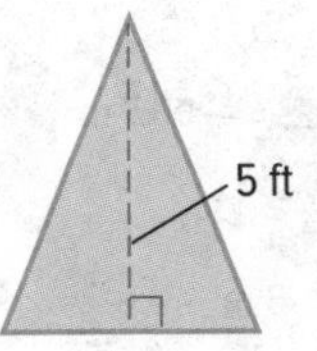

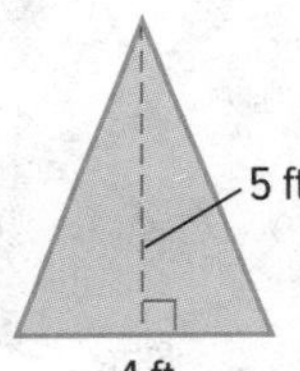

area of base: $\frac{1}{2}(4)(3.5) = 7$

area of each lateral face: $\frac{1}{2}(4)(5) = 10$

Add to find the surface area.
$7 + 10 + 10 + 10 = 37$ square feet

Got it? **Do these problems to find out.**

c.

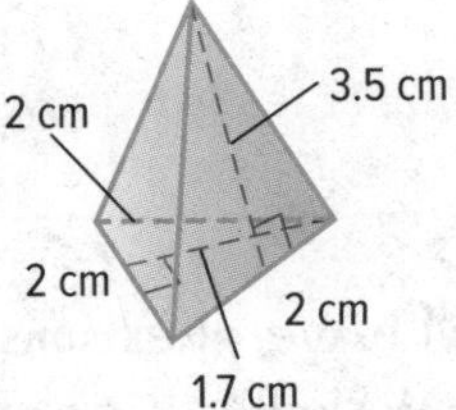

d.

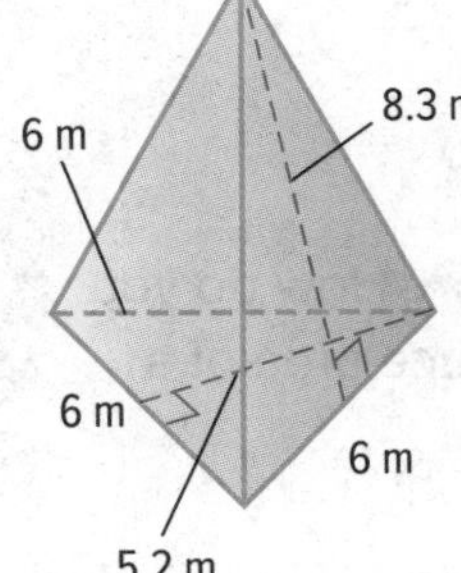

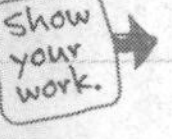

c. ______________

d. ______________

Example

3. A pyramid puzzle has all sides that are equilateral triangles. Each triangle has side lengths of 8 centimeters. The slant height is 6.9 centimeters. Find the surface area of the puzzle.

Create a net and then use it to find the surface area of the pyramid.

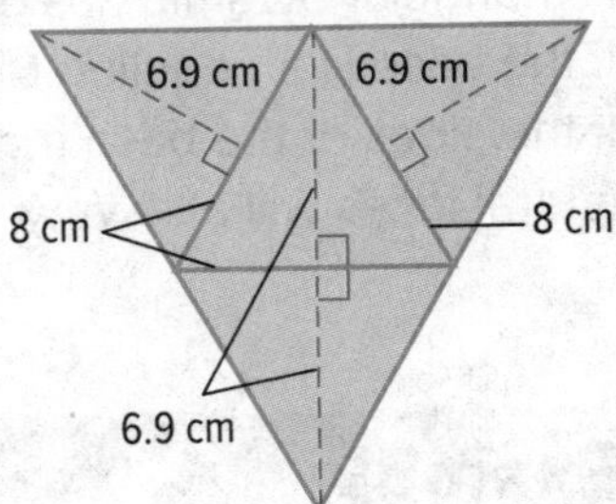

Each face has an area of $\frac{1}{2}(8)(6.9)$ or 27.6 square centimeters. So, the surface area of the puzzle is $4 \cdot 27.6$ or 110.4 square centimeters.

Guided Practice

Find the surface area of each pyramid. (Examples 1–2)

1. ______

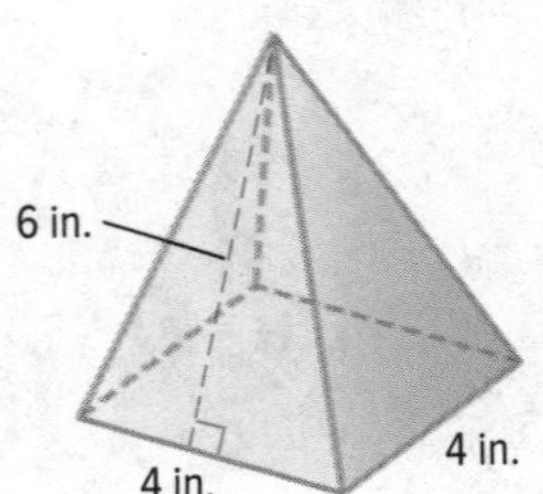

Show your work.

2. ______

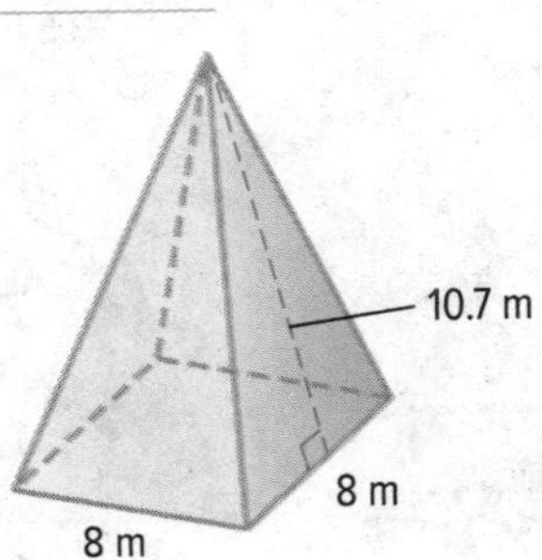

3. ______

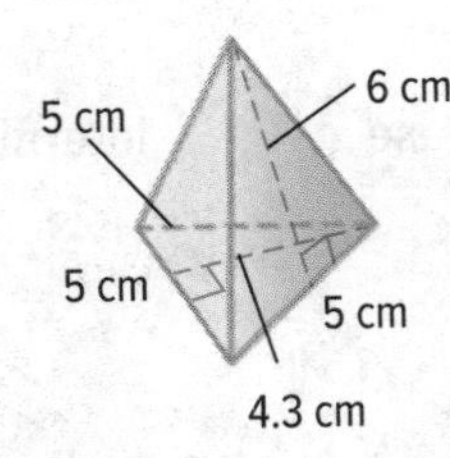

4. Pyramid-shaped gift boxes have square bases that measure 5 inches on each side. The slant height is 6.5 inches. How much cardboard is used to make each box? (Example 3)

______

5. **Building on the Essential Question** How do you use the area of a triangle to find the surface area of a triangular pyramid?

______

______

______

Rate Yourself!

☐ I understand surface area of pyramids.

Great! You're ready to move on!

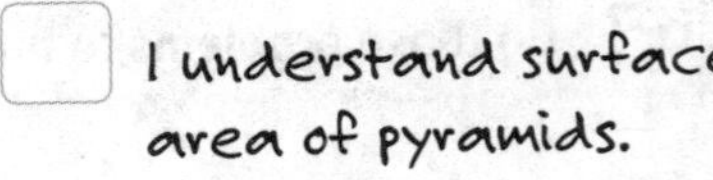

☐ I still have questions about surface area of pyramids.

No Problem! Go online to access a Personal Tutor.

FOLDABLES Time to update your Foldable!

Independent Practice

Go online for Step-by-Step Solutions

Find the surface area of each pyramid. (Examples 1–2)

1. ______

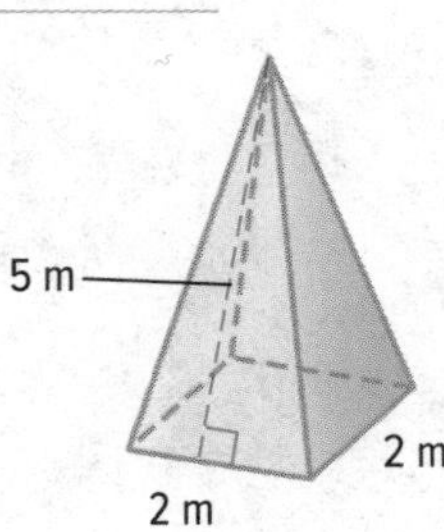

2. ______

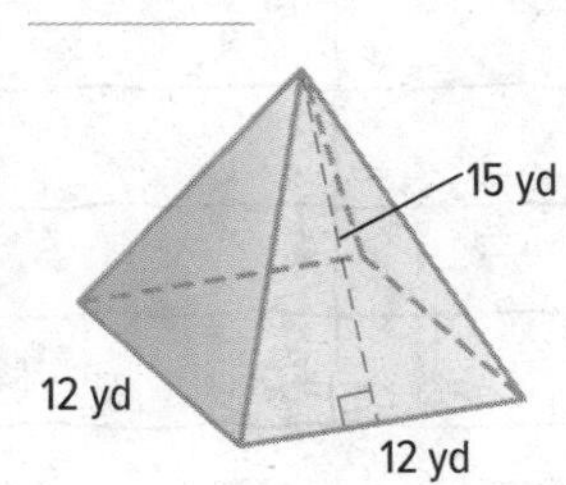

3. ______

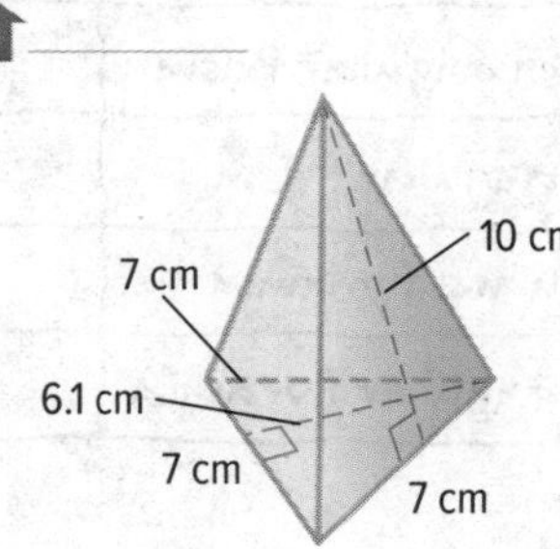

4. ______

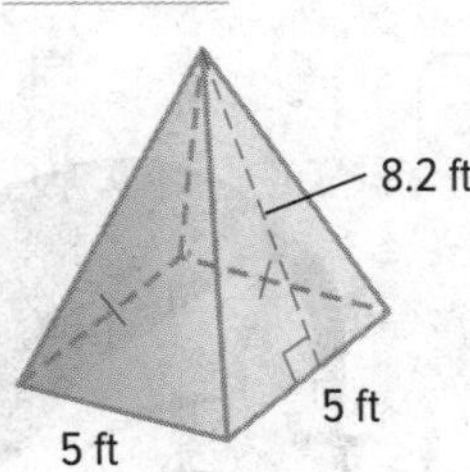

5. ______

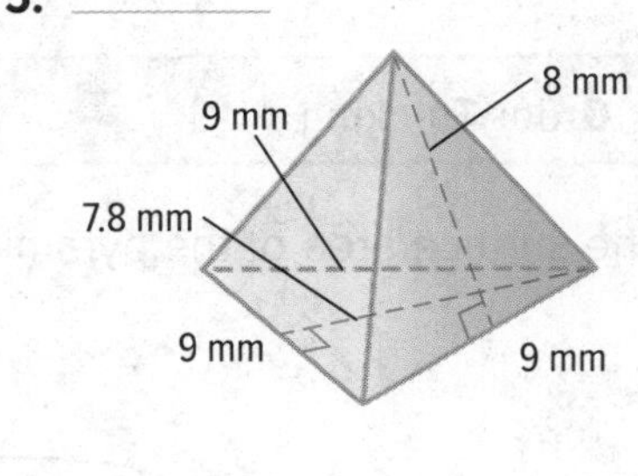

6. ______

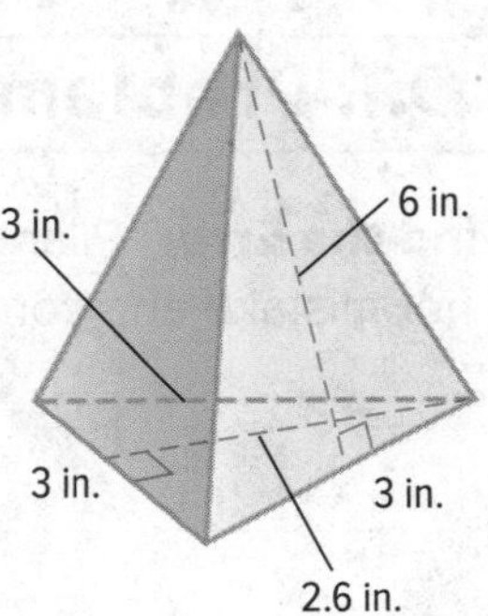

7. A tea bag is shaped like a square pyramid with the base measuring 4 centimeters on each side. The slant height is 4.5 centimeters. How much mesh is used to create the tea bag? (Example 3)

______

8. An earring design is shaped like a triangular pyramid. All the faces are equilateral triangles with side lengths of 14 millimeters. The slant height is 12.1 millimeters. What is the surface area of the earring? (Example 3)

______

9. An acting award is a square pyramid with a base that measures 6 inches on each side. The slant height is 8 inches. What is the surface area of the award? (Example 3)

______

10. **MP Identify Structure** Refer to the figures listed in the table. Determine the number of faces the figure has of each two-dimensional shape. Explain.

Figure	Rectangular Faces	Triangular Faces
Rectangular Prism		
Triangular Prism		
Square Pyramid		
Triangular Pyramid		

H.O.T. Problems Higher Order Thinking

11. **MP Find the Error** Pilar is finding the surface area of the pyramid shown. Find her mistake and correct it.

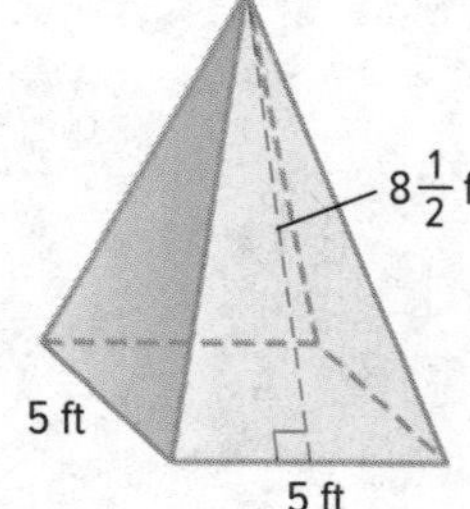

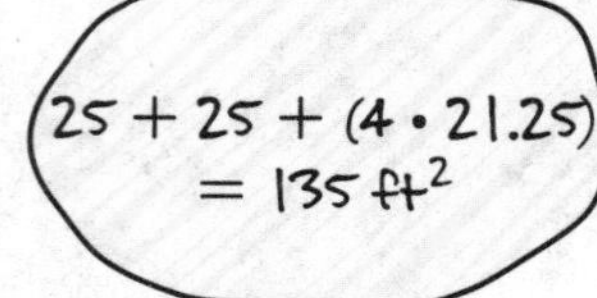

12. **MP Persevere with Problems** The *lateral surface area L.A.* of a pyramid is the area of its lateral faces. Use the square pyramid at the right to complete each step to find the lateral surface area of any pyramid.

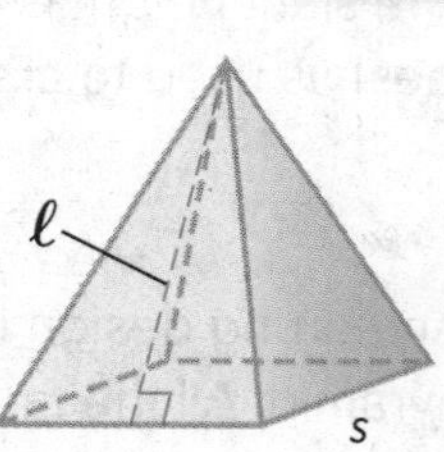

$L.A. = \frac{1}{2}s\ell +$ ______ Lateral surface area

$= \frac{1}{2}($ ______ $)\ell$ Distributive Property

$=$ ______ Perimeter of base: $P = s + s + s + s$

13. **MP Justify Conclusions** Suppose you could climb to the top of the Pyramid Arena in Memphis, Tennessee. Which path would be shorter, climbing a lateral edge or the slant height? Justify your response.

Extra Practice

Find the surface area of each pyramid.

14. 55 m²

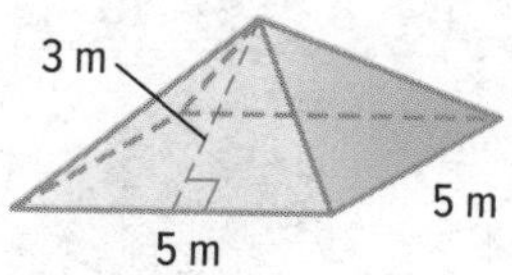

Homework Help

area of base: $5 \cdot 5 = 25$ m²

area of each face: $\frac{1}{2} \cdot 5 \cdot 3 = 7.5$ m²

surface area $= 25 + (4 \cdot 7.5)$

$= 25 + 30$ or 55 m²

15. 223.5 ft²

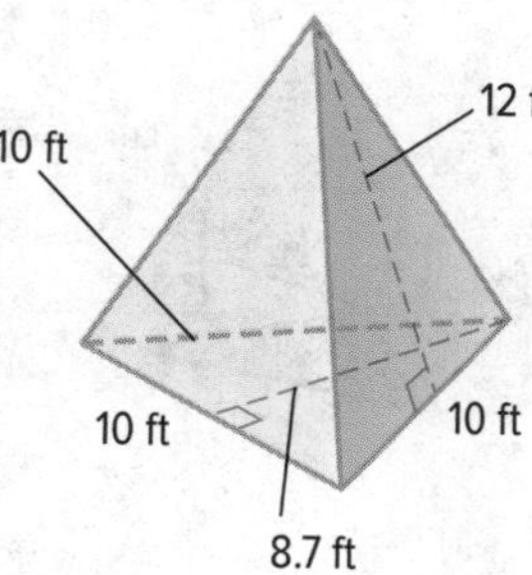

area of base: $\frac{1}{2} \cdot 10 \cdot 8.7 = 43.5$ ft²

area of each face: $\frac{1}{2} \cdot 10 \cdot 12 = 60$ ft²

surface area $= 43.5 + (3 \cdot 60)$

$= 43.5 + 180$ or 223.5 ft²

16. ______

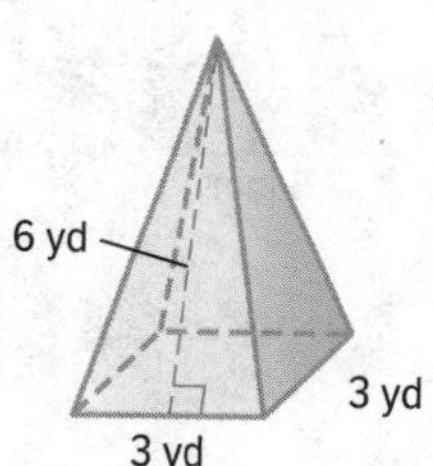

17. ______

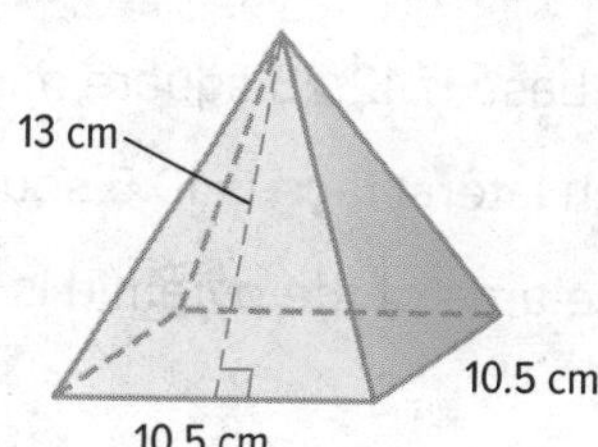

18. ______

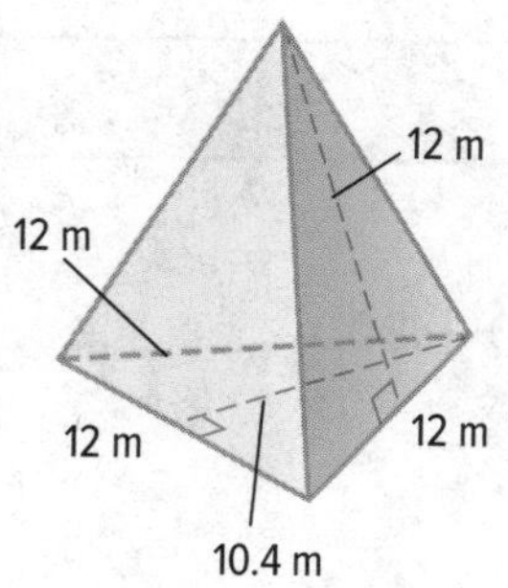

19. ______

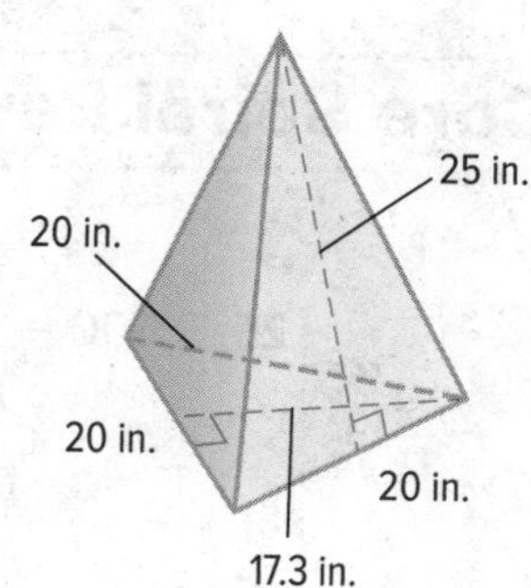

20. A paper model of the Khafre pyramid in Egypt has a square base 7.2 centimeters on each side. The slant height is 6 centimeters How much paper was used to make the model?

______

21. MP **Be Precise** A triangular pyramid has a surface area of 336 square inches. It is made up of equilateral triangles with side lengths of 12 inches. What is the slant height?

______

CCSS Power Up! Common Core Test Practice

22. A salt shaker is shaped like a square pyramid. The perimeter of the base is 16 cm, the height of the salt shaker is 10 cm, and the slant height is about 10.2 cm. Select values to label the net below with the correct dimensions.

2	10
4	10.2
8	16

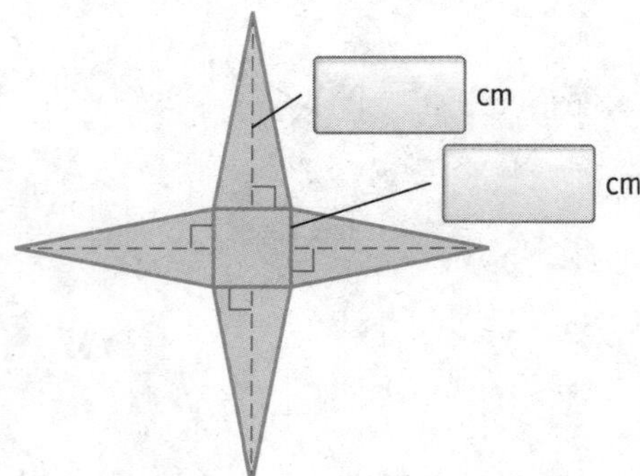

What is the surface area of the salt shaker? ______

23. A square pyramid has the dimensions shown. Determine if each statement is true or false.

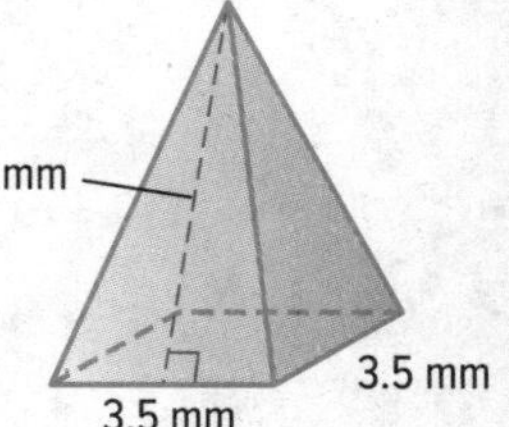

a. The pyramid has 1 base and 3 lateral faces. ☐ True ☐ False

b. The area of the base is 12.25 square millimeters. ☐ True ☐ False

c. The area of each lateral face is 10.5 square millimeters. ☐ True ☐ False

d. The total surface area of the pyramid is 54.25 mm^2. ☐ True ☐ False

CCSS Common Core Spiral Review

Divide. 5.NBT.6

24. $240 \div 10 =$ ______

25. $3,600 \div 36 =$ ______

26. $4,800 \div 80 =$ ______

27. Jalisa and two of her friends are sharing the cost of a taxi ride to the airport. The taxi ride costs $24.75. How much will each person pay? 5.NBT.7

______

28. How many centimeters are equal to 0.05 meters? 5.MD.1

______

21ST CENTURY CAREER in Design

Interior Designer

Do you like coming up with new ways to decorate your room, or are you always rearranging the furniture? You could have a career doing just that by becoming an interior designer. Interior designers plan the interior space and furnishings of homes, offices, and other places. Their designs are based on the client's specifications, tastes, and budget. Interior designers are responsible for recommending color schemes, furniture, lighting, and remodeling options. Many interior designers also develop their own product lines such as furniture, bedding, and accessories.

Is This the Career for You?

Are you interested in a career as an interior designer? Take some of the following courses in high school.

- Algebra
- Geometry
- Interior Design
- Intro to CAD

Turn the page to find out how math relates to a career in Design.

MP You be the Designer!

Use the labeled figures to solve each problem. Round to the nearest tenth if necessary.

1. A client wants to buy the rectangular ottoman with the most storage area inside. Which one should she choose? Explain your reasoning. ____________________

2. Find the volume of the paisley blanket chest. ____________________

3. What is the volume of the toy chest? How does it compare to the volume of the paisley blanket chest? ____________________

4. A designer is having the red ottoman reupholstered. If the bottom is not covered, estimate the amount of fabric needed. ____________________

5. How much fabric is needed to cover the purple ottoman? ____________________

6. How much greater is the surface area of the paisley blanket chest than the surface area of the toy chest? ____________________

Red Ottoman

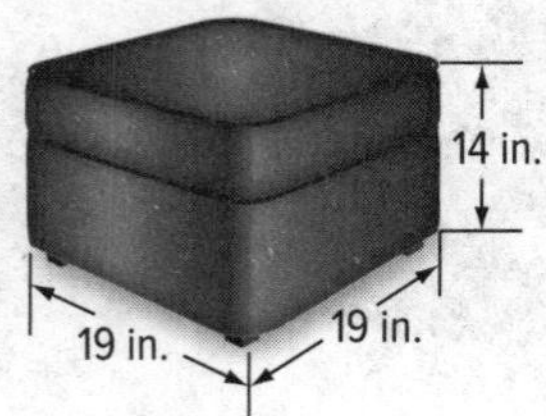

Purple Ottoman

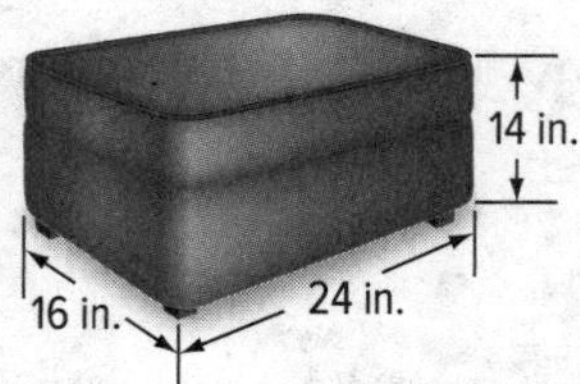

Blanket Chest

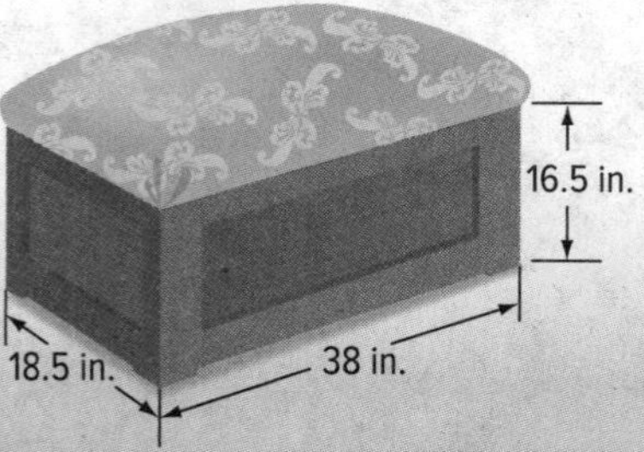

Toy Chest

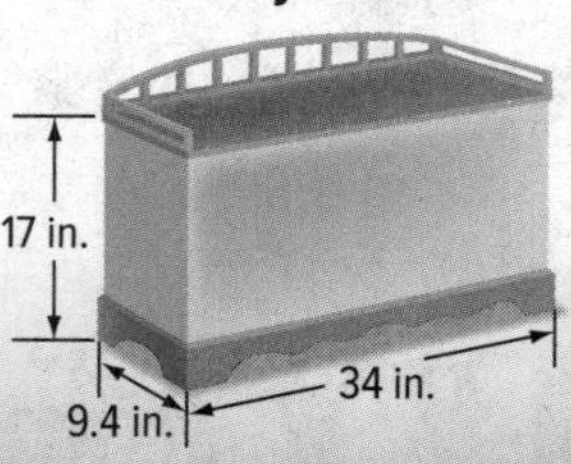

MP Career Project

It's time to update your career portfolio! Use grid paper to make a scale drawing of a room in your home. Model the furniture using squares, rectangles, and triangles drawn to scale. Cut out each shape and use them to create different room arrangements. Then, tape the pieces onto the grid paper. Describe the room's color scheme and style.

____

Do you think you would enjoy a career as an interior designer? Why or why not?

____

Chapter Review

Vocabulary Check

Complete each sentence using the vocabulary list at the beginning of the chapter. Then circle the word that completes the sentence in the word search.

1. A figure with length, width, and height is a ______________.

2. ______________ is the sum of the area of all the faces of a three-dimensional figure.

3. The amount of space inside a three-dimensional figure is its ______________.

4. A prism that has triangular bases is a ______________.

5. A ______________ is a prism that has rectangular bases.

6. Volume is measured in ______________.

7. The point where three or more faces intersect is the ______________.

8. The ______________ is the height of each lateral face.

9. Any face that is not a base is a ______________.

Key Concept Check

Use Your FOLDABLES®

Use your Foldable to help review the chapter.

Tape here

Tab 1

Real-World Examples

Formulas

Model

Tab 2

Tape here

Got it?

Use the figure provided to complete the cross number puzzle.

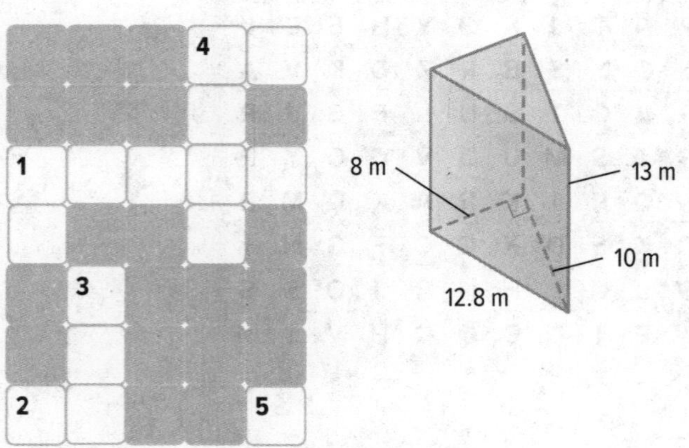

Across

1. surface area of the prism
2. height of the base triangle
4. height of the prism
5. length of the base triangle

Down

1. area of the base
3. volume of the prism
4. length of one side of the base triangle

Moving Time

The Davidsons are moving into a new house and have rented a trailer to transport boxes for the move. They bought cardboard boxes, like the one shown, in which they will pack their belongings. The trailer has 180 cubic feet of cargo space. The height of the trailer is 7.5 feet and the width is 4 feet. The perimeter of the base of the trailer is 20 feet.

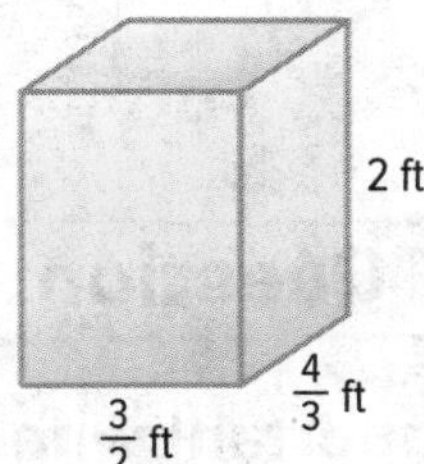

Write your answers on another piece of paper. Show all of your work to receive full credit.

Part A
The Davidsons need to know the dimensions of the trailer so they can maximize space while packing boxes. What is the length and the width of the trailer in feet?

Part B
If the boxes can be put in the trailer in any position, what is the greatest number of boxes that will fit in the trailer? How many boxes will fit in the trailer if each box must be loaded as shown in the image (with 2 feet as the height).

Part C
The family is taking three gift-wrapped presents (using the same boxes). How much wrapping paper will they need? Draw a net to represent one of the boxes.

Reflect

Answering the Essential Question

Use what you learned about volume and surface area to complete the graphic organizer.

Essential Question

HOW is shape important when measuring a figure?

	Draw it.	How do you find the volume?	How do you find the surface area?
rectangular prism			
triangular prism			

Answer the Essential Question. HOW is shape important when measuring a figure?

UNIT PROJECT

Watch

A New Zoo A zoo is a great place to explore wild animals and learn about their habitats. In this project you will:

- **Collaborate** with your classmates as you explore some animals at the zoo and design your own zoo.
- **Share** the results of your research in a creative way.
- **Reflect** on how you use different measurements to solve real-life problems.

By the end of this Project, you may be interested in working at the zoo or even working as a designer to help create new living areas for the animals.

Collaborate

Go Online Work with your group to research and complete each activity. You will use your results in the Share section on the following page.

1. Choose 10 zoo animals. Research various characteristics of each animal, such as average weight, lifespan, incubation period, and the temperature of its natural habitat. Write a brief summary for each animal that you choose.

2. Create a bar graph that shows the average weight, the average lifespan, and the average incubation period for the 10 animals you chose.

3. Organize the characteristics found in Exercise 1 for each animal in a table or spreadsheet. Then describe how you could use these characteristics to help you design the animals' living spaces.

4. Research the amount of living space needed for each animal. Use this information to design and draw your own zoo. Be sure to include the dimensions and area. Which animals have the largest living areas? Explain why.

5. Find the area of each animal's enclosure that you designed in Exercise 4. Also, find the volume and surface area of any buildings at the zoo you designed.

With your group, decide on a way to present what you have learned about designing a zoo. Some suggestions are listed below, but you can also think of other creative ways to present your information. Remember to show how you used mathematics to complete this project!

- Design a web page that can be used to describe the zoo. Some questions to consider are:
  - Which zoo attractions should be promoted to get more tourists to visit your zoo?
  - Include a map of your zoo.
- Design a living area for a giant panda exhibit. Be sure to include drawings and explanations of why you designed the exhibit the way you did.

Check out the note on the right to connect this project with other subjects.

Environmental Literacy

Research the living conditions for animals in zoos today compared to the living conditions of the past. How has it changed over time? Things to consider:

- size of living space
- average lifespan difference
- behavior changes

6. **Answer the Essential Question** How can you use different measurements to solve real-life problems?

   a. How did you use what you learned about area to solve real-life problems?

   b. How did you use what you learned about volume and surface area to solve real-life problems?

UNIT 5

Statistics and Probability

Essential Question

WHY is learning mathematics important?

Chapter 11
Statistical Measures

Statistical data has a distribution that can be described by its center or by its spread. In this chapter, you will find and use measures of center and measures of variation to describe sets of data.

Chapter 12
Statistical Displays

Statistical data can be represented in a variety of ways. In this chapter, you will represent and analyze data using line plots, histograms, and box plots.

Unit Project Preview

Let's Exercise Pediatricians recommend that children and teens do 60 minutes or more of physical activity each day to promote physical fitness. That includes riding bikes, skateboarding, dancing, and even just walking to school.

Survey twenty students about the sports or other physical activities they participate in each week. Then make a bar graph of the top five activities.

At the end of Chapter 12, you'll complete a project about physical fitness. Grab your sneakers as you get ready to run with this exciting task.

Participating in Physical Activities

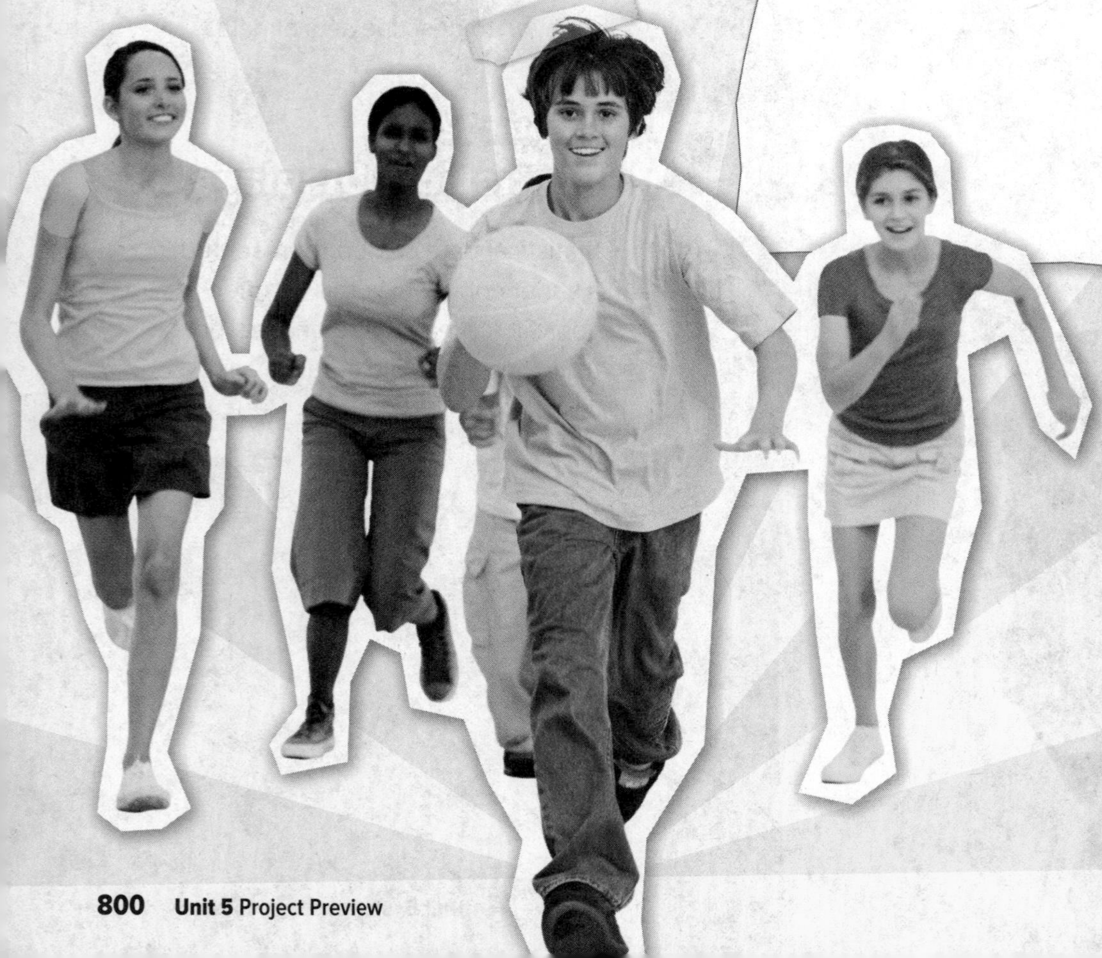

Chapter 11
Statistical Measures

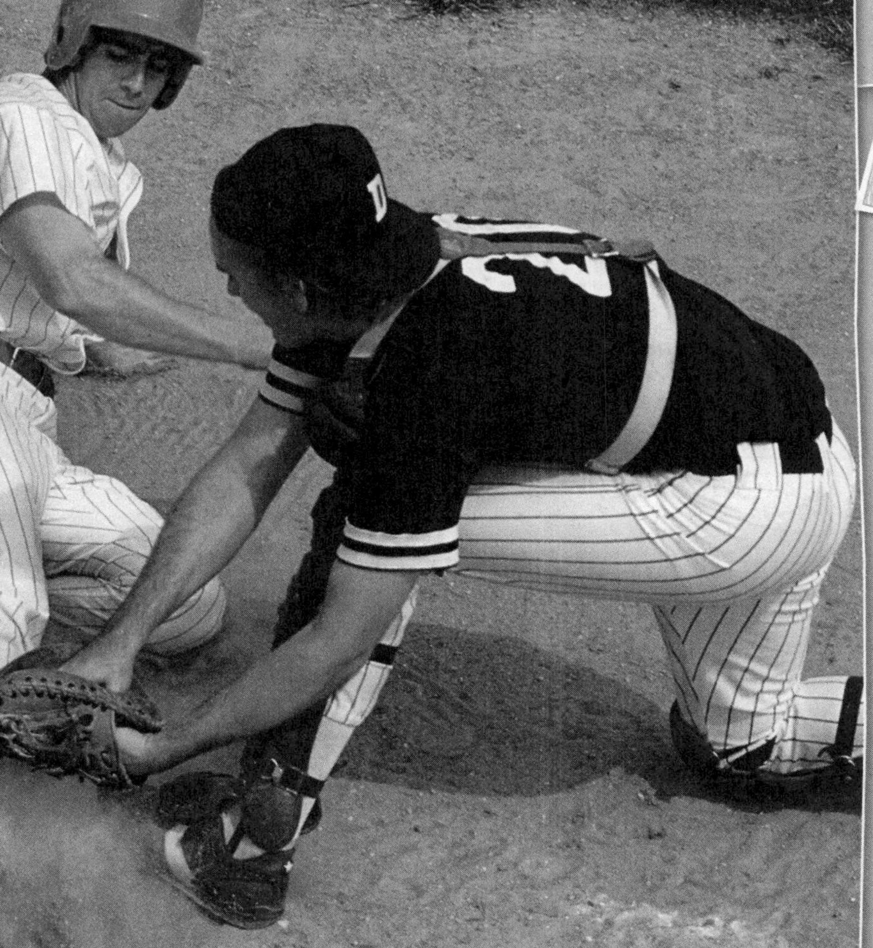

Essential Question

HOW are the mean, median, and mode helpful in describing data?

Common Core State Standards

Content Standards
6.SP.1, 6.SP.3, 6.SP.5, 6.SP.5b, 6.SP.5c, 6.SP.5d

MP Mathematical Practices
1, 2, 3, 4, 5, 6

Math in the Real World

Sports A baseball team scored 9, 6, 8, 16, and 5 points in 5 games. Plot the scores on the number line.

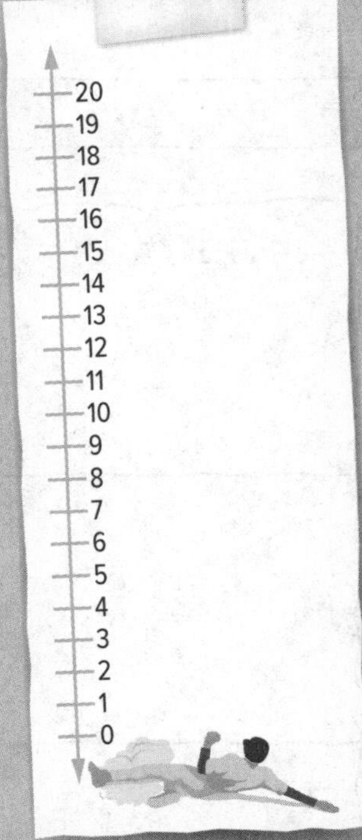

FOLDABLES® Study Organizer

1. Cut out the Foldable on page FL13 of this book.
2. Place your Foldable on page 856.
3. Use the Foldable throughout this chapter to help you learn about statistical measures.

What Tools Do You Need?

Vocabulary

- average
- first quartile
- interquartile range
- mean
- mean absolute deviation
- measure of center
- measures of variation
- median
- mode
- outliers
- quartiles
- range
- statistical question
- third quartile

Review Vocabulary

Graphic Organizer One way to remember vocabulary terms is to connect them to an opposite term or example. Use this information to complete the graphic organizer.

quotient

Definition

Opposite

Example

What Do You Already Know?

Read each statement. Decide whether you agree (A) or disagree (D). Place a checkmark in the appropriate column and then justify your reasoning.

Statistical Measures			
Statement	A	D	Why?
The median of a data set is the same thing as the average of the data set.			
The range is the difference between the least and greatest numbers in a data set.			
A measure of variation describes the change in values in a set of data.			
The measures of center include mean, median, and mode.			
A statistical question is a question that anticipates and accounts for a variety of answers.			
The first quartile is the same as the median of a data set.			

When Will You Use This?

Here are a few examples of how statistics are used in the real world.

Activity 1 What is your favorite sports team? Use the Internet to find the number of wins your team has had in each of the last five seasons. Compare your teams wins to the number of wins by another person's favorite team.

Activity 2 Go online at **connectED.mcgraw-hill.com** to read the graphic novel ***Baseball Challenge***. Did the two baseball teams ever have a season in which they had the same number of wins?

Are You Ready?

Try the Quick Check below.
Or, take the Online Readiness Quiz.

Quick Review

Common Core Review 5.NBT.7

Example 1

Find 12.53 + 9.87 + 16.24 + 22.12.

$$\begin{array}{r} \scriptstyle 2\,1\;\;1 \\ 12.53 \\ 9.87 \\ 16.24 \\ +\ 22.12 \\ \hline 60.76 \end{array}$$

Add.

Example 2

Michelle read 56.5 pages of her book on Monday and Tuesday. If she read the same amount of pages each day, how many pages did she average each day?

56.5 ÷ 2 = 28.25 Divide the total number of pages by the number of days.

Michelle averaged 28.25 pages per day.

Quick Check

Add Decimals **Find each sum.**

1. 6.20 + 31.59 + 11.11 + 19.85 =

2. 22.69 + 15.45 + 9.87 + 26.79 =

3. Sonya went to the baseball game. She paid $10.50 for admission. She bought a drink for $2.75, a bag of popcorn for $4.60, and a hot dog for $3.75. How much did she spend in total?

Divide Decimals **Find each quotient.**

4. 79.2 ÷ 6 =

5. 72.60 ÷ 3 =

6. 240.5 ÷ 13 =

7. The Chen family drove 345.6 miles on their vacation. They drove the same amount each of the 3 days. How many miles did they drive each day?

How Did You Do?

Which problems did you answer correctly in the Quick Check? Shade those exercise numbers below.

      7

Inquiry Lab
Statistical Questions

HOW are surveys created to collect and analyze data?

Content Standards
6.SP.1, 6.SP.3

Mathematical Practices
1, 3, 4

Anderson Advertising is collecting information for a pizza shop. They want to know the number of toppings most customers prefer on their pizza. They will use this information to determine the weekly special.

Hands-On Activity 1

Statistics deals with collecting, organizing, and interpreting pieces of information, or *data*. One way to collect data is by asking statistical questions. A **statistical question** is a question that anticipates and accounts for a variety of answers.

The table below gives some examples of statistical questions and questions that are *not* statistical questions.

Statistical Questions	Not Statistical Questions
How many text messages do you send each day?	What is the height in feet of the tallest mountain in Colorado?
What is the minimum driving age for each state in the United States?	How many people attended last night's jazz concert?

Create a survey similar to the one Anderson Advertising would use to survey your classmates. Consider a cheese pizza with no additional toppings as a pizza with one topping.

Step 1 Write a statistical question. *How many toppings do you like on your pizza?*

Step 2 Survey your classmates.

Step 3 Record the results in the table to the right. Add additional numbers of toppings to the table as necessary.

How Many Toppings Do You Like on Your Pizza?	
Number of Toppings	**Number of Responses**

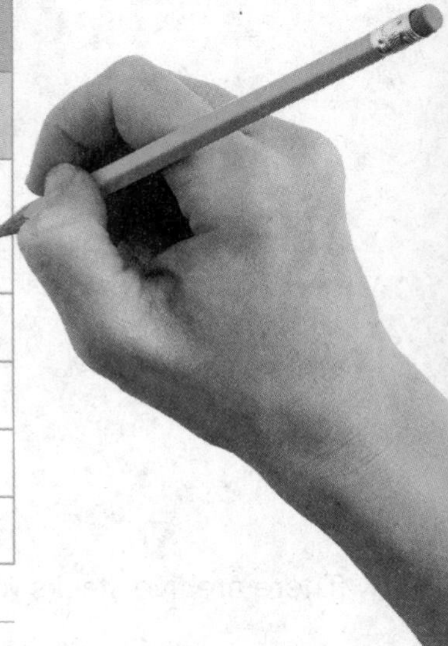

Why is the following a statistical question? *How many toppings do you like on your pizza?*

____

____

Hands-On Activity 2

Sometimes a set of data can be organized into intervals to more easily organize it. This often happens when the set of data has a wide range of values.

Suppose you want to determine the number of video games each of your math classmates has at home.

Step 1 Write the statistical question. *How many different video games do you own?*

Step 2 Survey your classmates.

Step 3 Record the results in the table to the right.

How Many Different Video Games Do You Own?	
Number of Video Games	**Number of Responses**
less than 5	
5–9	
10–14	
15 or more	

Hands-On Activity 3

You can use surveys to provide information about patterns in the responses.

Suppose you surveyed five students using the statistical question, *How many Web sites did you visit before school this morning?* The students said 4, 3, 5, 1, and 2 Web sites. If the total amount was equally distributed among all five students, how many Web sites did each student visit?

Step 1 Make a stack of centimeter cubes to represent the number of Web sites visited by each student as shown.

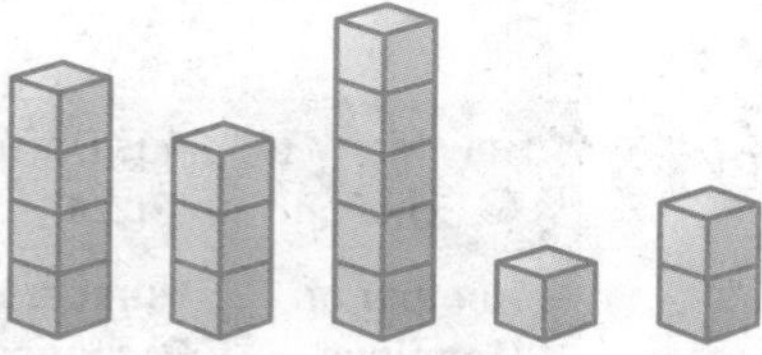

Step 2 Move the cubes so that each stack has the same number of cubes. Draw your models in the space below.

There are five stacks with ☐ cubes in each stack. So, if the responses were equally distributed, each student visited ☐ Web sites before school.

Work with a partner. State whether each question is a statistical question. Explain your reasoning.

1. Who was the first president of the United States?

____

2. How much time do the students in my school spend on the Internet each night?

____

3. What is the height of the tallest waterslide at Wild Rides Water Park?

____

4. What are the cabin rental prices for each of the state parks in Kentucky?

____

Work with a partner. Determine the equal share if the total number of centimeter cubes were equally distributed among the groups. Draw your models in the space provided.

5. 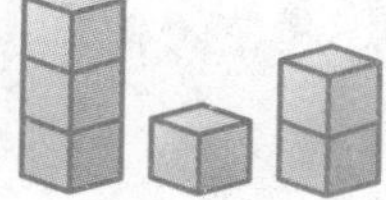

6. 

7. 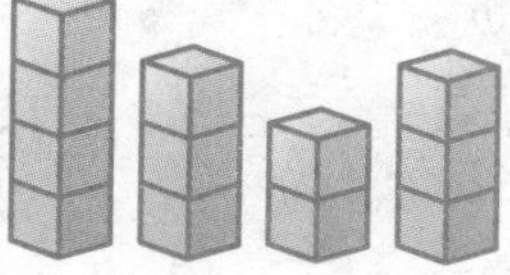

8. 

Analyze and Reflect

Collaborate

Work with a partner to determine the equal share for each exercise. Use centimeter cubes or counters if needed. The first one is done for you.

	Scenario	Responses	Response Total	Number of Responses	Equal Share
	Rainfall (inches)	7, 5, 2, 6	7 + 5 + 2 + 6 = 20	4	5
9.	Books Read	8, 7, 3			
10.	Eggs Hatched	5, 2, 3, 6			
11.	States Visited	1, 4, 2, 5, 3			
12.	Photos Taken	5, 3, 7, 2, 4, 3			
13.	Miles Hiked	11, 12, 8, 9			

14. MP **Reason Inductively** Compare the answers you provided in the table above. How does the response total and the number of responses relate to the equal share? Write a rule you can use to evenly distribute a data set without using centimeter cubes. ______

Create

On Your Own

15. MP **Model with Mathematics** Write a survey question that yields data without variability. Rewrite the question so that it yields data with variability.

16. MP **Model with Mathematics** Write a real-world problem that involves equal shares. Find the equal share of your data set.

17. Inquiry HOW are surveys created to collect and analyze data?

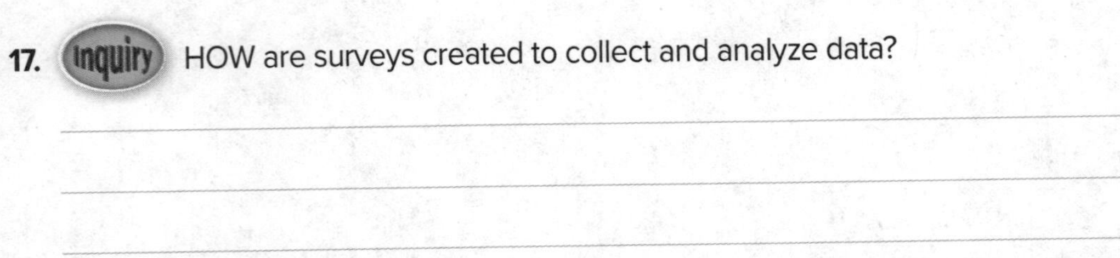

Lesson 1
Mean

Real-World Link

Music Tina and her friends downloaded songs for 6 weeks, as shown in the table.

Number of Songs Downloaded Each Week					
12	6	10	9	4	1

1. How many total songs were downloaded? ____________

2. On average, how many songs did they download each week?

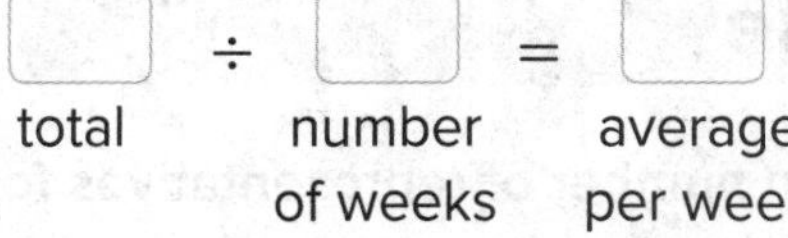

3. On the number line below, draw an arrow that points to the average. Plot the number of songs downloaded on the number line.

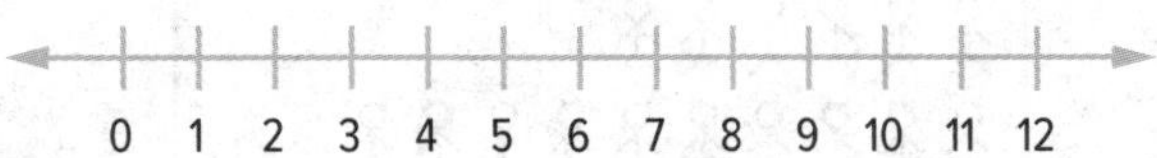

4. How far below the average is 1? 4? 6? How far above the average is 9? 10? 12? ____________

5. What is the sum of the distances between the average and the points below the average? above the average? ____________

6. Explain why the average is the balance point of the data.

____

____

____

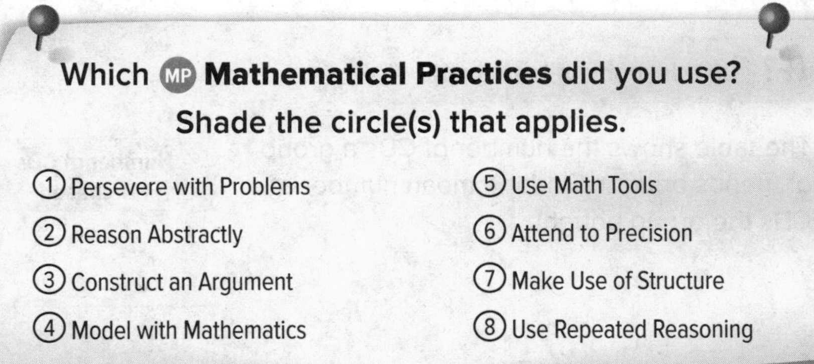

Which MP **Mathematical Practices** did you use? Shade the circle(s) that applies.

① Persevere with Problems
② Reason Abstractly
③ Construct an Argument
④ Model with Mathematics
⑤ Use Math Tools
⑥ Attend to Precision
⑦ Make Use of Structure
⑧ Use Repeated Reasoning

Essential Question

HOW are the mean, median, and mode helpful in describing data?

Vocabulary

mean
average

Common Core State Standards

Content Standards
6.SP.3

MP **Mathematical Practices**
1, 2, 3, 4, 6

Key Concept **Mean**

The **mean** of a data set is the sum of the data divided by the number of pieces of data. It is the balance point for the data set.

Work Zone

On the previous page, you found a single number to describe the number of songs downloaded each week. The **average**, or mean, summarizes the data using a single number.

You can find the mean of a set of data shown in different displays such as pictographs and dot plots.

Example

1. Find the mean number of representatives for the four states shown in the pictograph.

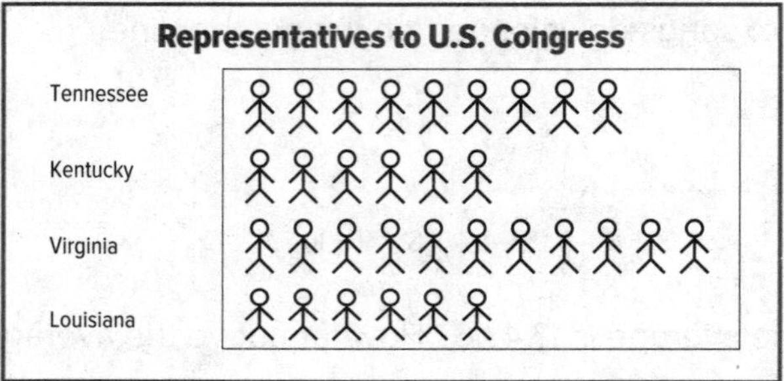

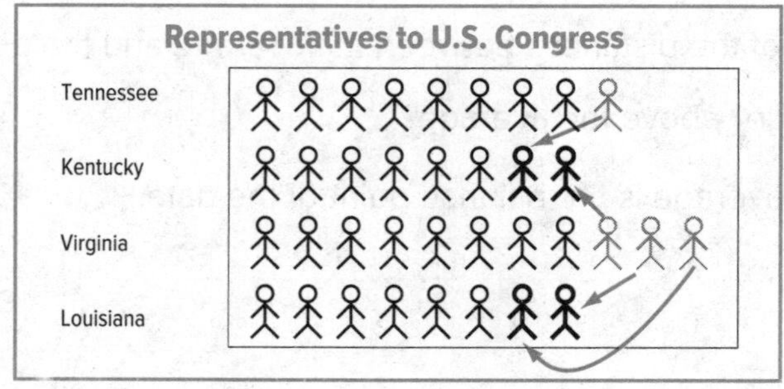

Move the figures to equally distribute the total number of representatives among the four states.

Each state has a mean or average of 8 representatives.

Including Data

Even if a data value is 0, it still should be counted in the total number of pieces of data.

Got it? Do this problem to find out.

a. ______

a. The table shows the number of CDs a group of friends bought. Find the mean number of CDs the group bought.

Number of CDs Purchased		
3	4	6
0	2	

Examples

2. **The dot plot shows the recorded high temperatures for six days in Little Rock, Arkansas. Find the mean temperature.**

High Temperatures

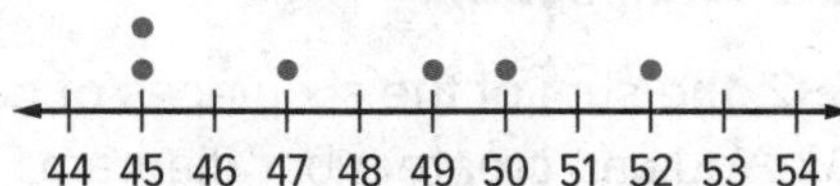

$$\text{mean} = \frac{45 + 45 + 47 + 49 + 50 + 52}{6}$$ ◄····· sum of the data / ◄····· number of data items

$$= \frac{288}{6} \text{ or } 48$$ Simplify.

The mean is 48 degrees. So, all of the data values can be summarized with a single number, 48.

3. **The dot plot shows the number of runs a baseball team had for each game of a 4-game series. Find the mean number of runs for the series.**

Number of Runs

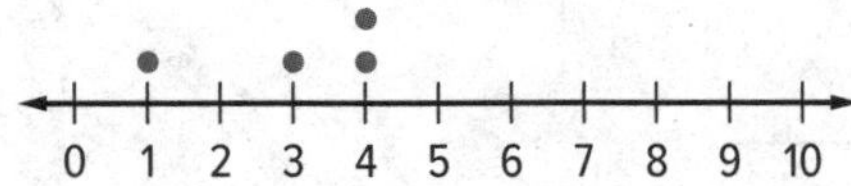

mean = $\frac{\square}{\square}$ ◄····· sum of the data / ◄····· number of data items

= $\frac{\square}{\square}$ or $\square$ Simplify.

The mean number of runs for the series is $\square$.

Got it? **Do this problem to find out.**

b. The dot plot shows the number of books Deanna read each week of a reading challenge. Find the mean number of books she read.

Books Read

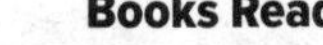

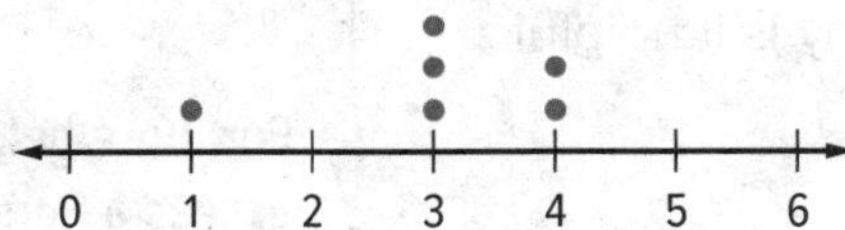

Dot Plots

In a dot plot, individual data values are represented as dots above a number line.

Show your work.

b. ____________

The mean is sometimes described as the balance point. Explain below what this means using the data set {2, 2, 3, 8, 10}.

Example

4. The number of minutes Marielle spent talking on her cell phone each month for the past five months were 494, 502, 486, 690, and 478. Suppose the mean for six months was 532 minutes. How many minutes did she talk on her cell phone during the sixth month?

If the mean is 532, the sum of the six pieces of data must be 532×6 or 3,192. You can create a bar diagram.

3,192					
494	502	486	690	478	?

$$3{,}192 - (494 + 502 + 486 + 690 + 478) = 3{,}192 - 2{,}650$$
$$= 542$$

Marielle talked 542 minutes during the sixth month.

Guided Practice

1. The dot plot shows the number of beads sold. Find the mean number of beads. (Examples 1–3)

______

Number of Beads

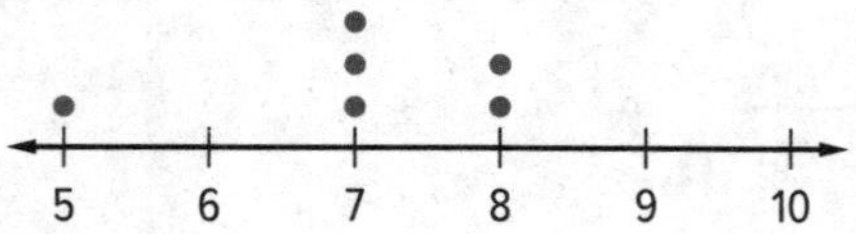

Dot plot: 5 (1 dot), 6 (0), 7 (3 dots), 8 (2 dots), 9 (0), 10 (0). Axis labels: 5, 6, 7, 8, 9, 10

2. The table shows the greatest depths of four of the five oceans in the world. If the average greatest depth is 8.094 kilometers, what is the greatest depth of the Southern Ocean? (Example 4) ______

Ocean	Greatest Depth (km)
Pacific	10.92
Atlantic	9.22
Indian	7.46
Arctic	5.63
Southern	■

3. **Building on the Essential Question** Why is it helpful to find the mean of a data set?

______

______

Rate Yourself!

How confident are you about finding the mean of a data set? Check the box that applies.

For more help, go online to access a Personal Tutor.

FOLDABLES Time to update your Foldable!

Name ________________ My Homework ________________

Independent Practice

Go online for Step-by-Step Solutions

Find the mean for each data set. (Examples 1–3)

1. ________

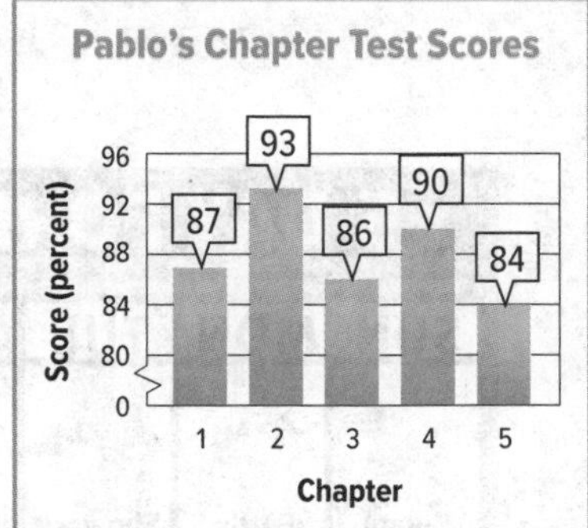

2. ________

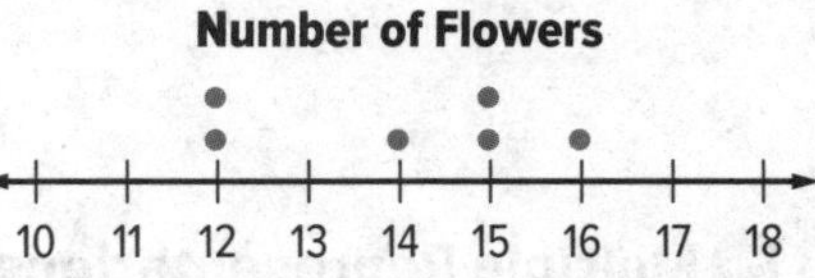

3. **Financial Literacy** Jamila babysat nine times. She earned \$15, \$20, \$10, \$12, \$20, \$16, \$80, and \$18 for eight babysitting jobs. How much did she earn the ninth time if the mean of the data set is \$24? (Example 4)

____

4. **Model with Mathematics** Refer to the graphic novel frame below for Exercises a–b.

a. What is the mean number of wins for the Cranes? for the Panthers?

____

____

b. Based on your answer for part **a**, is the mean a good measure for determining which team has the better record? Explain.

____

____

5. A *stem-and-leaf plot* is a display that organizes data from least to greatest. The digits of the least place value form the leaves, and the next place-value digits form the stems. The stem-and-leaf plot shows Marcia's scores on several tests. Find the mean test score.

Stem	Leaf
7	8
8	5 8 9
9	2 6

7|8 = 78

6. **MP Multiple Representations** The graphic shows the 5-day forecast.

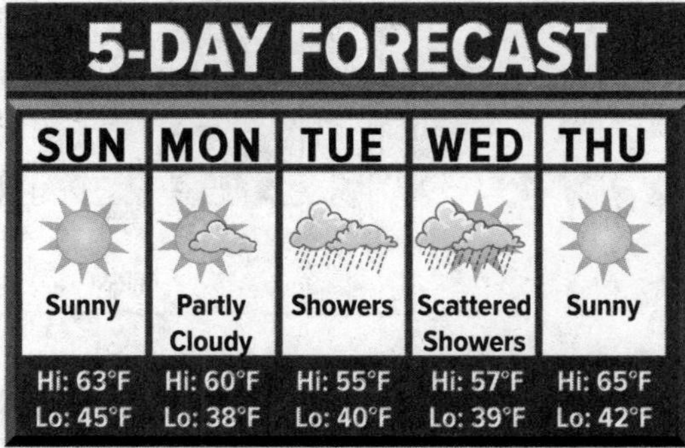

a. **Numbers** What is the difference between the mean high and mean low temperature for this 5-day period? Justify your answer.

b. **Graph** Make a double-line graph of the high and low temperatures for the 5-day period.

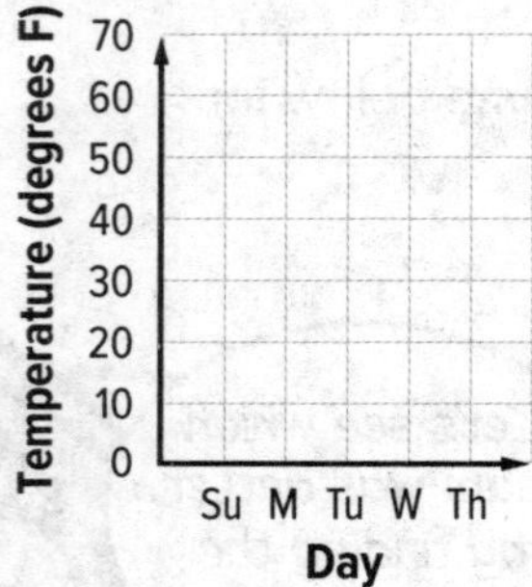

H.O.T. Problems Higher Order Thinking

7. **MP Reason Abstractly** Create a data set that has five values. The mean of the data set should be 34.

8. **MP Persevere with Problems** The mean of a set of data is 45 years. Find the missing numbers in the data set {40, 45, 48, ?, 54, ?, 45}. Explain the method or strategy you used.

9. **MP Reason Inductively** If 99 students had a mean quiz score of 82, how much is the mean score increased by the addition of a single score of 99? Explain.

Name ______________________ My Homework ______________

Extra Practice

Find the mean for each data set.

10. 8 bags

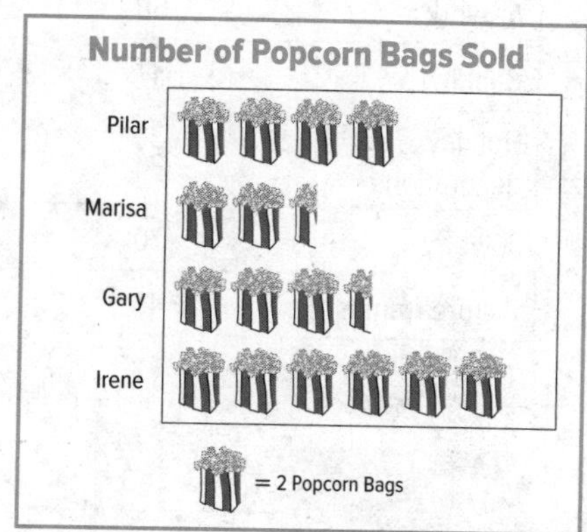

Homework Help

$\frac{8+5+7+12}{4} = 8$

11. ______

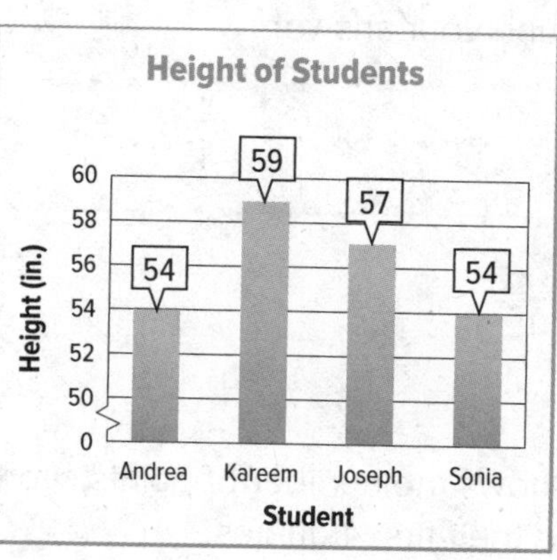

12. ______

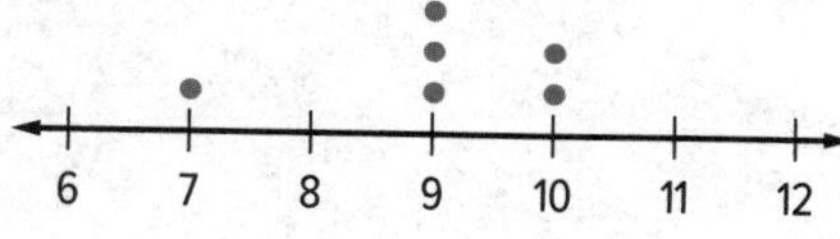

13. ______

Number of Tickets Sold

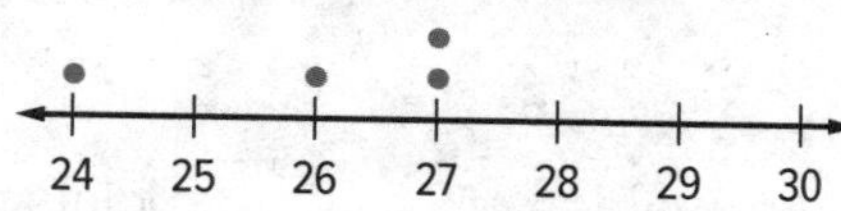

14. MP **Be Precise** The table shows the approximate heights of some of the tallest U.S. trees.

Tallest Trees in U.S.	
Tree	**Height (ft)**
Western Red Cedar	160
Coast Redwood	320
Monterey Cypress	100
California Laurel	110
Sitka Spruce	200
Port-Orford-Cedar	220

a. Find the mean of the data. ______

b. Find the mean if the Coast Redwood is not included in the data set. ______

c. How does the height of the Coast Redwood affect the mean of the data? ______

______

______

______

d. Suppose Blue Spruce was included in the list and the mean decreased to 165 feet. What is the height of the Blue Spruce?

______

CCSS Power Up! Common Core Test Practice

15. The table shows the money raised by each type of booth at a craft sale. The mean amount raised per booth was $59. How much money was raised by the T-shirt booth? Explain how you found your answer.

Northside Craft Sale	
Booth	**Money Raised ($)**
Artwork	58
Candies	47
Holiday decorations	54
Jewelry	70
Picture frames	45
T-shirts	?

16. The table shows the number of points scored by a football team during their first 4 games.

Game	1	2	3	4
Points Scored	24	30	22	28

Select values to complete the model below to find the mean number of points scored per game.

$$\frac{\square + \square + \square + \square}{\square} = \square$$

1	24
2	26
3	28
4	30
22	32

An average of ______ points were scored per game.

CCSS Common Core Spiral Review

Compare the numbers using < or >. 4.NBT.2

17. 18 ◯ 16

18. 65 ◯ 63

19. 22 ◯ 28

20. 34 ◯ 31

21. 75 ◯ 79

22. 67 ◯ 57

23. The table shows the distances from Louisville to several cities.

a. How much farther is it from Louisville to Charlotte than from Louisville to Lexington? 4.NBT.4 ______

b. Which city is the greatest distance from Louisville? 4.NBT.2

______

City	Distance (miles)
Charlotte	474
Cincinnati	100
Indianapolis	114
Lexington	75
St. Louis	265

Lesson 2

Median and Mode

Vocabulary Start-Up

A data set can also be described by its median or its mode. The mean, median, and mode are called **measures of center** because they describe the center of a set of data.

Find the definition of each term in the glossary. Then complete the graphic organizer.

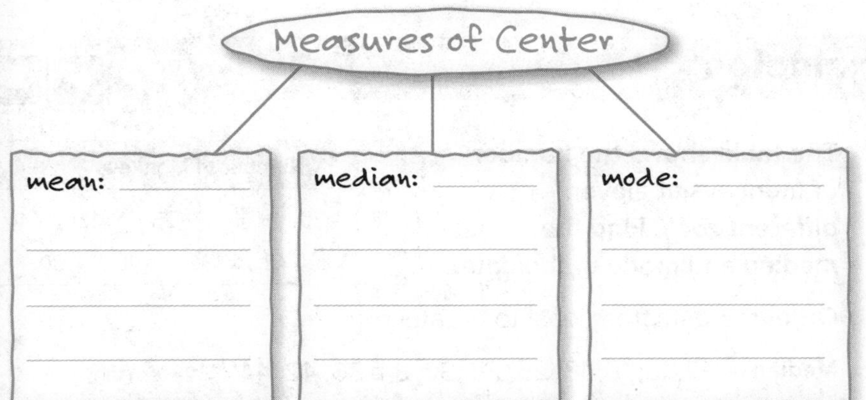

Essential Question

HOW are the mean, median, and mode helpful in describing data?

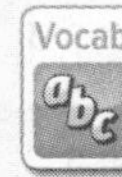

Vocabulary

measures of center
median
mode

Common Core State Standards

Content Standards
6.SP.3, 6.SP.5, 6.SP.5b, 6.SP.5c

MP Mathematical Practices
1, 3, 4, 5, 6

Real-World Link

Hurricanes The table shows the number of Atlantic hurricanes in different years.

Atlantic Hurricanes						
5	15	9	7	4	9	8

1. Order the data from least to greatest. Circle the number in the middle of your list. ______

2. Find the mean. Compare the middle number to the mean of the data. Round to the nearest hundredth if necessary.

______

Which MP **Mathematical Practices** did you use? Shade the circle(s) that applies.

① Persevere with Problems
② Reason Abstractly
③ Construct an Argument
④ Model with Mathematics
⑤ Use Math Tools
⑥ Attend to Precision
⑦ Make Use of Structure
⑧ Use Repeated Reasoning

Key Concept

Median and Mode

Work Zone

The **median** of a list of values is the value appearing at the center of a sorted version of the list, or the mean of the two central values, if the list contains an even number of values.

The **mode** is the number or numbers that occur most often.

Just as mean is one value used to summarize a data set, the median and mode also summarize a data set with a single number. If there is more than one number that occurs with the same frequency, a data set may have more than one mode.

Examples

1. The table shows the number of monkeys at eleven different zoos. Find the median and mode of the data.

Number of Monkeys					
28	36	18	25	12	44
18	42	34	16	30	

Order the data from least to greatest.

Median 12, 16, 18, 18, 25, (28), 30, 34, 36, 42, 44 — 28 is in the center.

Mode 12, 16, [18, 18], 25, 28, 30, 34, 36, 42, 44 — 18 occurs most often.

The median is 28 monkeys. The mode is 18 monkeys.

2. Dina recorded her scores on 7 tests in the table. Find the median and mode of the data.

Test Scores			
93	88	94	93
85	97	90	

Order the data from least to greatest.

___

Circle the number in the center. This is the median.

Circle the most frequently occurring numbers. This value is the mode.

The median is a score of ____. The mode is a score of .

Got it? **Do this problem to find out.**

a. ________

a. The list shows the number of stories in the 11 tallest buildings in Springfield. Find the median and mode of the data.

40, 38, 40, 37, 33, 30, 20, 24, 21, 17, 19

Examples

3. **Find the median and mode of the temperatures displayed in the graph.**

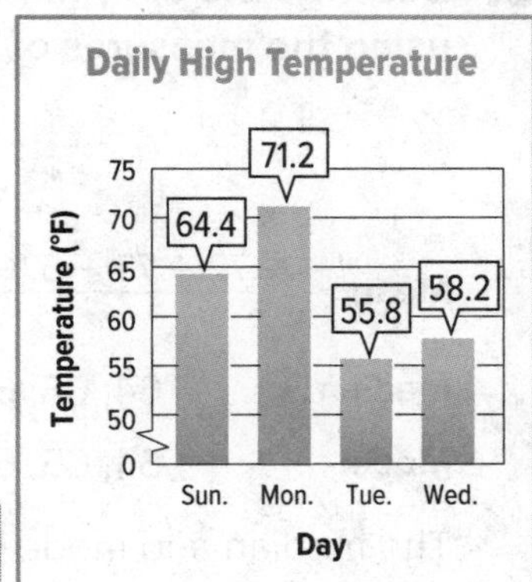

Median 55.8, 58.2, 64.4, 71.2

$$\frac{58.2 + 64.4}{2} = \frac{122.6}{2}$$
$$= 61.3°$$

There are an even number of data values. So, to find the median, find the mean of the two central values.

Mode There is no mode.

4. **Miguel researched the average precipitation in several states. Find and compare the median and mode of the average precipitation.**

State	Precipitation (in.)	State	Precipitation (in.)
Alabama	58.3	Louisiana	60.1
Florida	54.5	Maine	42.2
Georgia	50.7	Michigan	32.8
Kentucky	48.9	Missouri	42.2

Median 32.8, 42.2, 42.2, 48.9, 50.7, 54.5, 58.3, 60.1

$$\frac{48.9 + 50.7}{2} = \frac{99.6}{2}$$
$$= 49.8$$

Mode 32.8, 42.2, 42.2, 48.9, 50.7, 54.5, 58.3, 60.1

The median is 49.8 inches and the mode is 42.2 inches. The median is 7.6 inches greater than the mode.

Got it? **Do these problems to find out.**

b. Find the median and mode of the costs in the table.

Cost of Backpacks ($)			
16.78	48.75	31.42	18.38
22.89	51.25	28.54	26.79

c. Find and compare the median and mode of the costs in the table.

Cost of Juice ($)			
1.65	1.97	2.45	2.87
2.35	3.75	2.49	2.87

b. ______________

c. ______________

Example

5. Describe the daily high temperatures using the measures of center.

Daily High Temperature (°F)			
72	73	67	65
71	64	71	

Mean $\frac{72 + 73 + 67 + 65 + 71 + 64 + 71}{7} = \frac{483}{7}$ or 69°

Median 64, 65, 67, 71, 71, 72, 73

Mode 64, 65, 67, 71, 71, 72, 73

The median and mode both equal 71 degrees. They are both 2 degrees greater than the mean. The data follows the measures of center in that the temperatures are close to the measures of center.

d. ________

Got it? **Do this problem to find out.**

d. Describe the cost of CDs using the measures of center.

Cost of CDs ($)		
11.95	12.89	19.99
19.99	12.59	18.49

Guided Practice

1. Find and compare the median and mode for the following set of data. monthly spending: \$46, \$62, \$62, \$57, \$50, \$42, \$56, \$40 (Examples 1–4)

2. Describe the daily high temperatures using the measures of center. (Example 5)

Daily High Temperature (°F)			
34	35	31	36
31	24	33	

3. **Building on the Essential Question** How are mean and median similar?

Independent Practice

Go online for Step-by-Step Solutions

Find and compare median and mode for each set of data. (Examples 1–4)

1. math test scores: 97, 85, 92, 86 ______________________

2. 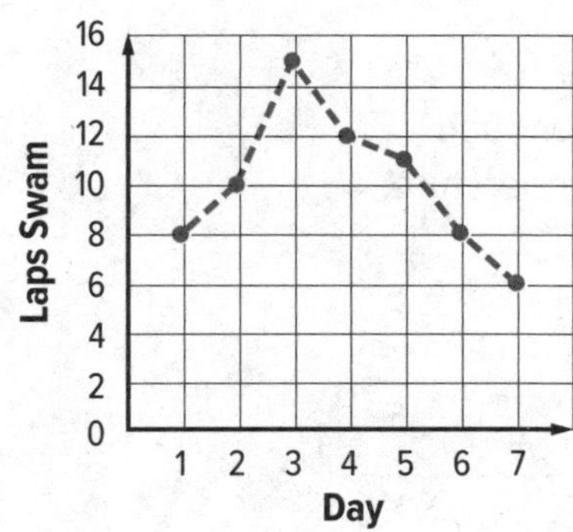

____

3. Describe the average speeds using the measures of center. (Example 5)

____

____

____

Average Speeds (mph)			
40	52	44	46
52	40	44	50
41	44	44	50

4. **MP Model with Mathematics** Refer to the graphic novel frame below for Exercises a–b.

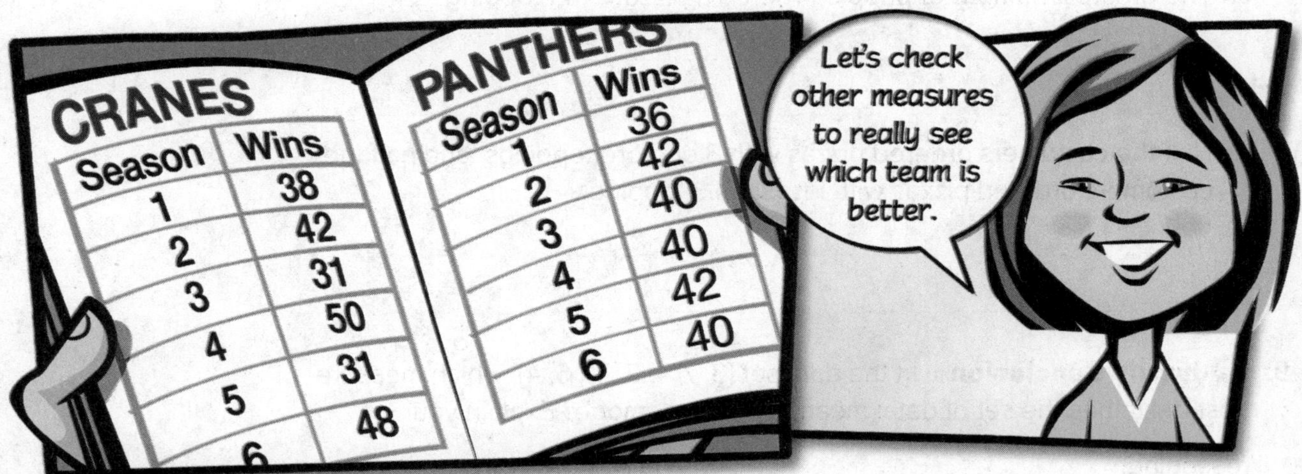

a. Find the median and mode for each team's wins.

____

____

b. Which team had the better record? Justify your response.

____

____

5. A Louisville newspaper claims that during seven days, the high temperature in Lexington was typically 6° warmer than the high temperature in Louisville. What measure was used to make this claim?

Justify your answer. ______________________

Daily High Temperatures (°F)	
Louisville	Lexington
75 50 80 72 70 84 70	80 73 75 74 71 76 76

6. **Use Math Tools** Use the Internet to find the high temperatures for each of the last seven days in a city near you. Then find the median high temperature.

H.O.T. Problems Higher Order Thinking

7. **Persevere with Problems** The ticket prices for a concert series were \$12, \$37, \$45, \$18, \$8, \$25, and \$18. What was the ticket price of the eighth and final concert in this series if the set of 8 prices had a mean of \$23, a mode of \$18, a median of \$19.50? ______________________

8. **Construct an Argument** One evening at a local pizzeria, the following number of toppings were ordered on each large pizza.

3, 0, 1, 1, 2, 5, 4, 3, 1, 0, 0, 1, 1, 2, 2, 3, 6, 4, 3, 2, 0, 2, 1, 3

Determine whether each statement is *true* or *false*. Explain your reasoning.

a. The greatest number of people ordered a pizza with 1 topping.

b. Half the customers ordered pizzas with 3 or more toppings, and half the customers ordered pizzas with less than 3 toppings.

9. **Justify Conclusions** In the data set {3, 7, 4, 2, 31, 5, 4}, which measure best describes the set of data: mean, median, or mode? Explain your reasoning. ______________________

10. **Model with Mathematics** Create a list of six values where the mean, median, and mode are 45, and only two of the values are the same.

Extra Practice

Find and compare median and mode for each set of data.

11. age of employees: 23, 22, 15, 44, 44 median: 23; mode: 44; The mode is 21 years more than the median.

Homework Help

Median: 15, 22, 23, 44, 44

Mode: 15, 22, 23, 44, 44

12. minutes spent on homework: 18, 20, 22, 11, 19, 18, 18

13.

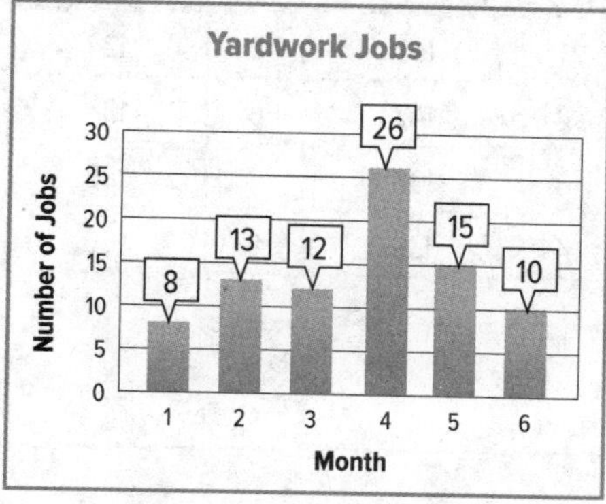

14. Describe the test grades using the measures of center.

Test Grades			
100	77	80	65
87	85	85	82
100	97	95	75

15. MP **Be Precise** Fill in the graphic organizer with the description. The first one is done for you.

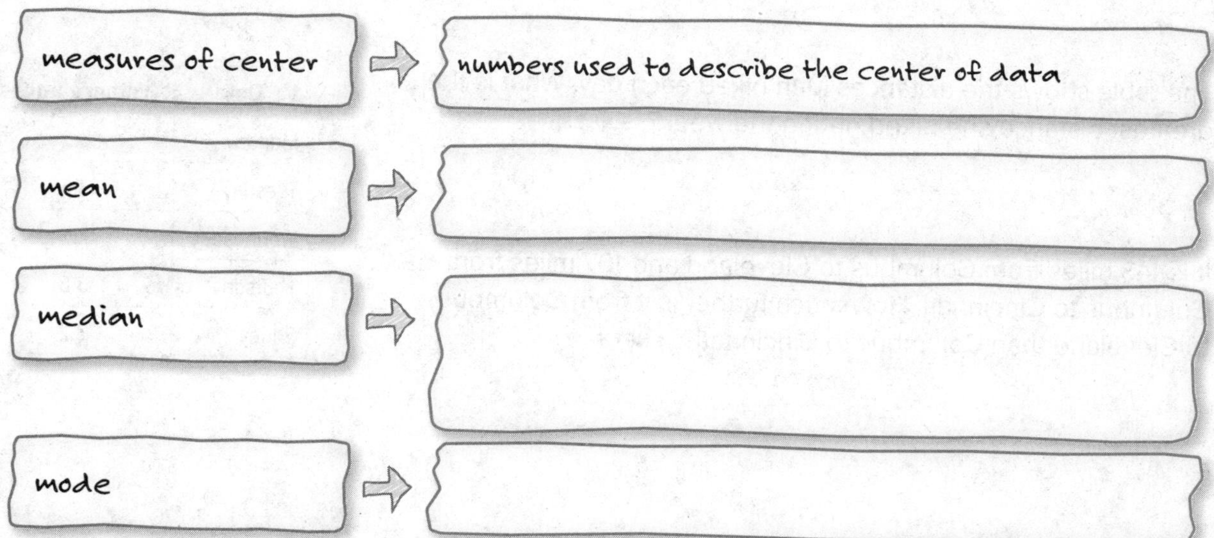

CCSS Power Up! Common Core Test Practice

16. The list of data shows the number of schools in 12 different counties.

Number of Schools in Different Counties			
4	3	6	10
3	14	8	5
7	11	7	8

Arrange the data values in order from least to greatest.

What are the two middle numbers in the data set?

What is the median number of schools in the 12 counties?

17. The table shows the number of concerts performed each year by a band. Determine if each statement is true or false.

Year	Number of Concerts	Year	Number of Concerts
1	142	5	124
2	142	6	138
3	136	7	136
4	136	8	150

a. The median is 135 concerts. ☐ True ☐ False

b. The mode is 136 concerts. ☐ True ☐ False

c. The mean is 138 concerts. ☐ True ☐ False

CCSS Common Core Spiral Review

Find the greatest number in the data set. 4.NBT.2

18. {23, 35, 31, 28, 26, 34}

19. {56, 58, 49, 50, 56, 57}

20. {78, 81, 79, 84, 82, 83}

Find the least number in the data set. 4.NBT.2

21. {62, 58, 56, 61, 59, 57}

22. {24, 29, 22, 26, 23, 24}

23. {56, 58, 52, 54, 53, 57}

24. The table shows the distances Mari biked each day. What is the greatest distance she biked during the week? 5.NBT.3b

Day	Distance (miles)
Monday	5.2
Tuesday	3.5
Wednesday	4.9
Thursday	3.8
Friday	3.2

25. It is 143 miles from Columbus to Cleveland and 107 miles from Columbus to Cincinnati. How much further is it from Columbus to Cleveland than Columbus to Cincinnati? 4.NBT.4

 Problem-Solving Investigation

Use Logical Reasoning

Content Standards
6.SP.1

Mathematical Practices
1, 3, 4

Case #1 Speak to Me

Amy surveyed 15 students with the statistical question, "Do you speak Spanish, French, both languages, or neither language?" Four students speak French, seven students speak Spanish, and two students speak both languages.

Use a Venn diagram to find how many students speak neither Spanish nor French.

Understand *What are the facts?*

- You know ☐ classmates speak Spanish and ☐ classmates speak French.
- You know that ☐ students speak both languages.

Plan *What is your strategy to solve this problem?*

Make a Venn diagram to organize the information. Use logical reasoning to find the answer.

Solve *How can you apply the strategy?*

Draw and label two overlapping circles to represent the two languages. Since 2 students speak both languages, place a 2 in the section that is part of both circles. Use subtraction to determine the number for each of the other sections.

only French: 4 – ☐ = ☐

only Spanish: 7 – ☐ = ☐

neither: 15 – ☐ – ☐ – ☐ = ☐

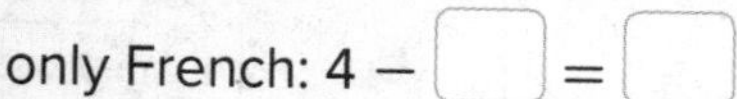

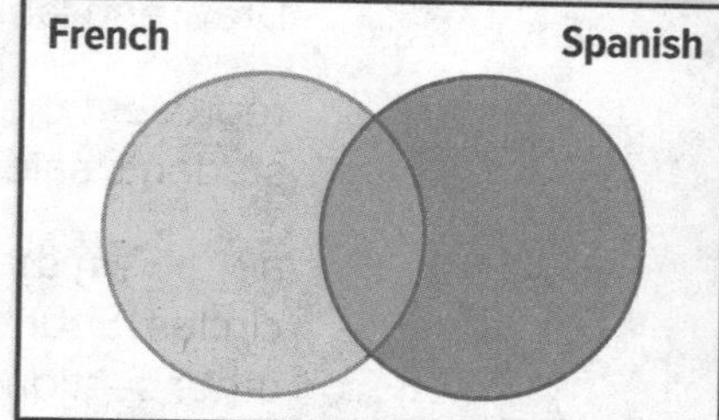

So, ☐ students speak neither French nor Spanish.

Check *Does the answer make sense?*

Check each circle to see if the appropriate number of students is represented.

Analyze the Strategy

Reason Inductively Explain why Amy's question, "Do you speak Spanish, French, both languages, or neither language?" is a statistical question.

Case #2 Battle of the Mascots

Nick conducted a survey of 85 students about a new school mascot. The results showed that 40 students liked Tigers, and 31 students liked Bears. Of those students, 12 liked both Tigers and Bears.

How many students liked neither Tigers nor Bears?

Understand

Read the problem. What are you being asked to find?

I need to find ______________________________

Underline key words and values in the problem. What information do you know?

☐ students were surveyed. In the survey, ☐ students said they liked Tigers, ☐ said they liked Bears, and ☐ said they liked both.

2 Plan

Choose a problem-solving strategy.

I will use the ______________________________ strategy.

3 Solve

Use your problem-solving strategy and a Venn diagram to solve the problem.

Draw and label two overlapping circles to represent the two mascots. Since ☐ students said they liked both mascots, place a ☐ in the section that is part of both circles. Subtract to find the numbers for the other sections.

Tigers | Bears

only tigers: ______________ only bears: ______________

neither tigers nor bears: ______________

So, ☐ students liked neither tigers nor bears as the school mascot.

4 Check

Use information from the problem to check your answer.

____

____

____

Work with a small group to solve the following cases. Show your work on a separate piece of paper.

Case #3 Marketing

A survey showed that 70 customers bought white bread, 63 bought wheat bread, and 35 bought rye bread. Of those who bought exactly two types of bread, 12 bought wheat and white, 5 bought white and rye, and 7 bought wheat and rye. Two customers bought all three.

How many customers bought only wheat bread?

Case #4 Pets

Dr. Poston is a veterinarian. One week she treated 20 dogs, 16 cats, and 11 birds. Some owners had more than one pet, as shown in the table.

How many owners had only a dog as a pet?

Pet	Number of Owners
dog and cat	7
dog and bird	5
cat and bird	3
dog, cat, and bird	2

Case #5 Sports

The Student Council surveyed a group of 24 students by asking the statistical question, “Do you like softball, basketball, both, or neither?” The results showed that 14 students liked softball, and 18 liked basketball. Of these, 8 liked both.

How many students liked just softball and how many liked just basketball?

Case #6 Money

Jorge has $138.22 in his savings account. He deposits $10.75 every week and withdraws $31.68 every four weeks.

What will his balance be in 8 weeks?

Mid-Chapter Check

Vocabulary Check

1. Define *mean*. Then determine the mean of the following data set {22, 18, 38, 6, 24, 18}. (Lesson 1)

2. Fill in the blank in the sentence below with the correct term. (Lesson 2)

   The ____________ is the number or numbers that occur most often in a set.

Skills Check and Problem Solving

Find the mean of each data set. (Lesson 1)

3. number of home runs by baseball players in a season: 43, 21, 35, 15, 35

4. number of different birds spotted: 7, 10, 13, 9, 12, 3

Find the median and mode for each set of data. (Lesson 2)

5. hours spent studying: 4, 2, 5, 7, 1

6. heights of buildings in feet: 35, 42, 40, 25, 42, 54, 50

7. MP **Use Math Tools** Use the table that shows the lengths of different lizards. Find and compare the median and mode of the data. (Lesson 2)

Lizard Length (cm)			
14	12	14	14
19	18	11	16
30	12	19	15

8. MP **Persevere with Problems** The table shows the number of minutes spent doing different exercises. The mean time spent exercising was 18.2 minutes. How many minutes were spent doing sit-ups? (Lesson 2)

Daily Exercises	
Exercise	**Time (min)**
Pull-ups	8
Push-ups	10
Running	38
Sit-ups	■
Weight lifting	20

Lesson 3

Measures of Variation

Vocabulary Start-Up

Measures of variation are used to describe the distribution, or spread, of the data. They describe how the values of a data set vary with a single number. A *quartile* is one measure of variation.

Look in a dictionary and find words that begin with *quar-*. Write two of the words and their definitions.

Word beginning with quar-	Definition

Based on the definitions you found, fill in the blank below.

Quartiles are values that divide a set of data into ______ equal parts.

Essential Question

HOW are the mean, median, and mode helpful in describing data?

Vocabulary

measures of variation
quartiles
first quartile
third quartile
interquartile range
range
outliers

Common Core State Standards

Content Standards
6.SP.3, 6.SP.5, 6.SP.5c

MP Mathematical Practices
1, 2, 3, 4, 5

Real-World Link

Surveys James asked his classmates how many hours of TV they watch on a typical day.

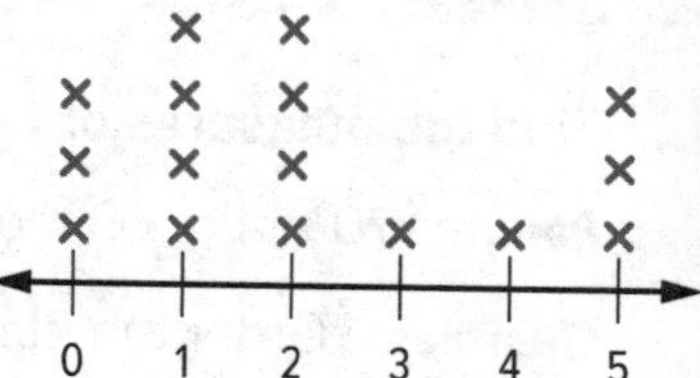

1. Divide the data into 4 equal parts. Draw a circle around each part.

2. How many data values are in each group? ______

Which MP **Mathematical Practices** did you use? Shade the circle(s) that applies.

① Persevere with Problems
② Reason Abstractly
③ Construct an Argument
④ Model with Mathematics
⑤ Use Math Tools
⑥ Attend to Precision
⑦ Make Use of Structure
⑧ Use Repeated Reasoning

Key Concept

Measures of Variation

Work Zone

Quartiles are values that divide the data set into four equal parts.

First and Third Quartiles

The first and third quartiles are the medians of the data values less than the median and the data values greater than the median, respectively.

Interquartile Range (IQR)

The distance between the first and third quartiles of the data set.

Range

The difference between the greatest and least data values.

Measures of variation of a data set are shown below.

Q_1 ↓ (at 2), median ↓ (between 3 and 4), Q_3 ↓ (at 6)

0, 0, 1, 1, 2, 2, 2, 3, 4, 5, 6, 6, 7, 7, 7, 8

The median of the data values less than the median is the first quartile or Q_1; in this case, 1.5.

The median of the data values greater than the median is the third quartile or Q_3; in this case, 6.5.

One fourth of the data lie below the first quartile and one fourth of the data lie above the third quartile. So, one half of the data lie between the first quartile and third quartile.

Example

1. Find the measures of variation for the data.

Animal	Speed (mph)
cheetah	70
lion	50
cat	30
elephant	25
mouse	8
spider	1

Range 70 − 1 or 69 mph

Quartiles Order the numbers.

Q_1 (8), median = 27.5 (between 25 and 30), Q_3 (50)

1 8 25 30 50 70

Interquartile Range 50 − 8 or 42 $Q_3 - Q_1$

The range is 69, the median is 27.5, the first quartile is 8, the third quartile is 50, and the IQR is 42.

Interquartile Range

If the interquartile range is low, the middle data are grouped closely together.

Got it? Do this problem to find out.

a. Determine the measures of variation for the data 64, 61, 67, 59, 60, 58, 57, 71, 56, and 62.

a. ________

Find Outliers and Analyze Data

An **outlier** is a data value that is either much *greater* or much *less* than the other values in a data set. If a data value is more than 1.5 times the value of the interquartile range beyond the quartiles, it is an outlier.

STOP and Reflect

Which measure of center would most likely be affected by an outlier? Explain below.

Example

2. The ages of candidates in an election are 23, 48, 49, 55, 57, 63, and 72. Name any outliers in the data.

Find the interquartile range: 63 − 48 = 15

Multiply the interquartile range by 1.5: 15 × 1.5 = 22.5

Subtract 22.5 from the first quartile and add 22.5 to the third quartile to find the limits for the outliers.

48 − **22.5** = 25.5 63 + **22.5** = 85.5

The only age beyond the limits is 23. So, it is the only outlier.

Got it? Do this problem to find out.

b. The lengths, in feet, of various bridges are 88, 251, 275, 354, and 1,121. Name any outliers in the data set.

b. ________________

Example

3. The table shows a set of scores on a science test in two different classrooms. Compare and contrast their measures of variation.

Room A	Room B
72	63
100	93
67	79
84	83
65	98
78	87
92	73
87	81
80	65

Find the measures of variation for both rooms.

	Room A	Room B
Range	100 − 65 = 35	98 − 63 = 35
Median	80	81
Q_3	$\frac{87 + 92}{2} = 89.5$	$\frac{87 + 93}{2} = 90$
Q_1	$\frac{67 + 72}{2} = 69.5$	$\frac{65 + 73}{2} = 69$
IQR	89.5 − 69.5 = 20	90 − 69 = 21

Both classrooms have a range of 35 points, but Room B has an interquartile range of 21 points while Room A's interquartile range is 20 points. There are slight differences in the medians as well as the third and first quartiles.

c. ______________

Got it? Do this problem to find out.

c. Temperatures for the first half of the year are given for Antelope, Montana, and Augusta, Maine. Compare and contrast the measures of variation of the two cities.

Month	Antelope, MT	Augusta, ME
January	21	28
February	30	32
March	42	41
April	58	53
May	70	66
June	79	75

Guided Practice

1. The average wind speeds for several cities in Pennsylvania are given in the table. (Examples 1 and 2)

Wind Speed	
Pennsylvania City	**Speed (mph)**
Allentown	8.9
Erie	11.0
Harrisburg	7.5
Middletown	7.7
Philadelphia	9.5
Pittsburgh	9.0
Williamsport	7.6

   a. Find the range of the data. ______________

   b. Find the median and the first and third quartiles.

   ____

   c. Find the interquartile range. ______________

   d. Identify any outliers in the data. ______________

2. The heights of several types of palm trees, in feet, are 40, 25, 15, 22, 50, and 30. The heights of several types of pine trees, in feet, are 60, 75, 45, 80, 75, and 70. Compare and contrast the measures of variation of both kinds of trees. (Example 3)

____

3. **Building on the Essential Question** Describe the difference between measure of center and measure of variation. ______________

Rate Yourself!

Are you ready to move on? Shade the section that applies.

YES ? NO

For more help, go online to access a Personal Tutor.

FOLDABLES Time to update your Foldable!

Name ______________________ My Homework ______________________

Independent Practice

Go online for Step-by-Step Solutions

eHelp

1. The table shows the number of golf courses in various states. (Examples 1 and 2)

Number of Golf Courses			
California	1,117	New York	954
Florida	1,465	North Carolina	650
Georgia	513	Ohio	893
Iowa	437	South Carolina	456
Michigan	1,038	Texas	1,018

a. Find the range of the data. ____________

b. Find the median and the first and third quartiles.

____

c. Find the interquartile range. ____________

d. Name any outliers in the data. ____________

For each data set, find the median, the first and third quartiles, and the interquartile range. (Example 1)

2. texts per day: 24, 53, 38, 12, 31, 19, 26

____

3. daily attendance at the water park: 346, 250, 433, 369, 422, 298

____

4. The table shows the number of minutes of exercise for each person. Compare and contrast the measures of variation for both weeks. (Example 3) ____________

Minutes of Exercise		
	Week 1	**Week 2**
Tanika	45	30
Tasha	40	55
Tyrone	45	35
Uniqua	55	60
Videl	60	45
Wesley	90	75

5. **STEM** The table shows the number of known moons for each planet in our solar system. Use the measures of variation to describe the data. ____________

Known Moons of Planets			
Mercury	0	Jupiter	63
Venus	0	Saturn	34
Earth	1	Uranus	27
Mars	2	Neptune	13

6. MP **Use Math Tools** The *double stem-and-leaf plot*, where the stem is in the middle and the leaves are on either side, shows the high temperatures for two cities in the same week. Use the measures of variation to describe the data in the stem-and-leaf plot.

Minneapolis		Columbus
5 3 1 0	2	5 7 9 9
6 4	3	7
3	4	8
	5	
	6	2

6|3 = 36° *2|5 = 25°*

H.O.T. Problems Higher Order Thinking

7. MP **Find the Error** Hiroshi was finding the measures of variation of the following set of data: 89, 93, 99, 110, 128, 135, 144, 152, and 159. Find his mistake and correct it.

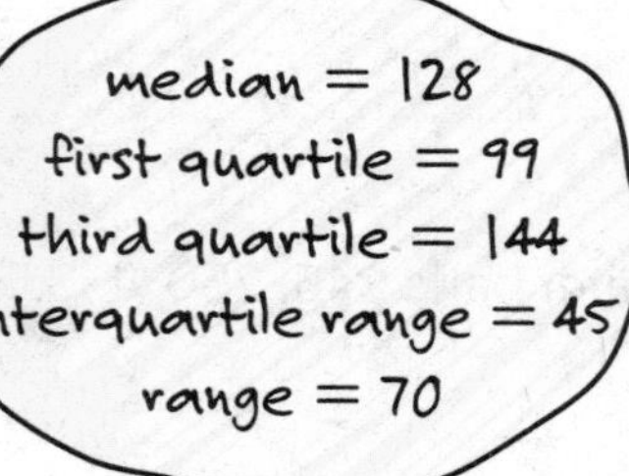

8. MP **Reason Abstractly** Create a list of data with at least six numbers that has an interquartile range of 15 and two outliers.

9. MP **Persevere with Problems** How is finding the first and third quartiles similar to finding the median?

10. MP **Reason Inductively** Explain why the median is not affected by very high or very low values in the data.

11. MP **Reason Inductively** Determine the range and IQR of each data set. Which measure of variation tells you more about the distribution of the data values? Explain.

Data Set A	Data Set B
1, 2, 2, 2, 3, 3, 4, 5, 5, 5, 6, 6, 17, 19, 21	1, 2, 9, 17, 17, 17, 17, 17, 17, 18, 18, 18, 19, 20, 21

Name ____________________ My Homework ____________________

Extra Practice

12. The table shows the countries with the most Internet users.

Millions of Internet Users	
China	99.8
Germany	41.88
India	36.97
Japan	78.05
South Korea	31.67
United Kingdom	33.11
United States	185.55

a. Find the range of the data.

153,880,000 185,550,000 − 31,670,000 = 153,880,000

b. Find the median and the first and third quartiles.

41,880,000; 33,110,000; 99,800,000

31.67 33.11 36.97 41.88 78.05 99.8 185.55

Q_1 Median Q_3

c. Find the interquartile range.

66,690,000 99,800,000 − 33,110,000 = 66,690,000

d. Name any outliers in the data. none

13. MP **Use Math Tools** The table shows teams in the National Football Conference (NFC) and the American Football Conference (AFC).

Penalties By NFL Teams			
NFC		AFC	
Dallas Cowboys	104	New England Patriots	78
Arizona Cardinals	137	Indianapolis Colts	67
Green Bay Packers	113	Jacksonville Jaguars	76
New Orleans Saints	68	San Diego Chargers	94
New York Giants	77	Cleveland Browns	114
Seattle Seahawks	59	Pittsburgh Steelers	80
Minnesota Vikings	86	Houston Texans	82

a. Which conference had a greater range of penalties? ____________________

b. Find the measures of variation for each conference. ____________________

c. Compare and contrast the measures of variation for each conference.

14. Find the median, the first and third quartiles, and the interquartile range for the cost of admission: $13.95, $24.59, $19.99, $29.98, $23.95, $28.99.

CCSS Power Up! Common Core Test Practice

15. The number of games won by 10 chess players is shown below.

13, 15, 2, 7, 5, 9, 11, 10, 12, 11

Which of the following statements are true? Select all that apply.

- ☐ Half of the players won more than 10.5 games and half won less than 10.5 games.
- ☐ The range of the data is 13 games.
- ☐ There are no outliers.
- ☐ Only one fourth of the players won more than 7 games.

16. The data at the right shows the number of dogs enrolled in different obedience classes.

Number of Dogs in Obedience Classes				
8	12	20	10	6
15	12	9	10	22

a. Order the values from least to greatest.

b. Find the range of the data.

c. Find the median and the first and third quartiles.

d. Find is the interquartile range?

CCSS Common Core Spiral Review

Divide. 5.NBT.6, 5.NBT.7

Show your work.

17. $160 \div 5 =$ ______

18. $188 \div 8 =$ ______

19. $133 \div 7 =$ ______

20. $87.5 \div 5 =$ ______

21. $136.5 \div 7 =$ ______

22. $74.4 \div 6 =$ ______

23. Refer to the table. How much farther did the Sing family drive on Friday than on Saturday? 4.NBT.4

Day	Distance (miles)
Thursday	68
Friday	193
Saturday	26
Sunday	95

24. Refer to the table. How many more hours did Koli work in week 2 than in week 3? 4.NBT.4

Week	Hours Worked
1	12
2	16
3	9

 Need more practice? Download more Extra Practice at **connectED.mcgraw-hill.com.**

Lesson 4

Mean Absolute Deviation

Real-World Link

Basketball The tables show the number of points two teams scored.

Ally's Team			
52	48	60	50
56	54	58	62

Lena's Team			
51	48	60	49
59	50	62	61

1. Plot each set of data on a number line.

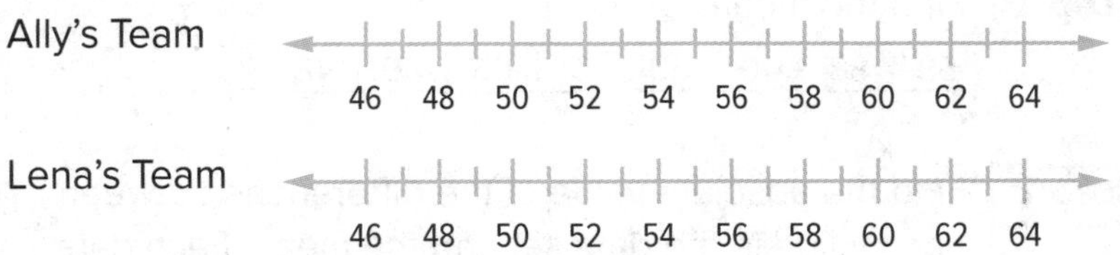

2. Find the mean of each set of data. Plot the means on the number lines with a star.

3. Find the range of each set of data. ______________________

____

4. Refer to the number lines. Compare and contrast each set of data.

____

____

____

Essential Question

HOW are the mean, median, and mode helpful in describing data?

Vocabulary

mean absolute deviation

Common Core State Standards

Content Standards
6.SP.5, 6.SP.5b, 6.SP.5c

MP Mathematical Practices
1, 2, 3, 4, 5, 6

Which MP Mathematical Practices did you use? Shade the circle(s) that applies.

① Persevere with Problems
② Reason Abstractly
③ Construct an Argument
④ Model with Mathematics
⑤ Use Math Tools
⑥ Attend to Precision
⑦ Make Use of Structure
⑧ Use Repeated Reasoning

Work Zone

Find Mean Absolute Deviation

You have used the interquartile range to describe the spread of a set of data. You can also use the mean absolute deviation. The **mean absolute deviation** of a set of data is the average distance between each data value and the mean.

Example

1. The table shows the maximum speeds of eight roller coasters. Find the mean absolute deviation of the set of data. Describe what the mean absolute deviation represents.

Maximum Speeds of Roller Coasters (mph)			
58	88	40	60
72	66	80	48

Step 1 Find the mean.

$$\frac{58 + 88 + 40 + 60 + 72 + 66 + 80 + 48}{8} = 64$$

Step 2 Find the absolute value of the differences between each value in the data set and the mean. Each data value is represented by an "x".

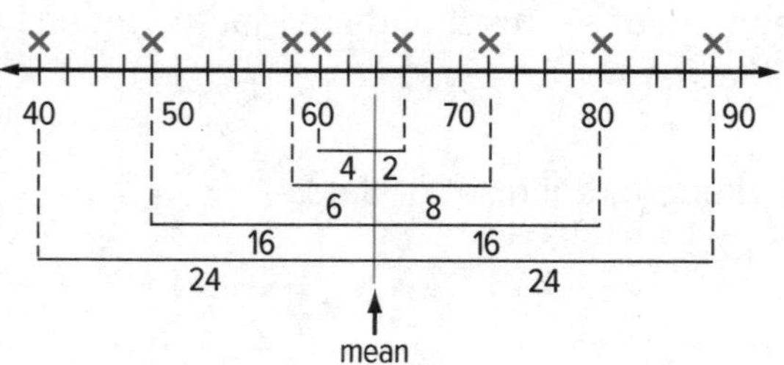

Step 3 Find the average of the absolute values of the differences between each value in the data set and the mean.

$$\frac{24 + 16 + 6 + 4 + 2 + 8 + 16 + 24}{8} = 12.5$$

The mean absolute deviation is 12.5. This means that the average distance each data value is from the mean is 12.5 miles per hour.

Got it? Do this problem to find out.

a. The table shows speeds of ten birds. Find the mean absolute deviation of the data. Round to the nearest hundredth. Describe what the mean absolute deviation represents.

Speeds of Top Ten Fastest Birds (mph)				
88	77	65	70	65
72	95	80	106	68

a. ________

Compare Variation

You can compare the mean absolute deviations for two data sets. A data set with a smaller mean absolute deviation has data values that are closer to the mean than a data set with a greater mean absolute deviation.

Example

2. The top five salaries and the bottom five salaries for the 2010 New York Yankees are shown in the table below. Salaries are in millions of dollars and are rounded to the nearest hundredth.

2010 New York Yankees Salaries (millions of $)	
Top Five Salaries	Bottom Five Salaries
33.00 24.29 22.60 20.63 16.50	0.45 0.44 0.43 0.41 0.41

a. Find the mean absolute deviation for each set of data. Round to the nearest hundredth.

Find the mean of the top five salaries.

$$\frac{33.00 + 24.29 + 22.60 + 20.63 + 16.50}{5} \approx 23.40$$

The mean is about $23.40 million.

Find the mean absolute deviation of the top five salaries.

$$\frac{9.60 + 0.89 + 0.80 + 2.77 + 6.90}{5} \approx 4.19$$

The mean absolute deviation is about $4.19 million.

Find the mean of the bottom five salaries.

$$\frac{0.45 + 0.44 + 0.43 + 0.41 + 0.41}{5} \approx 0.43$$

The mean is about $0.43 million.

Find the mean absolute deviation of the bottom five salaries.

$$\frac{0.02 + 0.01 + 0 + 0.02 + 0.02}{5} \approx 0.01$$

The mean absolute deviation is about $0.01 million.

b. Write a few sentences comparing their variation.

The mean absolute deviation for the bottom five salaries is much less than that for the top five salaries. The data for the bottom five salaries are closer together than the data for the top five salaries.

Mean Absolute Deviation

The absolute values of the differences between each data value and the mean for the top five salaries are calculated below.

$|33.00 - 23.40| = 9.60$

$|24.29 - 23.40| = 0.89$

$|22.60 - 23.40| = 0.80$

$|20.63 - 23.40| = 2.77$

$|16.50 - 23.40| = 6.90$

Got it? Do this problem to find out.

b. The table shows the running time in minutes for two kinds of movies. Find the mean absolute deviation for each set of data. Round to the nearest hundredth. Then write a few sentences comparing their variation.

Running Time for Movies (min)									
Comedy					Drama				
90	95	88	100	98	115	120	150	135	144

Guided Practice

1. Find the mean absolute deviation for the set of data. Round to the nearest hundredth if necessary. Then describe what the mean absolute deviation represents. (Example 1)

Number of Daily Visitors to a Web Site				
112	145	108	160	122

2. The table shows the height of waterslides at two different water parks. Find the mean absolute deviation for each set of data. Round to the nearest hundredth. Then write a few sentences comparing their variation. (Example 2)

Height of Waterslides (ft)									
Splash Lagoon					Wild Water Bay				
75	95	80	110	88	120	108	94	135	126

3. **Building on the Essential Question** What does the mean absolute deviation tell you about a set of data?

Rate Yourself!

☐ I understand how to find the mean absolute deviation.

Great! You're ready to move on!

☐ I still have questions about finding the mean absolute deviation.

No Problem! Go online to access a Personal Tutor. Tutor

FOLDABLES Time to update your Foldable!

Independent Practice

eHelp
Go online for Step-by-Step Solutions

Find the mean absolute deviation for each set of data. Round to the nearest hundredth if necessary. Then describe what the mean absolute deviation represents. (Example 1)

1.

Known Moons of Planets			
0	0	1	2
63	34	27	13

2.

Hard Drive (gigabytes)			
640	250	500	640
720	640	250	720

3. The table shows the lengths of the longest bridges in the United States and in Europe. Find the mean absolute deviation for each set of data. Round to the nearest hundredth if necessary. Then write a few sentences comparing their variation.

Longest Bridges (kilometers)									
United States					Europe				
38.4	36.7	29.3	24.1	17.7	17.2	11.7	7.8	6.8	6.6
12.9	11.3	10.9	8.9	8.9	6.1	5.1	5.0	4.3	3.9

For Exercises 4–7, refer to the table that shows the recent population, in millions, of the ten largest U.S. cities.

Population of Largest U.S. Cities (millions)				
1.5	3.8	1.3	1.6	2.9
1.4	0.9	2.3	8.4	1.3

4. Find the mean absolute deviation. Round to the nearest hundredth.

5. How many data values are closer than one mean absolute deviation away from the mean?

6. Which population is farthest from the mean? How far away from the mean is that population? Round to the nearest hundredth.

7. Are there any populations that are more than twice the mean absolute deviation from the mean? Explain.

MP Be Precise **For Exercises 8 and 9, look up the word *deviate* in a dictionary or online.**

8. What does the word *deviate* mean? How can it help you remember what the mean absolute deviation refers to? ______

9. How does the word *absolute* help you to remember how to calculate the mean absolute deviation? ______

H.O.T. Problems Higher Order Thinking

10. **MP Reason Abstractly** Create two sets of data, each with five values, that satisfy the following conditions.

The mean absolute deviation of Set A is less than the mean absolute deviation of Set B.

The mean of Set A is greater than the mean of Set B.

MP Persevere with Problems **For Exercises 11 and 12, refer to the table that shows the recorded speeds of several cars on a busy street.**

Recorded Speeds (mph)					
35	38	41	35	36	55

11. Calculate the mean absolute deviation both with and without the data value of 55. Round to the nearest hundredth if necessary.

12. Explain how including the value of 55 affects the mean absolute deviation.

13. **MP Construct an Argument** Explain why the mean absolute deviation is calculated using absolute value. ______

14. **MP Persevere with Problems** The table shows the high temperatures for 6 days. If the high temperature for day 7 is 61°F, how does the mean absolute deviation change?

High Temperature (°F)					
75	58	72	68	69	66

Name ________________ My Homework ________________

Extra Practice

MP Use Math Tools Find the mean absolute deviation for each set of data. Round to the nearest hundredth if necessary. Then describe what the mean absolute deviation represents.

15.

Digital Camera Prices ($)				
140	125	190	148	156
212	178	188	196	224

$26.76; The average distance each data value is from the mean is $26.76.

Homework Help

mean: $\frac{140+125+190+148+156+212+178+188+196+224}{10} = \$175.70$

mean absolute deviation: $\frac{35.7+50.7+14.3+27.7+19.7+36.3+2.3+12.3+20.3+48.3}{10} = 26.76$

16.

Grand Slam Singles Titles Won				
14	8	7	6	5
10	11	8	8	6

Copy and Solve Find the mean absolute deviation for each set of data. Round to the nearest hundredth. Then write a few sentences comparing their variation.

17. The table shows the amount of money raised by the homerooms for two grade levels at a middle school.

Money Raised ($)											
Sixth Grade						Seventh Grade					
88	116	94	108	112	124	144	91	97	122	128	132

18. The table shows the number of points scored each game for two different basketball teams.

Number of Points Scored											
Lakeside Panthers						Jefferson Eagles					
44	38	54	48	26	36	58	42	64	62	70	40

Power Up! Common Core Test Practice

19. Which of the following statements are true concerning the mean absolute deviation of a set of data? Select all that apply.

- ☐ It describes the variation of the data around the median.
- ☐ It describes the absolute value of the mean.
- ☐ It describes the variation of the data around the mean.
- ☐ It describes the average distance between each data value and the mean.

20. The table shows the prices for parking at three different beaches along the same coastline. Select the correct values to complete the model below to find the mean absolute deviation of the data.

Beach Parking ($)		
2.50	3.75	3.50

0.25	2.75	1
0.50	3.00	2
0.75	3.25	3
1.00	3.50	4
2.50	3.75	5

Find the mean:

$\dfrac{\square + \square + \square}{\square} = \square$

Find the absolute values of the differences between each data value and the mean:

$\square - \square = \square$

$\square - \square = \square$

$\square - \square = \square$

Find the mean of the absolute values of the differences:

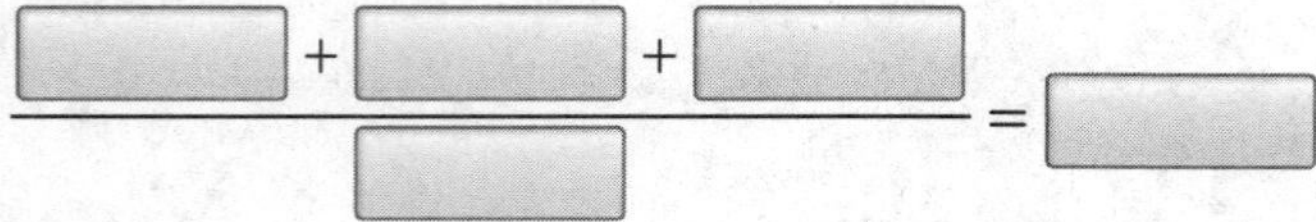

What is the mean absolute deviation of the data? ☐

Common Core Spiral Review

21. The table shows the number of different cones Delightful Dips ice cream shop sold in one afternoon. What is the total number of cones sold? **4.NBT.4** ________________

Flavor	Number of Cones
Chocolate	57
Cookie Crunch	49
Fudge Swirl	41
Strawberry	37
Vanilla	51

22. The hiking club wanted to cover a different trail each day for a week. On Monday they hiked 2.3 miles, on Tuesday they hiked 1.8 miles, on Wednesday they hiked 3.2 miles, on Thursday they hiked 1.4 miles, and on Friday they hiked 2.8 miles. What is the total distance they hiked? **5.NBT.7**

____

Lesson 5

Appropriate Measures

Real-World Link

Recycling The green committee had a recycling drive where they collected aluminum cans, plastic bottles, newspapers, and batteries. The weights collected on the first day are shown.

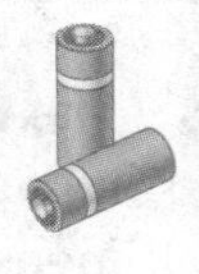

12.2 lb 11 lb 19.5 lb 13 lb

1. Find the mean weight collected. ____________

2. If the newspapers are not included, find the mean weight rounded to the nearest hundredth. ____________

3. How does the weight of the newspapers affect the mean?

____

____

4. What is the median for the data set? How does the median differ if the newspapers are not included?

____

____

Essential Question

HOW are the mean, median, and mode helpful in describing data?

Common Core State Standards

Content Standards
6.SP.5, 6.SP.5c, 6.SP.5d

MP Mathematical Practices
1, 3, 4

Which **MP Mathematical Practices** did you use? Shade the circle(s) that applies.

① Persevere with Problems
② Reason Abstractly
③ Construct an Argument
④ Model with Mathematics
⑤ Use Math Tools
⑥ Attend to Precision
⑦ Make Use of Structure
⑧ Use Repeated Reasoning

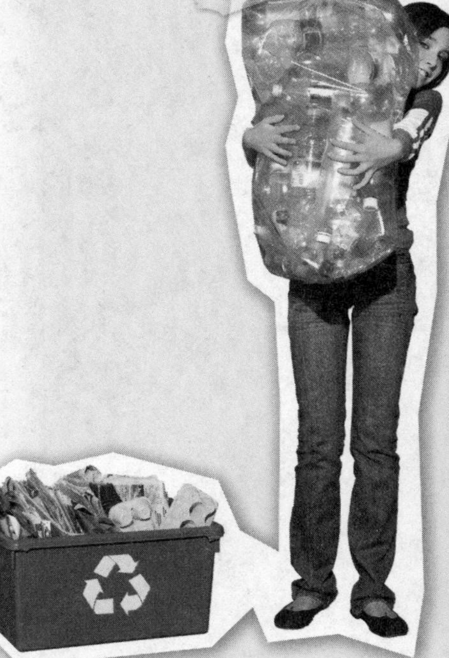

Key Concept

Using Mean, Median, and Mode

Measure	Most appropriate when...
mean	• the data have no extreme values.
median	• the data have extreme values. • there are no big gaps in the middle of the data.
mode	• data have many repeated numbers.

Work Zone

Sometimes, one measure is more appropriate than others to use to summarize a data set.

Examples

1. The table shows the number of medals won by the U.S. Which measure of center best represents the data? Then find the measure of center.

Year	1992	1996	2000	2004	2008
Number of Medals	112	101	97	103	110

Since the set of data has no extreme values or numbers that are repeated, the mean would best represent the data.

Mean $\frac{112 + 101 + 97 + 103 + 110}{5} = \frac{523}{5}$ or $104\frac{3}{5}$

The mean number of medals won is $104\frac{3}{5}$ medals.

2. The table shows the water temperature over several days. Which measure of center best represents the data? Then find the measure of center.

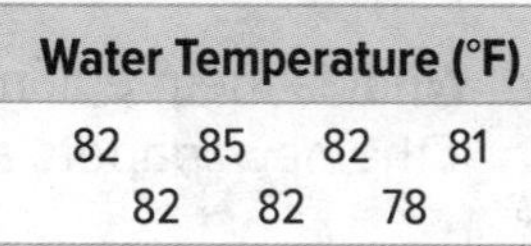

Water Temperature (°F)
82 85 82 81 82 82 78

In the set of data, there are no extreme values. There is a temperature repeated four times, so the mode 82° is the measure of center that best represents the data.

Got it? **Do this problem to find out.**

a. ______

a. The prices of several DVDs are \$22.50, \$21.95, \$25.00, \$21.95, \$19.95, \$21.95, and \$21.50. Which measure of center best represents the data? Justify your selection. Then find the measure of center.

Outliers and Appropriate Measure

Sometimes data sets contain outliers. Outliers are deviations from the majority of the data set. The outlier may affect the measures of center.

Examples

The table shows average life spans of some animals.

Average Life Span	
Animal	**Life Span (years)**
African elephant	35
Bottlenose dolphin	30
Chimpanzee	50
Galapagos tortoise	200
Gorilla	30
Gray whale	70
Horse	20

3. Identify the outlier in the data set.

Compared to the other values, 200 years is extremely high. So, it is an outlier.

4. Determine how the outlier affects the mean, median, and mode of the data.

Find the mean, median, and mode with and without the outlier.

With the outlier

Mean $\dfrac{35 + 30 + 50 + 200 + 30 + 70 + 20}{7} \approx 62$

Median 35

Mode 30

Without the outlier

Mean $\dfrac{35 + 30 + 50 + 30 + 70 + 20}{6} \approx 39$

Median 32.5

Mode 30

The mean life span decreased by 62 − 39 or 23 years. The median life span decreased by 35 − 32.5 or 2.5 years. The mode did not change.

5. Which measure of center best describes the data with and without the outlier? Justify your selection.

The mean was affected the most with the outlier. The median life span changed very little with and without the outlier, so it best describes the data in both cases. The mode does not describe the data very well since there were only two repeated numbers.

Outliers

In Example 3, 200 is an outlier.
IQR = 40
40 • 1.5 = 60
70 + 60 = 130
200 > 130
So, 200 is an outlier.

and Reflect

If a data set has an outlier, why might you use the median instead of the mean?

b. ____________

Got it? Do these problems to find out.

The prices of some new athletic shoes are shown in the table.

Price of Athletic Shoes			
$51.95	$47.50	$46.50	$48.50
$52.95	$78.95	$39.95	

b. Identify the outlier in the data set.

c. Determine how the outlier affects the mean, median, and mode of the data.

d. Tell which measure of center best describes the data with and without the outlier.

Guided Practice

1. The table shows the required temperatures for different recipes. (Examples 1–5)

Cooking Temperature (°F)			
175	325	325	350
350	350	400	450

a. Identify the outlier in the data set.

b. Determine how the outlier affects the mean, median, and mode of the data.

c. Tell which measure of center best describes the data with and without the outlier. Justify your selection.

2. **Building on the Essential Question** How does an outlier affect the mean, median, and mode of a data set?

Rate Yourself!

How well do you understand choosing the appropriate measure of center for a data set? Circle the image that applies.

Clear Somewhat Clear Not So Clear

For more help, go online to access a Personal Tutor.

Name ______________________ My Homework ______________________

Independent Practice

Go online for Step-by-Step Solutions

1. The number of minutes spent studying are: 60, 70, 45, 60, 80, 35, and 45. Find the measure of center that best represents the data. Justify your selection and then find the measure of center. (Examples 1 and 2)

2. The table shows monthly rainfall in inches for five months. Identify the outlier in the data set. Determine how the outlier affects the mean, median, and mode of the data. Then tell which measure of center best describes the data with and without the outlier. Round to the nearest hundredth. Justify your selection. (Examples 3–5)

Month	June	July	Aug.	Sept.	Oct.	Nov.
Rainfall (in.)	6.14	7.19	8.63	8.38	6.47	2.43

3. The table shows the average depth of several lakes.

Lake	Depth (ft)
Crater Lake	1,148
East Okoboji	10
Lake Gilead	43
Lake Erie	62
Great Salt Lake	14
Medicine Lake	24

a. Identify the outlier in the data set. ______________________

b. Determine how the outlier affects the mean, median, mode, and range of the data. ______________________

c. Tell which measure of center best describes the data with and without the outlier. ______________________

4. **MP Construct an Argument** Fill in the graphic organizer below.

Measure of Center	How can an outlier affect it?
mean	
median	
mode	

H.O.T. Problems Higher Order Thinking

5. MP **Find the Error** Pilar is determining which measure of center best describes the data set {12, 18, 16, 44, 15, 15}. Find her mistake and correct it.

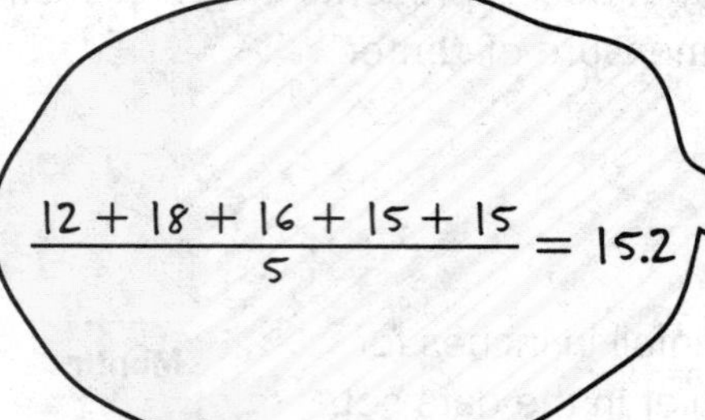

6. MP **Justify Conclusions** Determine whether the following statement is *true* or *false*. If true, explain your reasoning. If false, give a counterexample.

Of mean, median, and mode, the median will always be most affected by outliers.

7. MP **Persevere with Problems** Add three data values to the following data set so the mean increases by 10 and the median does not change.

42, 37, 32, 29, 20

8. MP **Model with Mathematics** Use the Internet to find some real-world data. Record your data in the space below.

a. Find the mean, median, and mode of your data set.

b. Are there any outliers? If so, how do they affect the measures of center?

c. Which measure of center best describes the data with and without the outlier?

Name ______ My Homework ______

Extra Practice

9. The number of songs downloaded per month by a group of friends were 8, 12, 6, 4, 2, 0, and 10. Find the measure of center that best represents the data. Justify your selection then find the measure of center. Since the set of data has no extreme values or numbers that are identical, the mean or median, 6 songs, would best represent the data.

There are no extreme values and no repeated numbers.

mean: $\frac{0 + 2 + 4 + 6 + 8 + 10 + 12}{7} = 6$

median: 0, 2, 4, (6), 8, 10, 12

10. The ages of participants in a relay race are 12, 15, 14, 13, 15, 12, 22, 16, and 11. Identify the outlier in the data set. Determine how the outlier affects the mean, median, and mode of the data. Then tell which measure of center best describes the data with and without the outlier. ______

______

______

______

______

11. MP **Justify Conclusions** The table shows the high temperatures during one week. Round to the nearest hundredth if necessary.

High Temperatures			
29°	27°	29°	25°
28°	29°	62°	

a. Identify the outlier in the data set. ______

b. Determine how the outlier affects the mean, median, mode, and range of the data. ______

______

______

c. Tell which measure of center best describes the data with and without the outlier. Explain your reasoning to a classmate. ______

______

______

______

Power Up! Common Core Test Practice

12. The table shows the number of points scored by a basketball team during their first 6 games. Determine if each statement is true or false.

Points Scored		
79	83	79
85	41	77

a. The median or mode is the best measure of center to represent the data. ☐ True ☐ False

b. The range is affected by the outlier. ☐ True ☐ False

c. The mean is the measure of center least affected by the outlier. ☐ True ☐ False

13. For each data set, select the most appropriate measure of center.

a. mp3 player prices: $45, $249, $77, $55, $24, $36, $60 ______

b. years of teaching experience: 19, 5, 7, 24, 20, 3, 28, 2, 16 ______

c. forecast high temperatures: 72°, 74°, 73°, 74°, 74°, 75°, 74° ______

mean

median

mode

Common Core Spiral Review

Find the total of each set of numbers. 4.NBT.4

14. {19, 16, 24, 22, 18} ______

15. {54, 48, 52, 57, 49} ______

16. {9, 5, 6, 7, 4, 11, 7} ______

17. {31, 36, 28, 34, 25} ______

18. Graph the numbers 15, 18, 22, 19, and 16 on the number line. 6.NS.6c

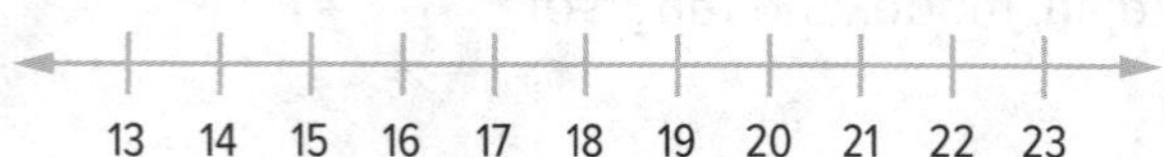

19. The table shows the number of tickets sold to the school musical on three days. How many total tickets were sold? 4.NBT.4

______

Day	Number of Tickets Sold
Wednesday	56
Thursday	79
Friday	68

21ST CENTURY CAREER in Marine Biology

Marine Biologist

Do all the unusual and amazing creatures in the ocean fascinate you? Do you think you would be good at coming up with your own experiments to test theories about them? If so, a career in marine biology might be something to think about! A marine biologist studies plants and animals that live in the ocean. These include everything from microscopic plankton to multi-ton whales. Marine biologists study organisms that live in the tiny layers of the surface and those that live thousands of meters below the surface.

Is This the Career for You?

If you would like to be a marine biologist, you may want to take some of the following courses in high school.

- Biology
- Calculus
- Chemistry
- Marine Science
- Statistics

Turn the page to find out how math relates to a career in Marine Biology.

Ready to Make Waves?

Use the information in the line plot and the table to solve each problem. Round to the nearest tenth if necessary.

1. Find the mean of the pipefish data. ______

2. Find the median and mode of the pipefish data. ______

3. What is the range of the pipefish data? Would you describe the data as spread out or close in value? Explain. ______

4. Identify the outlier in the artificial reef data. Find the mean with and without the outlier. ______

5. Describe how the outlier affects the mean in Exercise 4. ______

6. Find the median and mode of the artificial reef data. Which better represents the data? Explain. ______

Pipefish Specimens (cm)

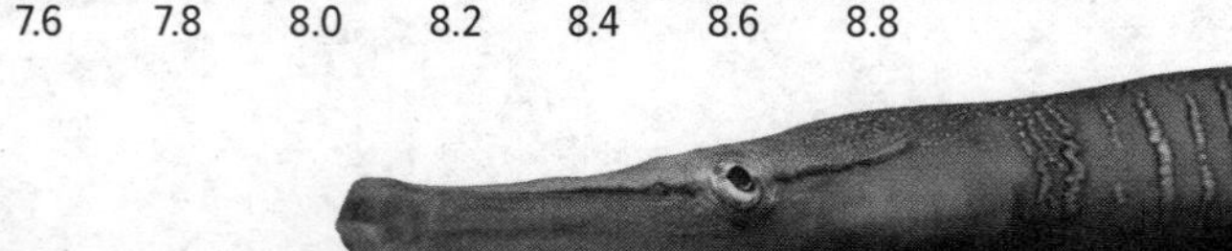

Number of Artificial Reefs in Florida Counties						
198	62	108	34	29	73	173
96	97	9	46	21	22	69
8	83	31	79	67	61	15
105	63	34	351	13	126	36
25	12	82	35	4		

Career Project

It's time to update your career portfolio! Use the Internet or another source to research several careers in marine biology. Write a brief summary comparing and contrasting the careers.

What subject in school is the most important to you? How would you use that subject in this career?

Chapter Review

Vocabulary Check

Reconstruct the vocabulary word and definition from the letters under the grid. The letters for each column are scrambled directly under that column.

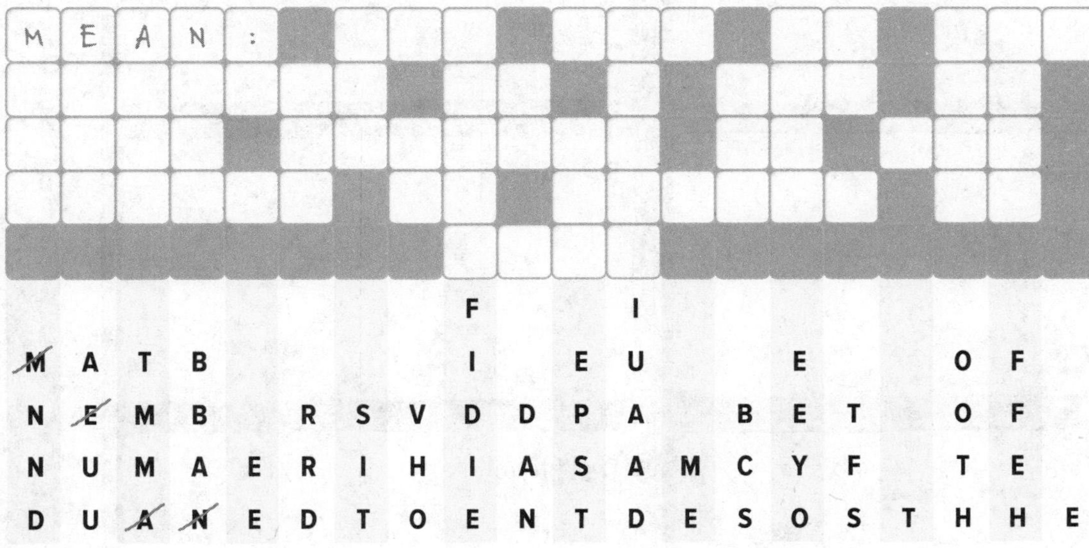

Complete each sentence using the vocabulary list at the beginning of the chapter.

1. The ________________ is the number(s) or item(s) that appear most often in a set of data.
2. Numbers that are used to describe the center of a set of data are ________________.
3. The difference between the greatest number and the least number in a set of data is the ________________.
4. The ________________ of a list of values is the value appearing at the center of a sorted version of the list, or the mean of the two central values, if the list contains an even number of values.
5. The ________________ is the distance between the first and third quartiles of a data set.
6. A value that is much higher or much lower than the other values of a data set is a(n) ________________.

Key Concept Check

Use Your FOLDABLES

Use your Foldable to help review the chapter.

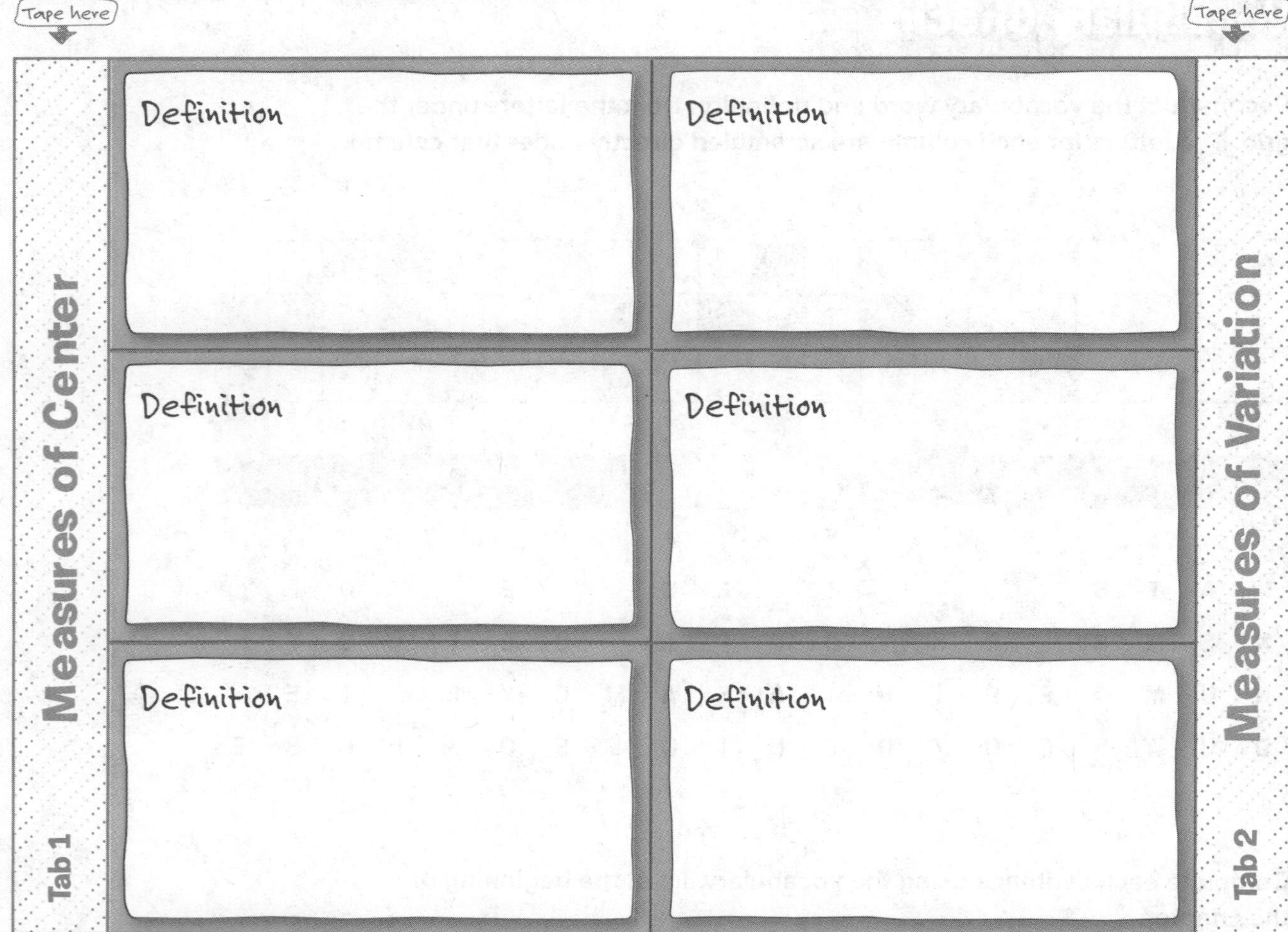

Got it?

Complete the cross number puzzle by finding the mean of each data set.

1		2		3		4
		5				
				6	7	
8					9	10
			11			
12						

Across

1. {563, 462, 490}
3. {260, 231, 248, 257}
5. {140, 163, 133, 116}
6. {21, 9, 18}
8. {145, 158, 182, 171}
9. {113, 82, 98, 91}
11. {7960, 8624, 8298, 8366}
12. {4625, 3989, 5465}

Down

1. {62, 58, 51, 41}
2. {5326, 5048, 4968}
3. {269, 293, 281}
4. {103, 89, 98, 98}
7. {720, 597, 756}
8. {142, 169, 150, 155}
10. {588, 615, 652, 653}
11. {70, 89, 90}

Athletic Awards

The local middle school athletic director has recorded the total points scored in each game by the school's basketball teams. She wants to give one of the teams a "most improved" award, but some of the game scores were lost.

Team	Game 1	Game 2	Game 3	Game 4	Game 5
7th Grade Boys	28	32	21	22	?
7th Grade Girls	17	21	20	24	?
8th Grade Girls	24	32	41	20	30
8th Grade Boys	43	39	46	50	52

Write your answers on another piece of paper. Show all of your work to receive full credit.

Part A

Find the missing point total for the 7th grade boys' fifth game if the mean of the first five games was 24.4 points. The median for the 7th grade girls' first five games was 20 points. Can you find the missing score with this information? Explain your answer.

Part B

Olga currently leads the 8th grade girls' team in scoring with a total of 50 points. Thomas leads the 8th grade boys' team with a total of 52 points. Which player should get the midseason MVP award based on their percentage of their team's total points?

Part C

Find the mean absolute deviation of the point totals for the 8th grade boys and for the 8th grade girls. Use your answers to figure out who gets the Most Consistent award between these two teams. Explain your answer.

Reflect

Answering the Essential Question

Use what you learned about mean, median, and mode to complete the graphic organizer.

Essential Question

HOW are the mean, median, and mode helpful in describing data?

	mean	median	mode
definition			
When is it appropriate to use?			
How does an outlier affect it?			

Answer the Essential Question. HOW are the mean, median, and mode helpful in describing data?

Chapter 12
Statistical Displays

Essential Question

WHY is it important to carefully evaluate graphs?

Common Core State Standards

Content Standards
6.SP.2, 6.SP.4, 6.SP.5, 6.SP.5a, 6.SP.5b, 6.SP.5c, 6.SP.5d

MP Mathematical Practices
1, 2, 3, 4, 5, 6, 7

Math in the Real World

Roller Coaster The table shows the drop of several different roller coasters.

Roller Coaster	Drop (ft)
Anaconda	144
Mind Eraser	95
Scorpion	60
Thunderbolt	70

Draw bars to represent the drop of each roller coaster.

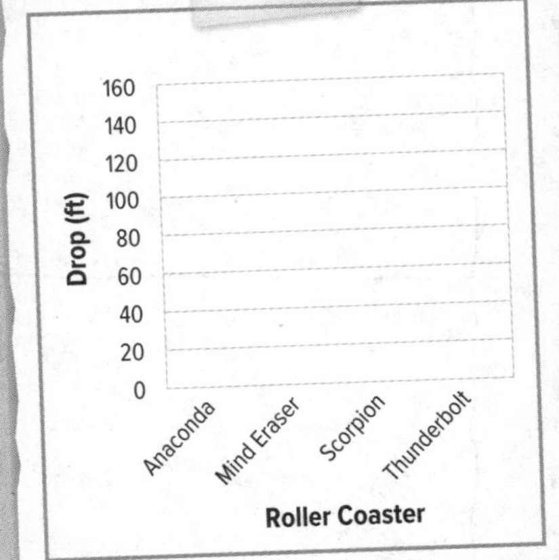

FOLDABLES® Study Organizer

Cut out the Foldable on page FL15 of this book.

Place your Foldable on page 922.

Use the Foldable throughout this chapter to help you learn about statistical displays.

What Tools Do You Need?

Vocabulary

box plot
cluster
distribution
dot plot
frequency distribution
gap
histogram
line graph
line plot
peak
symmetric

Review Vocabulary

Using a graphic organizer can help you remember important vocabulary terms. Fill in the graphic organizer for the word *graph*.

graph

Definition

Example

Picture

What Do You Already Know?

Place a checkmark below the face that expresses how much you know about each concept. Then scan the chapter to find a definition or example of it.

☹ I have no clue. 😐 I've heard of it. ☺ I know it!

Statistical Displays				
Concept	☹	😐	☺	Definition or Example
analyzing data distributions				
box plots				
dot plots				
histograms				
line graphs				
selecting appropriate displays				

When Will You Use This?

Here are a few examples of how statistical displays are used in the real world.

Activity 1 Find a bar graph in a newspaper, magazine, or on the Internet. Describe what information it is showing.

Activity 2 Go online at **connectED.mcgraw-hill.com** to read the graphic novel ***Glee Club Quest***. How many total tickets have been sold by each grade level?

Try the Quick Check below.
Or, take the Online Readiness Quiz.

Quick Review

Common Core Review 6.SP.4c

Example 1

Find the mean of the data set.

{15, 30, 20, 25, 30}

$15 + 30 + 20 + 25 + 30 = 120$ Add.

$\frac{120}{5} = 24$ Divide.

The mean is 24.

Example 2

Find the median of the data set.

{65, 57, 33, 41, 49}

33 41 (49) 57 65 Order the numbers from least to greatest.

The number in the middle is 49, so 49 is the median.

Quick Check

Mean **Find the mean of each data set.**

1. {8, 13, 21, 12, 29, 13}

2. {52, 76, 61, 58, 68}

3. {35, 18, 22, 20, 36, 31}

4. Jackson's social studies grades during one quarter are shown in the table. What is his mean score for the quarter?

Social Studies Grades (%)					
94	89	96	93	90	99
87	97	95	93	98	97

Median **Find the median of each data set.**

5. {56, 61, 54, 54, 58, 59}

6. {124, 131, 114, 148, 126}

7. {85, 79, 82, 90, 84, 87}

8. The table shows the high temperatures in a certain city for a week. What is the median temperature?

High Temperature (°F)						
71	64	56	52	62	62	66

How Did You Do?

Which problems did you answer correctly in the Quick Check? Shade those exercise numbers below.

8

Lesson 1

Line Plots

Real-World Link

Activities Students in Mr. Cotter's class were asked how many after-school activities they have. Their responses are shown in the table.

Step 1 Use the data to complete the frequency table.

Number of Activities			
0	2	1	3
1	1	3	4
2	1	0	1
2	3	2	1

→

Number of Activities	
Number	Tally
0	
1	
2	
3	
4	

Step 2 Turn the table so the number of activities is along the bottom on a number line. Instead of tally marks, place Xs above the number line. The Xs for 0 activities have been placed for you.

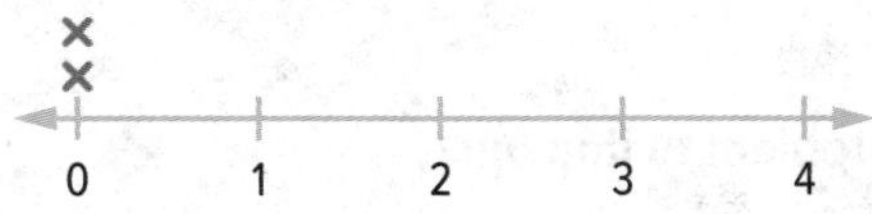

The data is now represented in a *line plot*.

Which MP Mathematical Practices did you use? Shade the circle(s) that applies.

① Persevere with Problems
② Reason Abstractly
③ Construct an Argument
④ Model with Mathematics
⑤ Use Math Tools
⑥ Attend to Precision
⑦ Make Use of Structure
⑧ Use Repeated Reasoning

Essential Question

WHY is it important to carefully evaluate graphs?

Vocabulary

line plot
dot plot

Common Core State Standards

Content Standards
6.SP.4, 6.SP.5, 6.SP.5a, 6.SP.5b, 6.SP.5c

MP Mathematical Practices
1, 3, 4

Work Zone

Make a Line Plot

One way to give a picture of data is to make a line plot. A **line plot** is a visual display of a distribution of data values where each data value is shown as a dot or other mark, usually an X, above a number line. A line plot is also known as a **dot plot**.

Example

1. Jasmine asked her class how many pets they had. The results are shown in the table. Make a line plot of the data. Then describe the data presented in the graph.

Number of Pets					
3	2	2	1	3	1
0	1	0	2	3	4
0	1	1	4	2	2
1	2	2	3	0	2

Step 1 Draw and label a number line.

0 1 2 3 4

Step 2 Place as many Xs above each number as there are responses for that number. Include a title.

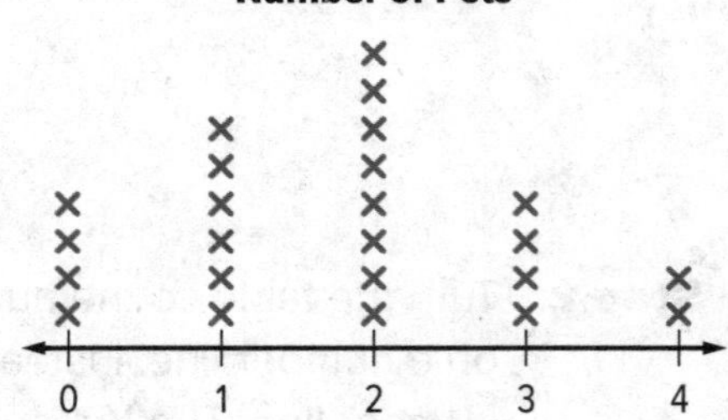

Step 3 Describe the data. 24 students responded to the question. No one has more than 4 pets. Four students have no pets. The response given most is 2 pets. This represents the mode.

Got it? **Do this problem to find out.**

a. ______

a. Javier asked the members of his 4-H club how many projects they were taking. The results are shown in the table. Make a line plot of the data. Then describe the data in the graph.

Number of Projects				
2	4	3	3	1
0	5	4	2	2
1	3	2	1	2

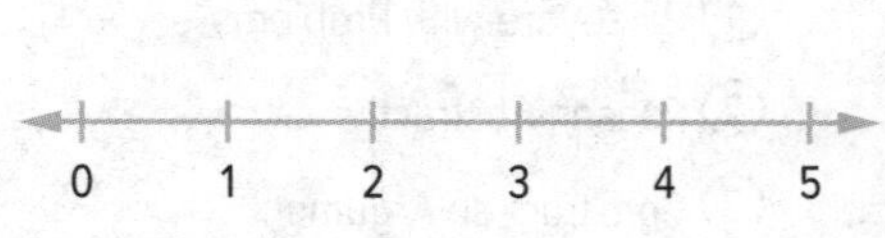

Analyze Line Plots

You can describe a set of data using measures of center as well as measures of variability. The range of the data and any outliers are also useful in describing the data.

Examples

The line plot shows the prices of cowboy hats.

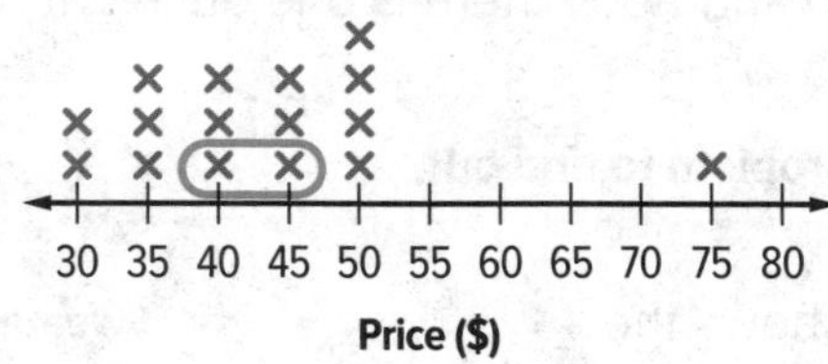

2. Find the median and mode of the data. Then describe the data using them.

There are 16 hat prices, in dollars, represented in the line plot. The median is between the 8th and 9th pieces of data.

The two middle numbers, shown on the line plot, are 40 and 45. So, the median is $42.50. This means that half of the cowboy hats cost more than $42.50 and half cost less than $42.50.

The number that appears most often is 50. So, the mode of the data is 50. This means that more cowboy hats cost $50 than any other price.

Suppose two sets of data have the same median but different ranges. What can you conclude about the sets? Explain below.

3. Find the range and any outliers of the data. Then describe the data using them.

The range of the prices is $75 − $30 or $45. The limits for the outlier are $12.50 and $72.50. So, $75 is an outlier.

Got it? Do this problem to find out.

b. The line plot shows the number of magazines each member of the student council sold. Find the median, mode, range, and any outliers of the data. Then describe the data using them.

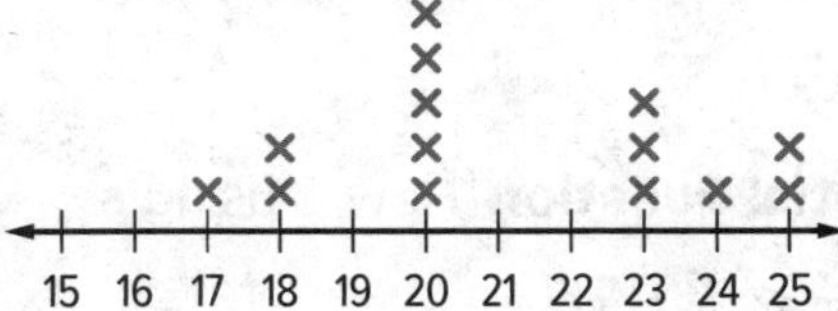

b. ____________

Example

4. The line plot shows the amount James deposited in his savings account each month. Describe the data. Include measures of center and variability.

Show your work.

Amount Saved ($)

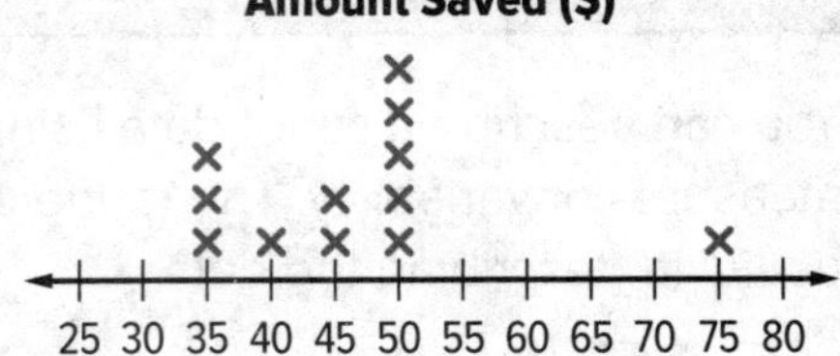

The mean is \$46.67. The median is \$47.50, and the mode is \$50. So, the majority of the data are close to the measures of center.

The range of the data is \$75 − \$35 or \$40. The interquartile range is $Q_3 - Q_1$, or \$50 − \$37.50 = \$12.50. So, half of the amounts are between \$37.50 and \$50. There is one outlier at \$75.

Got it? Do this problem to find out.

c. ____________

c. The line plot shows the prices of sweaters in a store. Describe the data. Include measures of center and variability.

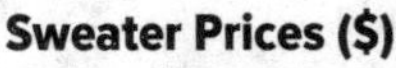

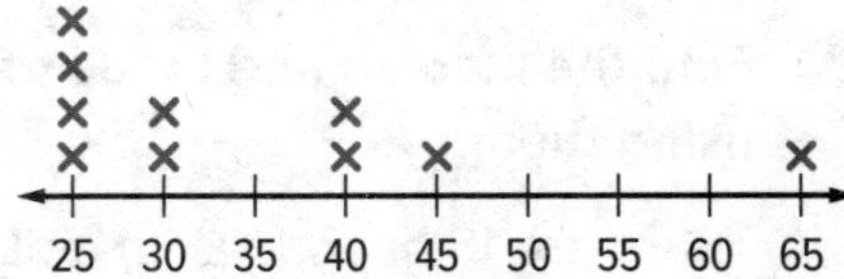

Guided Practice

1. Make a line plot for the set of data. Describe the data. Include measures of center and variability. (Examples 1–4)

Show your work.

Calories in Serving of Peanut Butter			
190	160	210	210
200	185	190	190
185	200	190	210
190	185	200	200

160 170 180 190 200 210 220

Calories

2. **Building on the Essential Question** How is using a line plot useful to analyze data?

Rate Yourself!

How confident are you about line plots? Check the box that applies.

For more help, go online to access a Personal Tutor.

Tutor

FOLDABLES Time to update your Foldable!

Independent Practice

Go online for Step-by-Step Solutions

Make a line plot for each set of data. Find the median, mode, range, and any outliers of the data shown in the line plot. Then describe the data using them. (Examples 1–3)

1. Length of summer camps in days:
7, 7, 12, 10, 5, 10, 5, 7, 10, 9, 7, 9, 6, 10, 5, 8, 7, and 8

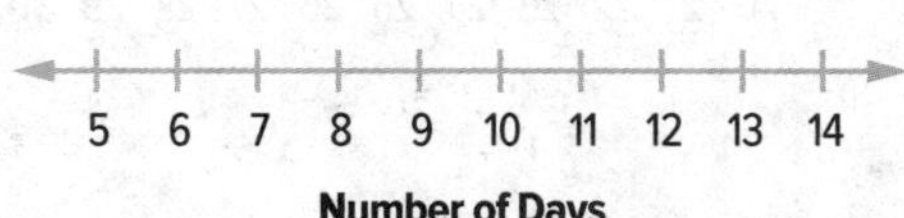

2.

Students' Estimates of Room Length (m)				
10	11	12	12	13
13	13	14	14	14
15	15	15	15	15
16	16	16	17	17
17	17	18	18	25

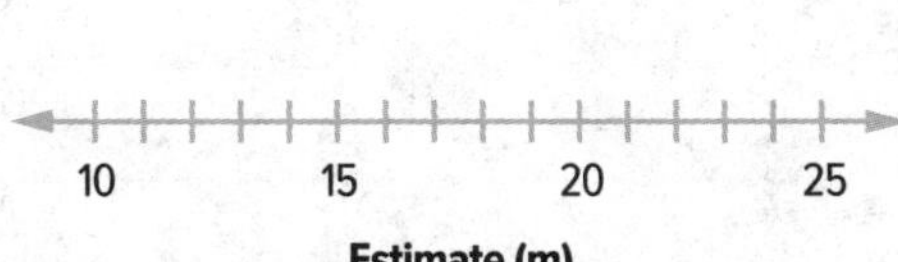

3. The line plot shows the number of songs in play lists. Describe the data. Include measures of center and variability. (Example 4)

Number of Songs in Play Lists

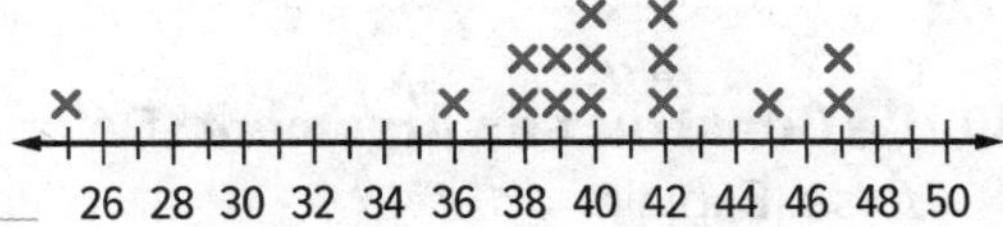

Inductive Reasoning The number of runs a softball team scored in their last five games is shown in the line plot. How many runs would the team need to score in the next game so that each statement is true?

4. The range is 10. ________

5. Another mode is 11. ________

6. The median is 9.5. ________

Runs Scored

5 6 7 8 9 10 11 12 13 14 15

H.O.T. Problems Higher Order Thinking

7. **MP Find the Error** Dwayne is analyzing the data in the line plot. Find his mistake and correct it.

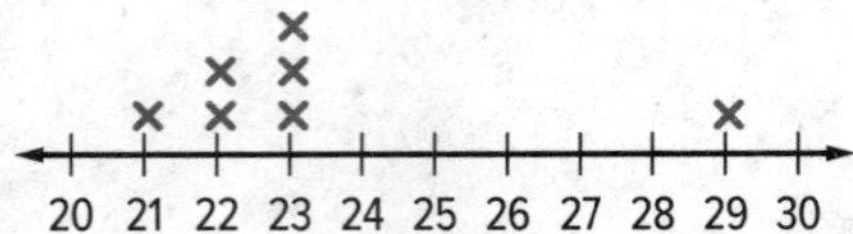

8. **MP Model with Mathematics** Write a survey question that has a numerical answer. Some examples are "How many CDs do you have?" or "How many feet long is your bedroom?" Ask your friends and family the question. Record the results and organize the data in a line plot. Use the line plot to make conclusions about your data. For example, describe the data using the measures of center and variability.

9. **MP Persevere with Problems** There are several sizes of flying disks in a collection. The range is 8 centimeters. The median is 22 centimeters. The smallest size is 16 centimeters. What is the largest disk in the collection?

10. **MP Construct an Argument** Determine whether the statement is *true* or *false*. Explain.

Line plots display individual data.

11. **MP Reason Inductively** The line plot shows the number of student visitors to the National Wildlife Refuge each day for two weeks. If the four Xs at 56 were not included in the data set, which measure of center would be most affected? Justify your response.

Number of Visitors

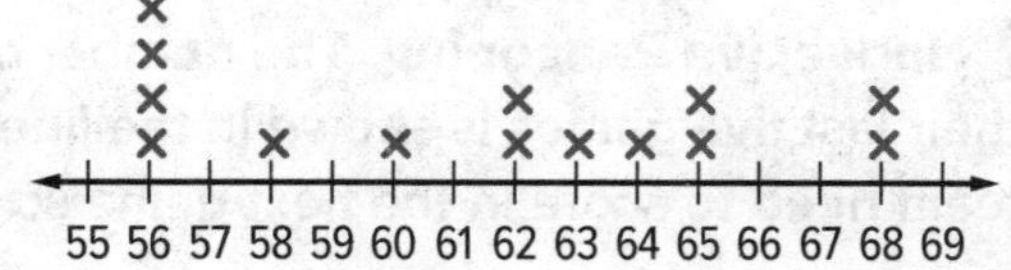

Name ______________________ My Homework ______________________

Extra Practice

Make a line plot for each set of data. Find the median, mode, range, and any outliers of the data shown in the line plot. Then describe the data using them.

12. Daily high temperatures in degrees Fahrenheit: 71, 72, 74, 72, 72, 68, 71, 67, 68, 71, 68, 72, 76, 75, 72, 73, 68, 69, 69, 73, 74, 76, 72, and 74

Homework Help

median: 72°F; mode: 72°F; range: 9°F; no outliers; The number of temperatures, in °F, represented is 24. The median means half the daily high temperatures are greater than 72°F and half are less. More days had a high of 72°F than any other temperature.

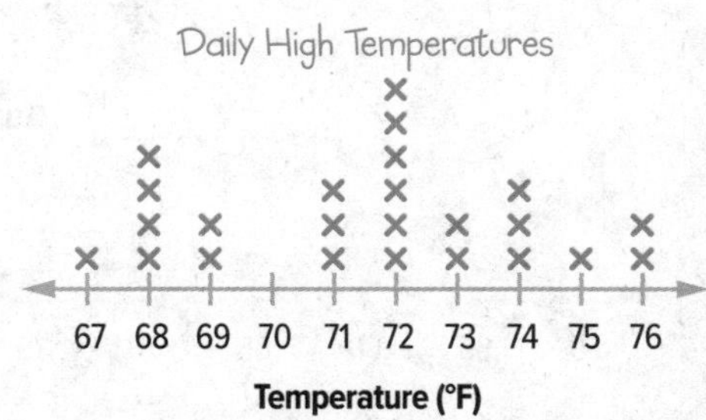

13.

Number of Tornadoes				
0	1	1	1	6
0	0	0	0	0
2	1	2	0	0

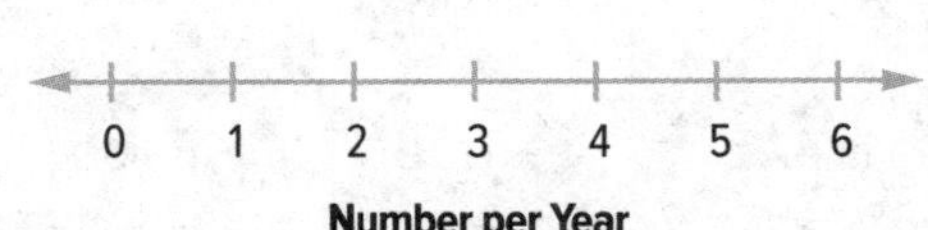

____

____

____

Copy and Solve **Describe the data in the line plots. Show your work on a separate piece of paper.**

14. The line plot shows the number of hours students spend watching TV each night. Describe the data. Include measures of center and variability. Round to the nearest tenth if necessary.

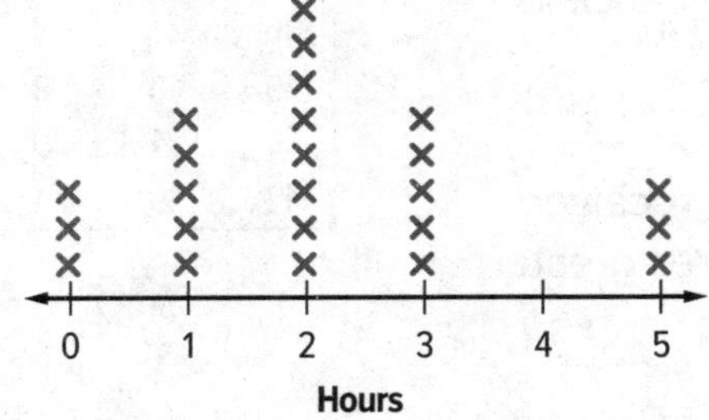

15. MP **Justify Conclusions** The line plot shows students' favorite pizza toppings. Which can you find using the line plot: the median, mode, range, or outlier(s)? Explain. Then write a sentence or two to describe the data set. Explain your reasoning to a classmate.

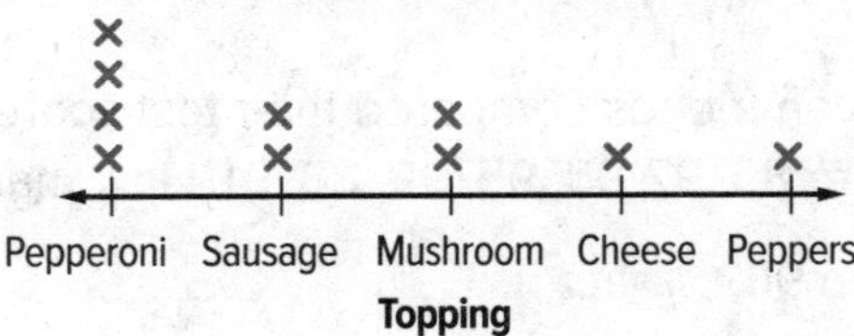

Power Up! Common Core Test Practice

16. The table shows the number of stories in 15 skyscrapers. Construct a line plot of the data.

Number of Stories				
54	88	80	88	70
78	101	69	88	85
102	73	88	110	80

Number of Stories

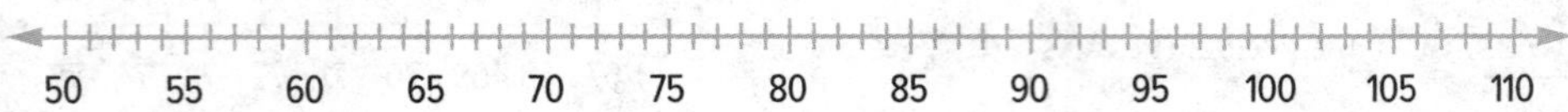

What are the median, first quartile, third quartile, and interquartile range of the data?

17. The line plot shows the number of weekly chores that some students have. Determine if each statement is true or false.

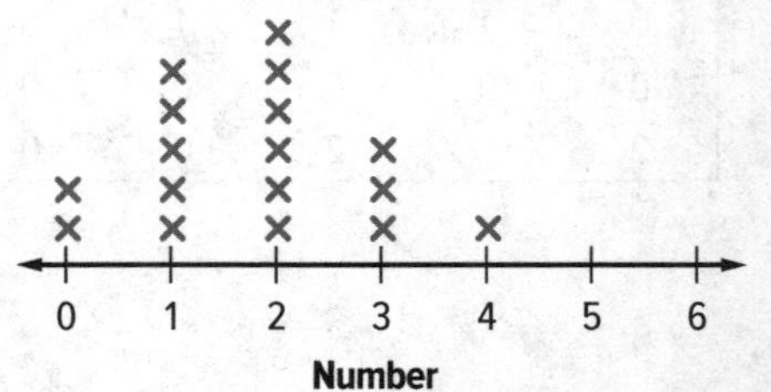

a. The median number of chores is 2. ☐ True ☐ False

b. The range of the data is 4. ☐ True ☐ False

c. The interquartile range of the data is 2. ☐ True ☐ False

Common Core Spiral Review

Fill in each ◯ with >, <, or = to make a true statement. **4.NBT.2, 5.NBT.3b**

18. 26 ◯ 19

19. 89 ◯ 92

20. 5.6 ◯ 6.5

21. 11.5 ◯ 105

22. 47 ◯ 44

23. 1.52 ◯ 14.8

24. The table shows the number of days several students attended an exercise class during a month. How many students attended a class less than 15 days? **4.NBT.2** ____________

Number of Days			
16	21	18	6
19	15	8	11
16	4	20	22
12	19	21	9

25. Seven friends compared their test scores. The scores they received were 89, 97, 93, 95, 90, 88, 91. How many people had scores greater than 90? **4.NBT.2** ____________

Lesson 2

Histograms

Real-World Link

Concerts Alicia researched the average price of concert tickets. The table shows the results.

Average Ticket Prices of Top 10 Money-Earning Concerts				
$83.87	$68.54	$51.53	$62.10	$59.58
$47.22	$66.58	$88.49	$50.63	$68.98

1. Fill in the tally column and frequency column on the frequency table.

Average Ticket Prices of Top 10 Money-Earning Concerts		
Price	Tally	Frequency
$25.00–$49.99		
$50.00–$74.99		
$75.00–$99.99		

2. What does each tally mark represent? ______

3. What is one advantage of using the frequency table?

______

4. What is one advantage of using the first table?

______

Which MP Mathematical Practices did you use? Shade the circle(s) that applies.

① Persevere with Problems
② Reason Abstractly
③ Construct an Argument
④ Model with Mathematics
⑤ Use Math Tools
⑥ Attend to Precision
⑦ Make Use of Structure
⑧ Use Repeated Reasoning

Essential Question

WHY is it important to carefully evaluate graphs?

Vocabulary

histogram
frequency distribution

Common Core State Standards

Content Standards
6.SP.4, 6.SP.5, 6.SP.5a, 6.SP.5b

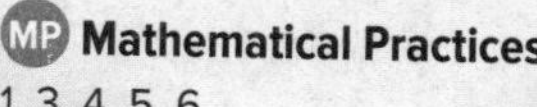

Mathematical Practices
1, 3, 4, 5, 6

Work Zone

Interpret Data

Data from a frequency table can be displayed as a histogram. A **histogram** is a type of bar graph used to display numerical data that have been organized into equal intervals. These intervals allow you to see the **frequency distribution** of the data, or how many pieces of data are in each interval.

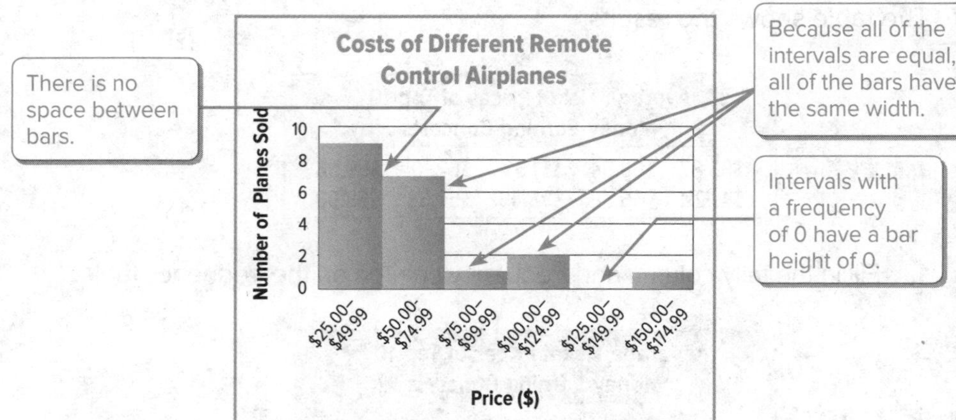

Example

1. Refer to the histogram above. Describe the histogram. How many remote control airplanes cost at least $100?

There are 9 + 7 + 1 + 2 + 1 or 20 prices, in dollars, recorded. More remote control airplanes had prices between $25.00 and $49.99 than any other range. There were no airplanes recorded with a price between $125.00 and $149.99.

Two remote control airplanes had prices between $100.00–$124.99 and one remote control airplane had a price between $150.00–$174.99. So, 2 + 1, or 3, remote control airplanes had prices that were at least $100.

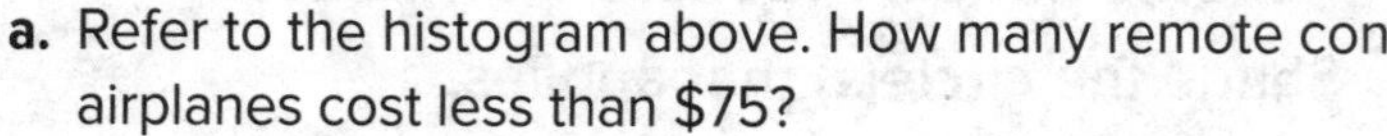

a. Refer to the histogram above. How many remote control airplanes cost less than $75?

a. ______

Construct a Histogram

You can use data from a table to construct a histogram.

Example

Watch Tutor

2. **The table shows the number of daily visitors to selected state parks. Draw a histogram to represent the data.**

Daily Visitors to Selected State Parks				
108	209	171	152	236
165	244	263	212	161
327	185	192	226	137
193	235	207	382	241

Step 1 Make a frequency table to organize the data. Use a scale from 100 through 399 with an interval of 50.

Daily Visitors to Selected State Parks		
Visitors	Tally	Frequency
100–149	||	2
150–199	~~||||~~ ||	7
200–249	~~||||~~ |||	8
250–299	|	1
300–349	|	1
350–399	|	1

Step 2 Draw and label a horizontal and vertical axis. Include a title. Show the intervals from the frequency table on the horizontal axis. Label the vertical axis to show the frequencies.

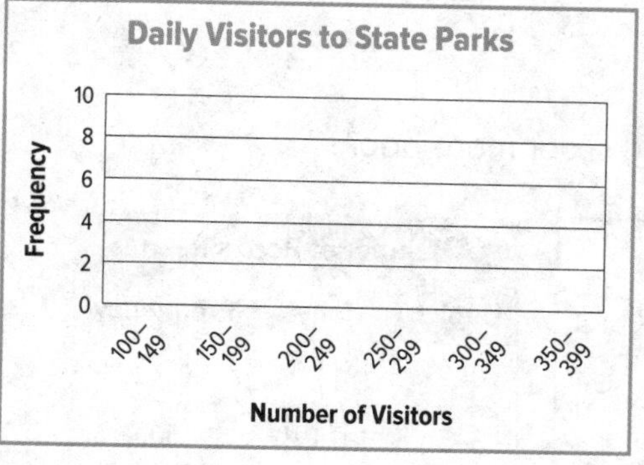

Step 3 For each interval, draw a bar whose height is given by the frequencies.

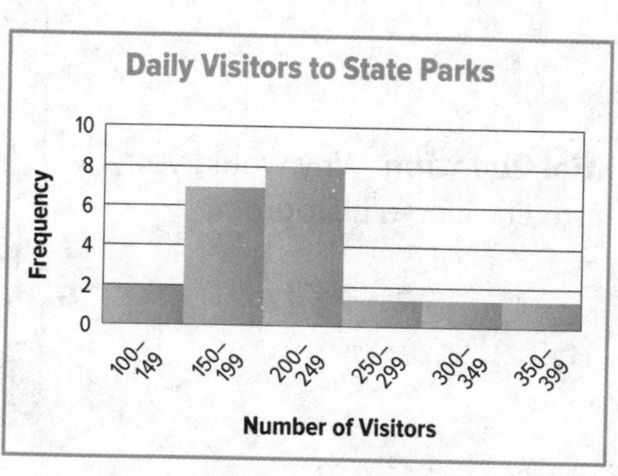

Scales and Intervals

It is important to choose a scale that includes all of the numbers in the data set. The interval should organize the data to make it easy to compare.

When is a histogram more useful than a table with individual data? Explain below.

Got it? Do this problem to find out.

b. The table at the right shows a set of test scores. Choose intervals, make a frequency table, and construct a histogram to represent the data.

Test Scores						
72	97	80	86	92	98	88
76	79	82	91	83	90	76
81	94	96	92	72	83	85
65	91	92	68	86	89	97

Test Scores		
Score	Tally	Frequency

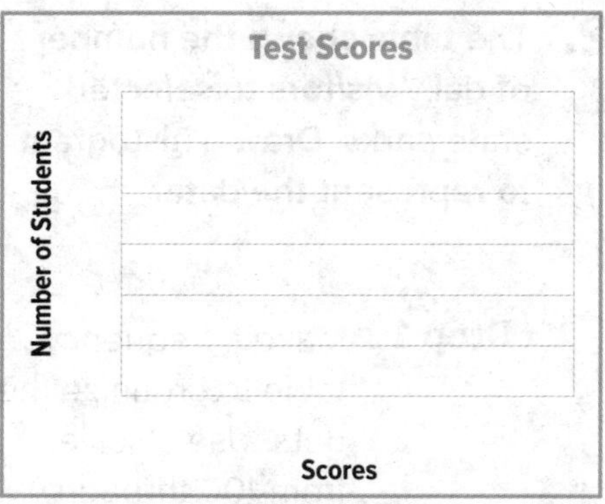

Guided Practice

1. The frequency table below shows the number of books read on vacation by the students in Mrs. Angello's class. (Examples 1 and 2)

   a. Draw a histogram to represent the data.

   b. Describe the histogram. ______________________

   c. How many students read six or more books? ______________

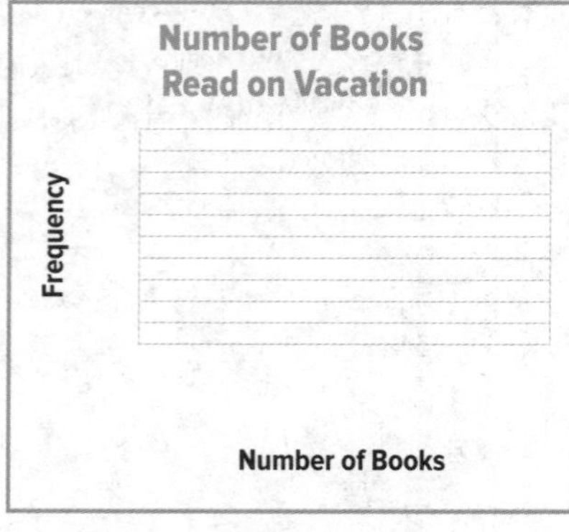

Number of Books Read		
Books	Tally	Frequency
0–2	𝍸 I	6
3–5	𝍸 𝍸	10
6–8	𝍸 II	7
9–11	III	3
12–14	IIII	4

2. **Building on the Essential Question** Why would you create a frequency table before creating a histogram?

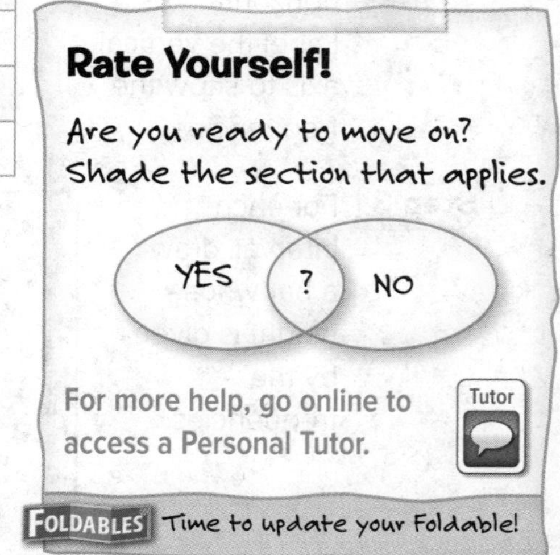

Name ________________ My Homework ________________

Independent Practice

Go online for Step-by-Step Solutions

For Exercises 1–4, use the histogram at the right. (Example 1)

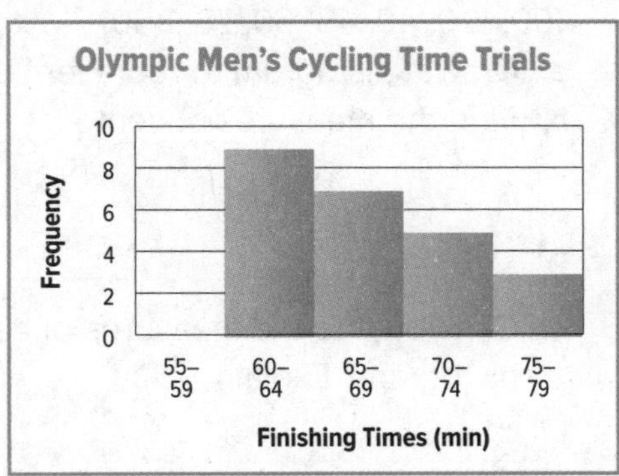

1. Describe the histogram. ________________

2. Which interval has 7 cyclists? ________________

3. Which interval represents the greatest number of cyclists?

____

4. How many cyclists had a time less than 70 minutes?

____

Draw a histogram to represent the set of data. (Example 2)

5.

Number of States Visited by Students in Marty's Class		
Number of States	Tally	Frequency
0–4	𝍸 ||||	9
5–9	|||	3
10–14	𝍸	5
15–19	|||	3
20–24	𝍸 |	6
25–29	|	1

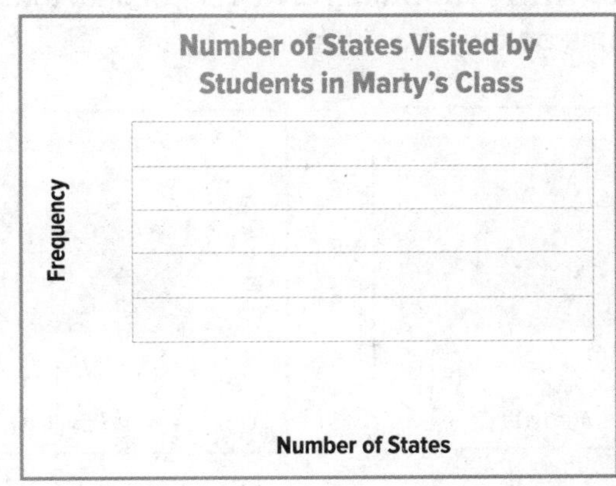

MP Use Math Tools For Exercises 6 and 7, refer to the histograms below.

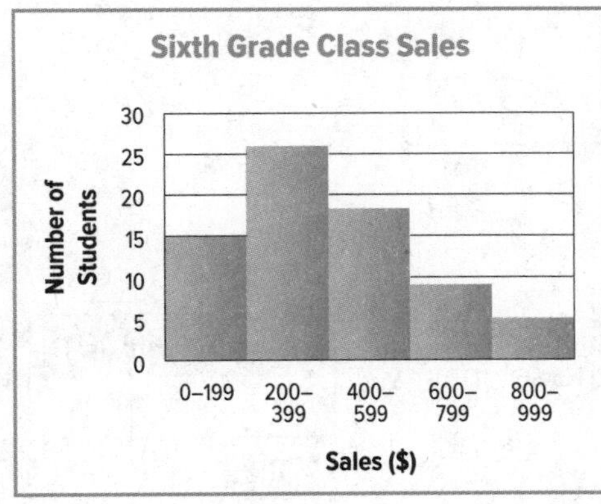

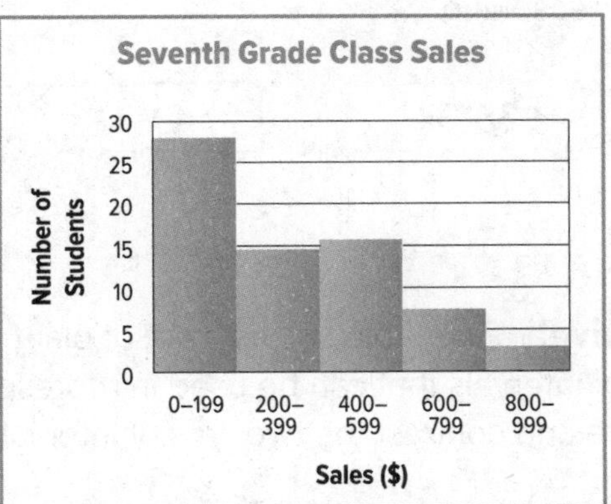

6. About how many students from both grades earned $600 or more?

____

7. Which grade had more students earn between $400 and $599?

____

8. **MP Be Precise** The following data provides the number of Calories of various types of frozen bars. {25, 35, 200, 280, 80, 80, 90, 40, 45, 50, 50, 60, 90, 100, 120, 40, 45, 60, 70, 350}

a. Draw a histogram to represent the data.

b. Find the measures of center.

Calories of Various Types of Frozen Bars

Number of Bars

Calories

c. Can you find the measures of center only from the histogram? Explain.

H.O.T. Problems Higher Order Thinking

9. **MP Persevere with Problems** Give a set of data that could be represented by both histograms below.

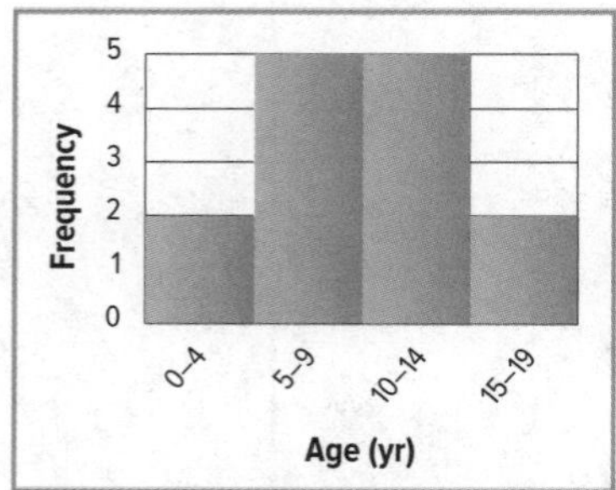

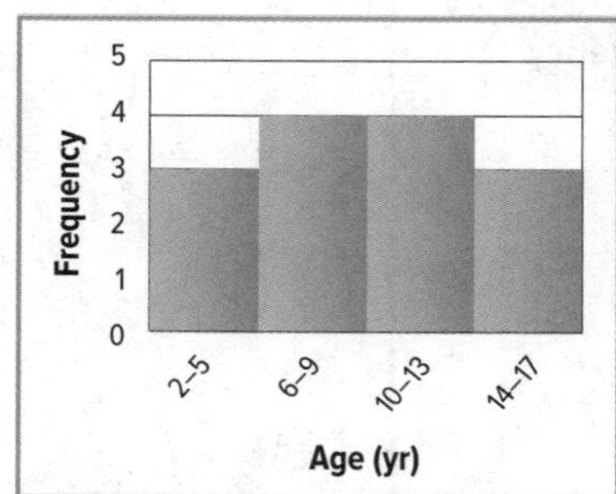

10. **MP Justify Conclusions** Identify the interval that is not equal to the other three. Explain your reasoning.

15-19 | 30-34 | 40-45 | 45-49

11. **MP Reason Inductively** The table shows a set of plant heights. Describe two different sets of intervals that can be used in representing the set in a histogram. Compare and contrast the two sets of intervals.

Plant Heights (in.)		
12	7	15
8	24	41
16	18	27
43	33	11
24	10	22

Name ______________________ My Homework ______________________

Extra Practice

For Exercises 12–16, use the histogram.

12. Describe the histogram. The ages of 30 players were collected. One player is older than 35, the rest are 35 or younger.

Homework Help: Add each of the frequencies to find the total players. $6 + 11 + 4 + 8 + 1 = 30$

Ages of Players on a Baseball Team

Age (yrs)	Frequency
20–23	6
24–27	11
28–31	4
32–35	8
36–39	1

13. Which interval represents the greatest number of players?

____

14. Which interval has 4 players? ______________________

15. How many players are younger than 28? ______________________

16. How many players have ages in the interval 32–35? ______________________

MP Model with Mathematics **Draw a histogram to represent the set of data.**

17.

Number of Homeruns in a Season		
Homeruns	**Tally**	**Frequency**
0–9	~~||||~~ ~~||||~~ ||	12
10–19	~~||||~~ ~~||||~~	10
20–29	~~||||~~ ||||	9
30–39	~~||||~~ ||||	9
40–49	~~||||~~		6

Number of Home Runs

Frequency

Number of Home Runs

18. **MP Find the Error** Pilar is analyzing the frequency table below. Find her mistake and correct it.

Distances from Home to School (mi)	Tally	Frequency
0.1–0.5	~~||||~~ ||	7
0.6–1.0	|||	3
1.1–1.5	~~||||~~	5
1.6–2.0	|||	3

15 people live less than 1.5 miles from school.

____

____

Power Up! Common Core Test Practice

19. The histogram shows the number of goals scored by the top players on a soccer team. Explain why there is not a bar for the interval of 30–44 goals.

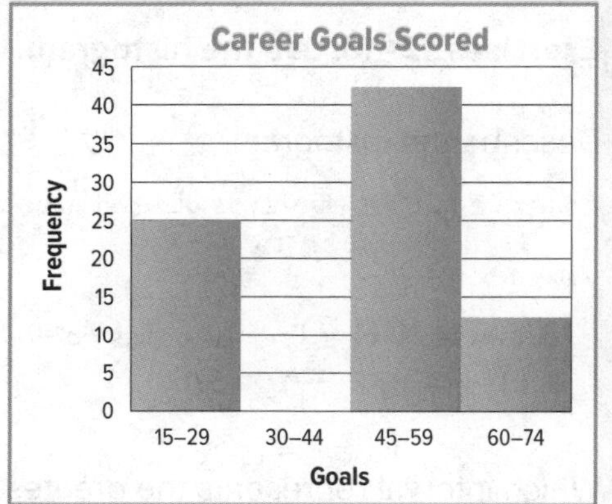

20. The table shows the number of sit-ups each member of a gym class completed in one minute. Choose an appropriate scale and intervals and construct a histogram of the data.

Number of Sit-Ups in One Minute				
30	15	34	22	28
20	25	26	31	29
27	30	19	22	28
32	31	27	23	26

Common Core Spiral Review

Divide. 4.NBT.6

21. $126 \div 3 =$ ______

22. $477 \div 9 =$ ______

23. $162 \div 6 =$ ______

24. $327 \div 5 =$ ______

25. $195 \div 2 =$ ______

26. $842 \div 4 =$ ______

27. Jamie, Tucker, and Lucinda bought a bag of apples. Jamie kept 0.25 of the apples, and Lucinda kept 0.5 of the apples. Who kept more of the apples?

5.NBT.3b ____________________

Lesson 3
Box Plots

Real-World Link

Football The table shows the number of touchdowns scored by each of the 16 teams in the National Football Conference in a recent year.

Number of Touchdowns							
47	41	35	38	28	54	49	24
49	44	27	34	37	44	26	36

1. Plot the data on a line plot.

Number of Touchdowns

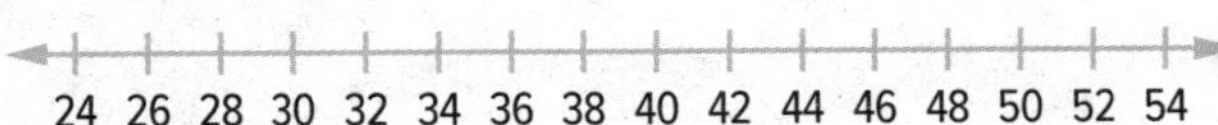

2. Find the median, lower extreme, upper extreme, first quartile and third quartile of the data. Place a star on the number line for each value.

median: ________ first quartile: ________

lower extreme: ________ third quartile: ________

upper extreme: ________

3. What percent of the teams scored less than 31 touchdowns?

____

4. What percent of the teams scored more than 37.5 touchdowns?

____

Which MP Mathematical Practices did you use? Shade the circle(s) that applies.

① Persevere with Problems
② Reason Abstractly
③ Construct an Argument
④ Model with Mathematics
⑤ Use Math Tools
⑥ Attend to Precision
⑦ Make Use of Structure
⑧ Use Repeated Reasoning

Essential Question

WHY is it important to carefully evaluate graphs?

Vocabulary

box plot

Common Core State Standards

Content Standards
6.SP.2, 6.SP.4, 6.SP.5, 6.SP.5b, 6.SP.5c

Mathematical Practices
1, 2, 3, 4, 7

Work Zone

Construct a Box Plot

A **box plot**, or box-and-whisker plot, uses a number line to show the distribution of a set of data by using the median, quartiles, and extreme values. A *box* is drawn around the quartile values, and the *whiskers* extend from each quartile to the extreme data points that are not outliers. The median is marked with a vertical line. The figure below is a box plot.

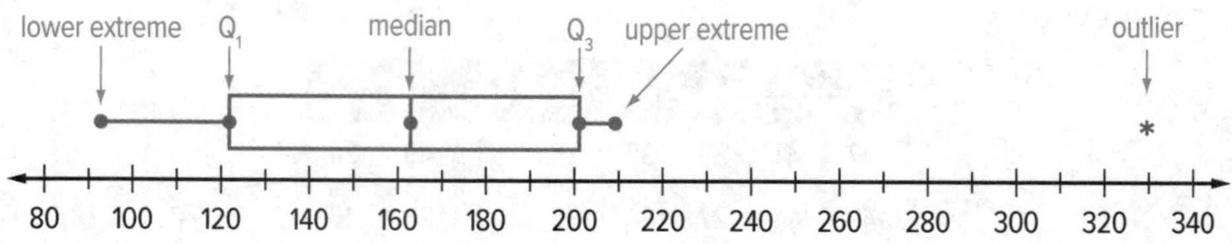

Box plots separate data into four parts. Even though the parts may differ in length, each contains 25% of the data. The box shows the middle 50% of the data.

Common Misconception
You may think that the median always divides the box in half. However, the median may not divide the box in half because the data may be clustered toward one quartile.

Example

1. Draw a box plot of the car speed data.

25 35 27 22 34 40 20 19 23 25 30

Step 1 Order the numbers from least to greatest. Then draw a number line that covers the range of the data.

Step 2 Find the median, the extremes, and the first and third quartiles. Mark these points above the number line.

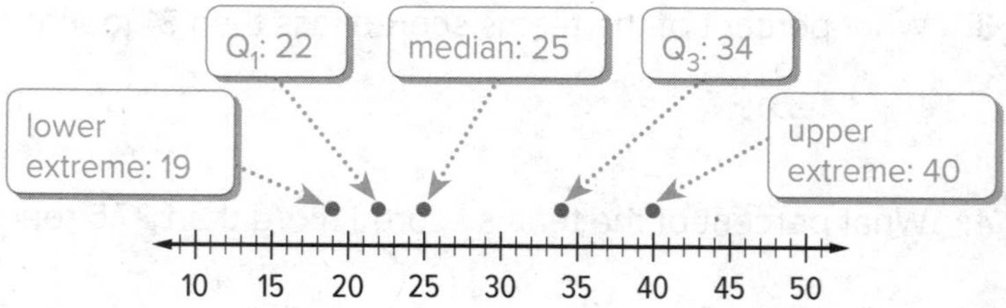

Step 3 Draw the box so that it includes the quartile values. Draw a vertical line through the box at the median value. Extend the whiskers from each quartile to the extreme data points. Include a title.

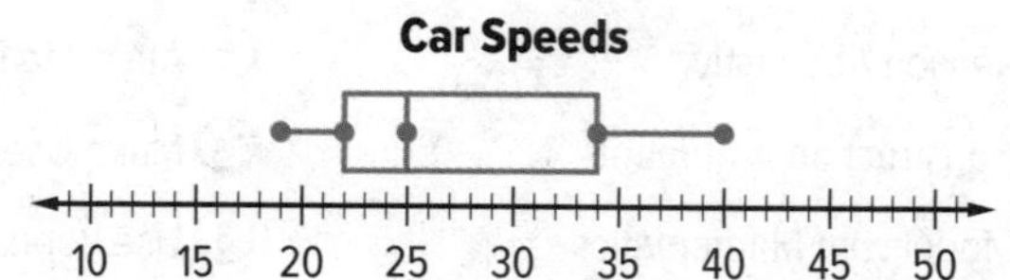

Got it? **Do this problem to find out.**

a. Draw a box plot of the data set below.
{$20, $25, $22, $30, $15, $18, $20, $17, $30, $27, $15}

a. ______

Interpret Data

Though a box plot does not show individual data, you can use it to interpret data.

Examples

Tutor

Refer to the box plot in Example 1.

2. Half of the drivers were driving faster than what speed?

Half of the 11 drivers were driving faster than 25 miles per hour.

3. What does the box plot's length tell about the data?

The length of the left half of the box plot is short. This means that the speeds of the slowest half of the cars are concentrated. The speeds of the fastest half of the cars are spread out.

Box Plots

- If the length of a whisker or the box is short, the values of the data in that part are concentrated.
- If the length of a whisker or the box is long, the values of the data in that part are spread out.

Got it? **Do this problem to find out.**

b. What percent were driving faster than 34 miles per hour?

b. ______

Example

Tutor

4. The box plot below shows the daily attendance at a fitness club. Find the median and the measures of variability. Then describe the data.

Fitness Club Attendance

45 50 55 60 65 70 75 80 85 90 95 100 105 110

The median is 72.5. The first quartile is 65 and the third quartile is 80. The range is 54 and the interquartile range is 15. There is an outlier at 110. Both whiskers are approximately the same size so the data, without the outlier, is spread evenly below and above the quartiles.

Outliers

If the data set includes outliers, then the whiskers will not extend to the outliers, just to the previous data point. Outliers are represented with an asterisk (*) on the box plot.

c. ______________

Got it? Do this problem to find out.

c. The number of games won in the American Football Conference in a recent year is displayed below. Find the median and the measures of variability. Then describe the data.

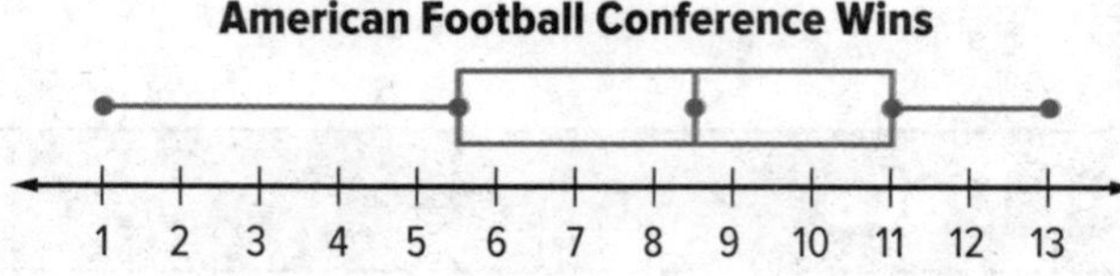

Guided Practice

1. Use the table. (Examples 1–3)

   a. Make a box plot of the data.

Depth of Recent Earthquakes (km)						
5	15	1	11	2	7	3
9	5	4	9	10	5	7

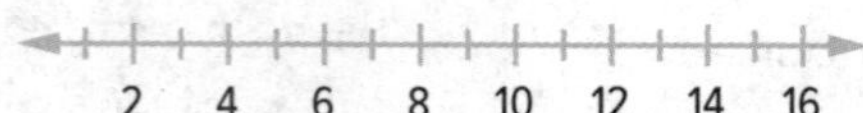

   b. What percent of the earthquakes were between 4 and 9 kilometers deep? ______________

   c. Write a sentence explaining what the length of the box plot means. ______________

2. Find the median and the measures of variability for the box plot shown. Then describe the data. (Example 4)

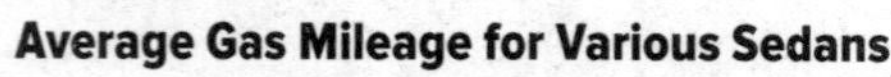

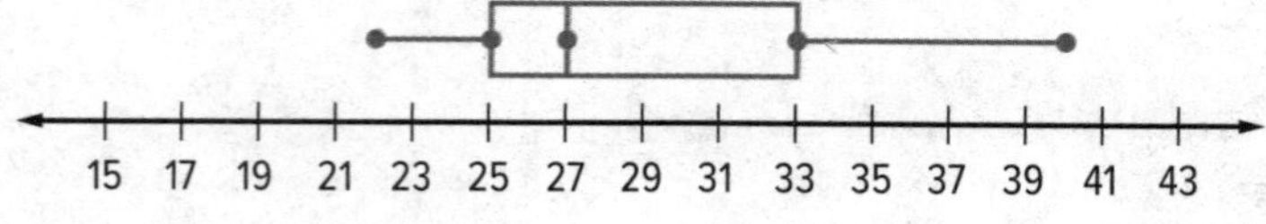

3. **Building on the Essential Question** How is the information you can learn from a box plot different from what you can learn from the same set of data shown in a line plot?

Rate Yourself!

How confident are you about making and interpreting box plots? Check the box that applies.

For more help, go online to access a Personal Tutor.

FOLDABLES Time to update your Foldable!

Name ____________________ My Homework ____________________

Independent Practice

Go online for Step-by-Step Solutions

Draw a box plot for each set of data. (Example 1)

1. {65, 92, 74, 61, 55, 35, 88, 99, 97, 100, 96}

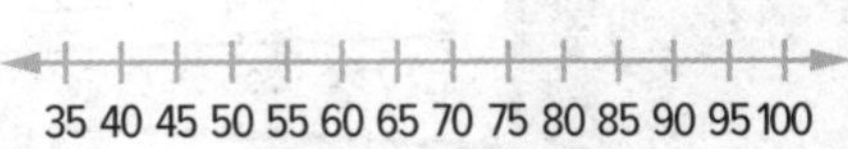

2.

Cost of MP3 Players ($)	
95	55
105	100
85	158
122	174
165	162

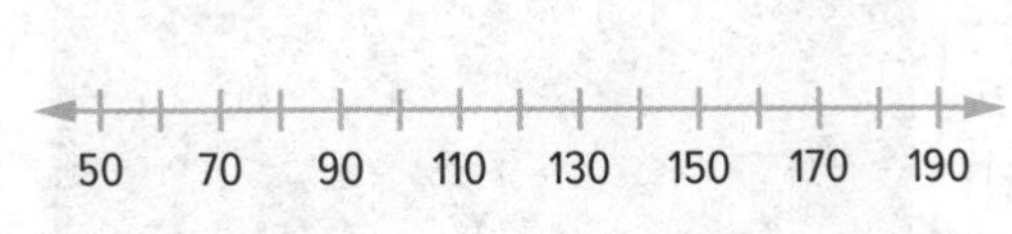

3. The table shows the length of coastline for the 13 states along the Atlantic Coast. (Examples 1–3)

Length of Coastline (mi)	
28	130
580	127
100	301
228	40
31	187
192	112
13	

a. Make a box plot of the data.

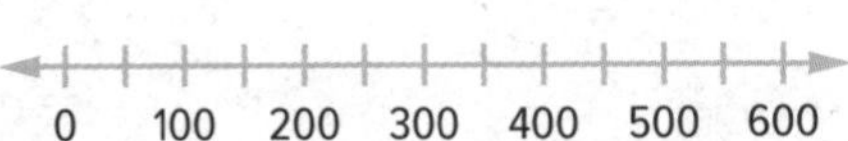

b. Half of the states have a coastline less than how many miles?

____

c. Write a sentence describing what the length of the box plot tells about the number of miles of coastline for states along the Atlantic coast.

____

____

____

4. The amount of Calories for a serving of certain fruits is displayed. Find the median and the measures of variability. Then describe the data. (Example 4)

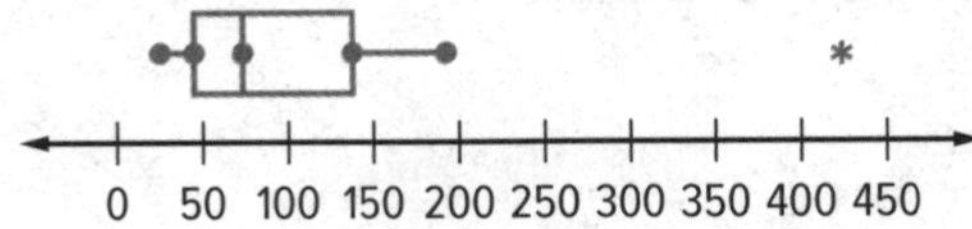

____

____

____

____

5. **MP Model with Mathematics** Refer to the graphic novel frame below for Exercises a–b.

a. Draw a box plot using the data for Grade 7.

b. Compare the box plots. Which grade sold more tickets? Explain.

H.O.T. Problems Higher Order Thinking

6. **MP Persevere with Problems** Write a set of data that contains 12 values for which the box plot has no whiskers. State the median, first and third quartiles, and lower and upper extremes.

7. **MP Reason Abstractly** Write a set of data that, when displayed in a box plot, will result in a long box and short whiskers. Draw the box plot.

8. **MP Reason Inductively** What can you conclude from a box plot where the length of the left box and whisker is the same as the length of the right box and whisker?

Extra Practice

Draw a box plot for each set of data.

9. {26, 22, 31, 36, 22, 27, 15, 36, 32, 29, 30}

15, 22, (22), 26, 27, (29), 30, 31, (32), 36, 36

median: 29; Q_1: 22; Q_3: 32

Mark the median, Q_1, Q_3, and extremes above the number line. Draw a box around the quartiles and a line through the center of the median. Connect the extremes to the box with a line.

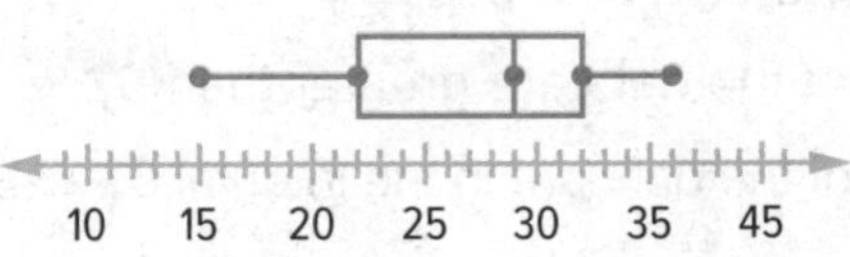

10.

Height of Waves (in.)		
80	51	77
72	55	65
42	78	67
40	81	68
63	73	59

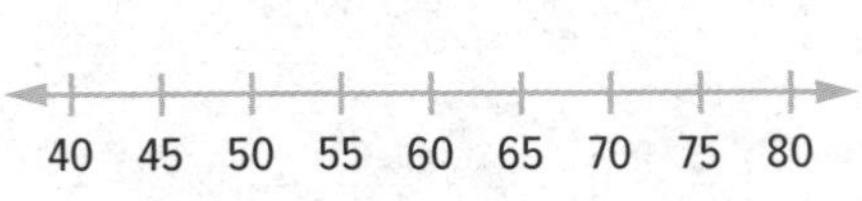

11. The box plot below summarizes math test scores.

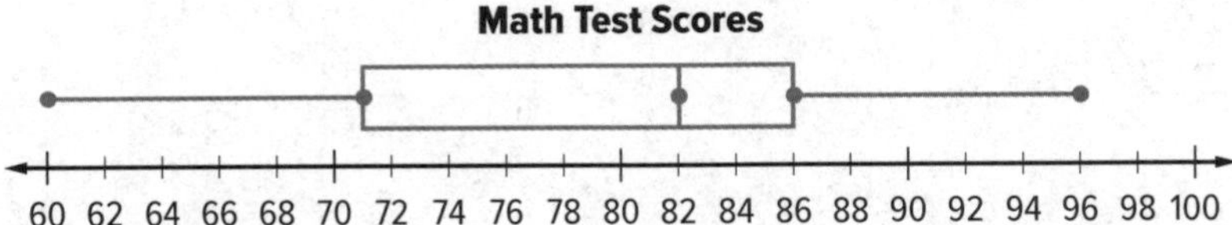

a. What was the greatest test score? ______________

b. Explain why the median is not in the middle of the box.

____

____

c. What percent of the scores were between 71 and 96? ______________

d. Half of the scores were higher than what score? ______________

12. MP **Identify Structure** Find the median, first and third quartiles, and the interquartile range for the set of data in the table. Create a box plot of the data.

Words Typed Per Minute		
80	42	65
72	63	81
67	73	40
51	68	59
77	55	78

0 10 20 30 40 50 60 70 80 90 100

Power Up! Common Core Test Practice

13. Which of the following statements are true about the box plot? Select all that apply.

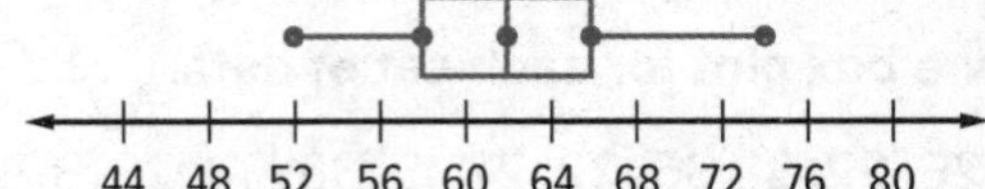

- ☐ Half of the data are greater than 62.
- ☐ Half of the data are in the interval 62–74.
- ☐ There are more data values in the interval 52–62 than there are in the interval 62–74.
- ☐ The value 74 is the maximum value.

14. The table shows the heights, in inches, of students in a classroom.

Height (in.)				
62	70	60	68	64
64	53	65	51	67
60	59	57	65	61

a. Construct a box plot of the data.

Height (in.)

48 50 52 54 56 58 60 62 64 66 68 70

b. What are the minimum, first quartile, median, third quartile, and maximum of the data?

Common Core Spiral Review

Find the total of each set of numbers. 4.NBT.4

15. {6, 8, 7, 9, 2, 4} ______

16. {15, 20, 35, 24, 31} ______

17. {16, 25, 35, 28, 31, 27} ______

18. {56, 58, 63, 51, 52} ______

19. {84, 106, 98, 88} ______

20. {34, 68, 23, 18, 57} ______

21. The table shows the number of raffle tickets each member of the drama club sold. How many members sold more than 50 raffle tickets? 4.NBT.2

Raffle Tickets Sold				
26	32	18	53	28
35	42	29	38	50
49	51	21	34	46
42	52	50	36	20

22. Rachel jumped rope for 6 minutes on Monday, 12 minutes on Tuesday, 7 minutes on Wednesday, 10 minutes on Thursday, and 8 minutes on Friday. Graph her times on a number line. 6.NS.6c

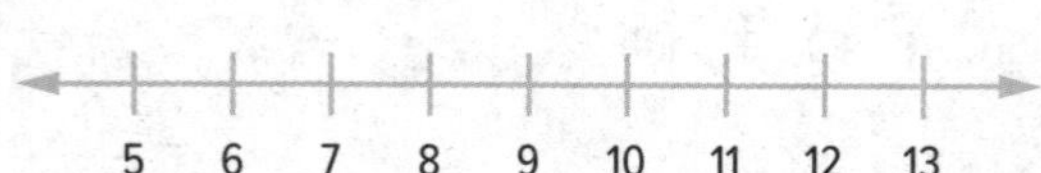

Problem-Solving Investigation
Use a Graph

Content Standards
6.SP.4, 6.SP.5, 6.SP.5c

Mathematical Practices
1, 3, 4

Case #1 Football

Finn's brother is on the football team and he is making a display of the number of points the team scored in each game last year. He uses the information in the table to make a line plot.

What score occurred most frequently?

Number of Points Scored			
35	35	43	21
49	35	21	24
34	35	21	

Understand *What are the facts?*

The range of the points is 49 − 21, or 28.

Plan *What is your strategy to solve this problem?*

Make a line plot to see which score occurs most frequently. Use the range to label the line plot from 20 to 50.

Solve *How can you apply the strategy?*

Plot each score on the line plot.

Number of Points Scored

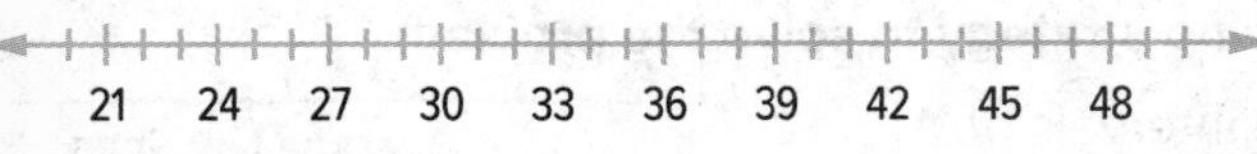

The score occurring most frequently is ☐.

Check *Does the answer make sense?*

The team scored 35 points four times. No other score occurred four or more times. So, the answer is reasonable.

Reason Inductively How would the results change if the team played a twelfth game and scored 21 points?

____

____

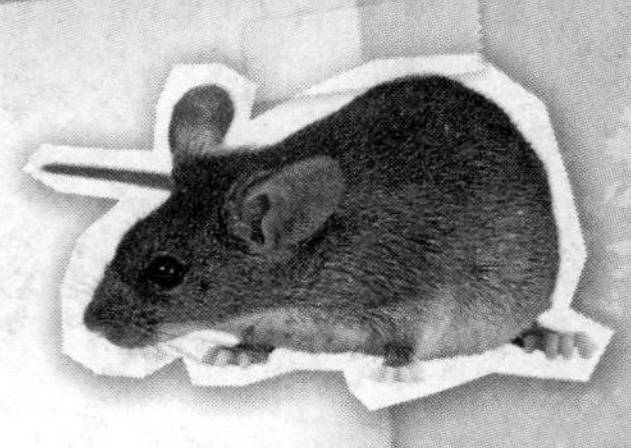

Case #2 Life Span

Different animals have different average life spans. The average life spans of several animals are shown in the table.

How many more animals have an average life span between 11 and 15 years than those that have an average life span between 1 and 5 years?

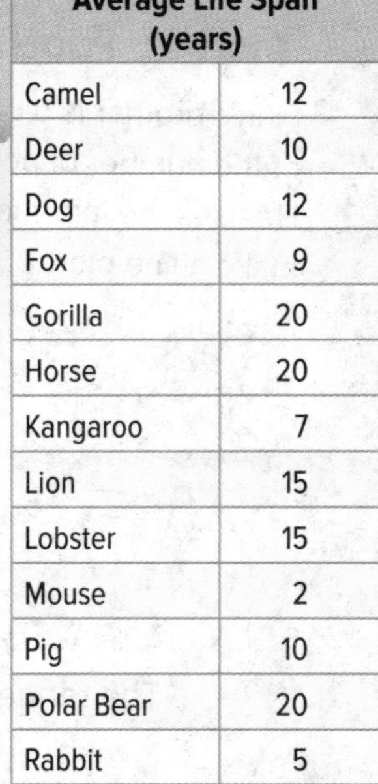

Average Life Span (years)	
Camel	12
Deer	10
Dog	12
Fox	9
Gorilla	20
Horse	20
Kangaroo	7
Lion	15
Lobster	15
Mouse	2
Pig	10
Polar Bear	20
Rabbit	5

Understand

Read the problem. What are you being asked to find?

I need to find ____

____.

What information do you know?

Animals with 11–15 year life span: ______________

Animals with 1–5 year life span: ______________

Plan

Choose a problem-solving strategy.

I will use the ______________________________ *strategy.*

3 Solve

Use your problem-solving strategy to solve the problem.

Make a histogram. Use intervals of

1–5 years, ________ years,

________ years, and 16–20 years.

So, there are ☐ more animals with an average life span between 11–15 years than with an average life span between 1–5 years.

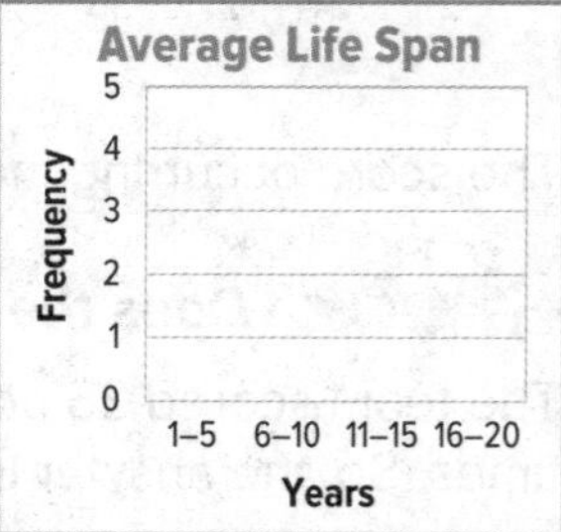

Check

Use information from the problem to check your answer.

There are four animals with an average life span between 11 and 15 years and two animals, mice and rabbits, with an average life span between 1 and 5 years.

Work with a small group to solve the following cases. Show your work on a separate piece of paper.

Case #3 Lawn Mowing

DeShawn mowed lawns over the summer to earn extra money. The number of lawns he mowed each week is shown in the line plot.

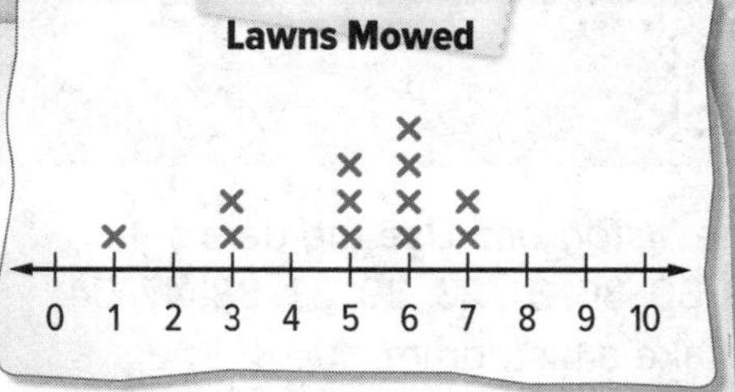

What is the mean number of lawns he mowed?

Case #4 Magazines

The box plot shows the number of magazines sold for a club fundraiser.

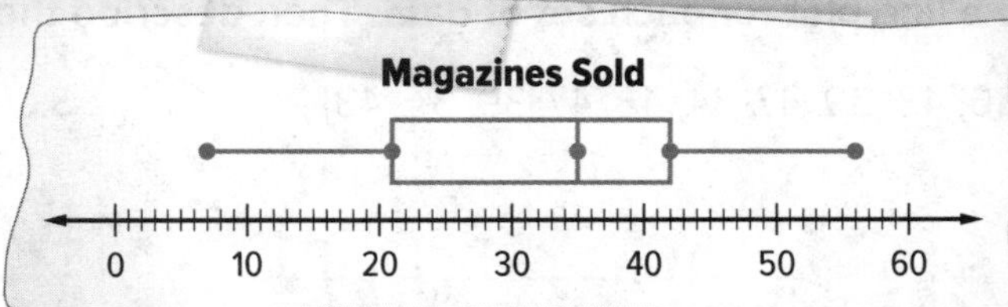

What is the difference between the median number of magazines sold and the most magazines sold?

Case #5 Quiz Scores

A teacher recorded quiz scores for a class in the table.

89	88	95	100
78	89	92	92
95	85	88	90
100	95	98	88
100	90	76	94

Make a line plot to determine the median quiz score.

Case #6 Exercise

To train for a marathon, Colleen plans to run four miles the first week and 150% the number of miles next week.

How many miles will Colleen run the next week?

Mid-Chapter Check

Vocabulary Check

1. **MP Be Precise** Define *histogram*. Use the data set {26, 37, 35, 49, 54, 53, 30, 36, 31, 28, 29, 33, 38, 47, 54, 50, 37, 26, 35, 51} to make a histogram. (Lesson 2)

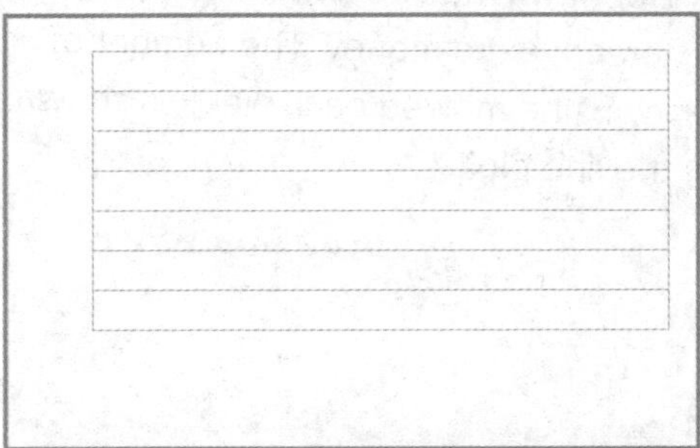

Skills Check and Problem Solving

Make a line plot for each set of data. Then describe the data. (Lesson 1)

2. {36, 43, 39, 47, 34, 43, 47, 39, 34, 43}

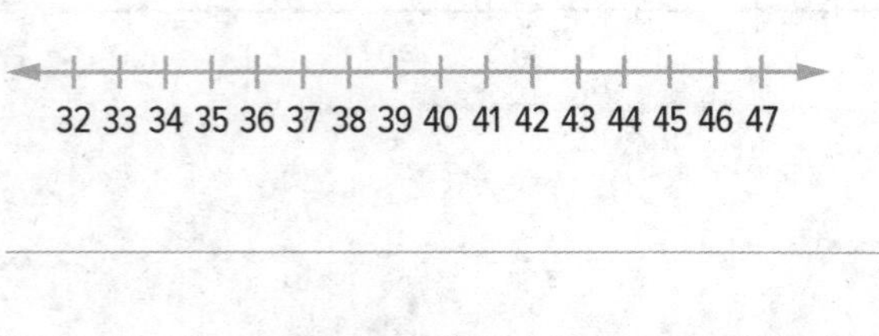

3. {63, 54, 57, 63, 52, 59, 52, 63, 61, 54}

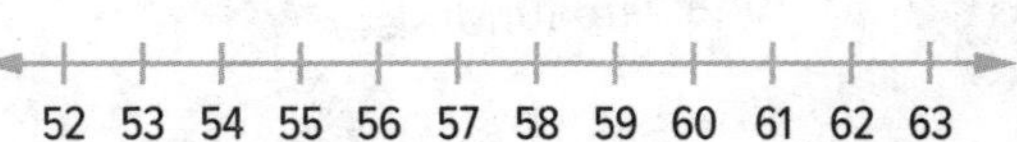

4. The histogram shows a movie theater's attendance each time a movie is shown. Describe the data in the histogram. (Lesson 2)

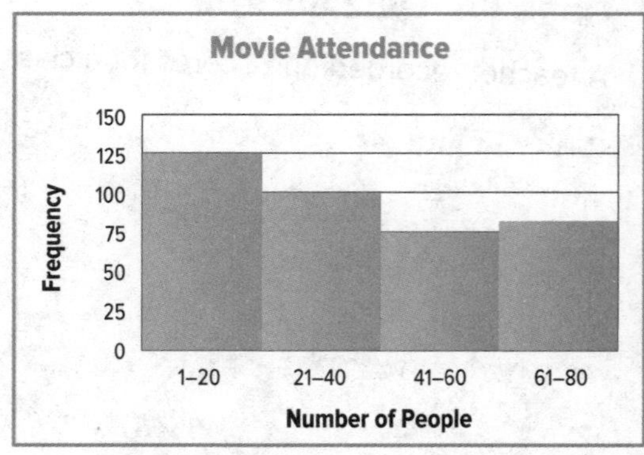

5. **MP Persevere with Problems** In a box plot, the first quartile, median, and third quartile are x, y, and 70, respectively. Give possible values for x and y according to each of the following conditions. (Lesson 3)

   a. The median separates the box into two quartiles, each with the same range.

   b. The box between the median and the third quartile is twice as long as the box between the median and the lower quartile.

Lesson 4

Shape of Data Distributions

Vocabulary Start-Up

The **distribution** of a set of data shows the arrangement of data values. The words below show some of the ways the distribution of data can be described. Match the words below to their definitions.

cluster	The left side of the distribution looks like the right side.
gap	The numbers that have no data value.
peak	The most frequently occurring values, or mode.
symmetry	Data that are grouped closely together.

Essential Question

WHY is it important to carefully evaluate graphs?

Vocabulary

distribution
symmetric distribution
cluster
gap
peak

Common Core State Standards

Content Standards
6.SP.2, 6.SP.5, 6.SP.5d

MP Mathematical Practices
1, 3, 4, 5, 7

Real-World Link

Parasailing The line plot shows the costs in dollars for parasailing for different companies on a certain beach.

Parasailing Costs ($)

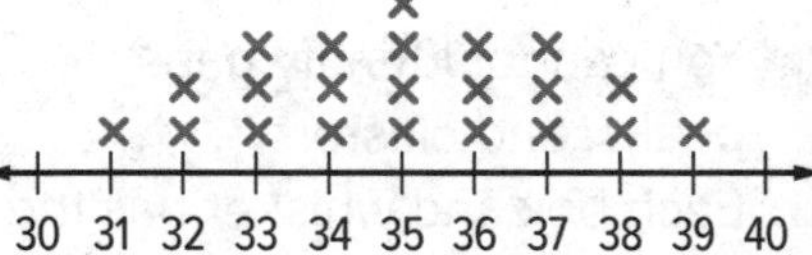

1. Draw a vertical line through the middle of the data. What do you notice?

______

2. Use one of the words shown above to write a sentence about the data.

______

Which MP **Mathematical Practices** did you use?
Shade the circle(s) that applies.

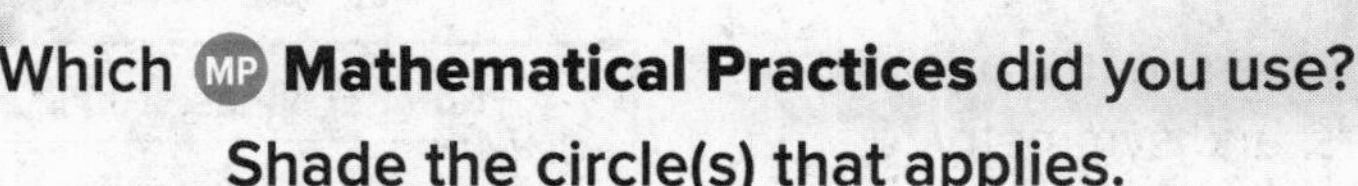

Work Zone

Describe the Shape of a Distribution

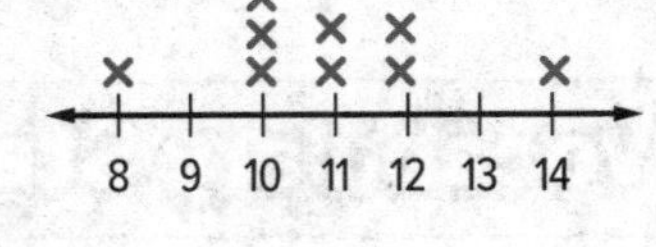

Data that are evenly distributed between the left side and the right side have a **symmetric distribution**. The distribution shown has a **cluster** of several data values within the interval 10–12. The **gaps** 9 and 13 have no data values. The value 10 is a **peak** because it is the most frequently-occurring value.

Examples

Describe the shape of each distribution.

1. The line plot shows the temperature in degrees Fahrenheit in a city over several days.

Temperature (°F)

15 16 17 18 19 20 21 22 23 24 25

You can use clusters, gaps, peaks, outliers and symmetry to describe the shape. The shape of the distribution is not symmetric because the left side of the data does not look like the right side of the data. There is a gap from 19–21. There are clusters from 16–18 and 22–25. The distribution has a peak at 22. There are no outliers.

2. The box plot shows the number of visitors to a gift shop in one month.

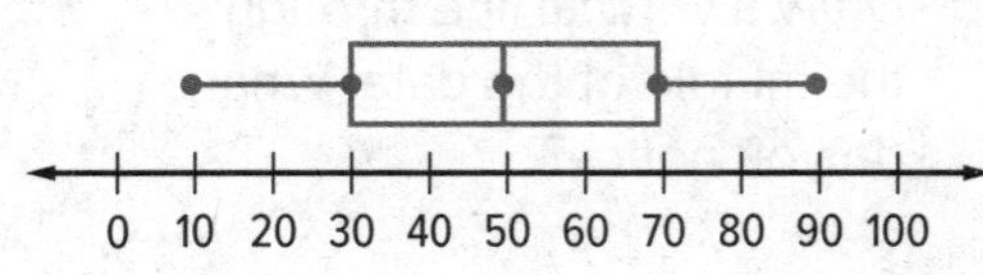

You cannot identify gaps, peaks, and clusters.
Each box and whisker has the same length. So, the data is evenly distributed. The distribution is symmetric since the left side of the data looks like the right side. There are no outliers.

Got it? **Do this problem to find out.**

a. ____________

a. Use clusters, gaps, peaks, outliers, and symmetry to describe the shape of the distribution at the right.

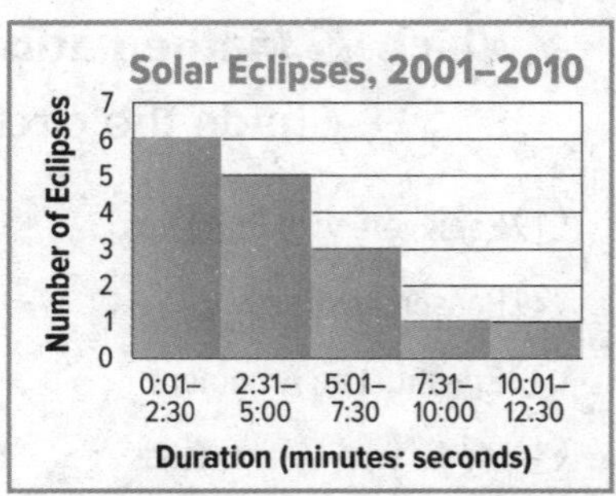

Measures of Center and Spread

Key Concept

Use the following flow chart to decide which measures of center and spread are most appropriate to describe a data distribution.

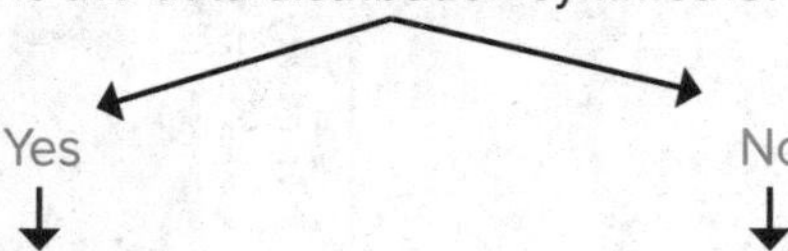

Use the **mean** to describe the center. Use the **mean absolute deviation** to describe the spread.

Use the **median** to describe the center. Use the **interquartile range** to describe the spread.

STOP and Reflect

Explain below which measures are most appropriate to describe the center and spread of a symmetric distribution.

If there is an outlier, the distribution is not usually symmetric.

Example

3. The line plot shows the number of states visited by students in a class.

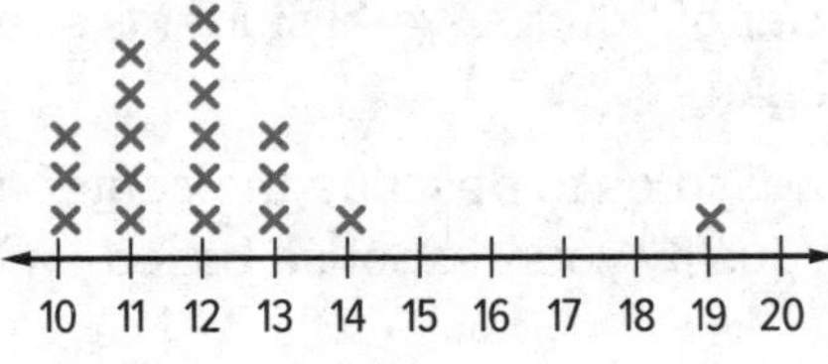

a. Choose the appropriate measures to describe the center and spread of the distribution. Justify your response based on the shape of the distribution.

The data are not symmetric and there is an outlier, 19. The median and interquartile range are appropriate measures to use.

b. Write a few sentences describing the center and spread of the distribution using the appropriate measures.

The median is 12 states. The first quartile is 11. The third quartile is 13. The interquartile range is 13–11, or 2 states.

The data are centered around 12 states. The spread of the data around the center is about 2 states.

Got it? Do this problem to find out.

b. Choose the appropriate measures to describe the center and spread of the distribution. Justify your response based on the shape of the distribution. Then describe the center and spread.

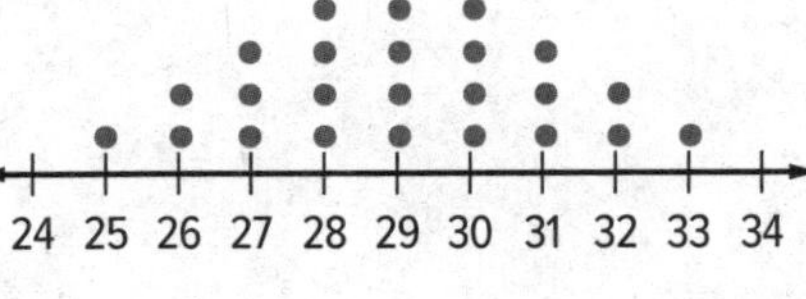

b. ______

Guided Practice

1. The histogram shows the wait times in minutes for entering a concert. Describe the shape of the distribution. (Example 1)

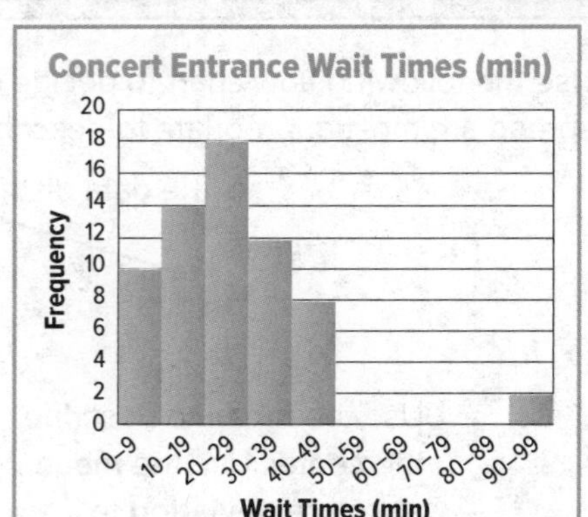

2. The box plot shows the weights in pounds of several dogs. Describe the shape of the distribution. (Example 2)

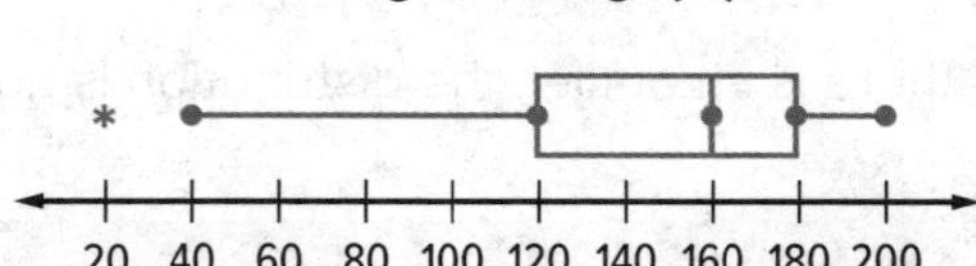

3. The line plot shows the number of hours several students spent on the Internet during the week. (Example 3)

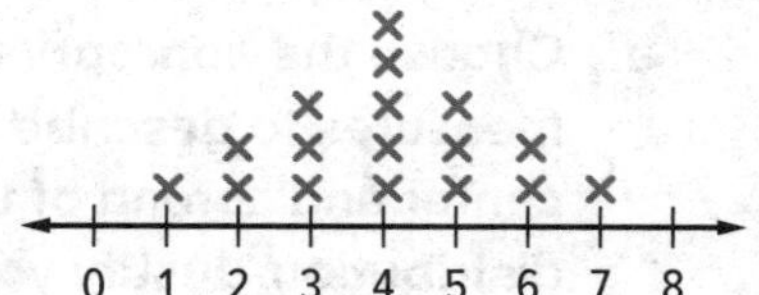

a. Choose the appropriate measures to describe the center and spread of the distribution. Justify your response based on the shape of the distribution.

b. Write a few sentences describing the center and spread of the distribution using the appropriate measures. Round to the nearest tenth if necessary.

4. **Building on the Essential Question** Why does the choice of measure of center and spread vary based on the type of data display?

Rate Yourself!

How well do you understand how to describe the shape of a distribution? Circle the image that applies.

Clear

Somewhat Clear

Not So Clear

For more help, go online to access a Personal Tutor.

Name ______________________ My Homework ______________________

Independent Practice

Go online for Step-by-Step Solutions 

1. The histogram shows the average animal speeds in miles per hour of several animals. Describe the shape of the distribution. (Example 1)

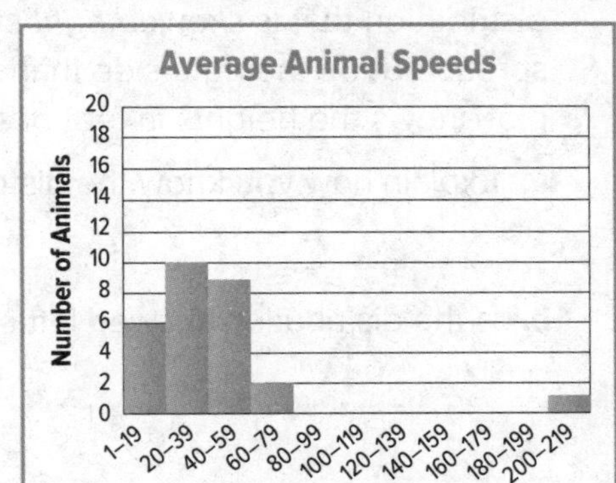

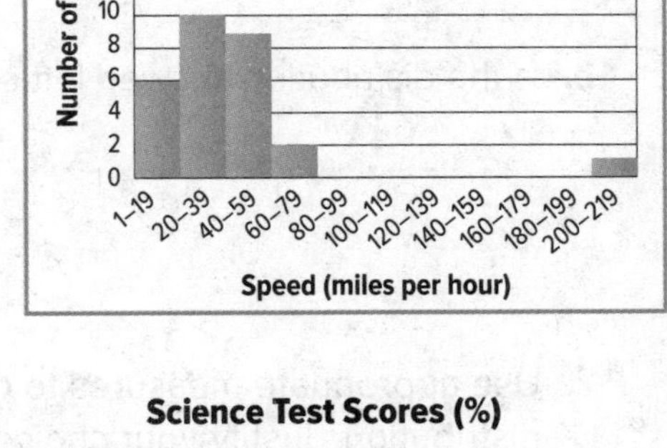

2. The box plot shows the science test scores for Mrs. Everly's students. Describe the shape of the distribution. (Example 2)

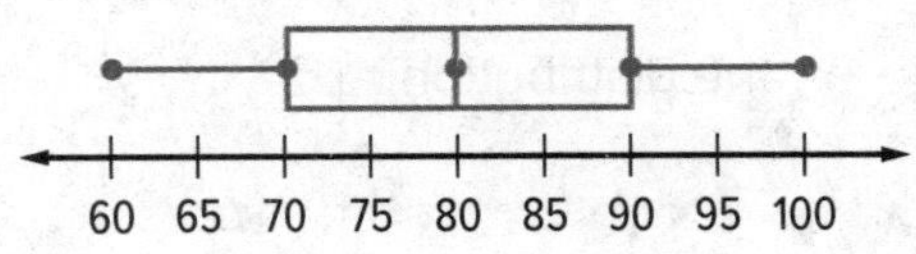

3. The line plot shows the number of text messages sent by different students in one day. (Example 3)

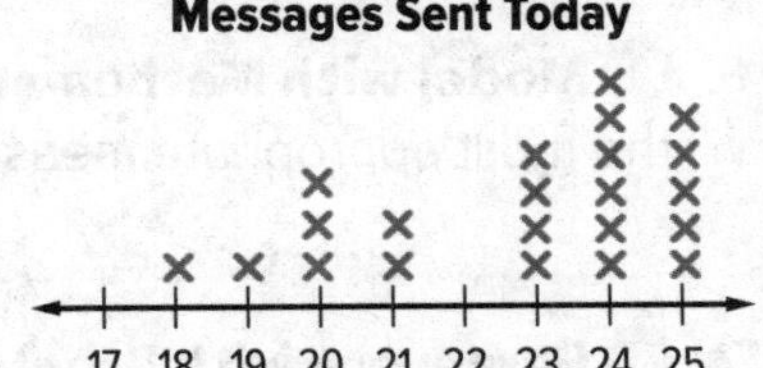

a. Choose the appropriate measures to describe the center and spread of the distribution. Justify your response based on the shape of the distribution.

b. Write a few sentences describing the center and spread of the distribution using the appropriate measures.

4. **MP Identify Structure** Fill in the graphic organizer to show when to use each measure regarding the shape of the distribution.

Measure	Symmetric or Not Symmetric
mean	
median	
interquartile range	
mean absolute deviation	

5. A distribution that is not symmetric is called *skewed.* A distribution that is *skewed left* shows data that is more spread out on the left side than on the right side. A distribution that is *skewed right* shows data that is more spread out on the right side than on the left side. The box plot shows the heights in feet of several trees.

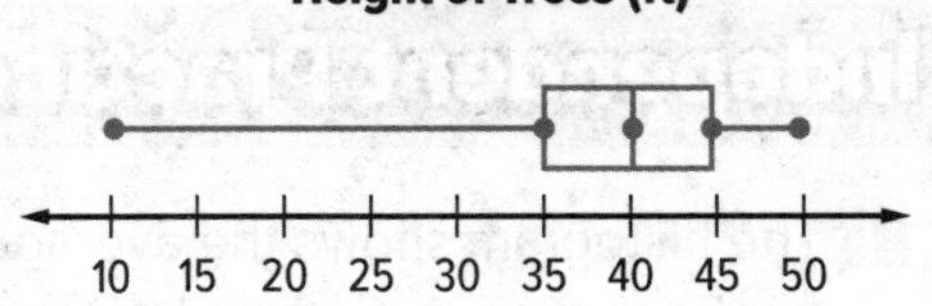

a. Explain how you know the distribution is not symmetric.

b. Is the distribution skewed left or skewed right? Explain.

c. Use appropriate measures to describe the center and spread of the distribution. Justify your choice of measure based on the shape of the distribution.

H.O.T. Problems Higher Order Thinking

6. **MP Model with Mathematics** Draw a line plot for which the median is the most appropriate measure to describe the center of the distribution.

7. **MP Persevere with Problems** Explain why you cannot describe the specific location of the center and spread of the box plot shown using the most appropriate measures.

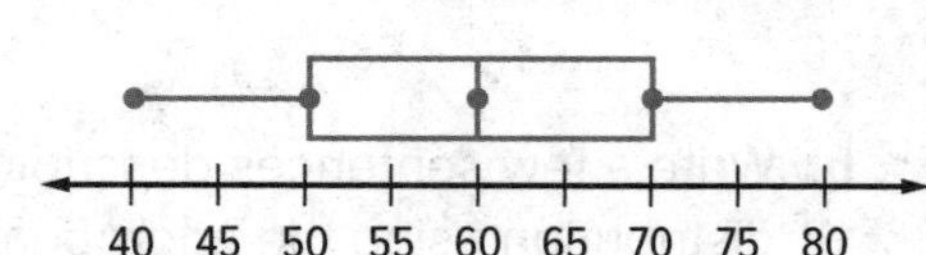

8. **MP Justify Conclusions** Tyra created the dot plot shown to represent the ages of the staff of the community pool. She concludes that since there is a peak at 19, the median is 19. She also concludes the two data values that are 25 to be outliers, so there are no gaps. Evaluate her conclusions.

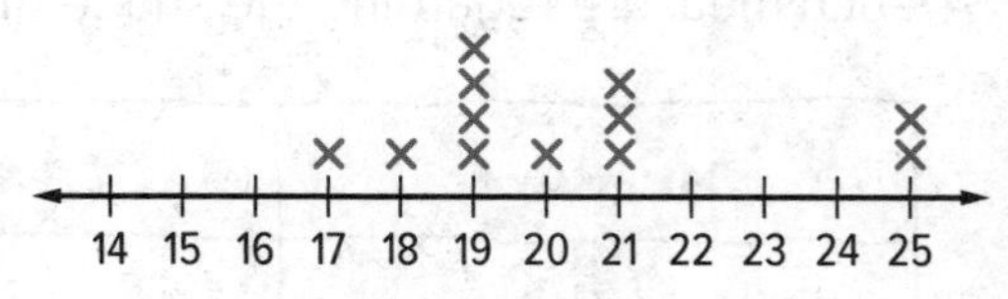

Name ______________ My Homework ______________

Extra Practice

9. The line plot shows the prices in dollars for several DVDs. Describe the shape of the distribution. Sample answer: The shape of the distribution is symmetric. The left side of the data looks like the right side. There is a cluster from $13–$15. There are no gaps in the data. The peak of the distribution is $14. There are no outliers.

Homework Help

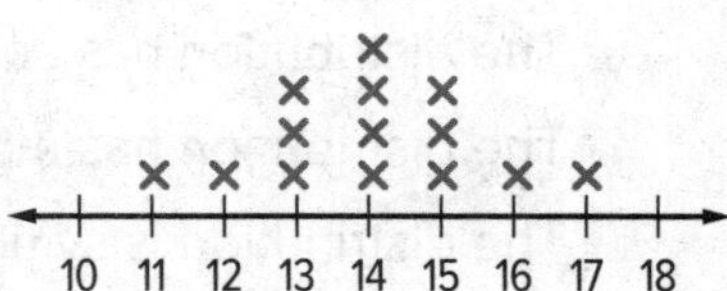

10. The box plot shows donations in dollars to charity. Describe the shape of the distribution.

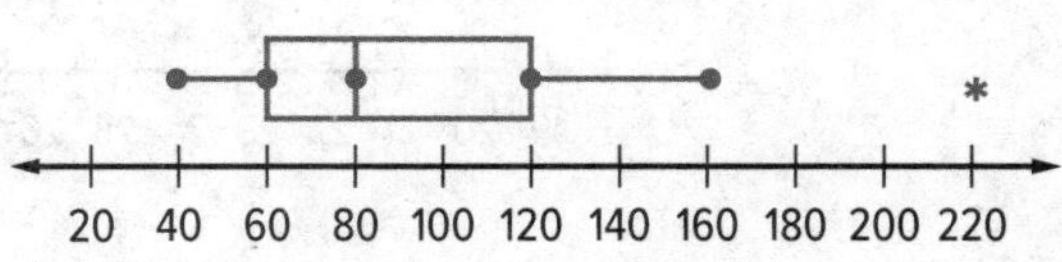

11. The line plot shows the number of miles Elisa ran each week.

a. Choose the appropriate measures to describe the center and spread of the distribution. Justify your response based on the shape of the distribution.

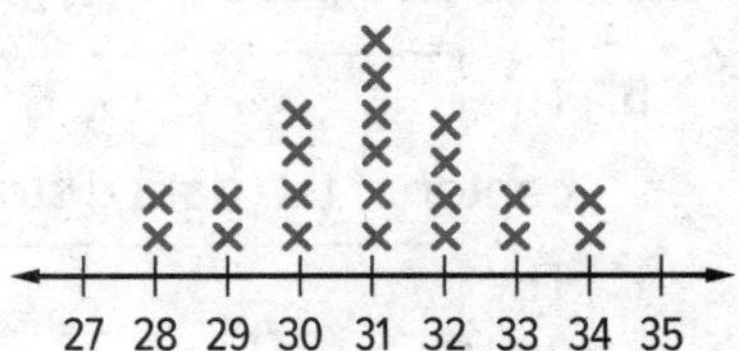

b. Write a few sentences describing the center and spread of the distribution using the appropriate measures. Round to the nearest tenth if necessary.

12. MP **Use Math Tools** The line plot shows the number of siblings for 18 students.

a. Explain how you know the distribution is not symmetric.

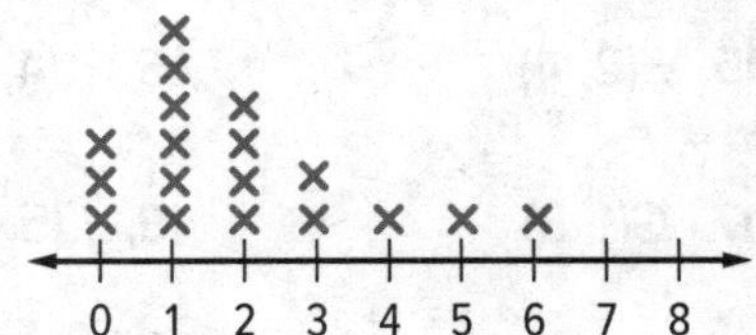

b. Is the distribution skewed left or skewed right? Explain.

c. Use appropriate measures to describe the center and spread of the distribution. Justify your choice of measure based on the shape of the distribution.

Power Up! Common Core Test Practice

13. Which of the following statements are true about the box plot? Select all that apply.

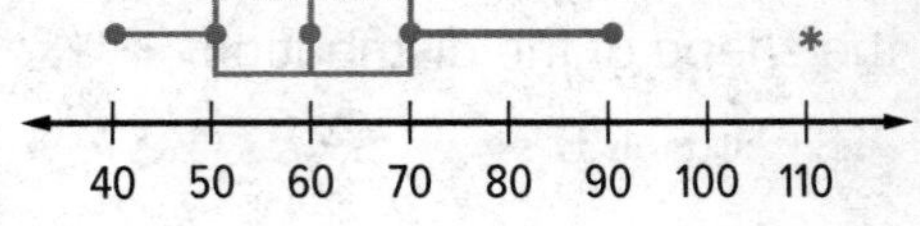

- ☐ The distribution has an outlier.
- ☐ The distribution has a gap of data.
- ☐ The distribution is symmetric.

14. The line plot shows the gas mileage of several different cars.

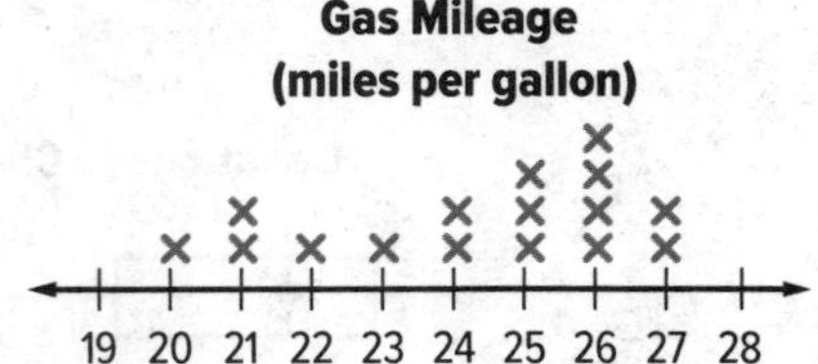

interquartile range
mean
mean absolute deviation
median
not symmetric
symmetric

Select the correct term to complete each statement.

a. The distribution is ________.

b. The ________ should be used to describe the center of the data distribution.

c. The ________ should be used to describe the spread of the data.

Common Core Spiral Review

Graph the points on the coordinate plane. 5.G.2

15. $F(2, 4)$
16. $K(4, 9)$
17. $G(1, 8)$
18. $L(5, 2)$
19. $H(2, 1)$
20. $M(9, 7)$
21. $I(8, 6)$
22. $N(5, 6)$

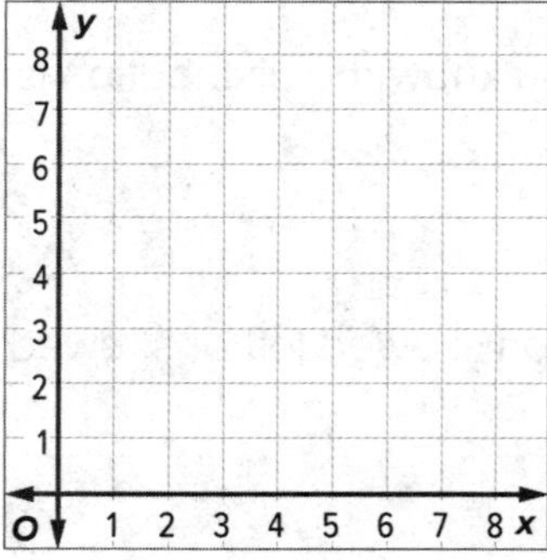

23. Callie is working on a small scrapbook. She completes 3 scrapbook pages each hour. How many pages will she complete in 12 hours? 4.NBT.5

Inquiry Lab
Collect Data

HOW do you answer a statistical question?

Content Standards
6.SP.4, 6.SP.5, 6.SP.5a, 6.SP.5b, 6.SP.5c, 6.SP.5d

Mathematical Practices
1, 3, 4

Aribelle surveyed students in the cafeteria lunch line. She asked the statistical question, *How many photos are currently stored in your cell phone?* She wants to organize the data and choose an appropriate way to display the results of her survey.

Hands-On Activity

You can collect, organize, display, and interpret data in order to answer a statistical question.

Step 1 Make a data collection plan. Aribelle chose to survey students in the cafeteria.

Step 2 Collect the data. The results of the survey are provided below.

55, 47, 58, 50, 66, 47, 54, 64, 47, 65,
43, 44, 51, 81, 54, 45, 57, 52, 58, 60

Step 3 Organize the data. Place the values in order from least to greatest.

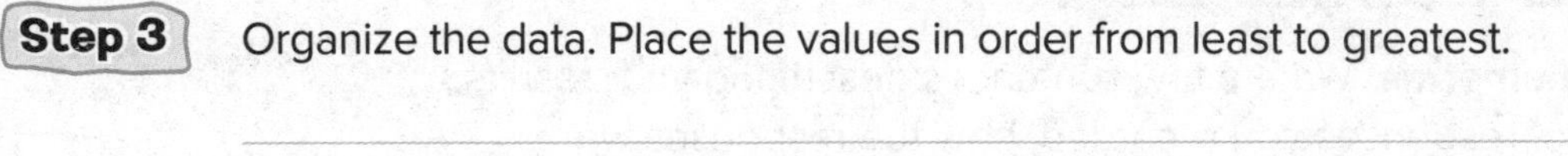

Step 4 Describe the data. There were a total of ☐ responses. The responses measure the number of ______. The data was collected using a ______. One attribute of the data is the median, which is ☐ photos. Another attribute is the interquartile range, which is ☐ photos. There is an outlier at ☐ photos.

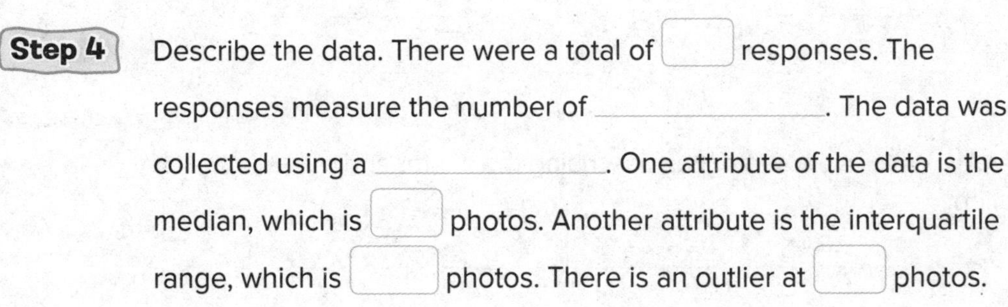

Step 5 Create a display of the data. Explain why a box plot would be an appropriate display of Aribelle's data. ______

Investigate

Work with a partner. Collect data in order to answer a statistical question.

1. Write a statistical question.

2. Collect the data and record the results in a table.

3. Create a display of the data.

Analyze and Reflect

4. **MP Model with Mathematics** Write a few sentences describing your results. Include the number of responses you recorded, how the responses were measured and/or gathered, and their overall pattern.

5. **MP Reason Inductively** Write a few sentences describing the center and spread of the distribution.

Create

6. **Inquiry** HOW do you answer a statistical question?

Lesson 5

Interpret Line Graphs

Real-World Link

Watch

Golf The table shows the prize money for winners of the Masters Tournament.

Money Won by Masters Tournament Winners	
Year	**Amount ($)**
2005	1,170,000
2006	1,225,000
2007	1,305,000
2008	1,305,000
2009	1,350,000
2010	1,350,000

1. Fill in the dollar difference between each consecutive year on the lines above.

2. If the data were plotted, would the points (year, amount) form a straight line? Explain.

3. The Masters Tournament is held once a year. If a *line graph* is made of these data, will there be any realistic data values between tournament dates? Explain.

Essential Question

WHY is it important to carefully evaluate graphs?

Vocabulary

line graph

Common Core State Standards

Content Standards
Extension of 6.SP.4

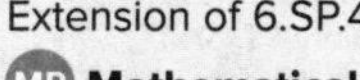

1, 3, 4

Which MP **Mathematical Practices** did you use? Shade the circle(s) that applies.

① Persevere with Problems
② Reason Abstractly
③ Construct an Argument
④ Model with Mathematics
⑤ Use Math Tools
⑥ Attend to Precision
⑦ Make Use of Structure
⑧ Use Repeated Reasoning

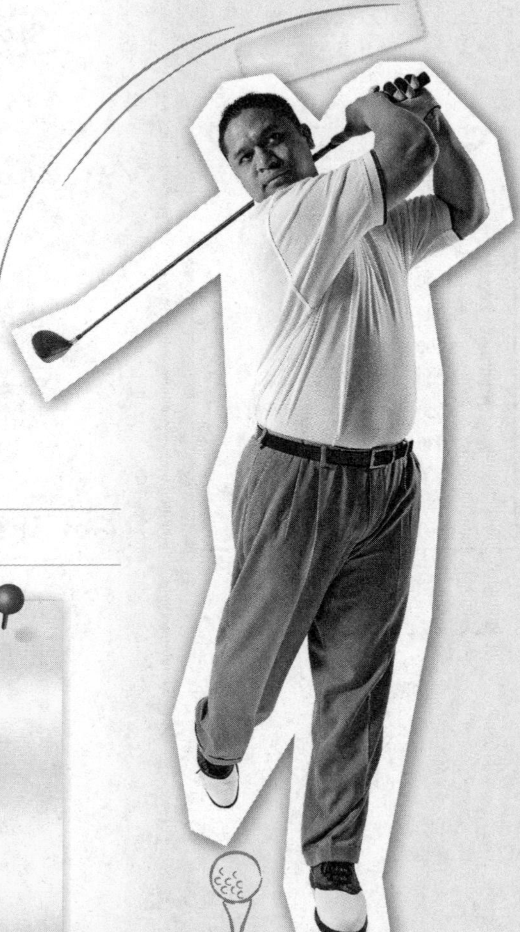

Work Zone

Make a Line Graph

A **line graph** is used to show how a set of data changes over a period of time. To make a line graph, decide on a scale and interval. Then graph pairs of data and draw a line to connect each point.

Example

Tutor

1. Make a line graph of the data of Earth's Population. Describe the change in Earth's population from 1750 to 2000.

Earth's Population						
Year	1750	1800	1850	1900	1950	2000
Population (millions)	790	980	1,260	1,650	2,555	6,080

Earth's Population

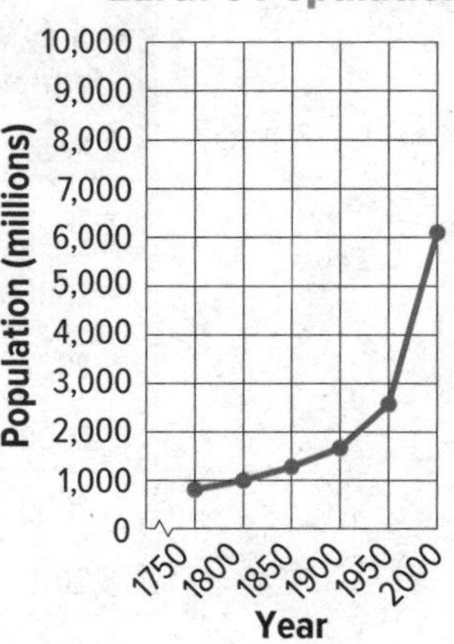

Line Graphs

The lines in a line graph are used to show the differences between data values and may not show precise values between data points.

Step 1 The data include numbers from 790 million to 6,080 million. So, a scale from 0 to 10,000 million and an interval of 1,000 million are reasonable.

Step 2 Let the horizontal axis represent the year. Let the vertical axis represent the population. Label the horizontal and vertical axes.

Step 3 Plot and connect the points for each year.

Step 4 Label the graph with a title.

Earth's population has increased drastically from 1750 to 2000.

Got it? **Do this problem to find out.**

a. Make a line graph of the data. Describe the change in the number of building permits filed from 2005 to 2010.

Number of Building Permits Filed in a Major City						
Year	2009	2010	2011	2012	2013	2014
Building Permits Filed	16,000	15,500	13,900	11,000	8,200	5,900

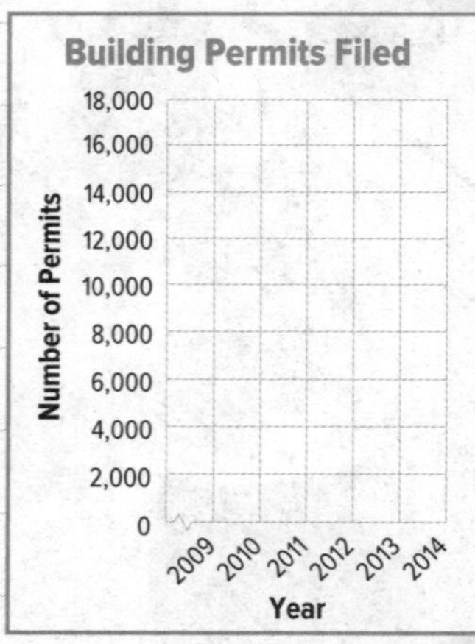

a. ______________

Interpret Line Graphs

By observing the upward or downward slant of the lines connecting the points, you can describe trends in the data and predict future events.

Example

2. The line graph below shows the cost of tuition at a college during several years. Describe the trend. Then predict how much tuition will cost in 2020.

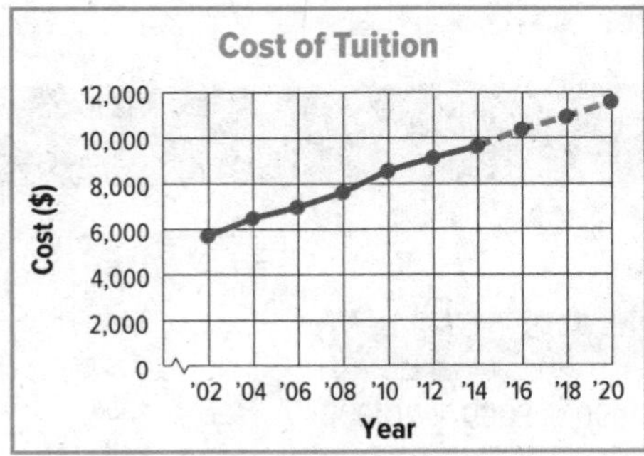

Notice that the increase from 2002 through 2012 is fairly steady. By extending the graph, you can predict that tuition in 2020 will cost a student about $11,500.

Got it? **Do this problem to find out.**

b. The line graph shows the growth of a plant over several weeks. Describe the trend. Then predict how tall the plant will be at 7 weeks.

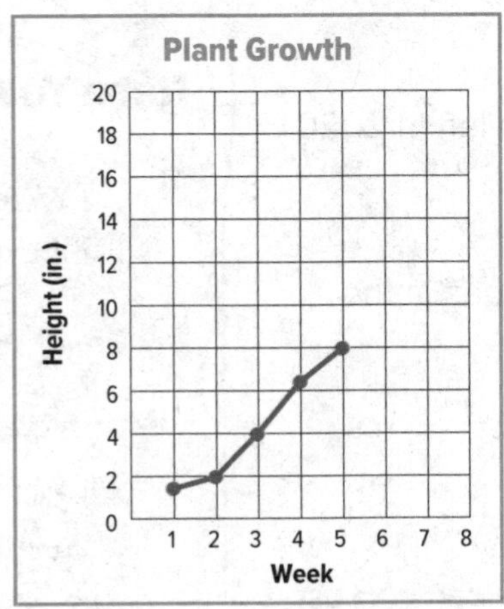

b. ________________

Example

3. What does the graph tell you about the popularity of skateboarding?

The graph shows that skateboard sales have been increasing each year. You can assume that the popularity of the sport is increasing.

Guided Practice

1. Make a line graph of the data. (Example 1)

World's Tropical Rainforests								
Year	1940	1950	1960	1970	1980	1990	2000	2010
Remaining Tropical Rainforests (millions of acres)	2,875	2,740	2,600	2,375	2,200	1,800	1,450	825

World's Tropical Rainforests

Remaining Tropical Rainforests (millions of acres)

3,000
2,750
2,500
2,250
2,000
1,750
1,500
1,250
1,000
750
500
250
0

1940 1950 1960 1970 1980 1990 2000 2010

Year

2. Describe the change in the world's remaining rainforests from 1940 to 2010. (Example 1) ______________________

____

3. Describe the trend in the remaining tropical rainforests. (Example 2) ______________________

4. Predict how many millions of acres there will be left in 2020. (Example 2) ______________________

5. What does the graph tell you about future changes in the remaining rainforests? (Example 3) ______________________

____

6. **Building on the Essential Question** How can you use line graphs to predict data?

____

____

Rate Yourself!

☐ I understand how to interpret line graphs.

Great! You're ready to move on!

☐ I still have some questions about interpreting line graphs.

No Problem! Go online to access a Personal Tutor.

FOLDABLES Time to update your Foldable!

Independent Practice

Go online for Step-by-Step Solutions

1. Make a line graph of the data. Then describe the change in the total amount Felisa saved from Week 1 to Week 5. (Example 1)

Felisa's Savings	
Week	**Total Amount ($)**
1	50
2	54
3	75
4	98
5	100

2. Use the graph at the right. (Examples 2–3)

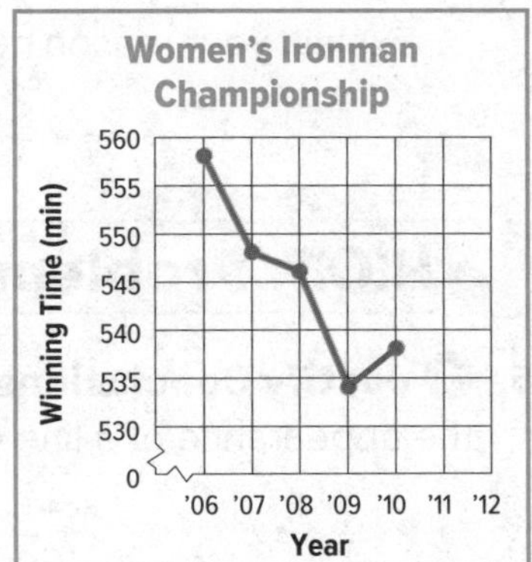

a. Describe the change in the winning times from 2006 to 2010.

b. Predict the winning time in 2015.

c. Predict when the winning time will be less than 500 minutes.

Copy and Solve **For Exercise 3, show your work on a separate piece of paper.**

3. **MP Model with Mathematics** Refer to the graphic novel frame below for Exercises a–b.

Year	Tickets Sold
2010	290
2011	360
2012	395
2013	450

a. Use the information in the table and draw a line graph to show the changes in ticket sales over the past four years.

b. Predict what the ticket sales will be in 2015.

4. Use the graph that shows the distance traveled by two cars on the same freeway headed in the same direction.

   a. Predict the distance traveled by Car A after 5 hours.

   b. Predict the distance traveled by Car B after 5 hours.

   c. How many miles do you think Car A will have traveled after 8 hours?

   d. Based on the graph, after how many hours will Car B have traveled about 360 miles?

   e. Based on the graph, which car will reach a distance of 500 miles first? Explain your reasoning.

Distance Traveled by Two Cars

H.O.T. Problems Higher Order Thinking

5. MP **Justify Conclusions** Can changes to the vertical scale or interval affect the appearance of a line graph? Justify your reasoning with examples.

6. MP **Persevere with Problems** Refer to the graph for Exercise 4. What can you conclude about the point at which the red and blue lines cross?

7. MP **Construct an Argument** Explain why line graphs are often used to make predictions.

8. MP **Model with Mathematics** Give an example of a set of data that is best represented in a line graph. Then make a line graph that could represent that data.

Name _______________ My Homework _______________

Extra Practice

9. MP **Model with Mathematics** Make a line graph of the data. Describe the change in the online sales of movie tickets for Weeks 1 to 5.

Online Sales of Movie Tickets	
Week	**Number of Tickets**
1	1,200
2	1,450
3	1,150
4	1,575
5	1,750

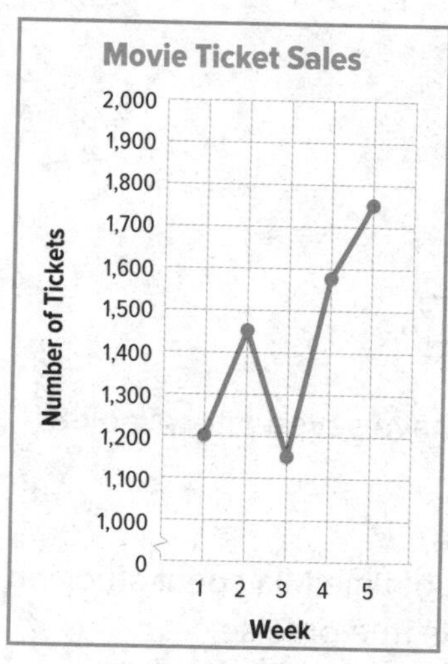

Homework Help

The online sales of movie tickets increased from Week 1 to Week 2, decreased in Week 3 and then increased again for Weeks 4 and 5.

10. Use the graph at the right.

   a. Describe the change in depth from 10 minutes to 35 minutes.

   ____

   b. Predict the depth at 45 minutes. ______________________________

   c. Predict when the depth will be more than 65 feet.

   ____

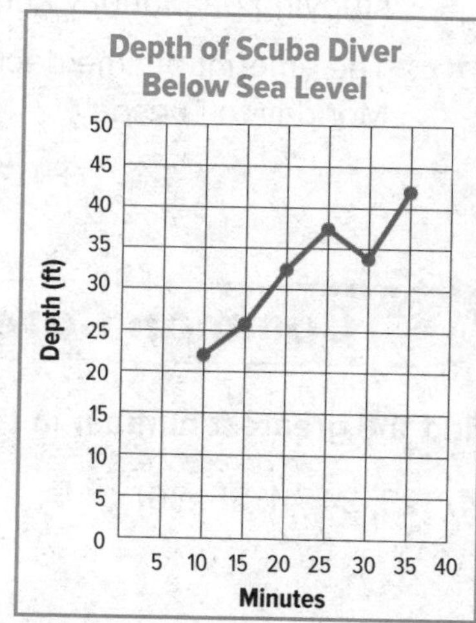

11. Use the line graph at the right.

   a. Between which years did the winning time change the most?

   Explain your reasoning. ______________________________

   b. Make a prediction of the winning time in the 2020 Olympics.

   Explain your reasoning. ______________________________

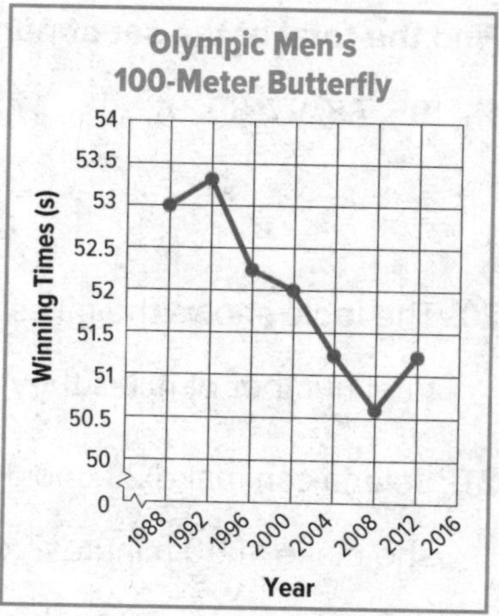

Power Up! Common Core Test Practice

12. The table shows the amount of money Kailey has saved after 5 weeks. Construct a line graph of the data.

Week	Amount Saved ($)
1	15
2	34
3	42
4	60
5	78

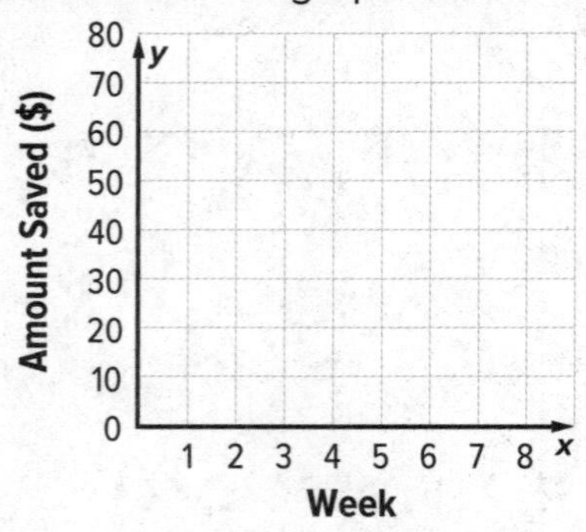

Predict how much Kailey will have saved after Week 8.

13. The graph shows the amount of time Mia spent studying last week. Determine if each statement is true or false.

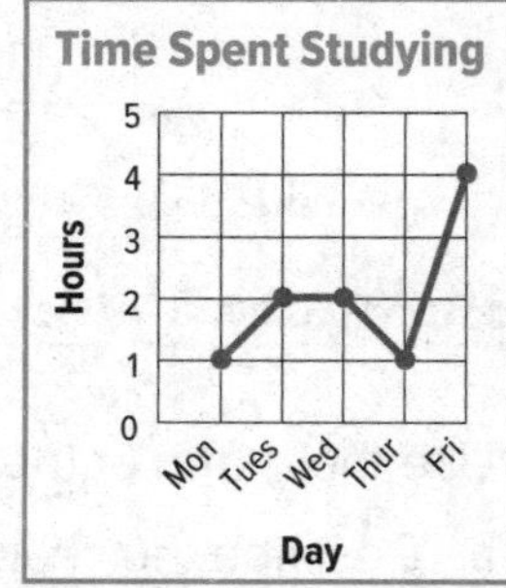

a. The amount of time increased the most from Thursday to Friday. ☐ True ☐ False

b. Mia spent the same amount of time studying on Monday and Wednesday. ☐ True ☐ False

c. The amount of time decreased from Monday to Tuesday. ☐ True ☐ False

Common Core Spiral Review

Find the greatest number in the set. 4.NBT.2

14. {23, 34, 41, 25, 36}

15. {65, 58, 64, 56, 62}

16. {18, 16, 22, 19, 24}

Find the total of the set of numbers. 4.NBT.4

17. {95, 88, 97, 89, 91}

18. {56, 71, 68, 62, 74}

19. {33, 36, 38, 29, 27}

20. The table shows the miles the Smythe family traveled each day. What is the total number of miles they traveled? 4.NBT.4 ______

Day	Miles
Saturday	125
Sunday	84
Monday	112

21. Selena can make 24 cookies in 30 minutes. At this rate, how many cookies can she make in 90 minutes? 6.RP.3b ______

Lesson 6

Select an Appropriate Display

Real-World Link

Animals The displays show the maximum speed of six animals.

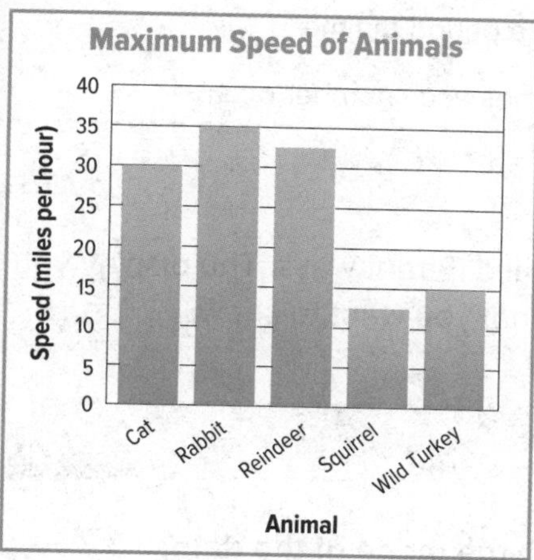

Animal Speeds	
Speeds	**Number of Animals**
1-5	
6-10	
11-15	
16-20	
21-25	
26-30	
31-35	

1. Use the bar graph to fill in the "Number of Animals" column in the table.

2. Which display allows you to find a rabbit's maximum speed?

____

3. In which display is it easier to find the number of animals with a maximum speed of 15 miles per hour or less? Explain.

____

____

____

Essential Question

WHY is it important to carefully evaluate graphs?

Common Core State Standards

Content Standards
Extension of 6.SP.4

MP Mathematical Practices
1, 3, 4, 5, 6

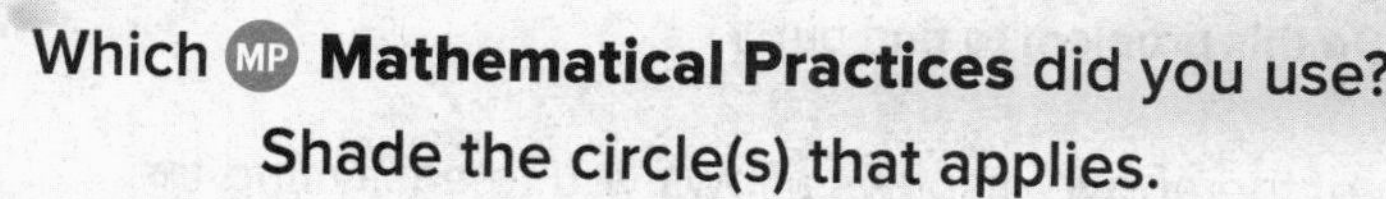

Which MP Mathematical Practices did you use? Shade the circle(s) that applies.

① Persevere with Problems
② Reason Abstractly
③ Construct an Argument
④ Model with Mathematics
⑤ Use Math Tools
⑥ Attend to Precision
⑦ Make Use of Structure
⑧ Use Repeated Reasoning

Key Concept — Statistical Displays

Work Zone

Type of Display	Best used to
Bar Graph	show the number of items in specific categories
Box Plot	show measures of variation for a set of data, also useful for very large sets of data
Histogram	show frequency of data divided into equal intervals
Line Graph	show change over a period of time
Line Plot	show how many times each number occurs

Data can often be displayed in several different ways. The display you choose depends on your data and what you want to show.

Example

1. Which display allows you to tell the mode of the data?

Lasagna Orders Each Night

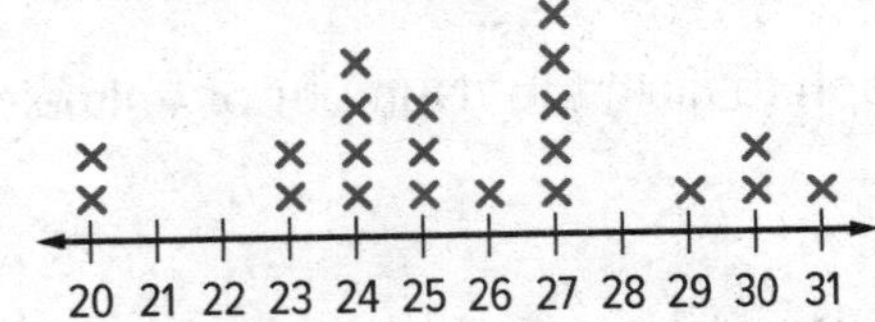

Lasagna Orders Each Night

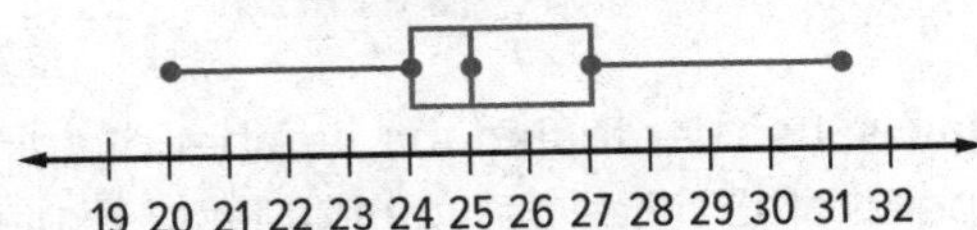

The line plot shows each night's data. The number of orders that occurs most frequently is 27. The box plot shows the spread of the data, but does not show individual data so it does not show the mode.

Got it? Do this problem to find out.

a. Which of the above displays allows you to easily find the median of the data?

a. ________

Examples

2. **A survey compared different brands of hair shampoo. The table shows the number of first-choice responses for each brand. Select an appropriate type of display to compare the number of responses. Justify your choice.**

Favorite Shampoo Survey			
Brand	**Responses**	**Brand**	**Responses**
A	35	D	24
B	12	E	8
C	42	F	11

These data show the number of responses for each brand. A bar graph would be the best display to compare the responses.

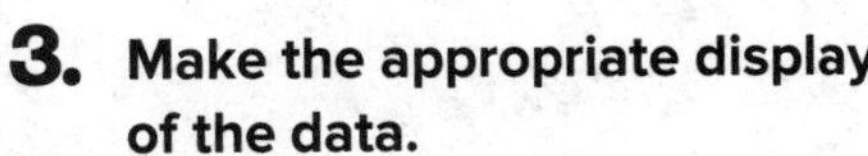

3. **Make the appropriate display of the data.**

Step 1 Draw and label horizontal and vertical axes. Add a title.

Step 2 Draw a bar to represent the number of responses for each brand.

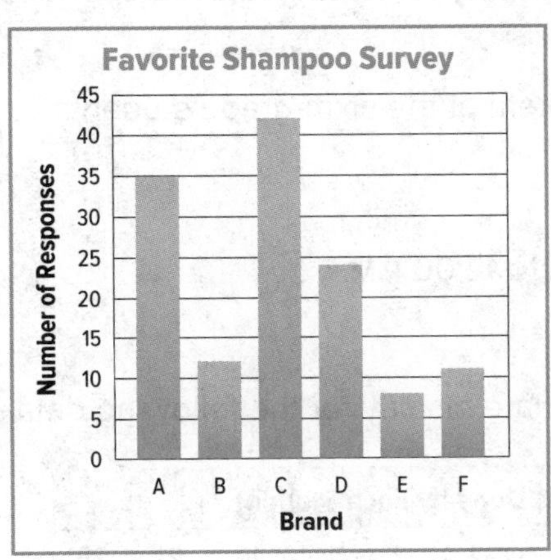

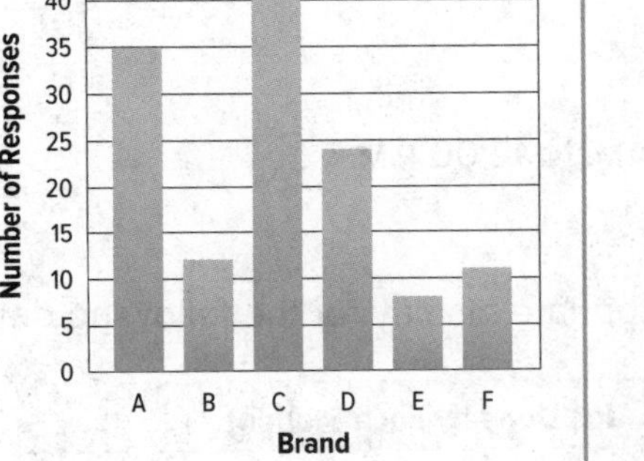

STOP and Reflect

What type of data are best represented in a bar graph? Explain below.

Got it? **Do these problems to find out.**

The table shows the quiz scores of Mr. Vincent's math class.

Math Quiz Scores											
70	70	75	80	100	85	85	65	75	85	95	90
90	100	85	90	90	95	80	85	90	85	90	75

b. Select an appropriate type of display to allow you to count the number of students with a score of 85. Explain your choice.

c. Make the appropriate display of the data.

b. ______________

Guided Practice

Check

1. Which display makes it easier to determine the greatest number of calendars sold? Justify your reasoning. (Example 1)

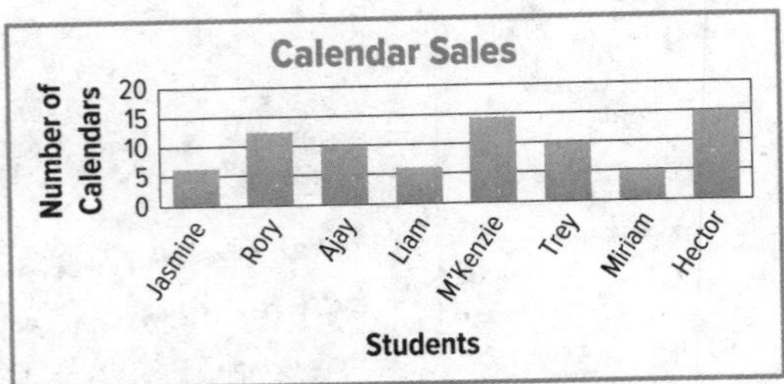

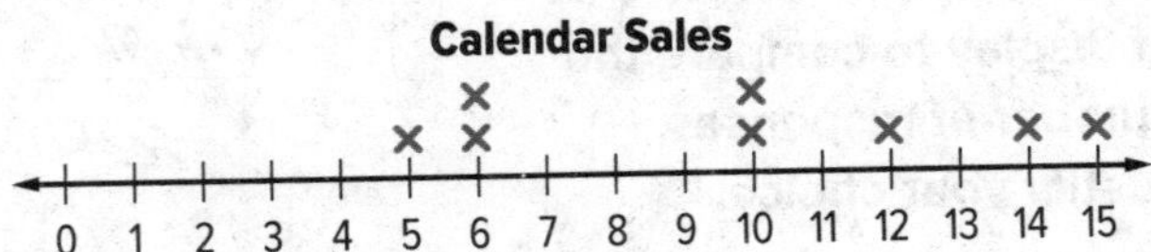

Select an appropriate type of display for data gathered about each situation. Justify your reasoning. (Example 2)

2. the favorite cafeteria lunch item of the sixth-grade students

3. the temperature from 6 A.M. to 12:00 P.M.

4. Select and make an appropriate display for the following data. (Example 3)

Number of Push-Ups Done by Each Student											
15	20	8	11	6	25	32	12	14	16	21	25
18	35	40	20	25	15	10	5	18	20	31	28

5. Building on the Essential Question Why is it important to choose the appropriate display for a set of data?

Rate Yourself!

How confident are you about selecting an appropriate display? Shade the ring on the target.

For more help, go online to access a Personal Tutor.

Name ____________________ My Homework ____________________

Independent Practice

Go online for Step-by-Step Solutions

1. Which display makes it easier to compare the maximum speeds of Top Thrill Dragster and Millennium Force? Justify your reasoning. (Example 1)

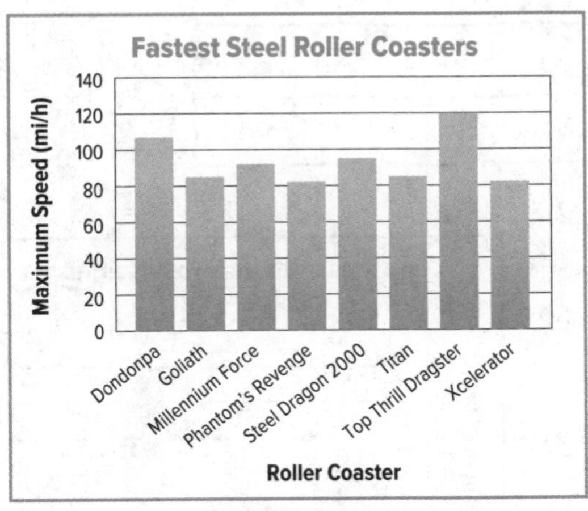

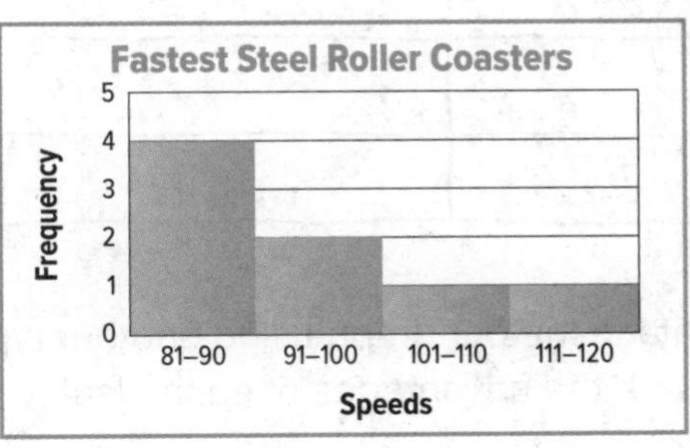

____

Select an appropriate type of display for data gathered about each situation. Justify your reasoning. (Example 2)

2. the test scores each student had on a language arts test

____

3. the median age of people who voted in an election

____

MP Use Math Tools Select and make an appropriate type of display for the situation. (Example 3)

Show your work.

4.

South American Country	Water Area (km^2)	South American Country	Water Area (km^2)
Argentina	47,710	Guyana	18,120
Bolivia	15,280	Paraguay	9,450
Chile	12,290	Peru	5,220
Ecuador	6,720	Venezuela	30,000

____

5. **MP Use Math Tools** Use the Internet or another source to find a set of data that is displayed in a bar graph, line graph, frequency table, or circle graph. Was the most appropriate type of display used? What other ways might these same data be displayed? ____________________

6. **MP Be Precise** Fill in the graphic organizer below.

Display	What it shows
line plot	
histogram	
box plot	
bar graph	

7. Display the data in the bar graph using another type of display. Compare the advantages of each display.

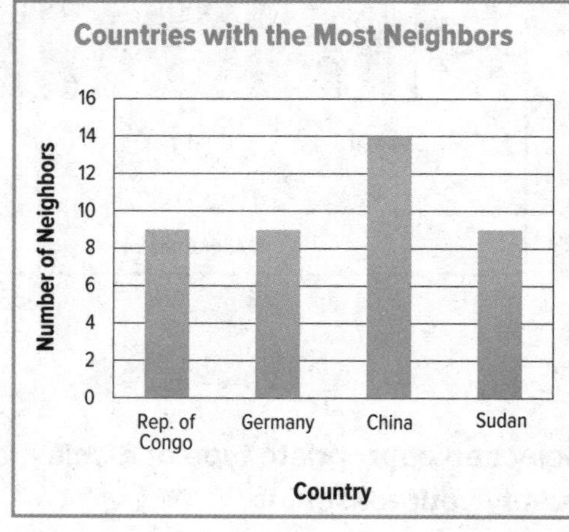

H.O.T. Problems Higher Order Thinking

8. **MP Construct an Argument** Determine whether the following statement is *true* or *false*. If true, explain your reasoning. If false, give a counterexample.

Any set of data can be displayed using a line graph.

9. **MP Persevere with Problems** Which type of display allows you to easily find the mode of the data? Explain your reasoning.

10. **MP Reason Inductively** The table shows the number of each type of plant at a botanical garden. The director of the garden would like to add cacti so that the relative frequency of the plant is 50%. How many cactus plants should the director add?

Type of Plant	Frequency
Rose	13
Cactus	18
Palm	4
Ferns	15

Extra Practice

11. Which display makes it easier to see the median distance? Justify your reasoning.

Winning Distance of Men's Olympic Javelin Throw Winners 1968–2008

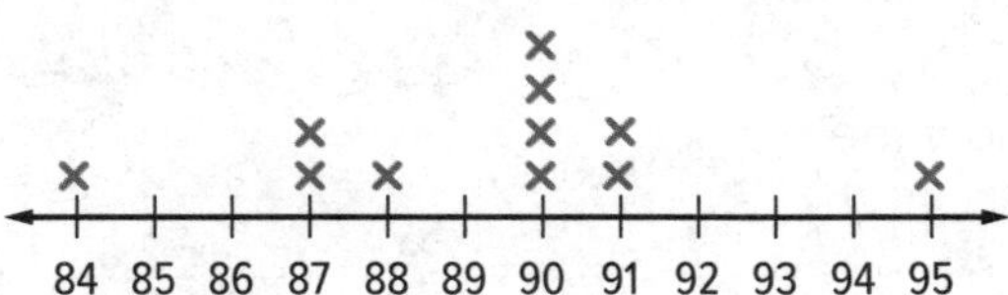

Winning Distances of Olympic Javelin Throw

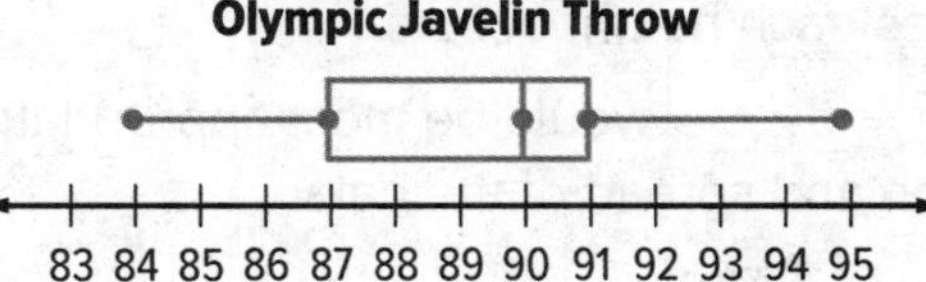

box plot; The median is easily seen on the box plot as the line in the box.

Select an appropriate type of display for data gathered about each situation. Justify your reasoning.

12. the amount of sales a company has over 6 months

13. the prices of five different brands of tennis shoes at an athletic store

14. the amount in a savings acount over a year

15. the shape of the distribution of a team's football scores for one season

MP **Model with Mathematics** **Select and make an appropriate type of display for the situation.**

16.

Number of Counties in Various Southern States	
67	67
95	82
33	64
63	29
46	100
75	77
95	105

CCSS Power Up! Common Core Test Practice

17. The table shows the heights of 15 different dogs. Complete each statement with the most appropriate type of data display.

Height of Dogs (in.)				
24	26	22	22	23
24	25	24	23	23
18	26	25	22	24

a. A ______ would be most appropriate to show the data divided into equal intervals.

b. A ______ would be most appropriate to show how many times each height occurs.

c. A ______ would be most appropriate to show the distribution and spread of the data.

18. Match each situation to the type of display that would best represent it.

- bar graph
- histogram
- line graph
- line plot

the favorite subject of the students in Mrs. Ling's homeroom ______

the weight a puppy gains in one year ______

the number of hits Dylan got in each game this baseball season ______

the number of each type of sandwich a deli sells during lunch ______

CCSS Common Core Spiral Review

Divide. 5.NBT.6

19. $36 \div 12 =$ ______

20. $108 \div 12 =$ ______

21. $138 \div 23 =$ ______

22. $204 \div 17 =$ ______

23. $192 \div 12 =$ ______

24. $390 \div 15 =$ ______

25. $324 \div 36 =$ ______

26. $540 \div 36 =$ ______

27. $792 \div 12 =$ ______

28. Measure the pencil below to the nearest centimeter. Then represent your measurement in meters. 5.MD.1 ______

Inquiry Lab

Use Appropriate Units and Tools

HOW do you determine a measureable attribute?

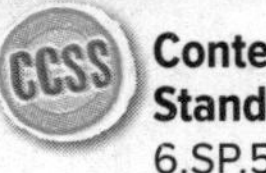

Content Standards
6.SP.5, 6.SP.5a, 6.SP.5b, 6.SP.5c

Mathematical Practices
1, 3, 4

Each item in a backpack has different attributes, such as color, size, and weight. Some of the attributes of the objects can be measured.

Hands-On Activity

You can choose the appropriate unit and tool to measure an object.

Step 1 Select an object in your classroom such as a desk, book, backpack, or trash can.

Step 2 List all of the measureable attributes of your object in the Step 3 table, for example length, weight or mass, time, or capacity.

Step 3 Select an appropriate tool and measure each attribute. Record each measure using appropriate units in the table below.

Object	Attribute	Tool	Measurement

Step 4 Choose a different object with at least one attribute that requires the use of a different tool to measure. Then repeat steps 1–3.

Object	Attribute	Tool	Measurement

Step 5 Write and solve a real-world problem in which one of your measurements is needed to solve the problem.

____

____

Collaborate

Investigate

Work with a partner. Choose an attribute common to several similar objects and use the appropriate unit and tool to measure.

1. Choose a set of objects and a measurable attribute.

____

2. Measure the attribute and record the results in a table. Then create a display of the data.

Collaborate

Analyze and Reflect

3. MP **Model with Mathematics** Write a few sentences describing your data. Include the number of observations, how the data was measured, and the overall pattern of the data. ____

4. MP **Make a Conjecture** Explain how the way you measured the objects influenced the shape of the display. ____

On Your Own

Create

5.  HOW do you determine a measureable attribute?

____

21ST CENTURY CAREER in Envionmental Science

Environmental Engineer

Are you concerned about protecting the environment? If so, you should think about a career in environmental science. Environmental engineers apply engineering principles along with biology and chemistry to develop solutions for improving the air, water, and land. They are involved in pollution control, recycling, and waste disposal. Environmental engineers also determine methods for conserving resources and for reducing environmental damage caused by construction and industry.

Is This the Career for You?

Are you interested in a career as an environmental engineer? Take some of the following courses in high school.

- Algebra
- Biology
- Environmental Science
- Environmental History

Turn the page to find out how math relates to a career in Environmental Science.

MP Thinking Green!

Use the information in the table to solve each problem. Round to the nearest tenth if necessary.

1. Find the mean, median, and mode of the percent of recycled glass data. ______

2. If Lee County is removed from the recycled aluminum cans data, which changes the most: the mean, median, or mode? Does this make sense? Explain your reasoning.

3. Find the range, quartiles, and interquartile range of the percent of recycled newspapers data. ______

4. Find any outliers in the percent of recycled plastic bottles data. ______

5. Make a box plot of the percent of recycled glass data.

6. Refer to the box plot you made in Exercise 5. Compare the parts of the box and the lengths of the whiskers. What does this tell you about the data? ______

Percent of Materials That Are Recycled				
County	Aluminum Cans	Glass	Newspapers	Plastic Bottles
Broward	15	13	41	7
Dade	4	17	28	15
Duval	31	17	81	7
Hillsborough	14	21	38	23
Lee	48	16	66	53
Orange	12	29	33	16
Polk	6	26	22	8

MP Career Project

It's time to update your career portfolio! Describe an environmental issue that concerns you. Explain how you, as an environmental engineer, would work to resolve this issue. Then research how the issue is being addressed by envionmental scientists today.

Choose your favorite school activity or volunteer job. Could it lead to a possible career? If so, what is it?

Chapter Review

Check

Vocabulary Check

Write the correct term for each clue in the crossword puzzle.

Across

4. the arrangement of a data set
6. a diagram that is constructed using five values
7. an empty space or interval in a set of data
9. a line plot using dots

Down

1. having one side of a distribution looking the same as the other side
2. a diagram that shows the frequency of data on a number line
3. a type of bar graph used to display numerical data that have been organized into equal intervals
5. data that are grouped closely together
8. the mode of the data

Key Concept Check

Use Your FOLDABLES

Use your Foldable to help review the chapter.

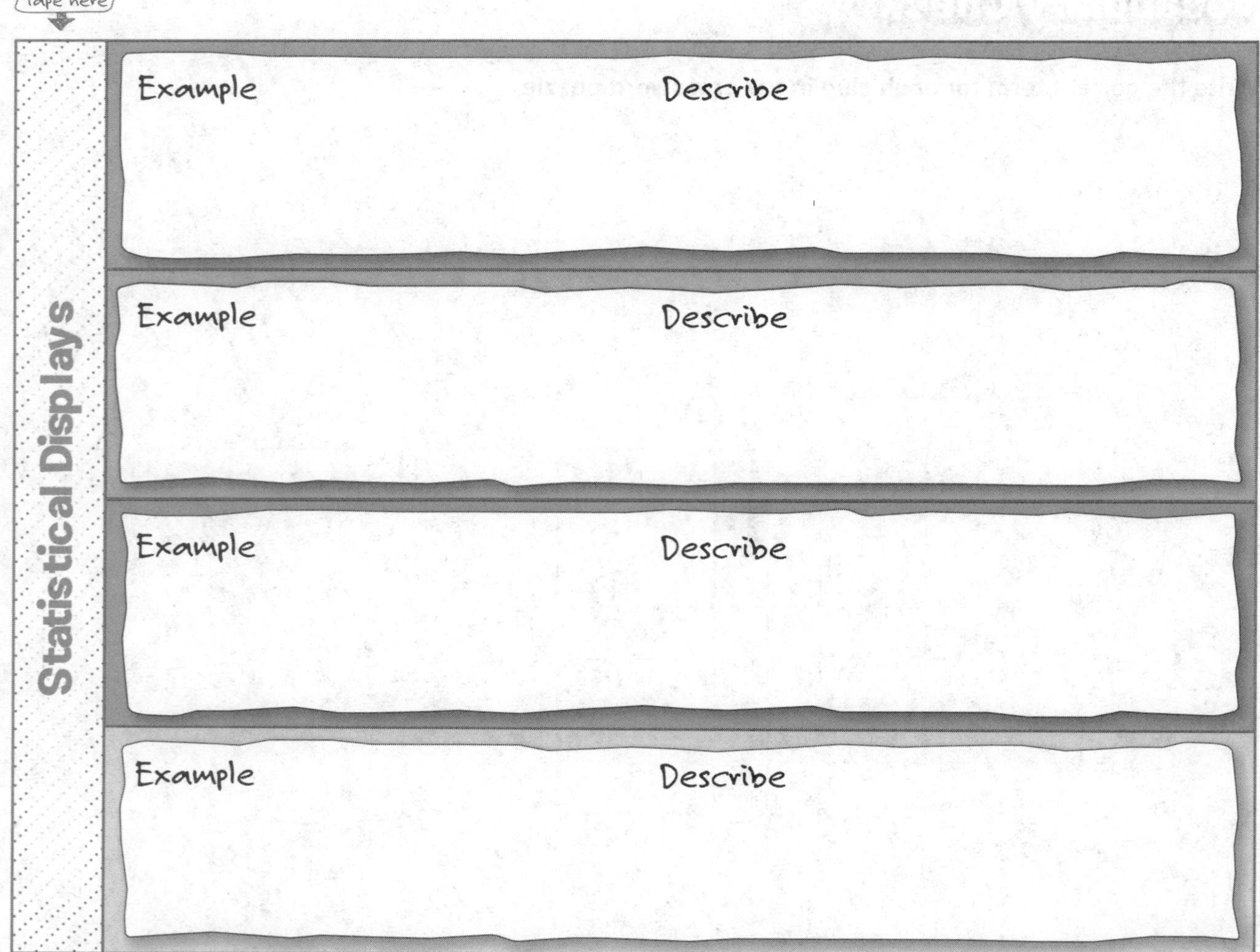

Got it?

Circle the correct term or number to complete each sentence.

1. It is best to use a (line plot, line graph) to show change over time.
2. A (cluster, gap) is the space on a graph that has no data values.
3. The median of a data set is easily seen in a (box plot, histogram).
4. A (line plot, box plot) will show the mode of the data set.
5. If a data set is symmetric, the spread should be described by the (interquartile range, mean absolute deviation).

Thanksgiving Meal

The local soup kitchen is attempting to budget for their annual Thanksgiving meal and need to predict how many people are going to come. The attendance numbers for the last several years are shown. The cost to prepare each meal is $3.

Year	1	2	3	4	5	6	7	8
Number of Meals	140	150	150	80	100	110	60	175

Write your answers on another piece of paper. Show all of your work to receive full credit.

Part A

Construct a box plot to display the information. Based on your plot, if an average number of people come to the dinner in Year 9 how much would it cost?

Part B

The graph shows the actual attendance for Years 1 through 9. How many dinners were served at the kitchen in Year 9? What was the food budget total? How close was the actual budget to the budget prediction you made in Part A? Explain your answer.

Number of Meals Eaten by Year

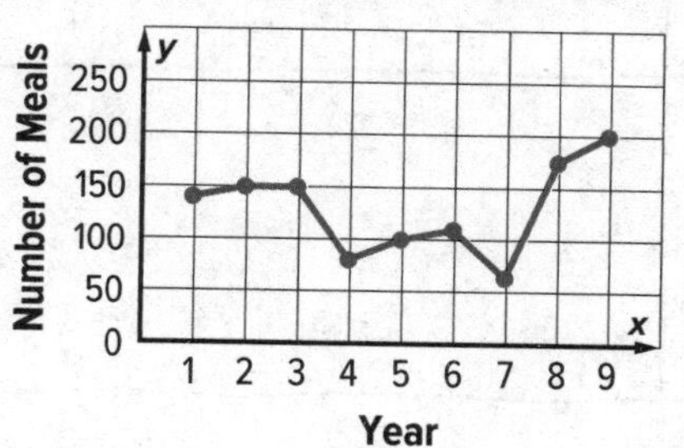

Part C

The box plot displays the number of meals served in Years 1 through 10. What can you determine about the number of dinners served in Year 10? Explain your answer.

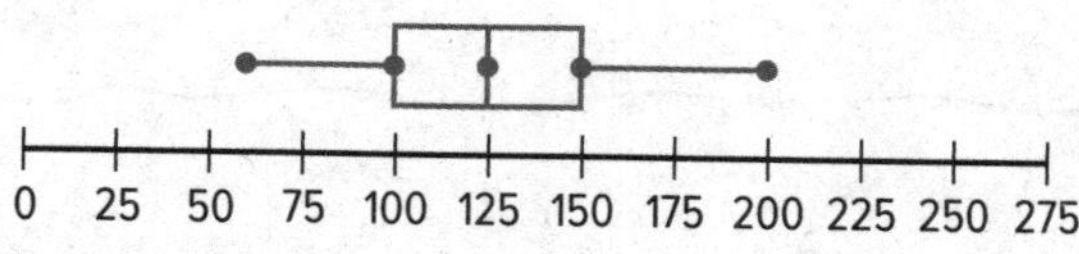

Reflect

Answering the Essential Question

Use what you learned about staistical displays to complete the graphic organizer.

Essential Question

WHY is it important to carefully evaluate graphs?

	When should I use it?
line graph	
histogram	
line plot	
box plot	

Answer the Essential Question. WHY is it important to carefully evaluate graphs?

UNIT PROJECT

Watch

Let's Exercise Regular physical activity not only keeps you fit, but helps you think clearly and improve your mood. In this project you will:

- **Collaborate** with your classmates as you research physical fitness.
- **Share** the results of your research in a creative way.
- **Reflect** on why learning mathematics is important.

By the end of this project, you just might be your family's personal trainer!

Collaborate

Go Online Work with your group to research and complete each activity. You will use your results in the Share section on the following page.

1. Survey at least ten students about the number of times they participate in sports or other physical activities each week. Find the mean. Then make a dot plot of the data.

2. Research 15 physical activities and the number of Calories burned per hour for each activity. Record the information and draw a box plot to represent the data.

3. Create a jogging schedule to train for a 5K run. Include the number of weeks needed to train and the increments of miles you would need to run. Calculate the number of Calories burned per run. Draw a line graph to represent the data.

4. Look up a fast food restaurant's menu that includes the number of Calories for each item. Record the number of Calories a person would consume if they ate at that restaurant for each meal in one day. Construct an appropriate graph to display your results.

5. Look at what the USDA considers a healthy diet. Based on what you learn, plan one day's worth of meals. Use a statistical display to compare this day's diet with the day's diet in Exercise 4.

With your group, decide on a way to share what you have learned about physical fitness. Some suggestions are listed below, but you can also think other creative ways to present your information. Remember to show how you used mathematics to complete each of the activities of this project!

- Write an article for the food or health section of an online magazine.
- Act as a pediatrician and create a digital presentation that promotes physical fitness.

Check out the note on the right to connect this project with other subjects.

connect with Language Arts

Health Literacy Suppose you are choosing a career as a fitness trainer. Make a brochure you can pass out to obtain clients. Include the following in your brochure:

- tables and graphs
- sample testimonials from satisfied customers.

6. **Answer the Essential Question** Why is learning mathematics important?

   a. How did you use what you learned about statistical measures to help you to understand why learning mathematics is important?

   b. How did you use what you learned about statistical displays to help you to understand why learning mathematics is important?

Glossary/Glosario

Go online for the eGlossary. 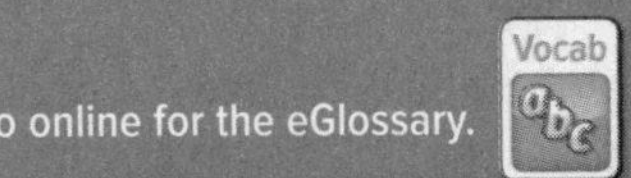

The eGlossary contains words and definitions in the following 13 languages:

Arabic	**Cantonese**	**Hmong**	**Spanish**	**Urdu**
Bengali	**English**	**Korean**	**Tagalog**	**Vietnamese**
Brazilian Portuguese	**Haitian Creole**	**Russian**		

English

Aa

absolute value The distance between a number and zero on a number line.

acute angle An angle with a measure greater than 0° and less than 90°.

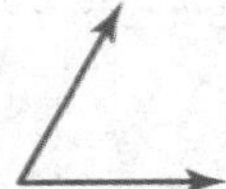

acute triangle A triangle having three acute angles.

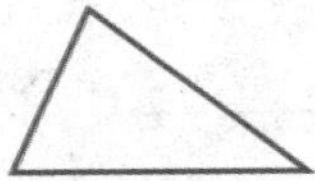

Addition Property of Equality If you add the same number to each side of an equation, the two sides remain equal.

algebra A mathematical language of symbols, including variables.

algebraic expression A combination of variables, numbers, and at least one operation.

analyze To use observations to describe and compare data.

angle Two rays with a common endpoint form an angle. The rays and vertex are used to name the angle.

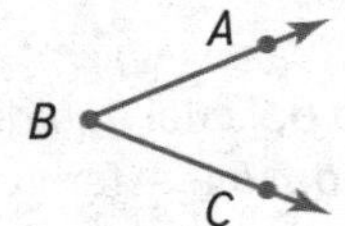

$\angle ABC$, $\angle CBA$, or $\angle B$

Español

valor absoluto Distancia entre un número y cero en la recta numérica.

ángulo agudo Ángulo que mide más de 0° y menos de 90°.

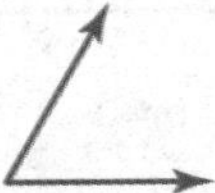

triángulo acutángulo Triángulo con tres ángulos agudos.

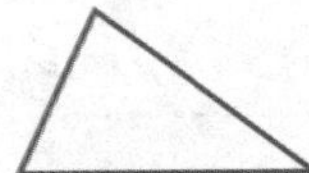

propiedad de adición de la igualdad Si sumas el mismo número a ambos lados de una ecuación, los dos lados permanecen iguales.

álgebra Lenguaje matemático que usa símbolos, incluyendo variables.

expresión algebraica Combinación de variables, números y, por lo menos, una operación.

analizar Usar observaciones para describir y comparar datos.

ángulo Dos rayos con un extremo común forman un ángulo. Los rayos y el vértice se usan para nombrar el ángulo.

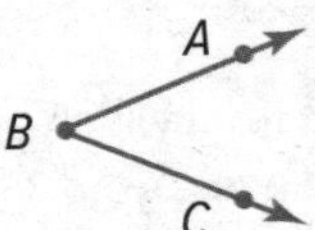

$\angle ABC$, $\angle CBA$ o $\angle B$

arithmetic sequence A sequence in which the difference between any two consecutive terms is the same.

sucesión aritmética Sucesión en la cual la diferencia entre dos términos consecutivos es constante.

Associative Property The way in which numbers are grouped does not change the sum or product.

propiedad asociativa La forma en que se agrupan tres números al sumarlos o multiplicarlos no altera su suma o producto.

average The sum of two or more quantities divided by the number of quantities; the mean.

promedio La suma de dos o más cantidades dividida entre el número de cantidades; la media.

Bb

bar notation A bar placed over digits that repeat to indicate a number pattern that repeats indefinitely.

notación de barra Barra que se coloca sobre los dígitos que se repiten para indicar el número de patrones que se repiten indefinidamente.

base Any side of a parallelogram.

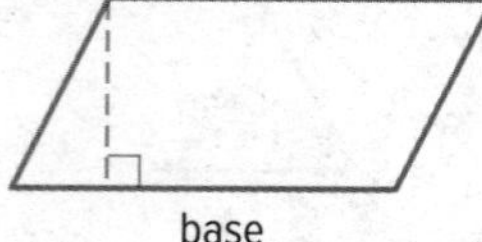

base Cualquier lado de un paralelogramo.

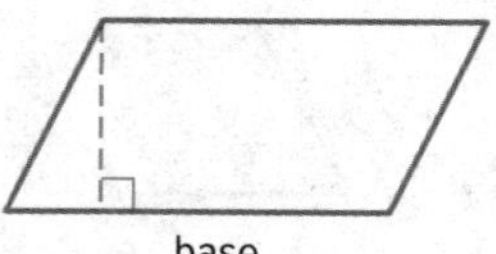

base One of the two parallel congruent faces of a prism.

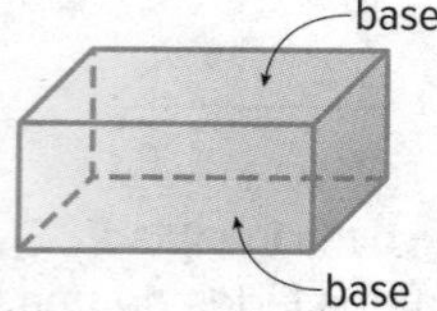

base Una de las dos caras paralelas congruentes de un prisma.

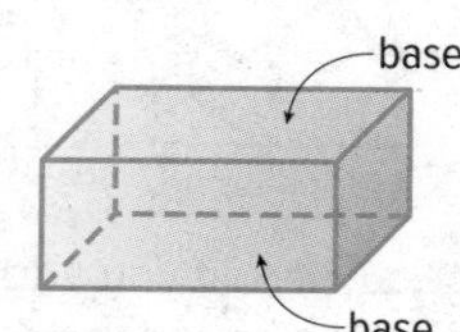

base In a power, the number used as a factor. In 10^3, the base is 10. That is, $10^3 = 10 \times 10 \times 10$.

base En una potencia, el número usado como factor. En 10^3, la base es 10. Es decir, $10^3 = 10 \times 10 \times 10$.

box plot A diagram that is constructed using five values.

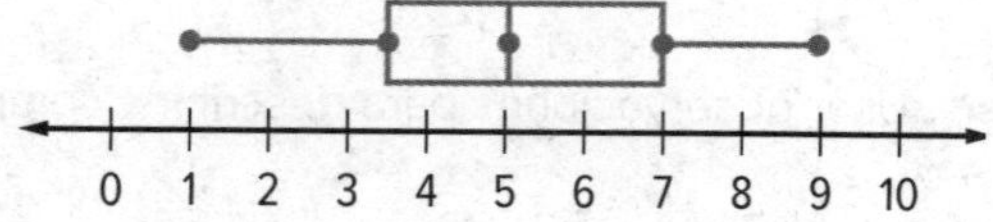

diagrama de caja Diagrama que se construye usando cinco valores.

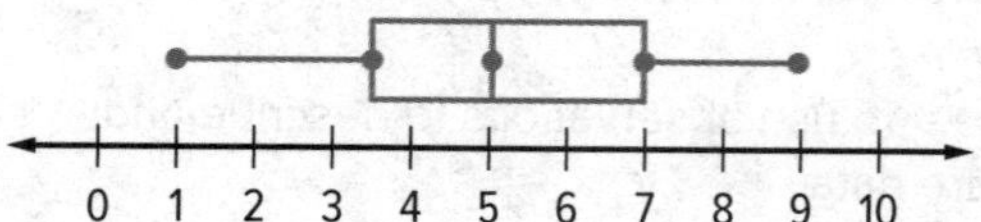

Cc

center The given point from which all points on a circle are the same distance.

centro Un punto dado del cual equidistan todos los puntos de un círculo o de una esfera.

circle The set of all points in a plane that are the same distance from a given point called the center.

círculo Conjunto de todos los puntos en un plano que equidistan de un punto dado llamado centro.

circle graph A graph that shows data as parts of a whole. In a circle graph, the percents add up to 100.

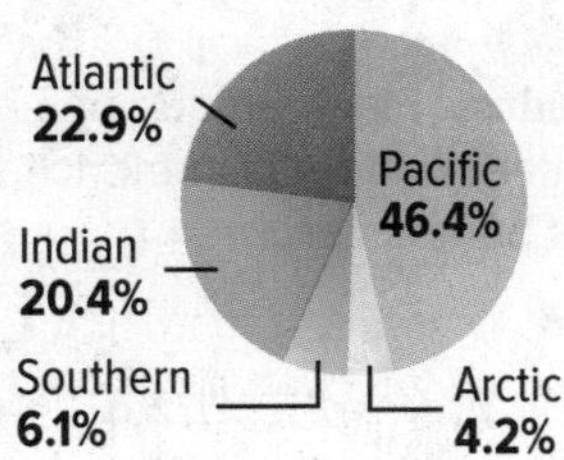

gráfica circular Gráfica que muestra los datos como partes de un todo. En una gráfica circular los porcentajes suman 100.

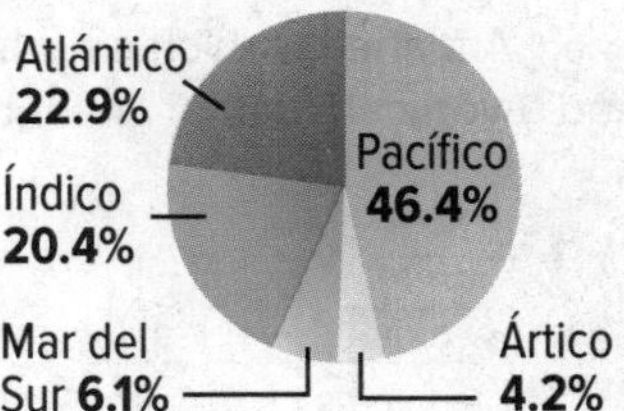

circumference The distance around a circle.

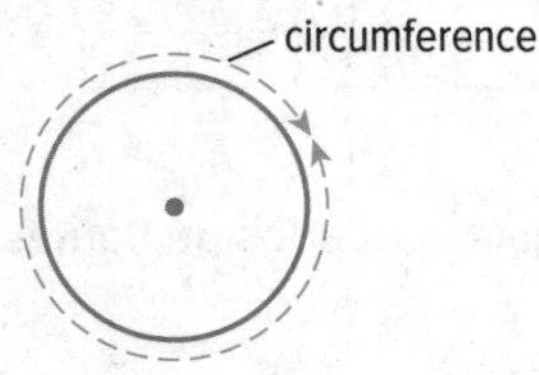

circunferencia La distancia alrededor de un círculo.

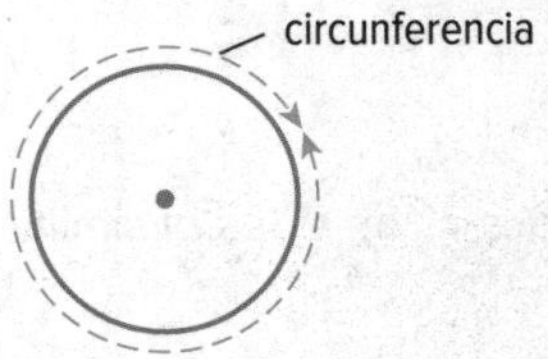

cluster Data that are grouped closely together.

agrupamiento Conjunto de datos que se agrupan.

coefficient The numerical factor of a term that contains a variable.

coeficiente El factor numérico de un término que contiene una variable.

Commutative Property The order in which numbers are added or multiplied does not change the sum or product.

propiedad commutativa La forma en que se suman o multiplican dos números no altera su suma o producto.

compatible numbers Numbers that are easy to use to perform computations mentally.

números compatibles Números que son fáciles de usar para realizar computations mentales.

complementary angles Two angles are complementary if the sum of their measures is 90°.

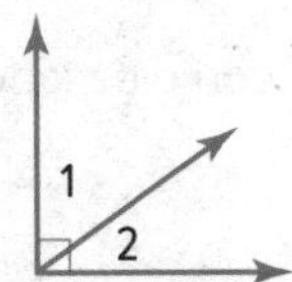

$\angle 1$ and $\angle 2$ are complementary angles.

ángulos complementarios Dos ángulos son complementarios si la suma de sus medidas es 90°.

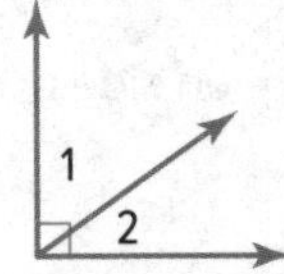

$\angle 1$ y $\angle 2$ son complementarios.

composite figure A figure made of triangles, quadrilaterals, semicircles, and other two-dimensional figures.

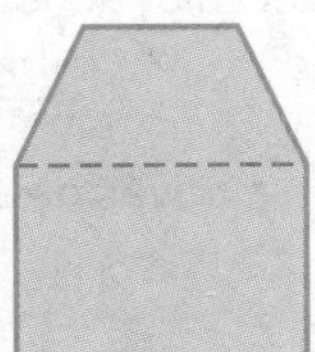

figura compuesta Figura formada por triángulos, cuadriláteros, semicírculos y otras figuras bidimensionales.

congruent Having the same measure.

congruente Ques tienen la misma medida.

congruent figures Figures that have the same size and same shape; corresponding sides and angles have equal measures.

figuras congruentes Figuras que tienen el mismo tamaño y la misma forma; los lados y los ángulos correspondientes con igual medida.

constant A term without a variable.

constante Un término sin una variable.

coordinate plane A plane in which a horizontal number line and a vertical number line intersect at their zero points.

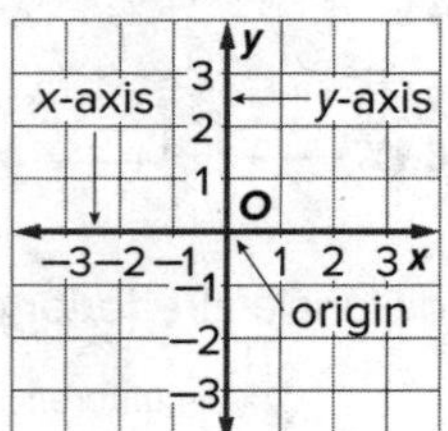

plano de coordenadas Plano en que una recta numérica horizontal y una recta numérica vertical se intersecan en sus puntos cero.

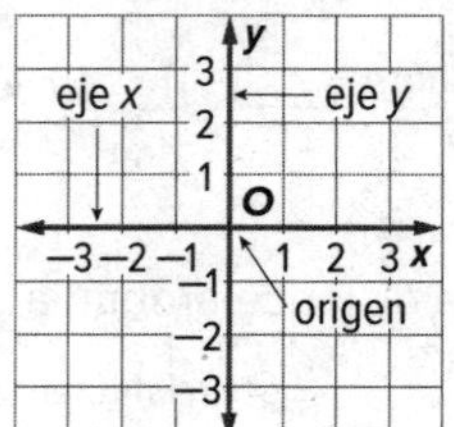

corresponding sides The sides of similar figures that "match."

lados correspondientes Lados de figuras semejantes que coinciden.

cubic units Used to measure volume. Tells the number of cubes of a given size it will take to fill a three-dimensional figure.

3 cubic units

unidades cúbicas Se usan para medir el volumen. Indican el número de cubos de cierto tamaño que se necesitan para llenar una figura tridimensional.

3 unidades cúbicas

Dd

data Information, often numerical, which is gathered for statistical purposes.

datos Información, con frecuencia numérica, que se recoge con fines estadísticos.

decagon A polygon having ten sides.

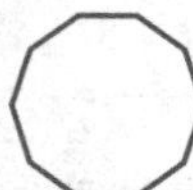

decágono Un polígono con diez lados.

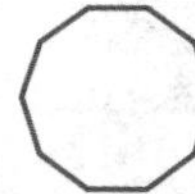

defining the variable Choosing a variable and deciding what the variable represents.

definir la variable Elegir una variable y decidir lo que representa.

dependent variable The variable in a relation with a value that depends on the value of the independent variable.

variable dependiente La variable en una relación cuyo valor depende del valor de la variable independiente.

diameter The distance across a circle through its center.

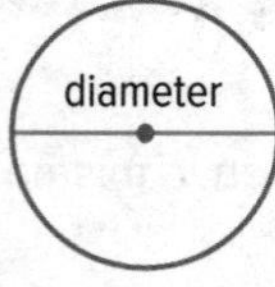

diámetro La distancia a través de un círculo pasando por el centro.

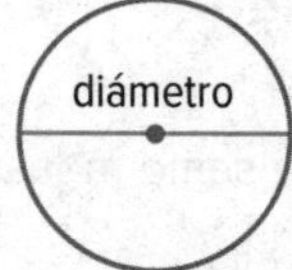

dimensional analysis The process of including units of measurement when you compute.

análisis dimensional Proceso que incluye las unidades de medida al hacer cálculos.

distribution The arrangement of data values.

distribución El arreglo de valores de datos.

Distributive Property To multiply a sum by a number, multiply each addend by the number outside the parentheses.

propiedad distributiva Para multiplicar una suma por un número, multiplica cada sumando por el número fuera de los paréntesis.

Division Property of Equality If you divide each side of an equation by the same nonzero number, the two sides remain equal.

propiedad de igualdad de la división Si divides ambos lados de una ecuación entre el mismo número no nulo, los lados permanecen iguales.

dot plot A diagram that shows the frequency of data on a number line. Also known as a line plot.

diagrama de puntos Diagrama que muestra la frecuencia de los datos sobre una recta numérica.

Ee

equals sign A symbol of equality, =.

signo de igualdad Símbolo que indica igualdad, =.

equation A mathematical sentence showing two expressions are equal. An equation contains an equals sign, =.

ecuación Enunciado matemático que muestra que dos expresiones son iguales. Una ecuación contiene el signo de igualdad, =.

equilateral triangle A triangle having three congruent sides.

triángulo equilátero Triángulo con tres lados congruentes.

equivalent expressions Expressions that have the same value.

expresiones equivalentes Expresiones que poseen el mismo valor, sin importer los valores de la(s) variable(s).

equivalent ratios Ratios that express the same relationship between two quantities.

razones equivalentes Razones que expresan la misma relación entre dos cantidades.

evaluate To find the value of an algebraic expression by replacing variables with numbers.

evaluar Calcular el valor de una expresión sustituyendo las variables por número.

exponent In a power, the number that tells how many times the base is used as a factor. In 5^3, the exponent is 3. That is, $5^3 = 5 \times 5 \times 5$.

exponente En una potencia, el número que indica las veces que la base se usa como factor. En 5^3, el exponente es 3. Es decir, $5^3 = 5 \times 5 \times 5$.

Ff

face A flat surface.

cara Una superficie plana.

factor the expression The process of writing numeric or algebraic expressions as a product of their factors.

factorizar la expresión El proceso de escribir expresiones numéricas o algebraicas como el producto de sus factores.

first quartile For a data set with median M, the first quartile is the median of the data values less than M.

primer cuartil Para un conjunto de datos con la mediana M, el primer cuartil es la mediana de los valores menores que M.

formula An equation that shows the relationship among certain quantities.

fórmula Ecuación que muestra la relación entre ciertas cantidades.

fraction A number that represents part of a whole or part of a set.

$\frac{1}{2}, \frac{1}{3}, \frac{1}{4}, \frac{3}{4}$

fracción Número que representa parte de un todo o parte de un conjunto.

$\frac{1}{2}, \frac{1}{3}, \frac{1}{4}, \frac{3}{4}$

frequency distribution How many pieces of data are in each interval.

distribución de frecuencias Cantidad de datos asociada con cada intervalo.

frequency table A table that shows the number of pieces of data that fall within the given intervals.

tabla de frecuencias Tabla que muestra el número de datos en cada intervalo.

function A relationship that assigns exactly one output value to one input value.

función Relación que asigna exactamente un valor de salida a un valor de entrada.

function rule An expression that describes the relationship between each input and output.

regla de funciones Expresión que describe la relación entre cada valor de entrada y de salida.

function table A table organizing the input, rule, and output of a function.

tabla de funciones Tabla que organiza las entradas, la regla y las salidas de una función.

gap An empty space or interval in a set of data.

laguna Espacio o intervalo vacío en un conjunto de datos.

geometric sequence A sequence in which each term is found by multiplying the previous term by the same number.

sucesión geométrica Sucesión en la cual cada término después del primero se determina multiplicando el término anterior por el mismo número.

graph To place a dot at a point named by an ordered pair.

graficar Colocar una marca puntual en el punto que corresponde a un par ordenado.

Greatest Common Factor (GCF) The greatest of the common factors of two or more numbers.

The greatest common factor of 12, 18, and 30 is 6.

máximo común divisor (MCD) El mayor de los factores comunes de dos o más números.

El máximo común divisor de 12, 18 y 30 es 6.

height The shortest distance from the base of a parallelogram to its opposite side.

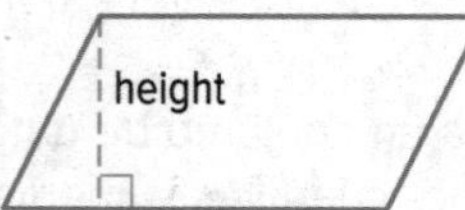

altura La distancia más corta desde la base de un paralelogramo hasta su lado opuesto.

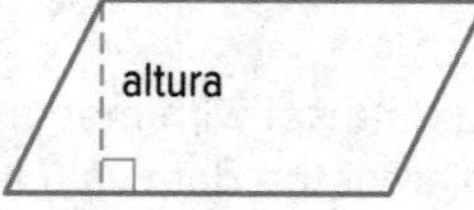

heptagon A polygon having seven sides.

heptágono Polígono con siete lados.

hexagon A polygon having six sides.

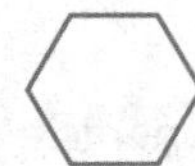

hexágono Polígono con seis lados.

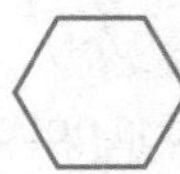

histogram A type of bar graph used to display numerical data that have been organized into equal intervals.

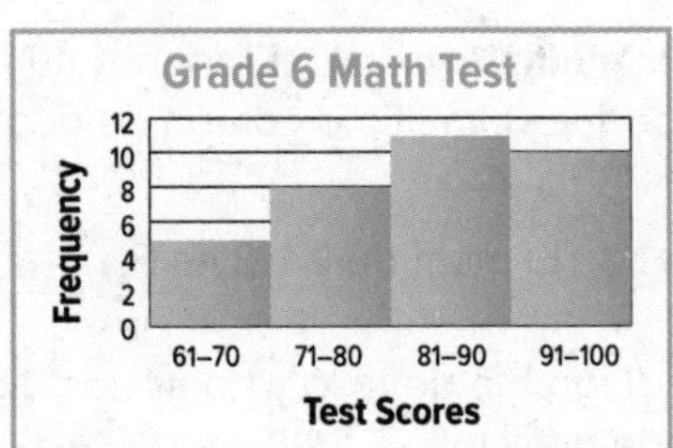

histograma Tipo de gráfica de barras que se usa para exhibir datos que se han organizado en intervalos iguales.

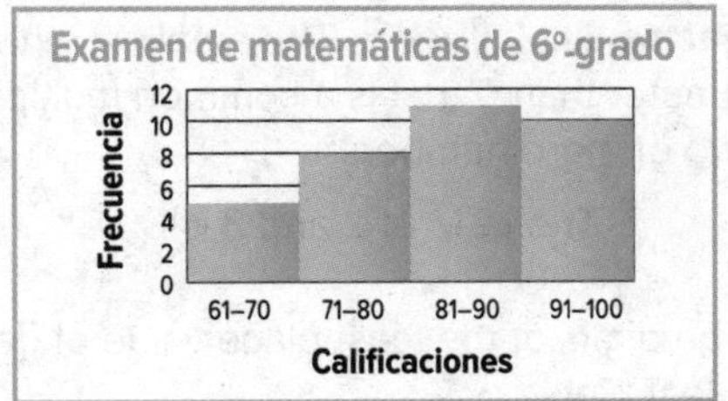

Ii

Identity Properties Properties that state that the sum of any number and 0 equals the number and that the product of any number and 1 equals the number.

propiedades de identidad Propiedades que establecen que la suma de cualquier número y 0 es igual al número y que el producto de cualquier número y 1 es iqual al número.

independent variable The variable in a function with a value that is subject to choice.

variable independiente Variable en una función cuyo valor está sujeto a elección.

inequality A mathematical sentence indicating that two quantities are not equal.

desigualdad Enunciado matemático que indica que dos cantidades no son iguales.

integer Any number from the set {... −4, −3, −2, −1, 0, 1, 2, 3, 4 ...} where ... means *continues without end.*

entero Cualquier número del conjunto {... −4, −3, −2, −1, 0, 1, 2, 3, 4 ...} donde ... significa que *continúa sin fin.*

interquartile range A measure of variation in a set of numerical data, the interquartile range is the distance between the first and third quartiles of the data set.

rango intercuartil El rango intercuartil, una medida de la variación en un conjunto de datos numéricos, es la distancia entre el primer y el tercer cuartil del conjunto de datos.

intersecting lines *Lines* that meet or cross at a common *point.*

rectas secantes *Rectas* que se intersectan o se cruzan en un *punto* común.

interval The difference between successive values on a scale.

intervalo La diferencia entre valores sucesivos de una escala.

inverse operations Operations which *undo* each other. For example, addition and subtraction are inverse operations.

operaciones inversas Operaciones que *se anulan* mutuamente. La adición y la sustracción son operaciones inversas.

isosceles triangle A triangle having at least two congruent sides.

triángulo isósceles Triángulo que tiene por lo menos dos lados congruentes.

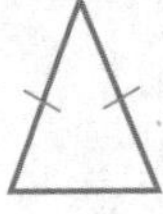

lateral face Any face that is not a base.

cara lateral Cualquier superficie plana que no sea la base.

least common denominator (LCD) The least common multiple of the denominators of two or more fractions.

mínimo común denominador (mcd) El menor múltiplo común de los denominadores de dos o más fracciones.

least common multiple (LCM) The smallest whole number greater than 0 that is a common multiple of each of two or more numbers.

The LCM of 2 and 3 is 6.

mínimo común múltiplo (mcm) El menor número entero, mayor que 0, múltiplo común de dos o más números.

El mcm de 2 y 3 es 6.

leaves The digits of the least place value of data in a stem-and-leaf plot.

hoja En un diagrama de tallo y hojas, los dígitos del menor valor de posición.

like terms Terms that contain the same variable(s) to the same power.

términos semejantes Términos que contienen la misma variable o variables elevadas a la misma potencia.

line A set of *points* that form a straight path that goes on forever in opposite directions.

recta Conjunto de *puntos* que forman una trayectoria recta sin fin en direcciones oputestas.

linear function A function that forms a line when graphed.

función lineal Función cuya gráfica es una recta.

line graph A graph used to show how a set of data changes over a period of time.

gráfica lineal Gráfica que se use para mostrar cómo cambian los valores durange un período de tiempo.

line of symmetry A line that divides a figure into two halves that are reflections of each other.

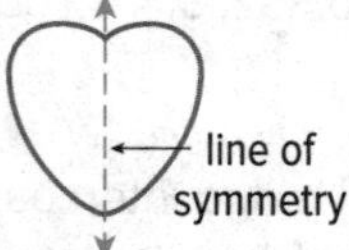

eje de simetría Recta que divide una figura en dos mitades especulares.

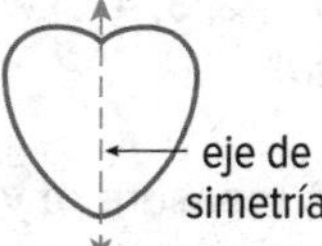

line plot A diagram that shows the frequency of data on a number line. Also known as a dot plot.

esquema lineal Diagrama que muestra la frecuencia de los datos sobre una recta numérica.

line segment A part of a *line* that connects two points.

segmento de recta Parte de una *recta* que conecta dos puntos.

line symmetry Figures that match exactly when folded in half have line symmetry.

simetría lineal Exhiben simetría lineal las figuras que coinciden exactamente al doblarse una sobre otra.

Mm

mean The sum of the numbers in a set of data divided by the number of pieces of data.

media La suma de los números en un conjunto de datos dividida entre el número total de datos.

mean absolute deviation A measure of variation in a set of numerical data, computed by adding the distances between each data value and the mean, then dividing by the number of data values.

desviación media absoluta Una medida de variación en un conjunto de datos numéricos que se calcula sumando las distancias entre el valor de cada dato y la media, y luego dividiendo entre el número de valores.

measures of center Numbers that are used to describe the center of a set of data. These measures include the mean, median, and mode.

medidas del centro Numéros que se usan para describir el centro de un conjunto de datos. Estas medidas incluyen la media, la mediana y la moda.

measures of variation A measure used to describe the distribution of data.

medidas de variación Medida usada para describir la distribución de los datos.

median A measure of center in a set of numerical data. The median of a list of values is the value appearing at the center of a sorted version of the list—or the mean of the two central values, if the list contains an even number of values.

mediana Una medida del centro en un conjunto de datos numéricos. La mediana de una lista de valores es el valor que aparece en el centro de una versión ordenada de la lista, o la media de los dos valores centrales si la lista contiene un número par de valores.

mode The number(s) or item(s) that appear most often in a set of data.

moda Número(s) de un conjunto de datos que aparece(n) más frecuentemente.

Multiplication Property of Equality If you multiply each side of an equation by the same nonzero number, the two sides remain equal.

propiedad de multiplicación de la igualdad Si multiplicas ambos lados de una ecuación por el mismo número no nulo, lo lados permanecen iguales.

Nn

negative integer A number that is less than zero. It is written with a − sign.

entero negativo Número que es menor que cero y se escribe con el signo −.

net A two-dimensional figure that can be used to build a three-dimensional figure.

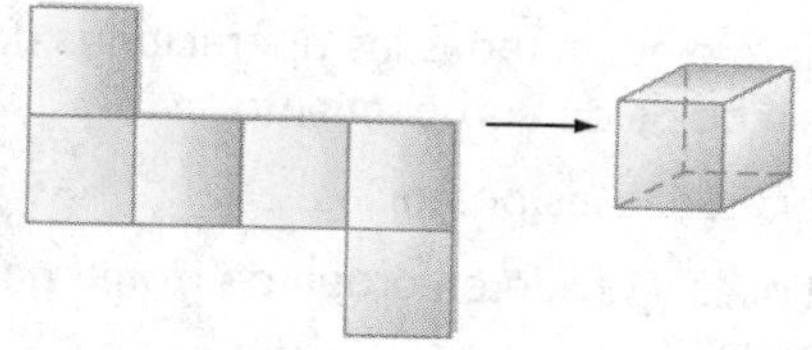

red Figura bidimensional que sirve para hacer una figura tridimensional.

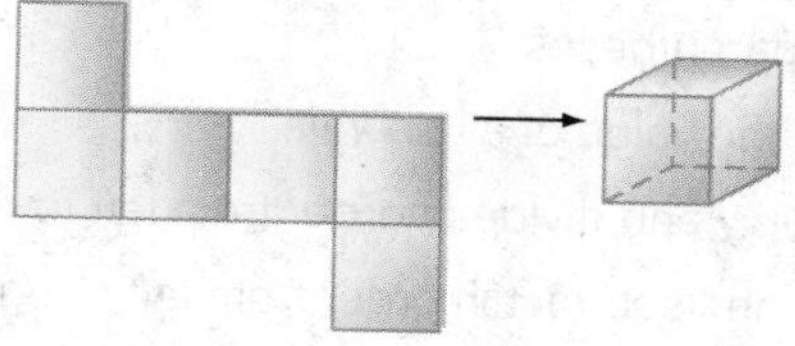

nonagon A polygon having nine sides.

enágono Polígono que tiene nueve lados.

numerical expression A combination of numbers and operations.

expresión numérica Una combinación de números y operaciones.

Oo

obtuse angle Any angle that measures greater than 90° but less than 180°.

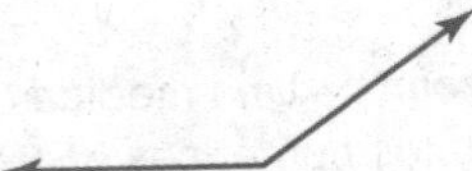

ángulo obtuso Cualquier ángulo que mide más de 90° pero menos de 180°.

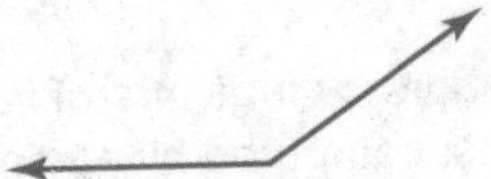

obtuse triangle A triangle having one obtuse angle.

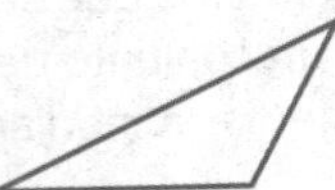

triángulo obtusángulo Triángulo que tiene un ángulo obtuso.

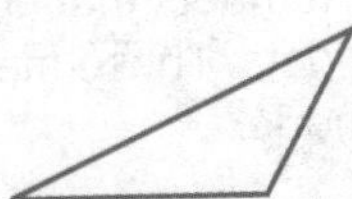

octagon A polygon having eight sides.

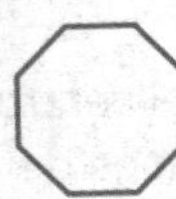

octágono Polígono que tiene ocho lados.

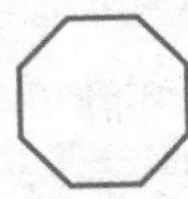

opposites Two integers are opposites if they are represented on the number line by points that are the same distance from zero, but on opposite sides of zero. The sum of two opposites is zero.

opuestos Dos enteros son opuestos si, en la recta numérica, están representados por puntos que equidistan de cero, pero en direcciones opuestas. La suma de dos opuestos es cero.

ordered pair A pair of numbers used to locate a point on the coordinate plane. The ordered pair is written in the form (*x*-coordinate, *y*-coordinate).

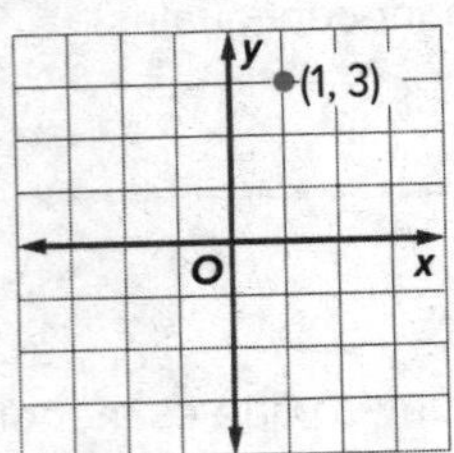

par ordenado Par de números que se utiliza para ubicar un punto en un plano de coordenadas. Se escribe de la forma (coordenada *x*, coordenada *y*).

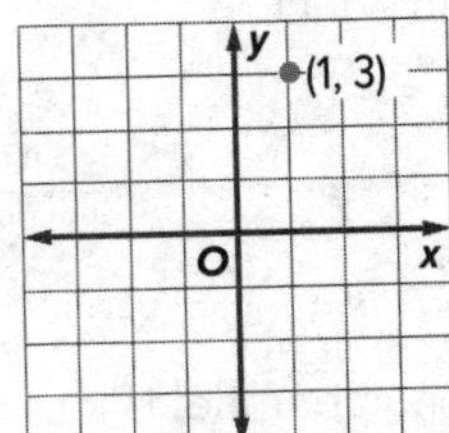

order of operations The rules that tell which operation to perform first when more than one operation is used.

1. Simplify the expressions inside grouping symbols, like parentheses.
2. Find the value of all powers.
3. Multiply and divide in order from left to right.
4. Add and subtract in order from left to right.

orden de las operaciones Reglas que establecen cuál operación debes realizar primero, cuando hay más de una operación involucrada.

1. Primero ejecuta todas las operaciones dentro de los símbolos de agrupamiento.
2. Evalúa todas las potencias.
3. Multiplica y divide en orden de izquierda a derecha.
4. Suma y resta en orden de izquierda a derecha.

origin The point of intersection of the *x*-axis and *y*-axis on a coordinate plane.

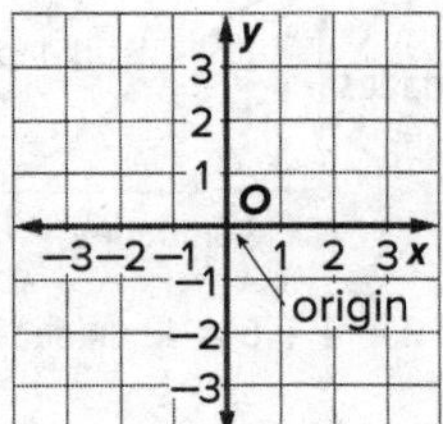

origen Punto de intersección de los ejes axiales en un plano de coordenadas.

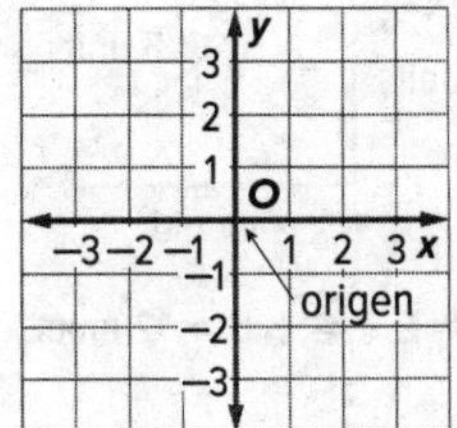

outlier A value that is much greater than or much less than than the other values in a set of data.

valor atípico Dato que se encuentra muy separado de los otros valores en un conjunto de datos.

parallel lines Lines in a plane that never intersect.

rectas paralelas Rectas en un plano que nunca se intersecan.

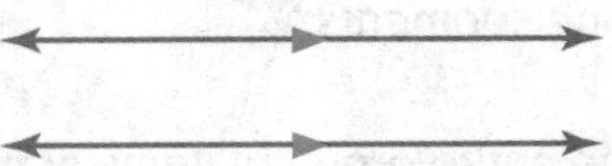

parallelogram A quadrilateral with opposite sides parallel and opposite sides congruent.

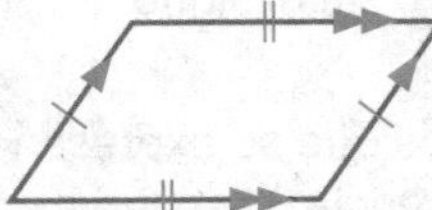

paralelogramo Cuadrilátero cuyos lados opuestos son paralelos y congruentes.

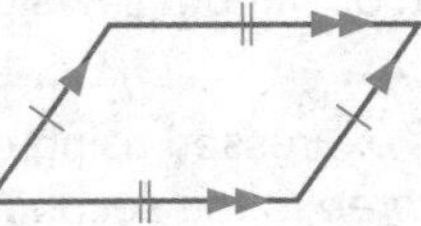

peak The most frequently occurring value in a line plot.

pico El valor que ocurre con más frecuencia en un diagrama de puntos.

pentagon A polygon having five sides.

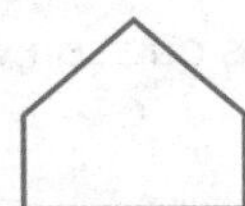

pentágono Polígono que tiene cinco lados.

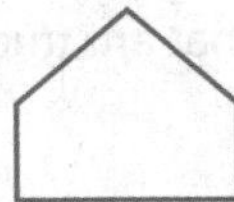

percent A ratio that compares a number to 100.

por ciento Razón en que se compara un número a 100.

percent proportion One ratio or fraction that compares part of a quantity to the whole quantity. The other ratio is the equivalent percent written as a fraction with a denominator of 100.

$$\frac{\text{part}}{\text{whole}} = \frac{\text{percent}}{100}$$

proporción porcentual Razón o fracción que compara parte de una cantidad a toda la cantidad. La otra razón es el porcentaje equivalente escrito como fracción con 100 de denominador.

$$\frac{\text{parte}}{\text{todo}} = \frac{\text{porcentaje}}{100}$$

perfect square Numbers with square roots that are whole numbers. 25 is a perfect square because the square root of 25 is 5.

cuadrados perfectos Números cuya raíz cuadrada es un número entero. 25 es un cuadrado perfecto porque la raíz cuadrada de 25 es 5.

perimeter The distance around a figure.

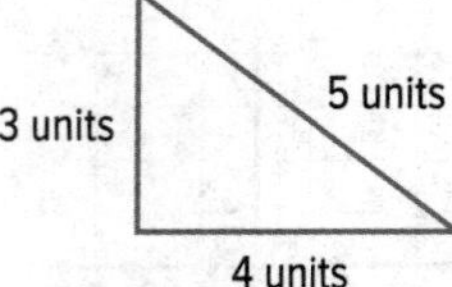

$P = 3 + 4 + 5 = 12$ units

perímetro La distancia alrededor de una figura.

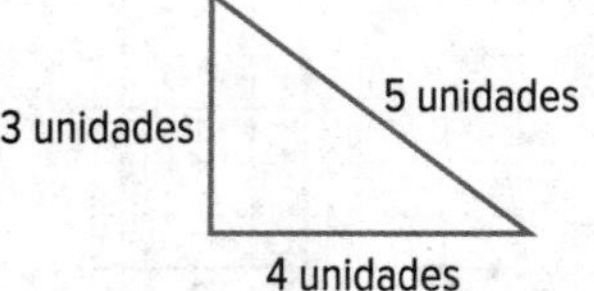

$P = 3 + 4 + 5 = 12$ unidades

pi The ratio of the circumference of a circle to its diameter. The Greek letter π represents this number. The value of pi is always 3.1415926....

pi Razón de la circunferencia de un círculo al diámetro del mismo. La letra griega π representa este número. El valor de pi es siempre 3.1415926....

plane A flat surface that goes on forever in all directions.

plano Superficie plana que se extiende infinitamente en todas direcciones.

point An exact location in space that is represented by a dot.

punto Ubicación exacta en el espacio que se representa con un marca puntual.

polygon A simple closed figure formed by three or more straight line segments.

polígono Figura cerrada simple formada por tres o más segmentos de recta.

population The entire group of items or individuals from which the samples under consideration are taken.

población El grupo total de individuos o de artículos del cual se toman las muestras bajo estudio.

positive integer A number that is greater than zero. It can be written with or without a + sign.

entero positivo Número que es mayor que cero y se puede escribir con o sin el signo +.

powers Numbers expressed using exponents. The power 3^2 is read *three to the second power,* or *three squared.*

potencias Números que se expresan usando exponentes. La potencia 3^2 se lee *tres a la segunda potencia* o *tres al cuadrado.*

prism A three-dimensional figure with at least three rectangular lateral faces and top and bottom faces parallel.

prisma Figura tridimensional que tiene por lo menos tres caras laterales rectangulares y caras paralelas superior e inferior.

properties Statements that are true for any number.

propiedades Enunciados que son verdaderos para cualquier número.

proportion An equation stating that two ratios or rates are equivalent.

proporción Ecuación que indica que dos razones o tasas son equivalentes.

pyramid A three-dimensional figure with at least three triangular sides that meet at a common vertex and only one base that is a polygon.

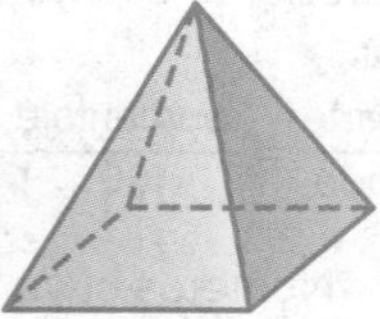

pirámide Una figura de tres dimensiones con que es en un un polígono y tres o mas caras triangulares que se encuentran en un vértice común.

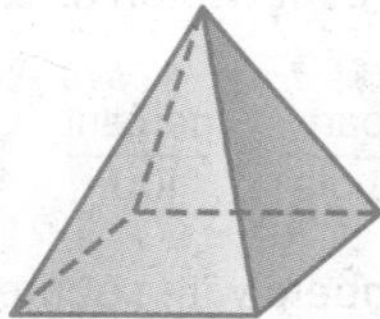

Qq

quadrants The four regions in a coordinate plane separated by the *x*-axis and *y*-axis.

cuadrantes Las cuatro regiones de un plano de coordenadas separadas por el eje *x* y el eje *y*.

quadrilateral A closed figure having four sides and four angles.

cuadrilátero Figura cerrada que tiene cuatro lados y cuatro ángulos.

quartiles Values that divide a data set into four equal parts.

cuartiles Valores que dividen un conjunto de datos en cuatro partes iguales.

Rr

radical sign The symbol used to indicate a nonnegative square root, $\sqrt{}$.

signo radical Símbolo que se usa para indicar una raíz cuadrada no negativa, $\sqrt{}$.

radius The distance from the center to any point on the circle.

radio Distancia desde el centro de un círculo hasta cualquier punto del mismo.

range The difference between the greatest number and the least number in a set of data.

rango La diferencia entre el número mayor y el número menor en un conjunto de datos.

rate A ratio comparing two quantities with different kinds of units.

tasa Razón que compara dos cantidades que tienen diferentes tipos de unidades.

rate of change A rate that describes how one quantity changes in relation to another. A rate of change is usually expressed as a unit rate.

tasa de cambio Tasa que describe cómo cambia una cantidad con respecto a otra. Por lo general, se expresa como tasa unitaria.

ratio A comparison of two quantities by division. The ratio of 2 to 3 can be stated as 2 out of 3, 2 to 3, 2 : 3, or $\frac{2}{3}$.

razón Comparación de dos cantidades mediante división. La razón de 2 a 3 puede escribirse como 2 de cada 3, 2 a 3, 2 : 3 ó $\frac{2}{3}$.

rational number A number that can be written as a fraction.

número racional Número que se puede expresar como fracción.

ratio table A table with columns filled with pairs of numbers that have the same ratio.

tabla de razones Tabla cuyas columnas contienen pares de números que tienen una misma razón.

ray A line that has one endpoint and goes on forever in only one direction.

rayo Recta con un extremo y la cual se extiende infinitamente en una sola dirección.

reciprocals Any two numbers that have a product of 1. Since $\frac{5}{6} \times \frac{6}{5} = 1$, $\frac{5}{6}$ and $\frac{6}{5}$ are reciprocals.

recíproco Cualquier par de números cuyo producto es 1. Como $\frac{5}{6} \times \frac{6}{5} = 1$, $\frac{5}{6}$ y $\frac{6}{5}$ son recíprocos.

rectangle A parallelogram having four right angles.

rectángulo Paralelogramo con cuatro ángulos rectos.

rectangular prism A prism that has rectangular bases.

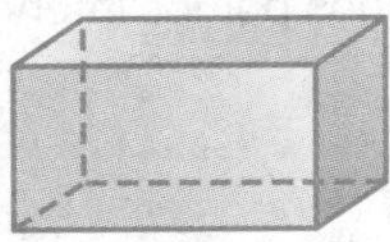

prisma rectangular Una prisma que tiene bases rectangulares.

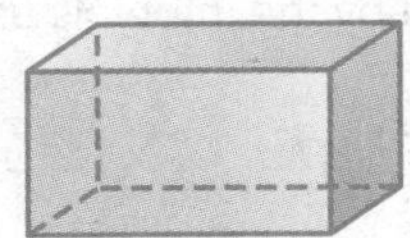

reflection The mirror image produced by flipping a figure over a line.

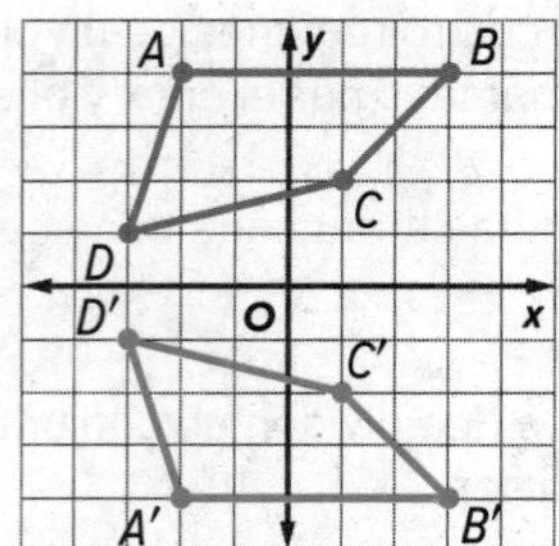

reflexión Transformación en la cual una figura se voltea sobre una recta. También se conoce como simetría de espejo.

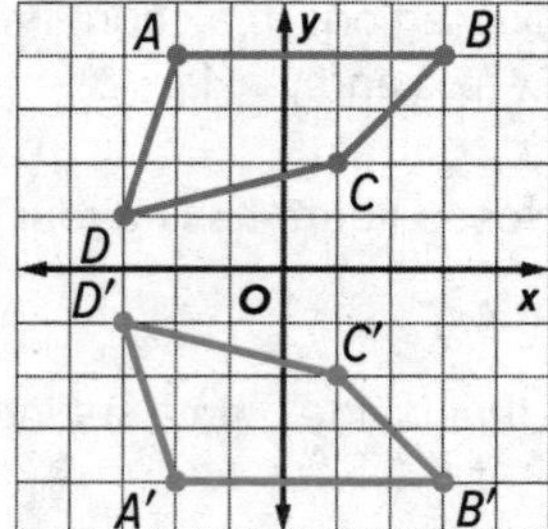

relation A set of ordered pairs such as (1, 3), (2, 4), and (3, 5). A relation can also be shown in a table or a graph.

relación Conjunto de pares ordenados como (1, 3), (2, 4) y (3, 5). Una relación también se puede mostrar en una tabla o una gráfica.

repeating decimal The decimal form of a rational number.

decimal periódico La forma decimal de un número racional.

rhombus A parallelogram having four congruent sides.

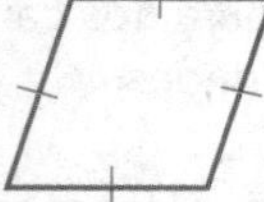

rombo Paralelogramo que tiene cuatro lados.

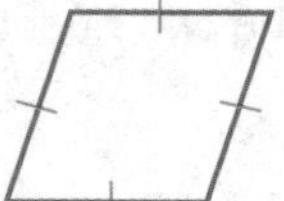

right angle An angle that measures exactly 90°.

ángulo recto Ángulo que mide exactamente 90°.

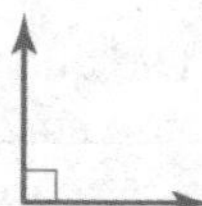

right triangle A triangle having one right angle.

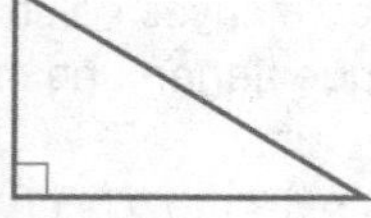

triángulo rectángulo Triángulo que tiene un ángulo recto.

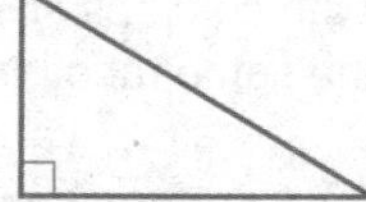

Ss

sample A randomly selected group chosen for the purpose of collecting data.

muestra Grupo escogido al azar o aleatoriamente que se usa con el propósito de recoger datos.

scale The set of all possible values of a given measurement, including the least and greatest numbers in the set, separated by the intervals used.

escala Conjunto de todos los valores posibles de una medida dada, incluyendo el número menor y el mayor del conjunto, separados por los intervalos usados.

scale The scale gives the ratio that compares the measurements of a drawing or model to the measurements of the real object.

escala Razón que compara las medidas de un dibujo o modelo a las medidas del objeto real.

scale drawing A drawing that is used to represent objects that are too large or too small to be drawn at actual size.

dibujo a escala Dibujo que se usa para representar objetos que son demasiado grandes o demasiado pequeños como para dibujarlos de tamaño natural.

scalene triangle A triangle having no congruent sides.

triángulo escaleno Triángulo sin lados congruentes.

scaling To multiply or divide two related quantities by the same number.

homotecia Multiplicar o dividir dos cantidades relacionadas entre un mismo número.

sequence A list of numbers in a specific order, such as 0, 1, 2, 3, or 2, 4, 6, 8.

sucesión Lista de números en un orden específico como, por ejemplo, 0, 1, 2, 3 ó 2, 4, 6, 8.

similar figures Figures that have the same shape but not necessarily the same size.

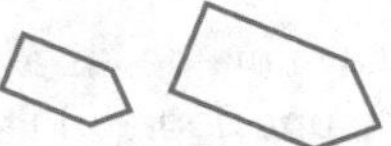

figuras semejantes Figuras que tienen la misma forma, pero no necesariamente el mismo tamaño.

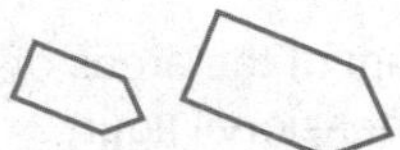

slant height The height of each lateral face.

altura oblicua Altura de cada cara lateral.

solution The value of a variable that makes an equation true. The solution of $12 = x + 7$ is 5.

solución Valor de la variable de una ecuación que hace verdadera la ecuación. La solución de $12 = x + 7$ es 5.

solve To replace a variable with a value that results in a true sentence.

resolver Reemplazar una variable con un valor que resulte en un enunciado verdadero.

square A rectangle having four right angles and four congruent sides.

cuadrado Rectángulo con cuatro ángulos rectos y cuatro lados congruentes.

square root The factors multiplied to form perfect squares.

raíz cuadrada Factores multiplicados para formar cuadrados perfectos.

statistical question A question that anticipates and accounts for a variety of answers.

cuestión estadística Una pregunta que se anticipa y da cuenta de una variedad de respuestas.

statistics Collecting, organizing, and interpreting data.

estadística Recopilar, ordenar e interpretar datos.

stem-and-leaf plot A system where data are organized from least to greatest. The digits of the least place value usually form the leaves, and the next place-value digits form the stems.

Stem	Leaf
1	2 4 5
2	
3	1 2 3 3 9
4	0 4 6 7

4 | 7 = 47

diagrama de tallo y hojas Sistema donde los datos se organizan de menor a mayor. Por lo general, los dígitos de los valores de posición menores forman las hojas y los valores de posición más altos forman los tallos.

Tallo	Hojas
1	2 4 5
2	
3	1 2 3 3 9
4	0 4 6 7

4 | 7 = 47

stems The digits of the greatest place value of data in a stem-and-leaf plot.

tallo Los dígitos del mayor valor de posición de los datos en un diagrama de tallo y hojas.

straight angle An angle that measures exactly 180°.

ángulo llano Ángulo que mide exactamente 180°.

Subtraction Property of Equality If you subtract the same number from each side of an equation, the two sides remain equal.

propiedad de sustracción de la igualdad Si sustraes el mismo número de ambos lados de una ecuación, los dos lados permanecen iguales.

supplementary angles Two angles are supplementary if the sum of their measures is 180°.

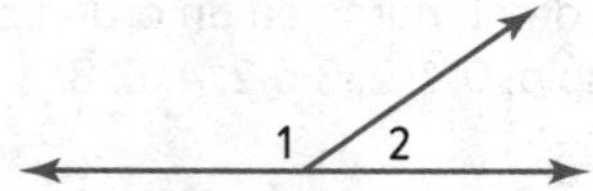

$\angle 1$ and $\angle 2$ are supplementary angles.

ángulos suplementarios Dos ángulos son suplementarios si la suma de sus medidas es 180°.

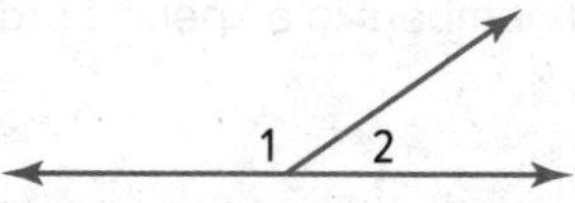

$\angle 1$ y $\angle 2$ son suplementarios.

surface area The sum of the areas of all the surfaces (faces) of a three-dimensional figure.
$S.A. = 2\ell h + 2\ell w + 2hw$

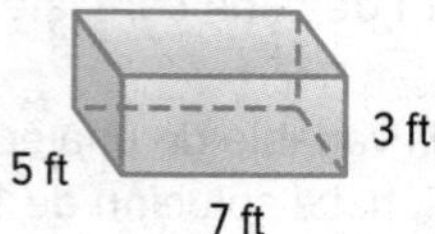

$S.A. = 2(7 \times 3) + 2(7 \times 5) + 2(3 \times 5)$
$= 142$ square feet

área de superficie La suma de las áreas de todas las superficies (caras) de una figura tridimensional.
$S.A. = 2\ell h + 2\ell w + 2hw$

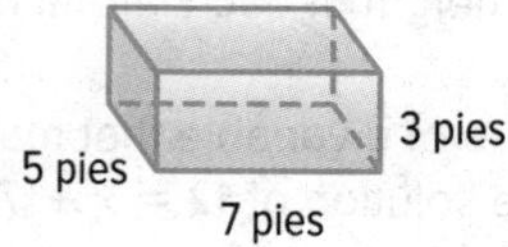

$S.A. = 2(7 \times 3) + 2(7 \times 5) + 2(3 \times 5)$
$= 142$ pies cuadrados

survey A question or set of questions designed to collect data about a specific group of people, or population.

encuesta Pregunta o conjunto de preguntas diseñadas para recoger datos sobre un grupo específico de personas o población.

symmetric distribution Data that are evenly distributed.

distribución simétrica Datos que están distribuidos.

term Each number in a sequence.

término Cada uno de los números de una sucesión.

term Each part of an algebraic expression separated by a plus or minus sign.

término Cada parte de un expresión algebraica separada por un signo más o un signo menos.

terminating decimal A decimal is called terminating if its repeating digit is 0.

decimal finito Un decimal se llama finito si el dígito que se repite es 0.

third quartile For a data set with median M, the third quartile is the median of the data values greater than M.

tercer cuartil Para un conjunto de datos con la mediana M, el tercer cuartil es la mediana de los valores mayores que M.

three-dimensional figure A figure with length, width, and height.

figura tridimensional Una figura que tiene largo, ancho y alto.

trapezoid A quadrilateral with one pair of parallel sides.

trapecio Cuadrilátero con un único par de lados paralelos.

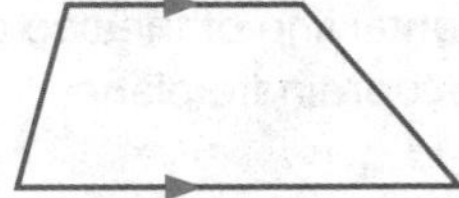

triangle A figure with three sides and three angles.

triángulo Figura con tres lados y tres ángulos.

triangular prism A prism that has triangular bases.

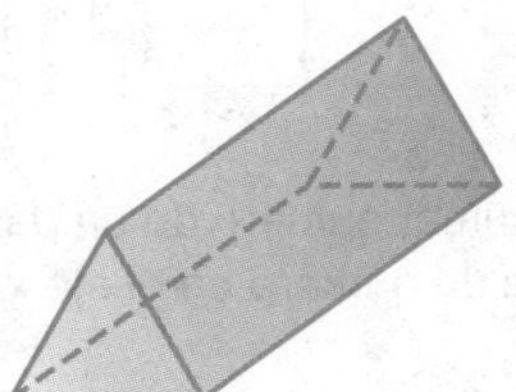

prisma triangular Prisma con bases triangulares.

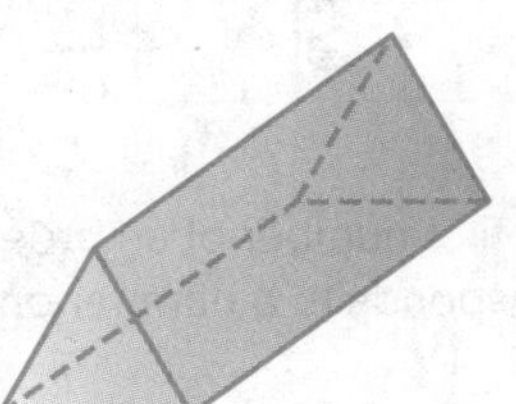

Uu

unit price The cost per unit.

precio unitario El costo por cada unidad.

unit rate A rate that is simplified so that it has a denominator of 1.

tasa unitaria Tasa simplificada para que tenga un denominador igual a 1.

unit ratio A unit rate where the denominator is one unit.

razón unitaria Tasa unitaria en que el denominador es la unidad.

Vv

variable A symbol, usually a letter, used to represent a number.

variable Un símbolo, por lo general, una letra, que se usa para representar un número.

vertex The point where three or more faces intersect.

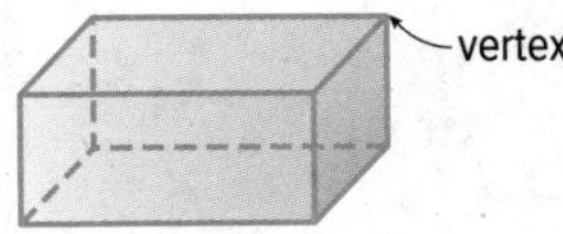

vértice El punto en que se intersecan dos o más caras del prisma.

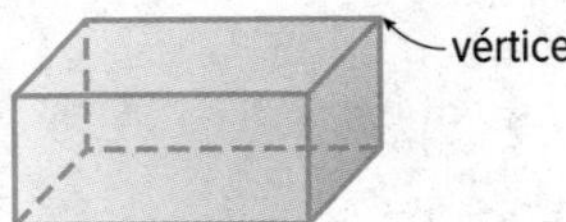

volume The amount of space inside a three-dimensional figure. Volume is measured in cubic units.

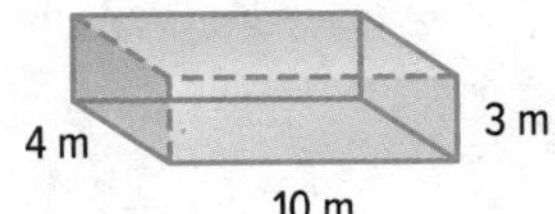

$V = 10 \times 4 \times 3 = 120$ cubic meters

volumen Cantidad de espacio dentro de una figura tridimensional. El volumen se mide en unidades cúbicas.

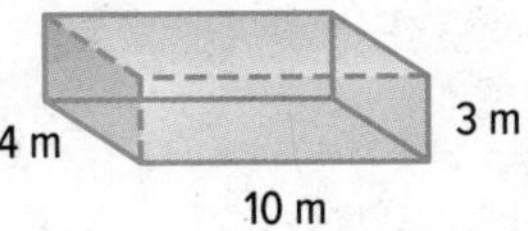

$V = 10 \times 4 \times 3 = 120$ metros cúbicos

Xx

***x*-axis** The horizontal line of the two perpendicular number lines in a coordinate plane.

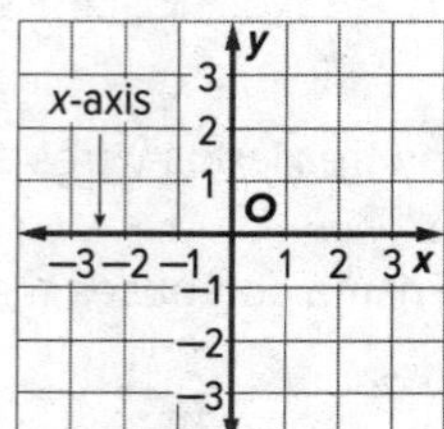

eje *x* La recta horizontal de las dos rectas numéricas perpendiculares en un plano de coordenadas.

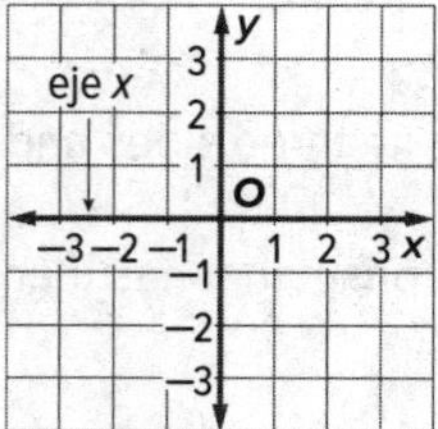

***x*-coordinate** The first number of an ordered pair. The *x*-coordinate corresponds to a number on the *x*-axis.

coordenada *x* El primer número de un par ordenado, el cual corresponde a un número en el eje *x*.

Yy

***y*-axis** The vertical line of the two perpendicular number lines in a coordinate plane.

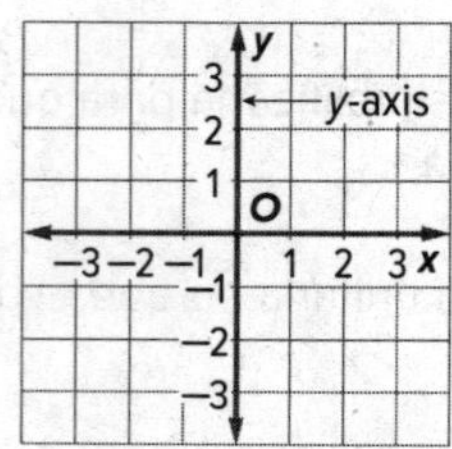

eje *y* La recta vertical de las dos rectas numéricas perpendiculares en un plano de coordenadas.

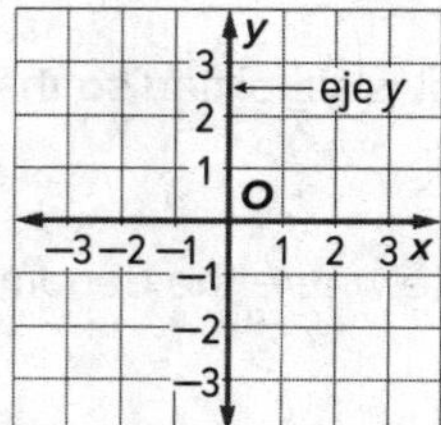

***y*-coordinate** The second number of an ordered pair. The *y*-coordinate corresponds to a number on the *y*-axis.

coordenada *y* El segundo número de un par ordenado, el cual corresponde a un número en el eje *y*.

Chapter 6 Expressions

Page 428 Chapter 6 Are You Ready?

1. 343 **3.** 6,561 **5.** $1\frac{5}{9}$ **7.** $\frac{1}{20}$

Pages 437–438 Lesson 6-1 Independent Practice

1. 6^2 **3.** 5^6 **5.** 27^4 **7.** $6 \times 6 \times 6 \times 6$; 1,296 **9.** $\frac{1}{8} \times \frac{1}{8} = \frac{1}{64}$ **11.** 1.0625 **13.** 1,100.727 **15a.** The next values are found by dividing the previous power by 2. **15b.** The next values are found by dividing the previous power by 4. **15c.** The next values are found by dividing the previous power by 10. **15d.** Any nonzero number with an exponent of 0 has a value of 1.

Powers of 2	Powers of 4	Powers of 10
$2^4 = 16$	$4^4 = 256$	$10^4 = 10{,}000$
$2^3 = 8$	$4^3 = 64$	$10^3 = 1{,}000$
$2^2 = 4$	$4^2 = 16$	$10^2 = 100$
$2^1 = 2$	$4^1 = 4$	$10^1 = 10$
$2^0 = 1$	$4^0 = 1$	$10^0 = 1$

17. 10^8; Sample answer: $10^7 = 10{,}000{,}000$ and $10^8 = 100{,}000{,}000$. 100,000,000 is much closer to 230,000,000 than 10,000,000.

Pages 439–440 Lesson 6-1 Extr a Practice

19. 10^3 **21.** 9^2 **23.** 13^5 **25.** 0.06×0.06; 0.0036 **27.** 8,100 square feet **29.** 42.875 miles **31.** There are 25 counters in the 5th figure; There are 81 counters in the 9th figure; There are 121 counters in the 11th figure. **33.** 8 **35.** $87

Pages 445–446 Lesson 6-2 Independent Practice

1. 9 **3.** 106 **5.** 117 **7.** 112 **9.** $5 \times \$7 + 5 \times \$5 + 5 \times \$2$; $70 **11.** $3 \times 10 + 2 \times 5$; 40 rolls **13a.** $(34 - 12) \div 2 + 7$ **13b.** Sample answer: $34 - (12 \div 2) + 7 = 34 - 6 + 7 = 28 + 7 = 35$ **15a.** $7 + 3 \times (2 + 4) = 25$ **15b.** $8^2 \div (4 \times 8) = 2$ **15c.** parentheses not needed

Pages 447–448 Lesson 6-2 Extra Practice

17. 6 **19.** 61 **21.** 35 **23.** 22 **25.** $3 \times 16 + 8^2$; $112 **27.** 4($0.50) + 3($2.25); $8.75 **29.** 9 **31.** 14 **33.** $37

Pages 453–454 Lesson 6-3 Independent Practice

1. 12 **3.** 18 **5.** 1 **7.** 20 **9.** $\frac{1}{8}$ m^3 **11.** $415.80 **13.** 29 **15.** 7 ft^2 **17.** Sample answer: Both numerical expressions and algebraic expressions use operations. An algebraic expression, such as $6 + a$, includes numbers and variables, where a numerical expression, such as, $6 + 3$ only includes numbers.

Pages 455–456 Lesson 6-3 Extra Practice

19. 24 **21.** 14 **23.** 3 **25.** 22 **27.** $117 **29.** $81\frac{1}{2}$ **31.** 180 **33.** 6 ft **35.** = **37.** < **39.** $14 + 8 = 22$

Pages 465–467 Lesson 6-4 Independent Practice

1. w = the width; $w - 6$ **3.** t = Tracey's age; $t - 6$ **5.** s = the number in the Senate; $4s + 35$; 435 members **7.** $2.54x$; 30.48 cm **9.** m = Marcella's age; $\frac{1}{3}m + 2$; Justin is 23 years old and Aimee is 42 years old. **11.** c = total customer order; $2 + 0.2c$ **13.** sometimes; Sample answer: $x - 3$ and $y - 3$ represent the sample values only when $x = y$.

Pages 467–468 Lesson 6-4 Extra Practice

15. a = the number of apples; $4 \times a$ or $4a$ **17.** j = the cost of James' dinner; $j - 5$ **19.** b = the cost of one game; $3b + 2$; $14 **21.** s = number of songs in Damian's library; $2s + 17$; 27 songs **23a.** False **23b.** True **25.** 7.8 **27.** 14.5

Page 471 Problem-Solving Investigation Act It Out

Case 3. Team 3 **Case 5.** 352 people

Pages 477–478 Lesson 6-5 Independent Practice

1. yes; Associative Property **3.** no; The first expression is equal to 17 and the second is equal to 1. **5.** No; the first expression is equal to 32, not 0. **7.** 75,000 • 5 and 5 • 75,000 **9.** $42r$ **11.** 3 **13.** Sample answer: $12 + (8 + 5)$ and $(12 + 8) + 5$ **15.** Sample answer: $24 \div 12 = 2$ and $12 \div 24 = 0.5$ **17.** Sample answer: Rewrite $48 + 82$ as $48 + (52 + 30)$. By using the Associative Property, $48 + (52 + 30) = (48 + 52) + 30$. So $48 + 82 = 130$.

Pages 479–480 Lesson 6-5 Extra Practice

19. yes; Identity Property **21.** yes; Commutative Property **23.** Sample answer: $(12 + 24) + 6$ and $12 + (24 + 6)$ **25.** $x + 6$ **27.** $8n$ **29.** $m + 15$ **31.** $2 \times (12 \times 25) + 15 \times 20$; $(2 \times 12) \times 25 + 15 \times 20$; $15 \times 20 + (2 \times 12) \times 25$ **33.** $10 + 5$ **35.** $200 + 9$ **37.** 80 cents; Since 3 dimes + 5 dimes = 8 dimes and 8 dimes × 10 cents = 80 cents, the value of the money is 80 cents.

Pages 489–490 Lesson 6-6 Independent Practice

1. $9(40) + 9(4) = 396$ **3.** $7(3) + 7(0.8) = 26.6$ **5.** $66 + 6x$ **7.** $6(43) - 6(35) = 6(43 - 35)$; 48 mi **9.** $6(9 + 4)$ **11.** $11(x + 5)$ **13.** $7(11x + 3)$ **15.** 0.37; Sample answer: $0.1(3.7) = 0.1(3) + 0.1(0.7) = 0.3 + 0.07 = 0.37$ **17.** Sample answer: The friend did not multiply 5 and 2. The expression $5(x + 2) = 5x + 10$.

Pages 491–492 Lesson 6-6 Extra Practice

19. 152 **21.** 11.7 **23.** $3x + 21$ **25.** $9(2.50 + 4) = 9(2.50) + 9(4)$; $58.50 **27.** $3(9 + 4)$ **29.** $4(4 + 5)$ **31.** $6(5 + 2x)$ **33a.** No **33b.** Yes **33c.** Yes **33d.** No **35.** 12.23 **37.** 3.6 **39.** 384 fluid ounces

Pages 499–500 Lesson 6-7 Independent Practice

1. $11x$ **3** $45x$ **5.** $21x + 35y$ **7** $6(4x + 3y)$ **9.** $4(x + 6) + 4x$; $\$8x + \$24$ **11** $6(3t + 2c) = 18t + 12c$ **13.** 9 **15a.** $3(x + 0.75) + 2x$; $\$5x + \$2.25$ **15b.** $6(x + 3) + 2x$; $\$8x + \$18$ **15c.** $2(x + 1.50) + 3x$; $\$5x + \$3$ **17.** Sample answer: The expressions are equivalent because they name the same number regardless of which number stands for y. **19.** $6x + 33$

Pages 501–502 Lesson 6-7 Extra Practice

21. $9x$ **23.** $21x$ **25.** $28x + 20y$ **27.** $5(2x + 3y)$ **29.** $4(x + 3 + 2)$; $\$4x + \$20$ **31.** $4(5t + 3j) = 20t + 12j$ **33.** terms: $2x$, $3y$, x, 7; like terms: $2x$, x; coefficients: 1, 2, 3; constant: 7 **35.** $2x + 3(x + 3) + (x + 6)$; $6x + 15$ **37.** $\frac{2}{7}$ **39.** 28

Page 505 Chapter Review Vocabulary Check

Across

1. algebraic **7.** powers **9.** base **13.** coefficient

Down

3. perfect squares **5.** like terms **11.** variable

Page 506 Chapter Review Key Concept Check

1. $12x + 12$ **3.** $3x - 6$ **5.** $2(x + 3)$

Chapter 7 Equations

Page 512 Chapter 7 Are You Ready?

1. 1.11 **3.** 2.69 **5.** $\frac{1}{3}$ **7.** $\frac{13}{40}$ mi

Pages 517–518 Lesson 7-1 Independent Practice

1 25 **3.** 5 **5.** 13 **7.** 3 **9.** 11 **11.** 5 games **13** 35 students **15.** Sample answer: $m + 8 = 13$ **17.** true; Since $m + 8$ is not equal to any specific value, there are no restrictions placed upon the value of m. **19.** Sample answer: $14 + x$ is an algebraic expression. $14 + x = 20$ is an algebraic equation.

Pages 519–520 Lesson 7-1 Extra Practice

21. 8 **23.** 7 **25.** 6 **27.** 5 **29.** 18 **31.** 8 cookies **33.** 8 ft **35.** $35 + d = 80$; 45 years **37.** 63 **39.** 115 **41.** 93

Pages 529–530 Lesson 7-2 Independent Practice

1 3 **3.** 2 **5** $m + 22 = 118$; 96 in. **7.** $\frac{2}{5}$ **9** $\frac{1}{4}$ **11.** Sample answers: $56 = 44 + x$; $36 = 24 + m$ **13.** $x + 9 = 11$; The solution for the other equations is 3. **15.** The value of y decreases by 4.

Pages 531–532 Lesson 7-2 Extra Practice

17. 3 **19.** 5 **21.** 5 **23.** $9 + x = 63$; 54 inches **25.** $\frac{1}{10}$ **27.** $\frac{1}{2}$ **29a.** $x + 15 = 85$ **29b.** $70 **31.** 38 **33.** 19 **35.** 17

Pages 539–540 Lesson 7-3 Independent Practice

1. 9 **3** 4 **5.** 3.4 **7.** $a - 6 = 15$; 21 years old **9.** 21 **11.** 1 **13** $x - 56 = 4$; $60 **15.** Elisa did not perform the inverse operation. Add 6 to each side to undo subtracting 6. **17.** Sample answer: I would use what I know about fact families to rewrite the equation $b + 7 = 16$. The solution is 9.

Pages 541–542 Lesson 7-3 Extra Practice

19. 6 **21.** 4 **23.** 14.7 **25.** $15 = v - 12$; 27 votes **27.** 19 **29.** $\frac{1}{2}$ **31.** $x - 12 = 3$; $15 **33a.** True **33b.** False **33c.** True **35.** 114 **37.** 104 **39.** 63

Page 545 Problem-Solving Investigation Guess, Check, and Revise

Case 3. five problems worth 2 points each and two problems worth 4 points each **Case 5.** $3 \times 4 + 6 \div 1 = 18$

Pages 555–556 Lesson 7-4 Independent Practice

1 6 **3.** 6 **5.** 2 **7.** $4e = 58$; $14.50 **9.** $\frac{1}{2}$

11 **a.** $25p = 2{,}544$; 107.76 points **b.** $20p = 2{,}150$; 107.5 points

13.

distance	=	rate	×	time
272 miles		r		4 hours

68

15. $4b = 7$; The solution for the other equations is 4. **17.** Sample answer: The Walkers traveled 240 miles in 4 hours. What was their average speed?; 60 miles per hour; The Walkers traveled an average of 60 miles per hour.

Pages 557–558 Lesson 7-4 Extra Practice

19. 5 **21.** 4 **23.** 2 **25.** 7 **27.** $1{,}764 = 28r$; 63 mph **29.** 5 **31.** 3 **33.** 4 **35a.** False **35b.** True **35c.** False **37.** 23 **39.** 52 **41.** 9 **43.** 21 bags

Pages 565–566 Lesson 7-5 Independent Practice

1\. 20 3. 15.04 5. $\frac{x}{4} = 3$; 12 dozen

7\.

+	−
Subtraction Property of Equality	Addition Property of Equality
×	÷
Division Property of Equality	Multiplication Property of Equality

9. True; Sample answer: Dividing by 3 is the same as multiplying by $\frac{1}{3}$. **11a.** $d = 50t$

11b.

Time (days)	1	2	3	4	5
Distance (miles)	50	100	150	200	250

11c. 50 days

Pages 567–568 Lesson 7-5 Extra Practice

13. 84 **15.** 169 **17.** 56 **19.** 3 **21.** $\frac{x}{3} = 2$; 6 eggs **23.** $\frac{r}{4} = 16$; 64 in. **25.** $\frac{x}{6} = 8$; \$48 **27.** > **29.** = **31.** > **33.** 60 in.

Page 571 Chapter Review Vocabulary Check

Across

1. division property **5.** solution **7.** addition property

Down

3. inverse operations

Page 572 Chapter Review Key Concept Check

1. $x = 16$ **3.** $x = 24$ **5.** $x = 68$

Chapter 8 Functions and Inequalities

Page 578 Chapter 8 Are You Ready?

1. > **3.** < **5.** 46 **7.** 3

Pages 583–584 Lesson 8-1 Independent Practice

1\.

Input (x)	3x + 5	Output
0	3(0) + 5	5
3	3(3) + 5	14
9	3(9) + 5	32

3.

Input (x)	x + 2	Output
0	0 + 2	2
1	1 + 2	3
6	6 + 2	8

5\.

Number of Guests (x)	30 ÷ x	Cupcakes per Guest (y)
6	30 ÷ 6	5
10	30 ÷ 10	3
15	30 ÷ 15	2

Cupcakes Per Guest
y
(6, 5)
(10, 3)
(15, 2)
x
Number of Guests

7. 56 miles

9.

Years (x)	223 million × \$10 × x
1	\$2,230,000,000
2	\$4,460,000,000
3	\$6,690,000,000

11. any number between 0 and 1; Sample answer: When you divide by a fraction, you multiply by the reciprocal. If the fraction is between 0 and 1, the reciprocal is greater than 1. **13.** Sample answer: Natalie is tying quilts for a charity. She has 48 yards of fabric to make quilts. Make a table that shows the number of quilts she can make that use 2, 3, and 4 yards of fabric.

Pages 585–586 Lesson 8-1 Extra Practice

15.

Input (x)	4x + 2	Output
1	4(1) + 2	6
3	4(3) + 2	14
6	4(6) + 2	26

17.

Input (x)	2x − 6	Output
3	2(3) − 6	0
6	2(6) − 6	6
9	2(9) − 6	12

19.

Hours (x)	$55x$	Miles (y)
3	55(3)	165
4	55(4)	220
5	55(5)	275

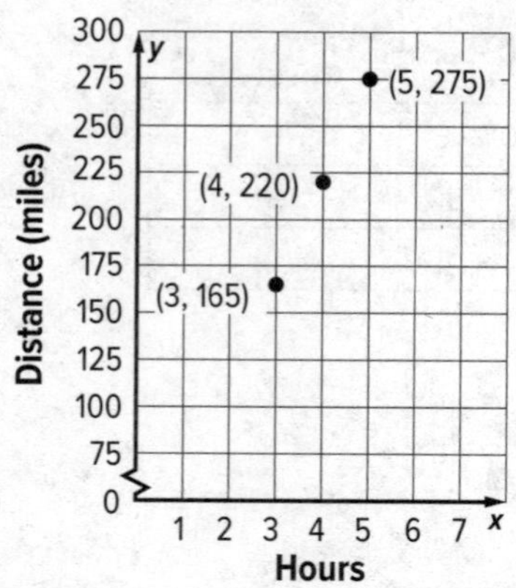

21a. False **21b.** True **21c.** True **23.** 2 **25.** 56 **27.** 72
29. Abby has twice as much money each month.

Pages 591–592 Lesson 8-2 Independent Practice

1 add 9 to the position number; $n + 9$; 21 **3.** Sample answer: This is a geometric sequence. Each term is found by multiplying the previous term by 3; 486, 1,458, 4,374
5. add 12; 52, 64 **7.** add $\frac{1}{2}$; $4\frac{1}{4}$, $4\frac{3}{4}$ **9.** 29.6
11. arithmetic sequence; 4.75, 5.75 **13** arithmetic sequence; Each term is found by adding 2 to the previous term.; $10 + 2 = 12$; 12 boxes **15.** The value of each term is the square of its position; n^2; 10,000.

Pages 593–594 Lesson 8-2 Extra Practice

17. subtract 4 from the position number; $n - 4$; 8 **19.** Each term is found by multiplying the previous term by 3; 324, 972, 2,916 **21.** add 3; 13, 16 **23.** add $1\frac{1}{2}$; $7\frac{1}{2}$, 9 **25.** 19.3
27. 2; 96; geometric **29.** 84 **31.** \$13.50

Pages 599–600 Lesson 8-3 Independent Practice

1. $y = 6x$

3

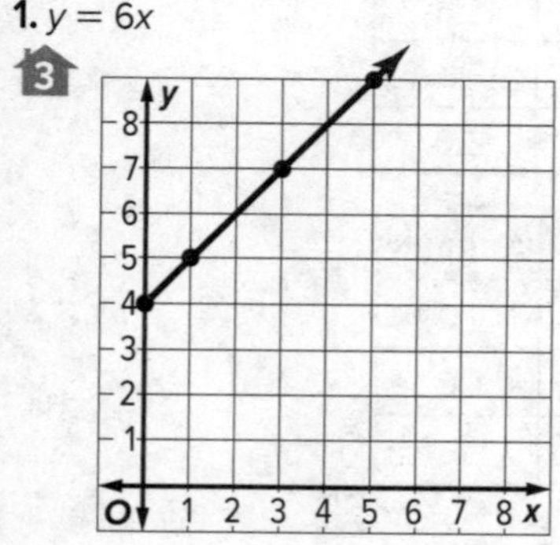

5.

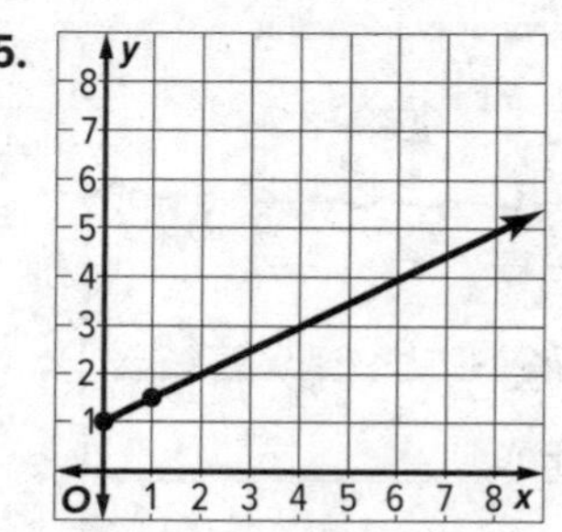

7

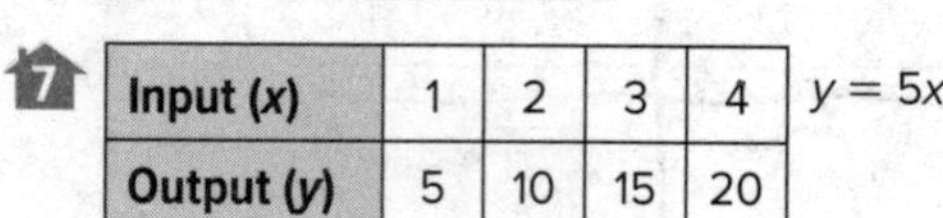

Input (x)	1	2	3	4
Output (y)	5	10	15	20

$y = 5x$

9. Sample answer: Ray is saving \$7 per week to buy a new DVD player. The variable y represents the total amount he has saved. The variable x represents the number of weeks.
11. Sample answer:

$y = x + 3$			
Input (x)	1	2	3
Output (y)	4	5	6

Inverse of $y = x + 3$			
Input (x)	4	5	6
Output (y)	1	2	3

$y = x - 3$

Pages 601–602 Lesson 8-3 Extra Practice

13. $y = 10x$
15.

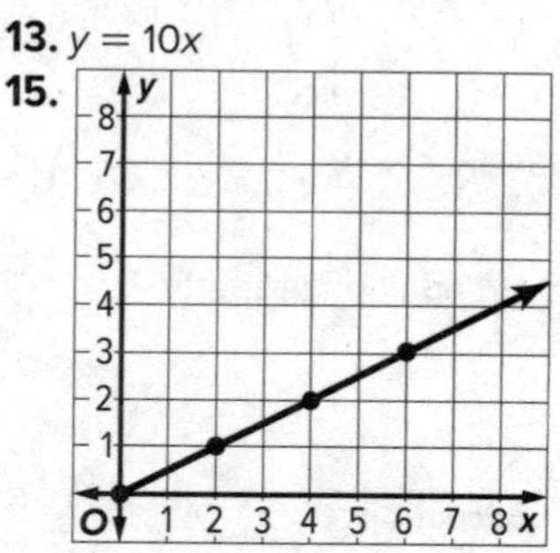

17.

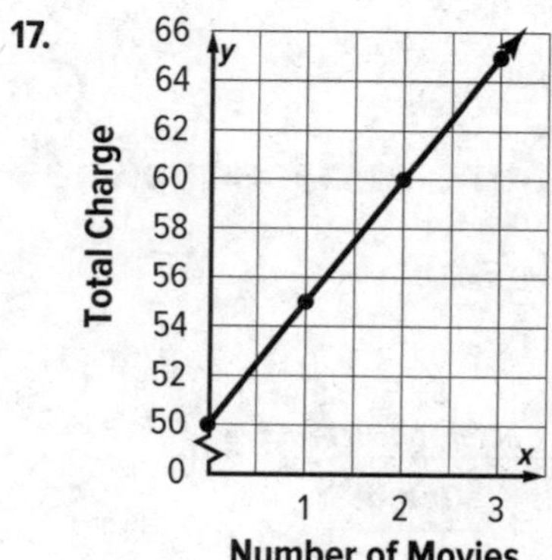

19a. True **19b.** True **19c.** False

21–27. 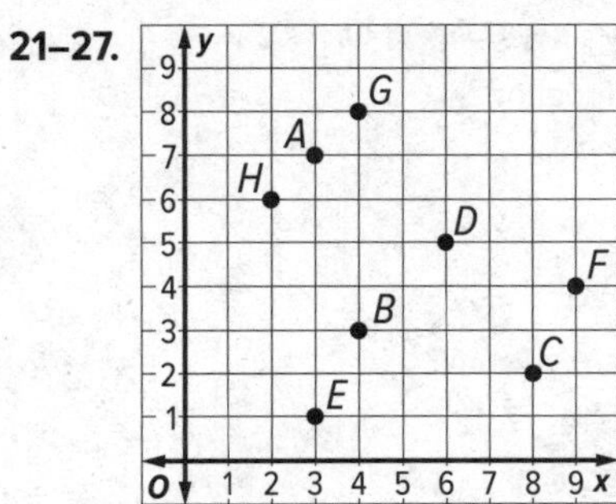

29.

Day	Time Studied (min)
Monday	20
Tuesday	45
Wednesday	30
Thursday	45

2 hours and 20 minutes

Pages 607–608 Lesson 8-4 Independent Practice

1 **a.** $v = 400d$

b.

Number of Days, *d*	1	2	3
Pounds Eaten, *v*	400	800	1200

c. 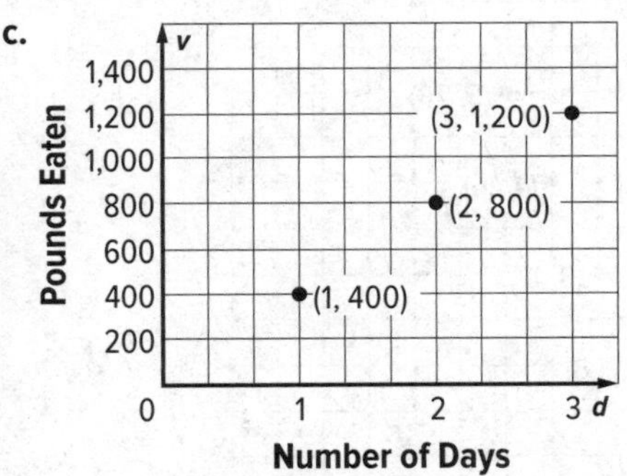

The graph is a line because with each day the amount of vegetation increases by 400.

3 **a.** $t = 3 + 1.75c$; where *t* represents the total earned and *c* represents the number of chores

b.

Number of Chores, *c*	1	2	3
Total Earned ($), *t*	4.75	6.50	8.25

c. 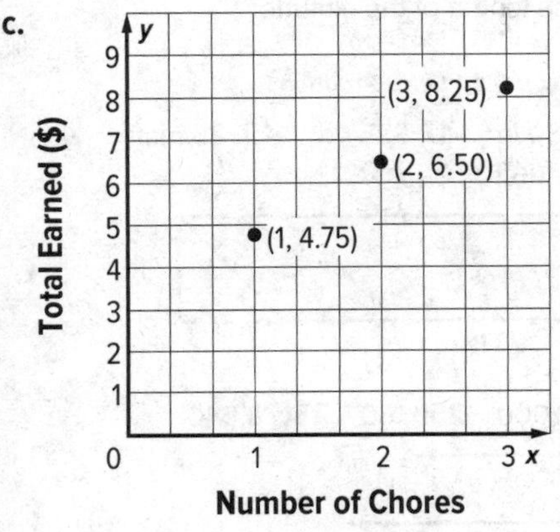

d. $11.75 **e.** The independent variable is the number of chores and the dependent variable is the total earned. **5.** no; the graphs of the lines will never meet other than at zero hours. **7.** $c = 25 + 2m$

Pages 609–610 Lesson 8-4 Extra Practice

9. Music Man: $t = 45n$; Road Tunes: $t = 35n$; where *t* represents the total cost and *n* represents the number of hours **11.** $m = 0.18b + 4$; $10.30 **13.** < **15.** < **17.** > **19.** Wednesday

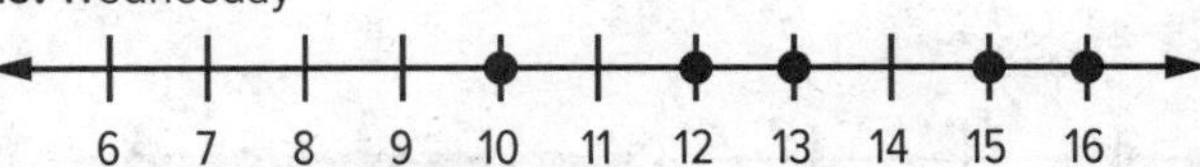

Page 613 Problem-Solving Investigation Make a Table

Case 3. 35 cubes **Case 5.** 50 and 36

Pages 621–622 Lesson 8-5 Independent Practice

1 5 **3.** yes **5.** stand up or suspended **7** Jan. and Feb.; $0.75 **9.** Sample answer: 0, 1, and 2 **11.** $a > c$; Sample answer: if $a > b$, then it is to the right of *b* on the number line. If $b > c$, then it is to the right of *c* on the number line. Therefore, *a* is to the right of *c* on the number line. **13a.** 5 and 6 **13b.** −3, −2, and −1 **13c.** 4 **13d.** none

Pages 623–624 Lesson 8-5 Extra Practice

15. 0 **17.** no **19.** Carmen, Eliot, and Ryan **21.** Jupiter; Saturn **23.** 5 + 3 **25.** 5 × 8 **27.** 6; 4

Pages 629–630 Lesson 8-6 Independent Practice

1. $p \leq 35$ **3** $p < 437$

5. 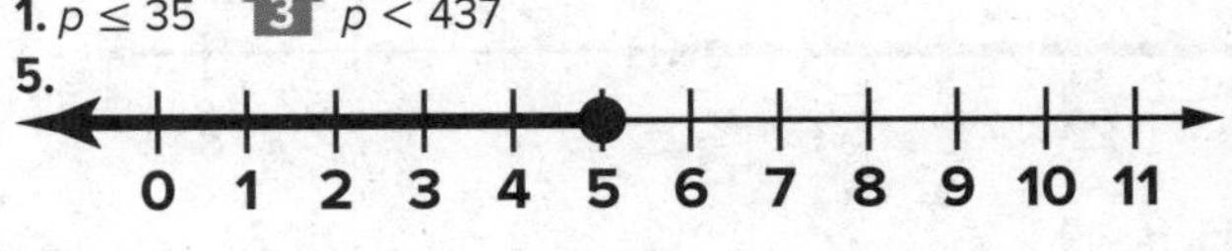

17 $s < 20$

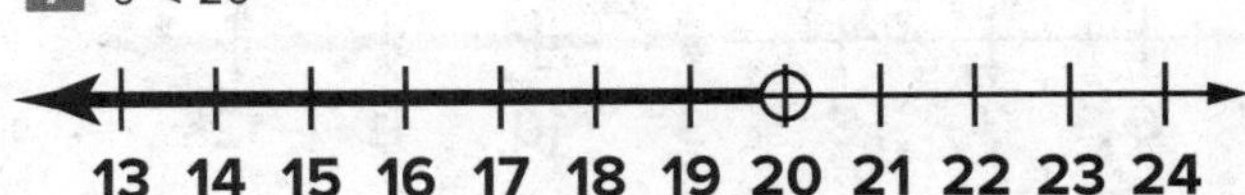

9. She used the incorrect symbol. "at least" means the values will be larger than 10, but include 10; $c \geq 10$ **11.** Sample answer: When an inequality uses the greater than or less than symbols, it does not include the number given. So, $x > 5$ and $x < 7$ do not include 5 or 7 respectively. When the greater than or equal to and less than or equal to symbols are used, the given numbers are included. So, $x \geq 5$ and $x \leq 7$ include 5 and 7, respectively.

Pages 631–632 Lesson 8-6 Extra Practice

13. $s \leq 50$ **15.** $h > 200$

17. **19.** $t < 4$

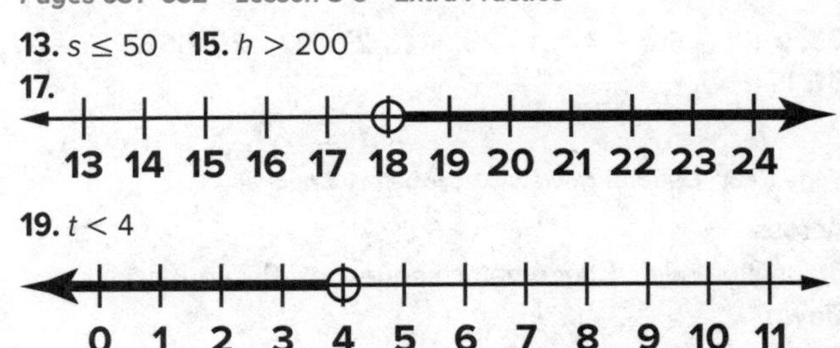

21a. True **21b.** False **21c.** True **23.** 5 **25.** 8 **27.** 5

29.

Pages 639–640 Lesson 8-7 Independent Practice

1. $y \leq 1$

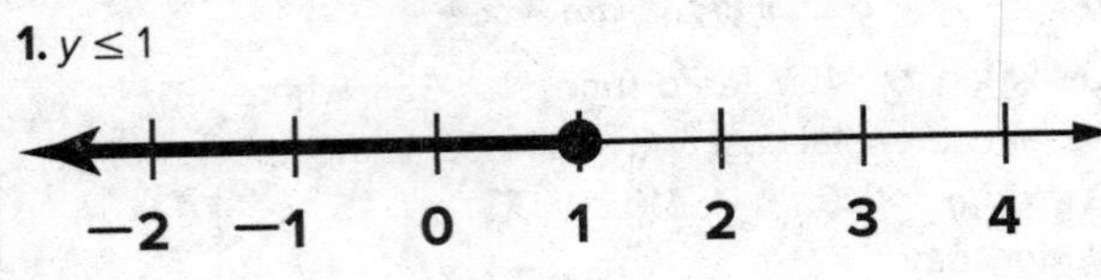

3 $x > 8$

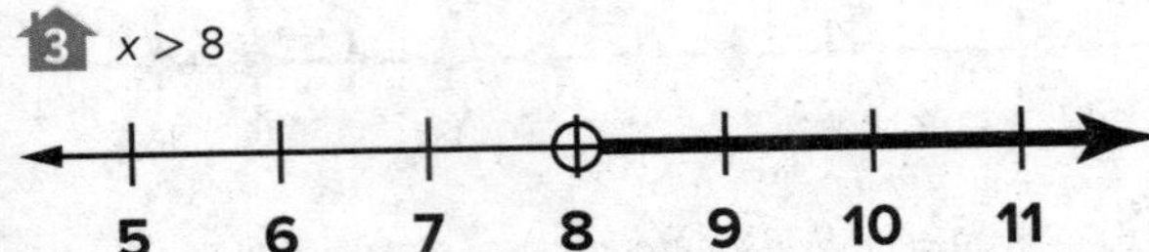

5 $0.1x \leq 5.00$; $x \leq 50$; The maximum is 50 letters

7 $p > \frac{53}{60}$

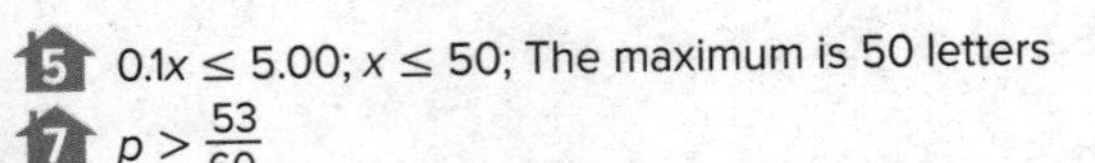

9. Sample answer: An airplane can hold 53 passengers and there are currently 32 passengers on board. How many more passengers can board the airplane? **11.** yes; Sample answer: $x > 5$ is not the same relationship as $5 > x$. However, $x > 5$ is the same relationship as $5 < x$.

Pages 641–642 Lesson 8-7 Extra Practice

13. $a < 5$

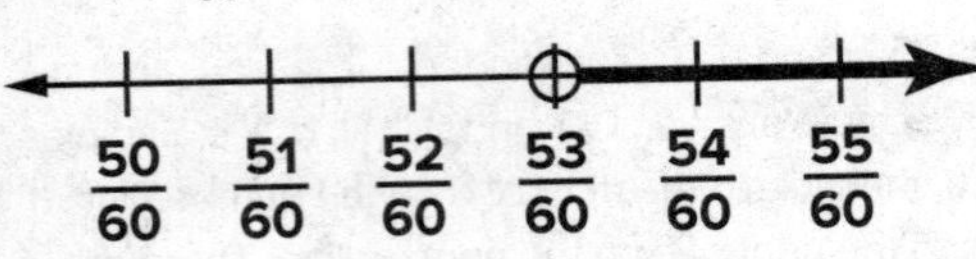

15. $d \geq 9$

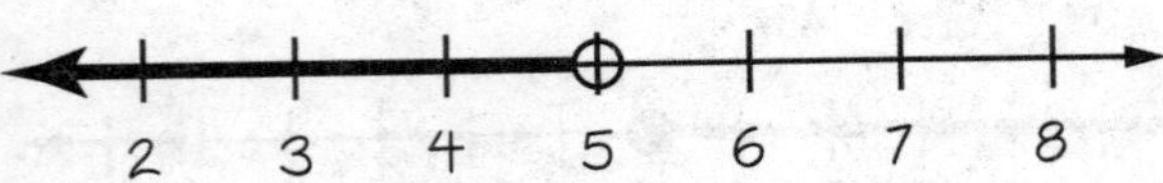

17. $g < 12$

19. $25b \geq 5{,}000$; $b \geq 200$; They need to sell a minimum of 200 backpacks

21. $n \geq \frac{3}{14}$

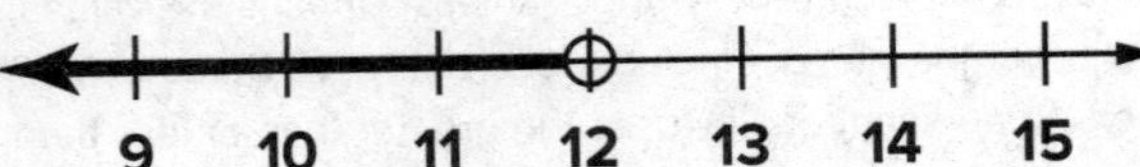

23. $y + 1 > 6$; $z - 4 > 1$ **25.** 144 **27.** 192 **29.** 66

31. 15 ft^2

Page 645 Chapter Review Vocabulary Check

Across

3. function rule **5.** geometric sequence **9.** sequence

Down

1. arithmetic sequence **7.** inequality

Page 646 Chapter Review Key Concept Check

1. 24 **3.** geometric **5.** function

Chapter 9 Area

Page 656 Chapter 9 Are You Ready?

1. 32 cm^2 **3.** 18 cm^2 **5.** 14 **7.** 12

Pages 665–666 Lesson 9-1 Independent Practice

1. 9 $units^2$ **3** 72 cm^2 **5.** $166\frac{1}{2}$ ft^2

7 no; In order for the area of the first floor to be 20,000 ft^2 and the base 250 feet, the height must be 20,000 ÷ 250 or 80 feet.

9a. Sample answers are given.

Base (cm)	Height (cm)	Area (cm^2)
1	4	4
2	4	8
3	4	12
4	4	16
5	4	20

9b.

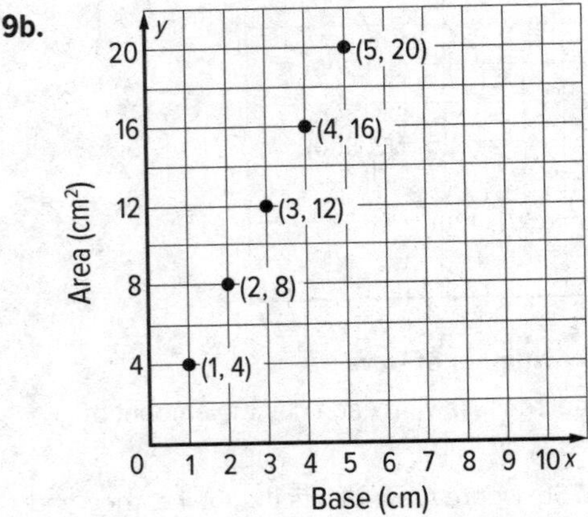

9c. It appears to form a line. **11.** Sample answer: Both parallelograms and rectangles have bases and heights. So, the formula $A = bh$ can be used for both figures. The height of a rectangle is the length of one of its sides while the height of a parallelogram is the length of the altitude.

Pages 667–668 Lesson 9-1 Extra Practice

13. 20 $units^2$ **15.** 180 in^2 **17.** 325 yd^2 **19.** 25 mm

21. Sample answer: 196 ft^2

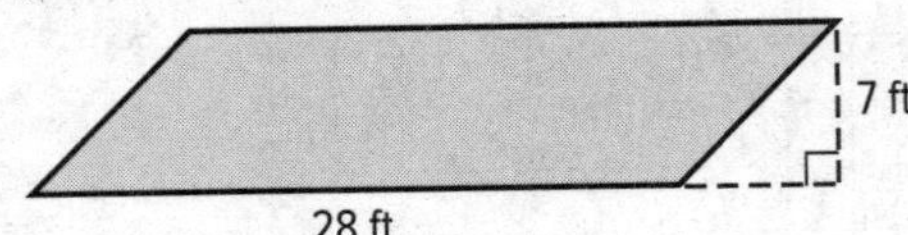

23. 84 cm^2 **25a.** 9,000 **25b.** 60 **25c.** 8,940

27.

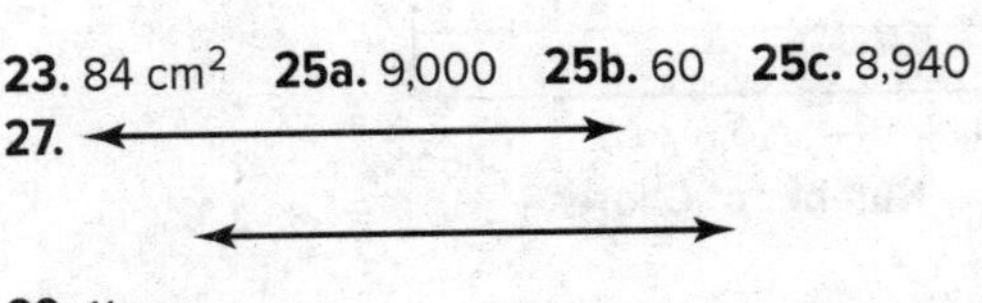

29. 11 songs

Pages 677–678 Lesson 9-2 Independent Practice

1. 24 units^2 **3.** 747 ft^2 **5.** 19 cm

7. a. $\frac{5x}{2}$ **b.**

c. The points appear to form a line.

9. The formula is $\frac{1}{2}bh$, not bh.

$$100 = \frac{b \cdot 20}{2}$$
$$b = 10 \text{ m}$$

11. Sample answer:

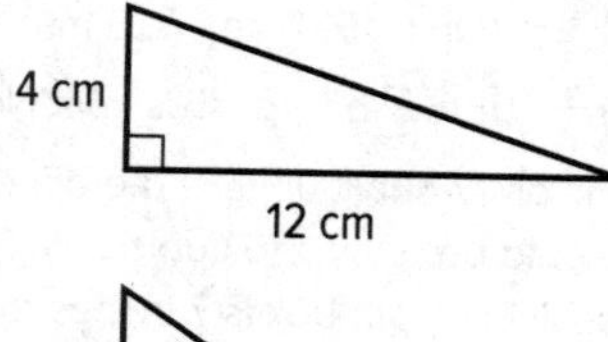

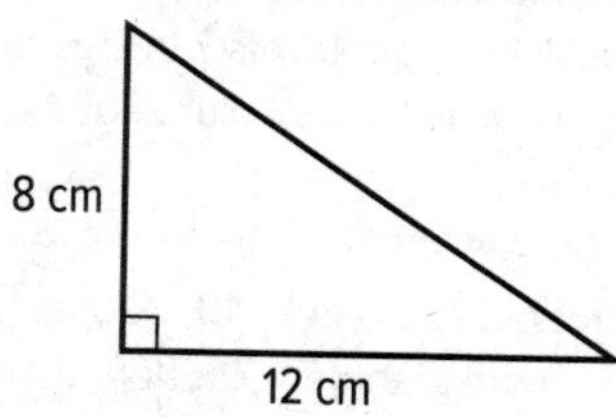

Area of first triangle is 24 cm^2; Area of second triangle is 48 cm^2; 1:2 or $\frac{1}{2}$.

Pages 679–680 Lesson 9-2 Extra Practice

13. $7\frac{1}{2}$ units^2 **15.** 87.5 m^2 **17.** 21 m **19.** 47.3 cm

21a. 27 ft^2 **21b.** 3 bags **23.** $\frac{7x}{2}$; Sample answer: The area is the product of the height (7), the base (x), and $\frac{1}{2}$, or $\frac{7x}{2}$.

25. trapezoid **27.** rhombus **29.** 3 lines

Pages 689–690 Lesson 9-3 Independent Practice

1. 168 yd^2 **3.** 112 m^2 **5.** 16 mm **7. a.** 7,000 ft^2
b. 4 bags
9. Sample answer:

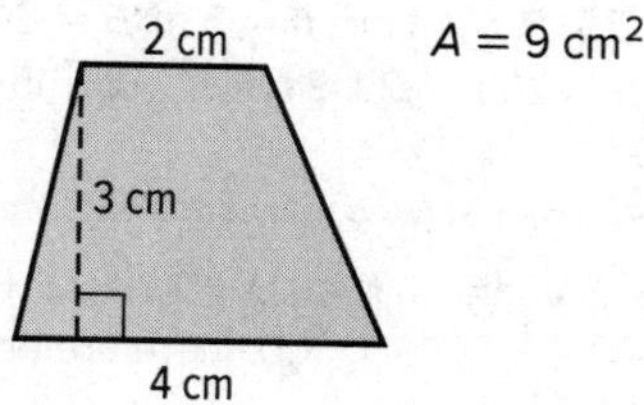

$A = 9$ cm^2

11. Sample answer: The lengths of the bases can be rounded to 20 m and 30 m, respectively. The area can be rounded to 250 m^2. Divide 250 by (20 + 30) or 50, and then multiply by 2. The height h is about 10 m. **13.** Sample answer: By knowing the formula for the area of a parallelogram is $A = bh$, I can draw two congruent trapezoids and rotate one so they create a parallelogram. After multiplying the base and height, I can divide by 2 to find the area of the trapezoid.

Pages 691–692 Lesson 9-3 Extra Practice

15. 121 cm^2 **17.** 187.6 ft^2 **19.** 3 miles **21.** 1,904 in^2
23. 100 cm^2 **25a.** True **25b.** False **27.** 256 **29.** 24 in.

Page 695 Problem-Solving Investigation Draw a Diagram

Case 3. 25 balloons **Case 5.** 272 customers

Pages 701–702 Lesson 9-4 Independent Practice

1. The perimeter is 4 times greater. The perimeter of the original figure is 36 cm and the perimeter of the new figure is 144 cm; 144 cm ÷ 36 cm = 4. **3.** The area is multiplied by $\frac{1}{3} \cdot \frac{1}{3}$ or $\frac{1}{9}$ the original area. The area of the original figure is 315 yd^2 and the area of the new figure is 35 yd^2; 35 yd^2 ÷ 315 yd^2 = $\frac{1}{9}$. **5.** Use the area and the length to find the width of the queen-size bed. The width of the bed is 4,800 ÷ 80, or 60 inches. So, the width of the dollhouse bed is $60 \cdot \frac{1}{12}$, or 5 inches. The length of the dollhouse bed is $80 \cdot \frac{1}{12}$ or $6\frac{2}{3}$ inches. **7.** Sample answer:

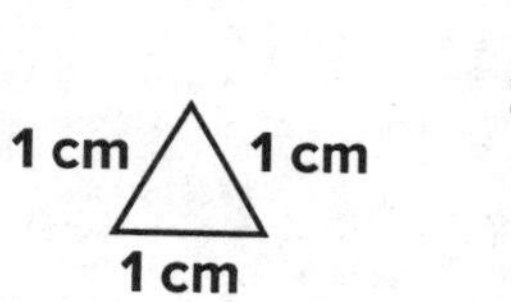

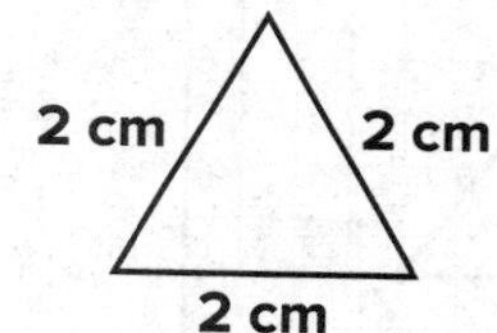

9. larger square: 12 units; smaller square: 6 units; Sample answer: The length of the sides for squares are equal. Divide 48 by 4 to get a side length of 12. The side length of the smaller square is half as big, so 6 units.

Pages 703–704 Lesson 9-4 Extra Practice

11. The perimeter is 6 times greater. The perimeter of the original figure is 30 ft and the perimeter of the new figure is 180 ft; 180 ft ÷ 30 ft = 6. **13.** The perimeter is $\frac{1}{4}$ the original perimeter. The perimeter of the original figure is 80 m and the perimeter of the new figure is 20 m; $\frac{1}{4} \cdot 80$ m = 20 m. The area is $\frac{1}{4} \cdot \frac{1}{4}$ or $\frac{1}{16}$ the original area. The area of the original figure is 240 m^2 and the area of the new figure is 15 m^2; 15 m^2 ÷ = $\frac{1}{16}$. **15a.** 4 **15b.** 4 **15c.** 25

17.
19.

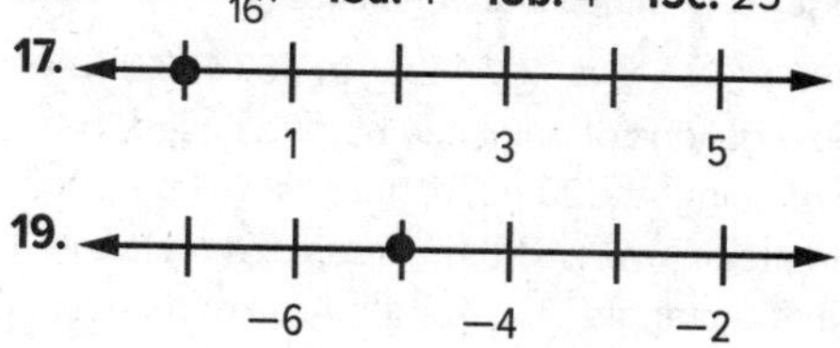

21. 15 yd; 10 yd

Pages 709–710 Lesson 9-5 Independent Practice

1. $DE = 5$ units, $EF = 3$ units, $FG = 5$ units, $GD = 3$ units; 16 units **3.** 120 cm **5.** 28 square units

7. rectangle; 45 units^2

11. Sample answer: Subtract the x-coordinates of the points with the same y-coordinates to find the length of 2 of the sides and then subtract the y-coordinates of the points with the same x-coordinates to find the length of the other 2 sides. Then find the sum of all 4 sides to find the perimeter.

Pages 711–712 Lesson 9-5 Extra Practice

13. $AB = 2$ units, $BC = 3$ units, $CD = 2$ units, $DA = 3$ units; 10 units **15.** 54 feet **17.** 24 square units

19. right triangle; 6 units^2

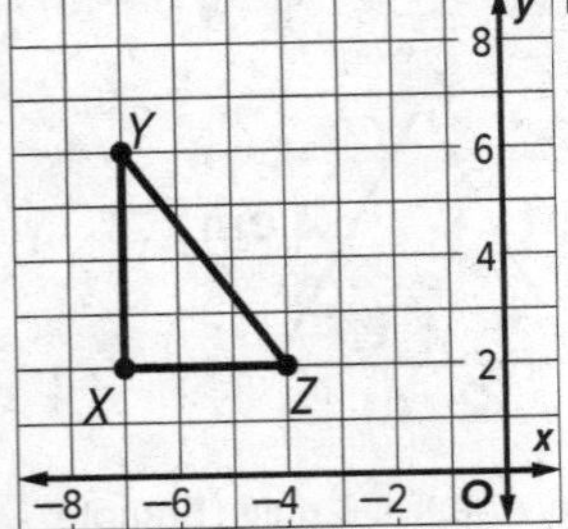

21. isosceles triangle; $A = 12$ units^2

23. one set of parallel sides; four vertices; two acute angles **25.** No sides are congruent. One pair of opposite sides is parallel. **27.** rectangle

Pages 721–722 Lesson 9-6 Independent Practice

1. 58.6 in^2 **3.** 189 ft^2 **5. a.** 467.4 ft^2 **b.** $467.4 \div 350 \approx 1.34$; Since only whole gallons of paint can be purchased, you will need 2 gallons of paint. At \$20 each, the cost will be $2 \times \$20$ or \$40. **7.** Sample answer: Add the areas of a rectangle and a triangle. Area of rectangle: $3 \times 4 = 12$; Area of triangle: $\frac{1}{2} \times 3 \times 3 = 4.5$; $12 + 4.5 = 16.5$. So, an approximate area is $16.5 \times 2{,}400$ or 39,600 mi^2. **9.** The area is multiplied by 4. Original area: 159.9 cm^2; new area: 639.6 cm^2

Pages 723–724 Lesson 9-6 Extra Practice

11. 66.2 m^2 **13.** 10,932 ft^2 **15a.** False **15b.** True **15c.** False **17.** 432 **19.** 14,400 **21.** 864 Calories

Page 727 Chapter Review Vocabulary Check

1. polygon **3.** parallelogram **5.** rhombus **7.** composite figure

Page 728 Chapter Review Key Concept Check

1. $A = \frac{1}{2}h(b_1 + b_2)$ **3.** $A = \frac{1}{2}(9.8)(7 + 12)$ **5.** $A = 93.1$

Chapter 10 Volume and Surface Area

Page 734 Chapter 10 Are You Ready?

1. 214.5 **3.** 172.8 **5.** 44 **7.** 101

Pages 743–744 Lesson 10-1 Independent Practice

1. 132 m^3 **3.** 171 in^3 **5.** 17 m **7.** 3 mm

9. a. $50\frac{5}{8}$ in^3 **b.** $16\frac{7}{8}$ in^3 **c.** 75% **11.** No; the volume of the figure is 3^3 or 27 cubic units. If the dimensions doubled, the volume would be 6^3 or 216 cubic units, eight times greater. **13.** Sample answer: A gift box is 7 inches long, 9 inches wide, and 4 inches tall. What is the volume of the gift box?; 252 in^2

Pages 745–746 Lesson 10-1 Extra Practice

15. 1,430 ft^3 **17.** 2,702.5 in^3 **19.** 360 mi^3 **21a.** 2,520; 14; 9 **21b.** 20 **23.** acute triangle **25.** right triangle **27.** isosceles; Sample answer: The triangle has two congruent sides.

Pages 751–752 Lesson 10-2 Independent Practice

1. 336 m^3 **3.** 104.0 cm^3 **5.** 108 in^3 **7.** 8 in. **9.** 10 yd **11.** To find the base area, Amanda should have multiplied by $\frac{1}{2}$. The base area of the prism is 6 cm^2, not 12 cm^2. So, the volume of the prism is 42 cm^3. **13.** The rectangular prism will hold more mints than the triangular prism. The rectangular prism has a volume of 144 in^3 while the triangular prism has a volume of 72 in^3.

Pages 753–754 Lesson 10-2 Extra Practice

15. 346.5 ft^3 **17.** 380 in^3 **19.** 10,395 in^3 **21.** 15 m **23.** 48 ft^3 **25.** $B = 48$ m^2, $h = 5$ m; $B = 24$ m^2, $h = 10$ m; $B = 12$ m^2, $h = 20$ m **27.** 9 units^2 **29.** 15 units^2

Page 757 Problem-Solving Investigation Make a Model

Case 3. yes; Sample answer: $8 + 10 + 12 + 14 + 16 + 18 + 20 = 98$; Since $98 < 100$, there are enough chairs. **Case 5.** 16 boxes

Pages 767–768 Lesson 10-3 Independent Practice

1. 2,352 yd^2 **3.** 3,668.94 m^2 **5.** 1,162 cm^2

7. Package A: 492 in^2; Package B: 404 in^2; Package A has a greater surface area. No, the volume of Package B is greater. **9.** 48 in^2; 144 in^2

Pages 769–770 Lesson 10-3 Extra Practice

13. 324 m^2 **15.** 384.62 cm^2 **17a.** 316.5 in^2 **17b.** 534 in^2 **17c.** 207.75 in^2 **19.** the amount of wrapping paper needed to cover a box; the amount of paint needed to cover a statue **21.** 218 **23.** equilateral; Sample answer: All three sides measure 15 inches.

Pages 777–778 Lesson 10-4 Independent Practice

1. 1,152 yd^2 **3** 13.6 m^2 **5** about 21.4 yd^2 **7.** 279.2 in^2 **9.** 7.5 in. **11.** Sample answer: Prism A with bases that are right triangles that measure 3 by 4 by 5 and with a height of 1. Prism B with bases that are right triangles that measure 1 by 1 by 1.4 and with a height of 10. Prism A has a greater volume while Prism B has a greater surface area.

Pages 779–780 Lesson 10-4 Extra Practice

13. 537 ft^2 **15.** 70.8 in^2 **17.** 282.7 cm^2 **19.** 428.1 cm^2 **21a.** False **21b.** False **21c.** True **23.** obtuse **25.** right

Pages 787–788 Lesson 10-5 Independent Practice

1. 24 m^2 **3** 126.35 cm^2 **5.** 143.1 mm^2 **7** 52 cm^2 **9.** 132 in^2 **11.** 110 ft^2; Sample answer: A pyramid has only one square base. To find the surface area, add $25 + (4 \cdot 21.25)$. **13.** It would be shorter to climb up the slant height. The bottom of the slant height is closer to the center of the base of the pyramid than the bottom of the lateral edge.

Pages 789–790 Lesson 10-5 Extra Practice

15. 223.5 ft^2 **17.** 383.25 cm^2 **19.** 923 in^2 **21.** 14 in. **23a.** False **23b.** True **23c.** True **23d.** True **25.** 100 **27.** $8.25

Page 793 Chapter Review Vocabulary Check

1. three-dimensional figure **3.** volume **5.** rectangular prism **7.** vertex **9.** lateral face

Page 794 Chapter Review Key Concept Check

Across
1. 480.4 **5.** 8
Down
1. 40 **3.** 520

Chapter 11 Statistical Measures

Page 804 Chapter 11 Are You Ready?

1. 68.75 **3.** $21.60 **5.** 24.20 **7.** 115.2 miles

Pages 813–814 Lesson 11-1 Independent Practice

1 88% **3** $25 **5.** 88 **7.** Sample answer: pages read: 27, 38, 26, 39, 40 **9.** 0.17; Sample answer: The sum of the scores for the 99 students must be 82×99 or 8,118. Adding the score of 99, the sum of the 100 students is 8,217. The new mean is then 82.17. The mean increased by $82.17 - 82$ or 0.17.

Pages 815–816 Lesson 11-1 Extra Practice

11. 56 in. **13.** 26 tickets **15.** $80; Sample answer: Multiply 59 by 6 and subtract the amounts given in the table. **17.** > **19.** < **21.** < **23a.** 399 miles **23b.** Charlotte

Pages 821–822 Lesson 11-2 Independent Practice

1 89; none; There is no mode to compare. **3.** The values are close. The median and mode are equal, 44 mph, and the mean of about 45.6 mph, is slightly more. The data follows the measures of center in the way that they are close to the measure of center. **5** Mode; The mode of the temperatures in Louisville is 70° and the mode for Lexington's temperatures is 76°. Since 76° − 70° = 6°, the mode was used to make this claim. **7.** $21 **9.** Sample answer: The median or mode best represents the data. The mean, 8, is greater than all but one of the data values.

Pages 823–824 Lesson 11-2 Extra Practice

11. median: 23; mode: 44; The mode is 21 years more than the median. **13.** median: 12.5; mode: none; There is no mode to compare. **15.** Sample answers are given.

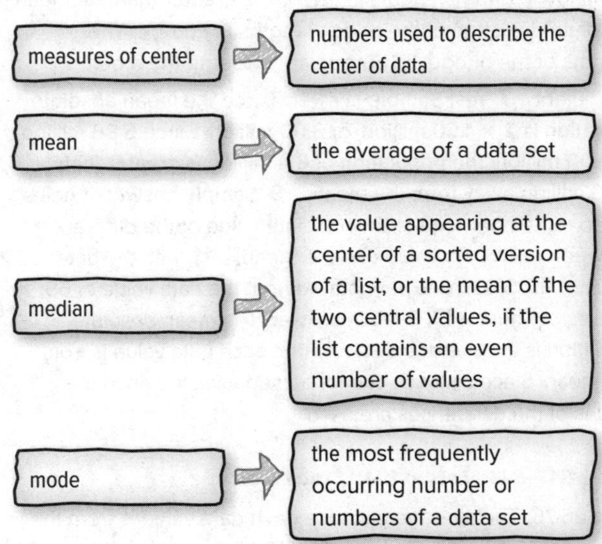

17a. False **17b.** True **17c.** True **19.** 58 **21.** 56 **23.** 52 **25.** 36 miles

Page 827 Problem-Solving Investigation Use Logical Reasoning

Case 3. 42 customers **Case 5.** 6 students; 10 students

Pages 833–834 Lesson 11-3 Independent Practice

1 **a.** 1,028 **b.** 923.5; 513; 1,038 **c.** 525 **d.** none
3 median: 357.5, Q_1: 298, Q_3: 422, IQR: 124
5. range: 63, median: 7.5, Q_3: 30.5, Q_1: 0.5, IQR: 30; Sample answer: The number of moons for each planet varies greatly. The IQR and range are both large. **7.** Sample answer: The median is correct, but Hiroshi included it when finding the third and first quartiles. The first quartile is 96, the third quartile is 148, and the interquartile range is 52. **9.** Sample answer: The third quartile is the median of the upper half of the data and the first quartile is the median of the lower half of the data.

11. Set A-range: 20; IQR: 4. Set B-range: 20; IQR: 1. Sample answer: The IQR tells more information, specifically that the middle half of the data in Set B are closer together than the middle half of the data in Set A.

Pages 835–836 Lesson 11-3 Extra Practice

13a. NFC **13b.** NFC—median: 86, Q_3: 113, Q_1: 68, IQR: 45; AFC—median: 80, Q_3: 94, Q_1: 76, IQR: 18 **13c.** Sample answer: The AFC had a median of 80 penalties and the NFC had a median of 86 penalties. The AFC had an IQR of 18 penalties while the NFC had an IQR of 45 penalties. The ranges were 47 penalties for the AFC and 78 penalties for the NFC. **15.** Half of the players won more than 10.5 games and half won less than 10.5 games; The range of the data is 13 games; There are no outliers. **17.** 32 **19.** 19 **21.** 19.5 **23.** 167 miles

Pages 841–842 Lesson 11-4 Independent Practice

1 17.88 moons; Sample answer: The average distance each data value is from the mean is 17.88 moons. **3.** United States: 9.77 km; Europe: 2.87 km; Sample answer: The mean absolute deviation in bridge lengths in the U.S. is greater than the mean absolute deviation of the bridge lengths in Europe. The lengths of the bridges in Europe are closer to the mean. **5** eight **7.** yes; Sample answer: Twice the mean absolute deviation is 2 × 1.50 million, or 3.00 million. Since 5.86 million > 3.00 million, the population of 8.4 million is greater than 3.00 million away from the mean. **9.** Sample answer: It helps me to remember to take the absolute value of the difference between each data value and the mean. **11.** with the data value of 55: 5.33 miles per hour; without the data value of 55: 2 miles per hour **13.** Sample answer: The mean absolute deviation is the average distance that each data value is from the mean. Since distance cannot be negative, the absolute values of the differences are used.

Pages 843–844 Lesson 11-4 Extra Practice

15. $26.76; The average distance each data value is from the mean is $26.76. **17.** Sixth grade: $10.67; Seventh grade: $16.67; Sample answer: The mean absolute deviation of the money raised by sixth grade homerooms is less than the mean absolute deviation of the money raised by seventh grade classrooms. The amounts of the money raised by the sixth grade homerooms are closer to the mean. **19.** It describes the variation of the data around the mean; It describes the average distance between each data value and the mean **21.** 235 cones

Pages 849–850 Lesson 11-5 Independent Practice

1 The mean best represents the data. There are no extreme values. mean: 56.4 minutes **3** **a.** 1,148 **b.** With the outlier, the mean is 216.83 ft, the median is 33.5 ft, there is no mode, and the range is 1,138 ft. Without the outlier, the mean is 30.6 ft, the median is 24 ft, there is no mode, and the range is 52 ft. **c.** With the outlier, the best measure is the median depth; without the outlier, the best measure is the mean. **5.** Pilar did not include the outlier. The mean is 20. The median, which is 15.5, best describes the data because the outlier affects the mean more than it affects the median. **7.** Sample answer: 125, 32, and 19

Pages 851–852 Lesson 11-5 Extra Practice

9. Since the set of data has no extreme values or numbers that are identical, the mean or median, 6 songs, would best represent the data. **11a.** 62° **11b.** With the outlier, the mean is 32.71°, the median is 29°, the mode is 29° and the range is 37°. Without the outlier, the mean is 27.83°, the median is 28.5°, the mode is 29°, and the range is 4°. **11c.** Sample answer: With the outlier, the best measure is the mode; without the outlier, the best measure is the mode; the outlier does not affect the mode, but affects the mean and median. **13a.** median **13b.** mean **13c.** mode **15.** 260 **17.** 154 **19.** 203 tickets

Page 855 Chapter Review Vocabulary Check

1. mode **3.** range **5.** interquartile range

Page 856 Chapter Review Key Concept Check

Across

1. 505 **3.** 249 **5.** 138 **9.** 96 **11.** 8312

Down

1. 53 **3.** 281 **7.** 691 **11.** 83

Chapter 12 Statistical Displays

Page 862 Chapter 12 Are You Ready?

1. 16 **3.** 27 **5.** 57 **7.** 84.5

Pages 867–868 Lesson 12-1 Independent Practice

1

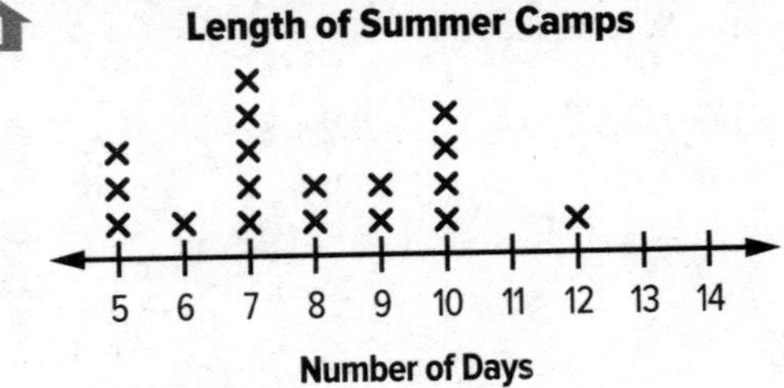

median: 7.5; mode: 7; range: 7; no outlier; There are a total of 18 summer camps represented. The median means that one half of the summer camps are longer than 7.5 days and one half are less. More camps are 7 days than any other number of days. **3** Sample answer: There are 15 play lists represented. mean: 40; median: 40; modes: 40 and 42; So, the majority of the data is close to the measures of center. Q_1: 38; Q_3: 42; IQR: 4, which means half the playlists have between 38 and 42 songs; there is an outlier at 25. **5.** 11 **7.** The outlier of the data set is 29°F, not 20°F. **9.** 24 cm **11.** mode; Sample answer: With the four values, the mean is 61.35, the median is 62, and the mode is 56. Without the four values, the mean is 63.5, the median is 63.5, and the modes are 62, 65, and 68. Not including the four values changes the mode more drastically.

Pages 869–870 Lesson 12-1 Extra Practice

13.

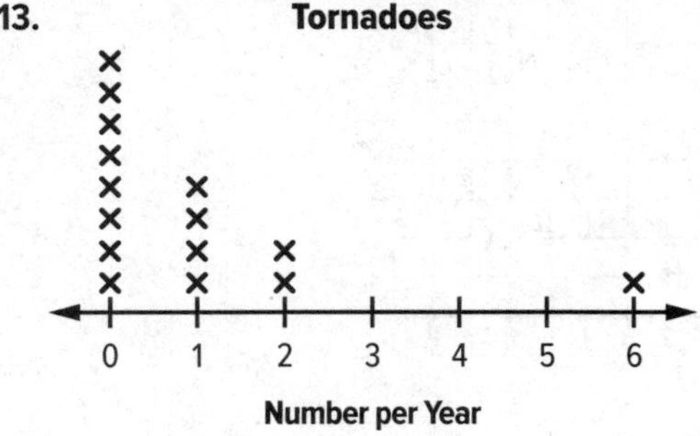

median: 0; mode: 0; range: 6; outlier: 6 There were 15 tornadoes represented. The median means that half the number of tornadoes was greater than zero and half the number of tornadoes was zero.
15. Sample answer: The median, range, and outliers do not exist because the data are not numerical. The mode is pepperoni, because more students prefer pepperoni than any other topping. The plot shows responses for 10 people. There are five different toppings. Two topping preferences were chosen by only one person. **17a.** True **17b.** True **17c.** False **19.** < **21.** < **23.** < **25.** 4

Pages 875–876 Lesson 12-2 Independent Practice

1. Sample answer: 24 cyclists participated. No one finished with a time lower than 60 minutes. **3** 60–64 minutes
5.

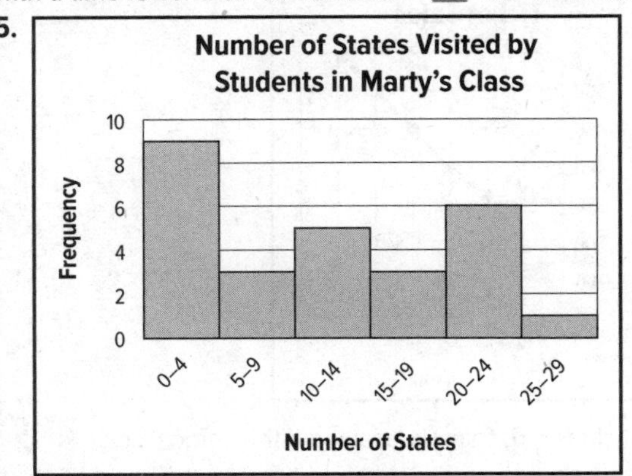

7 6th grade **9.** Sample answer: ages of students at summer camp: 3, 4, 5, 7, 7, 8, 8, 10, 10, 11, 13, 14, 15, 15 **11.** Sample answer: One set of intervals would be from 0 to 45, with intervals of 5. Another set would be from 0 to 50 with intervals of 10. If smaller intervals are used, less data values will be in each interval, therefore making the bars of the histogram shorter.

Pages 877–878 Lesson 12-2 Extra Practice

13. 24–27 yrs **15.** 17
17.

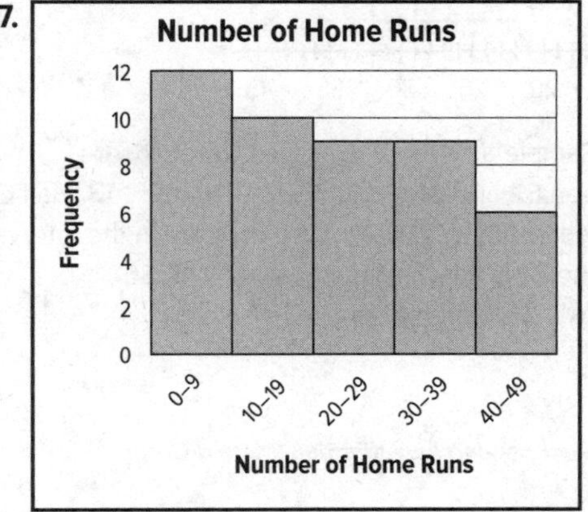

19. Sample answer: There were no players who scored between 30 and 44 goals in their career. **21.** 42 **23.** 27 **25.** 97.5 **27.** Lucinda

Pages 883–884 Lesson 12-3 Independent Practice

1

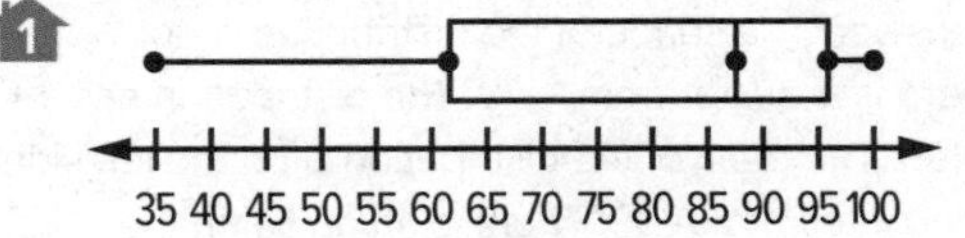

3 **a.**

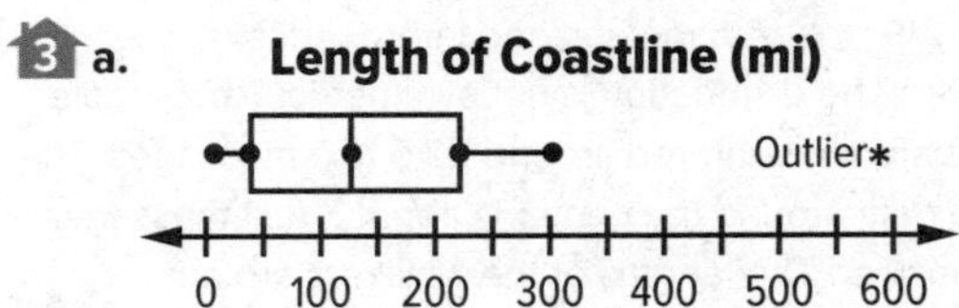

3b. 127 mi **c.** Sample answer: The length of the box plot shows that the number of miles of coastline for the top 25% of states varies greatly. The number of miles of coastline for the bottom 25% of states is concentrated.
5a.

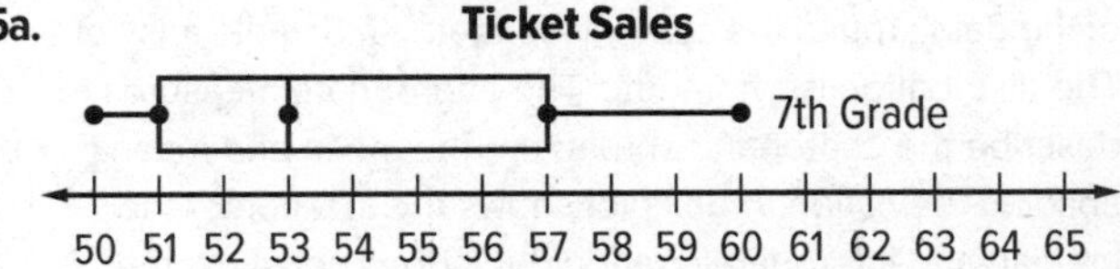

5b. Grade 6; Sample answer: The median, upper extreme, and first and third quartiles are higher for the grade 6 data.
7. Sample answer: {28, 30, 52, 68, 90, 92}

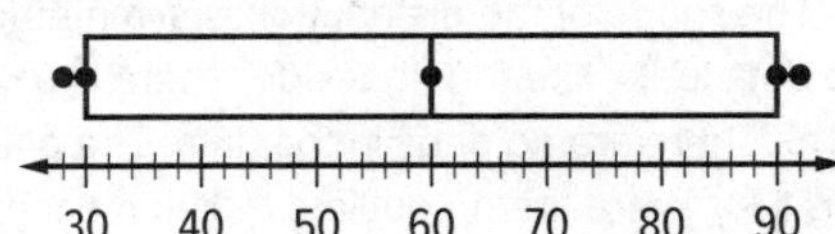

Selected Answers

Pages 885–886 Lesson 12-3 Extra Practice

9.

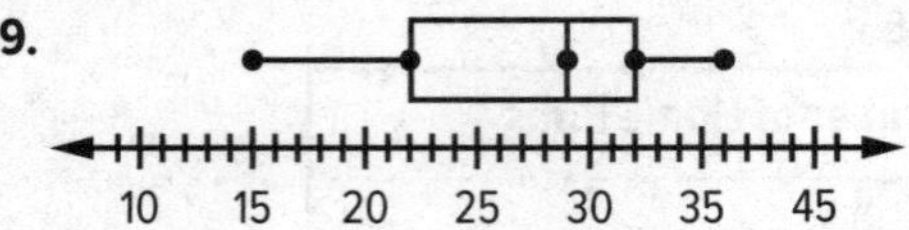

11a. 96 **11b.** Sample answer: The scores were closer together between 82 and 86. **11c.** 75% **11d.** 82 **13.** Half of the data are greater than 62; Half of the data are in the interval 62–74; The value 74 is the maximum value. **15.** 36 **17.** 162 **19.** 376 **21.** 3 members

Page 889 Problem-Solving Investigation Use a Graph

Case 3. 5 lawns
Case 5. 91

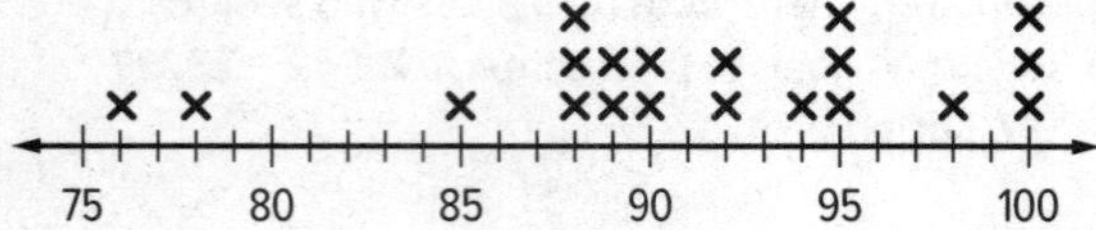

Pages 895–896 Lesson 12-4 Independent Practice

1 Sample answer: The shape of the distribution is not symmetric. There is a cluster from 1–79. The distribution has a gap from 80–199. The peak of the distribution is on the left side of the data in the interval 20–39. There is an outlier in the interval 200–219. **3** **a.** median and interquartile range; Sample answer: The distribution is not symmetric. **b.** Sample answer: The data are centered around 23.5 text messages. The spread of the data around the center is about 3 text messages. **5a.** Sample answer: The lengths of the whiskers are not equal. **5b.** skewed left; Sample answer: The data are more spread out on the left side due to the long left whisker. **5c.** Sample answer: Use the median and interquartile range to describe the center and spread since the distribution is not symmetric. The data are centered around 40 feet. The spread of the data around the center is 10 feet. **7.** Sample answer: The distribution is symmetric. The appropriate measures to describe the center and spread are the mean and mean absolute deviation. A box plot shows the location of the median and interquartile range but it does not show the location of the mean or the mean absolute deviation.

Pages 897–898 Lesson 12-4 Extra Practice

9. Sample answer: The shape of the distribution is symmetric. The left side of the data looks like the right side. There is a cluster from \$13–\$15. There are no gaps in the data. The peak of the distribution is \$14. There are no outliers. **11a.** mean and mean absolute deviation; Sample answer: The distribution is symmetric and there are no outliers. **11b.** Sample answer: The data are centered around 31 miles. The spread of the data around the center is about 1.3 miles. **13.** The distribution has an outlier; The distribution has a gap of data.

15–21.

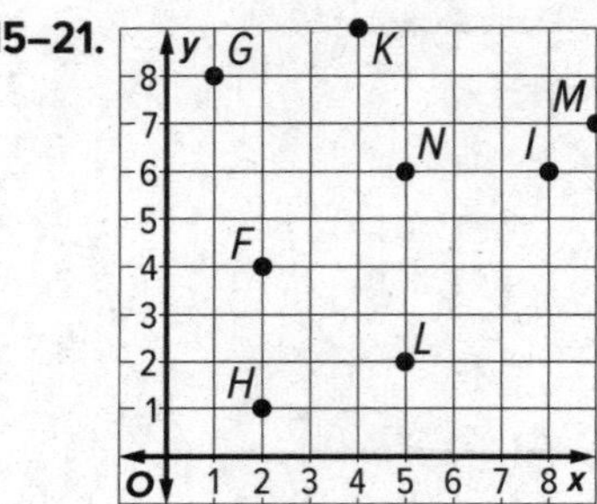

23. 36 pages

Pages 905–906 Lesson 12-5 Independent Practice

1

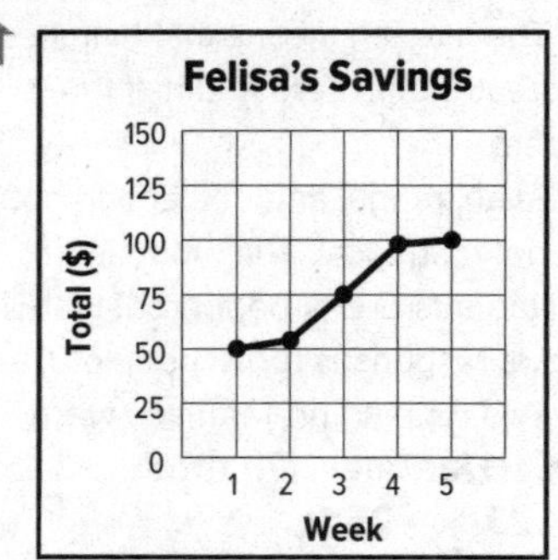

Sample answer: Felisa's total savings increased slowly for Weeks 1 and 2, then increased more dramatically for Weeks 3 and 4 with a slower increase for Week 5.

3a.

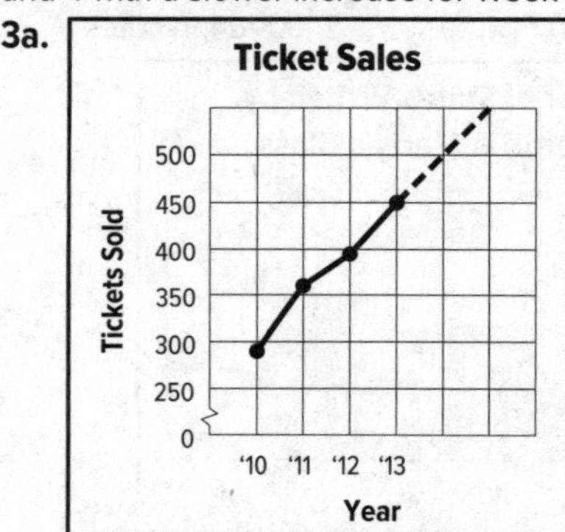

3b. 500 tickets **5.** Sample answer: If the vertical scale is much higher than the highest value, it makes the graph flatter. Changing the interval does not affect the graph. **7.** Sample answer: Line graphs are often used to make a prediction because they show changes over time and they allow the viewer to see data trends and make predictions.

Pages 907–908 Lesson 12-5 Extra Practice

9.

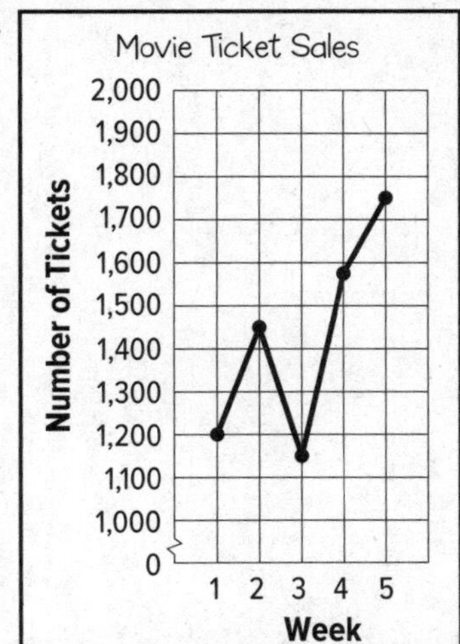

The online sales of movie tickets increased from Week 1 to Week 2, decreased in Week 3 and then increased again for Weeks 4 and 5. **11a.** 1992 and 1996; The winning time decreased by about 1 second. **11b.** Sample answer: 48.50 seconds; Based on the trend from 1992 to 2008, the winning time decreased. **13a.** True **13b.** False **13c.** False **15.** 65 **17.** 460 **19.** 163 **21.** 72 cookies

Pages 913–914 Lesson 12-6 Independent Practice

1 bar graph; The bar graph shows the maximum speeds, not just the interval in which the data occurs. **3.** box plot; A box plot easily displays the median.

7

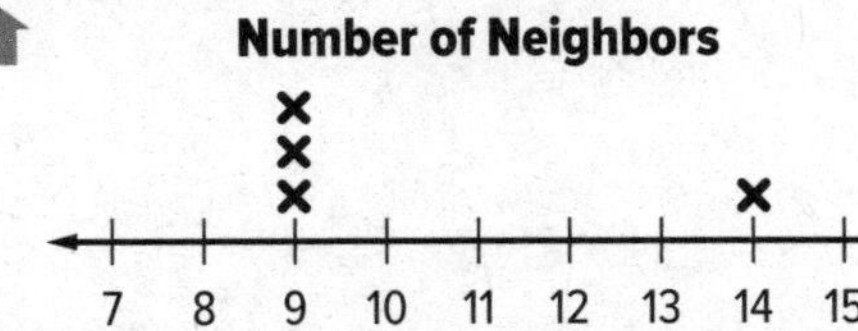

Sample answer: The line plot allows you to easily see how many countries have a given number of neighbors. The bar graph, however, allows you to see the number of neighbors for each given country. **9.** Sample answer: line plot; You can easily locate the values with the most Xs to find the mode.

Pages 915–916 Lesson 12-6 Extra Practice

11. box plot; The median is easily seen on the box plot as the line in the box. **13.** bar graph; A bar graph allows for the prices to be compared. **15.** Sample answer: box plot; A box plot easily shows the spread of data. **17a.** histogram **17b.** line plot **17c.** box plot **19.** 3 **21.** 6 **23.** 16 **25.** 9 **27.** 66

Page 921 Chapter Review Vocabulary Check

Across
7. gap **9.** dot plot
Down
1. symmetric **3.** histogram **5.** cluster

Page 922 Chapter Review Key Concept Check

1. line graph **3.** box plot **5.** mean absolute deviation

Index

Ff

Gg

Hh

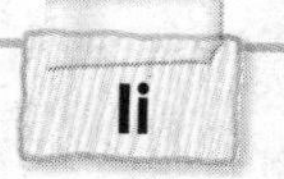
Ii

Kk

Ll

Mm

Index

Ss

Tt

Uu

Vv

Ww

Xx

Yy

Zz

Name ____________________

=

Work Mats

Name ______________________________

Name ____

Work Mats

Name ____

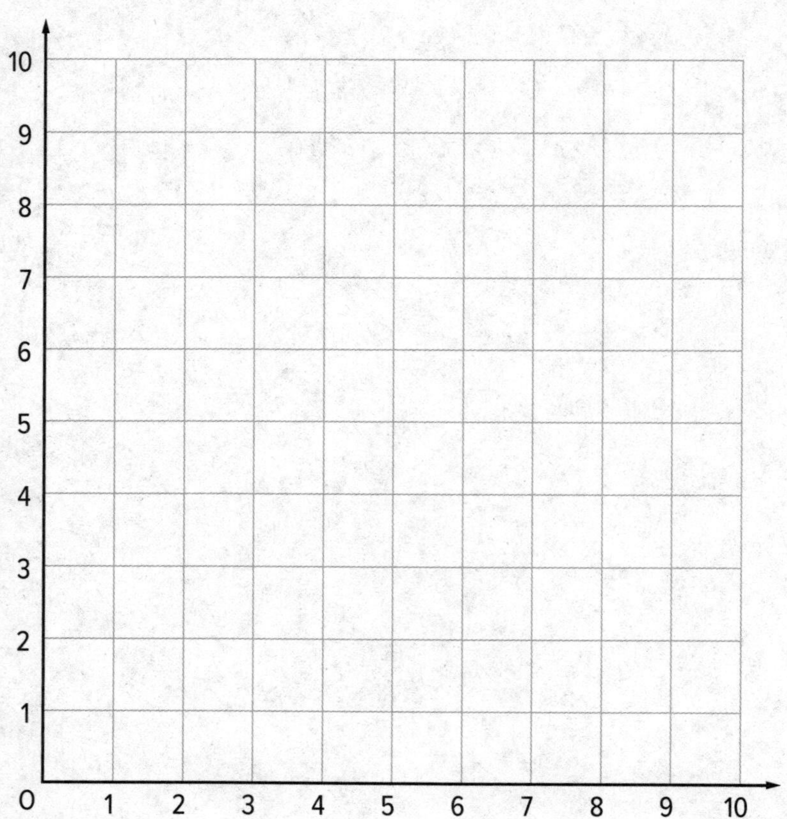

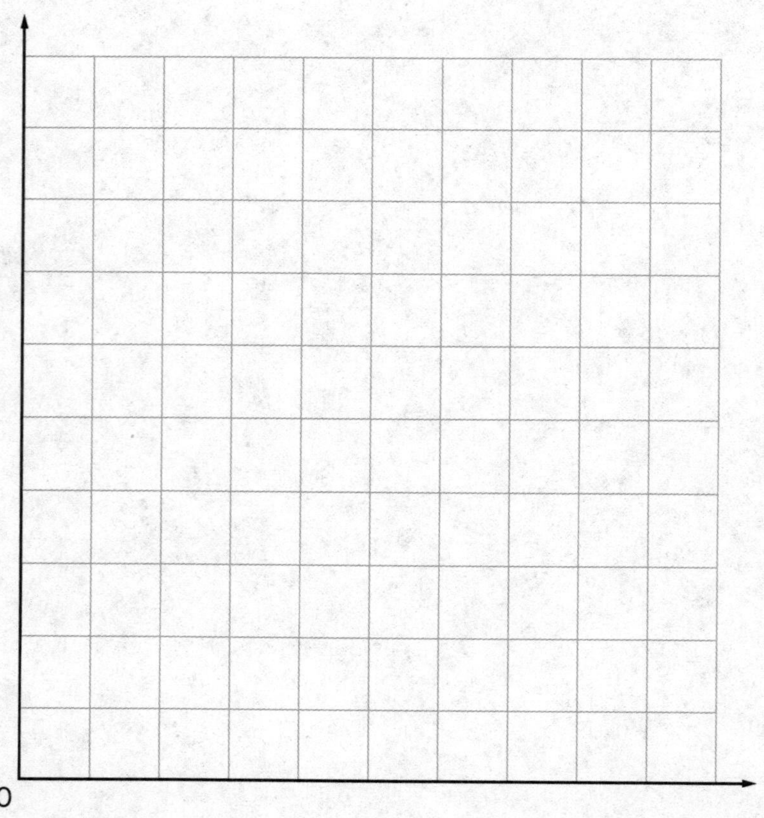

FOLDABLES® Study Organizers

What Are Foldables and How Do I Create Them?

Foldables are three-dimensional graphic organizers that help you create study guides for each chapter in your book.

Step 1 Go to the back of your book to find the Foldable for the chapter you are currently studying. Follow the cutting and assembly instructions at the top of the page.

Step 2 Go to the Key Concept Check at the end of the chapter you are currently studying. Match up the tabs and attach your Foldable to this page. Dotted tabs show where to place your Foldable. Striped tabs indicate where to tape the Foldable.

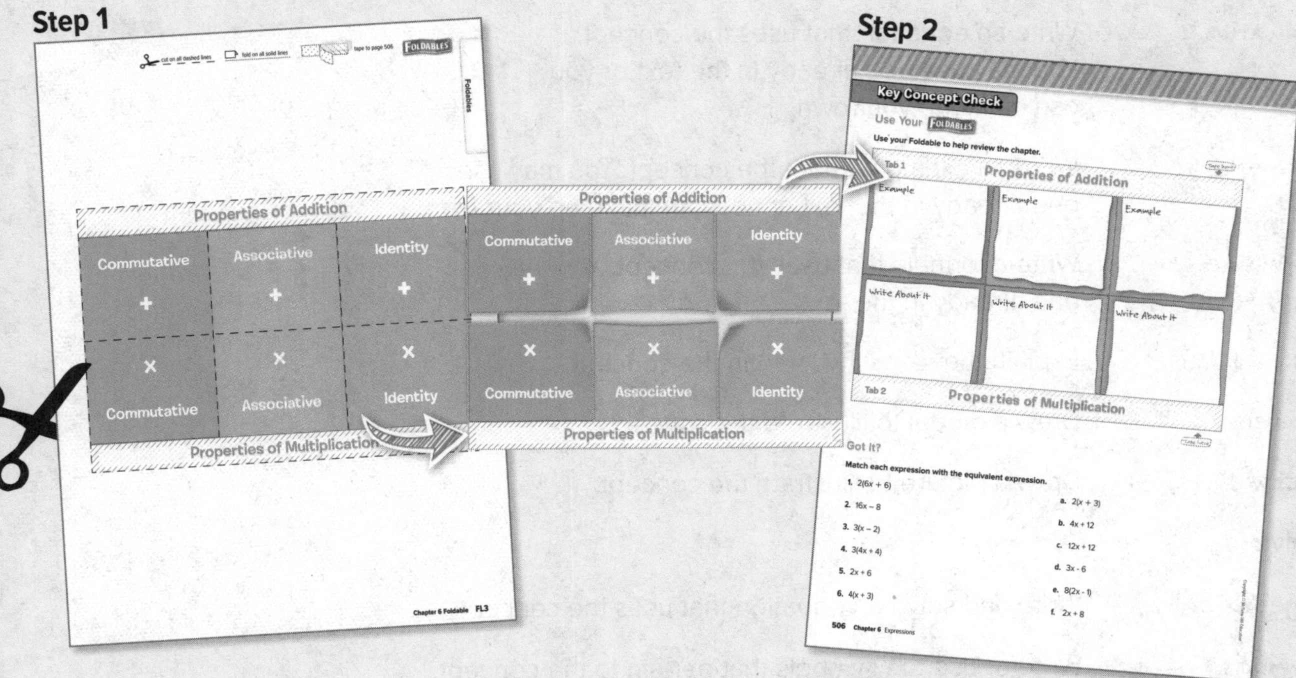

How Will I Know When to Use My Foldable?

When it's time to work on your Foldable, you will see a Foldables logo at the bottom of the **Rate Yourself!** box on the Guided Practice pages. This lets you know that it is time to update it with concepts from that lesson. Once you've completed your Foldable, use it to study for the chapter test.

How Do I Complete My Foldable?

No two Foldables in your book will look alike. However, some will ask you to fill in similar information. Below are some of the instructions you'll see as you complete your Foldable. **HAVE FUN** learning math using Foldables!

Instructions and what they mean

Best Used to... — Complete the sentence explaining when the concept should be used.

Definition — Write a definition in your own words.

Description — Describe the concept using words.

Equation — Write an equation that uses the concept. You may use one already in the text or you can make up your own.

Example — Write an example about the concept. You may use one already in the text or you can make up your own.

Formulas — Write a formula that uses the concept. You may use one already in the text.

How do I ...? — Explain the steps involved in the concept.

Models — Draw a model to illustrate the concept.

Picture — Draw a picture to illustrate the concept.

Solve Algebraically — Write and solve an equation that uses the concept.

Symbols — Write or use the symbols that pertain to the concept.

Write About It — Write a definition or description in your own words.

Words — Write the words that pertain to the concept.

Meet Foldables Author Dinah Zike

Dinah Zike is known for designing hands-on manipulatives that are used nationally and internationally by teachers and parents. Dinah is an explosion of energy and ideas. Her excitement and joy for learning inspires everyone she touches.

cut on all dashed lines fold on all solid lines tape to page 506

FOLDABLES®

Properties of Addition

Commutative +	Associative +	Identity +
× Commutative	× Associative	× Identity

Properties of Multiplication

cut on all dashed lines

fold on all solid lines

tape to page 506

FOLDABLES

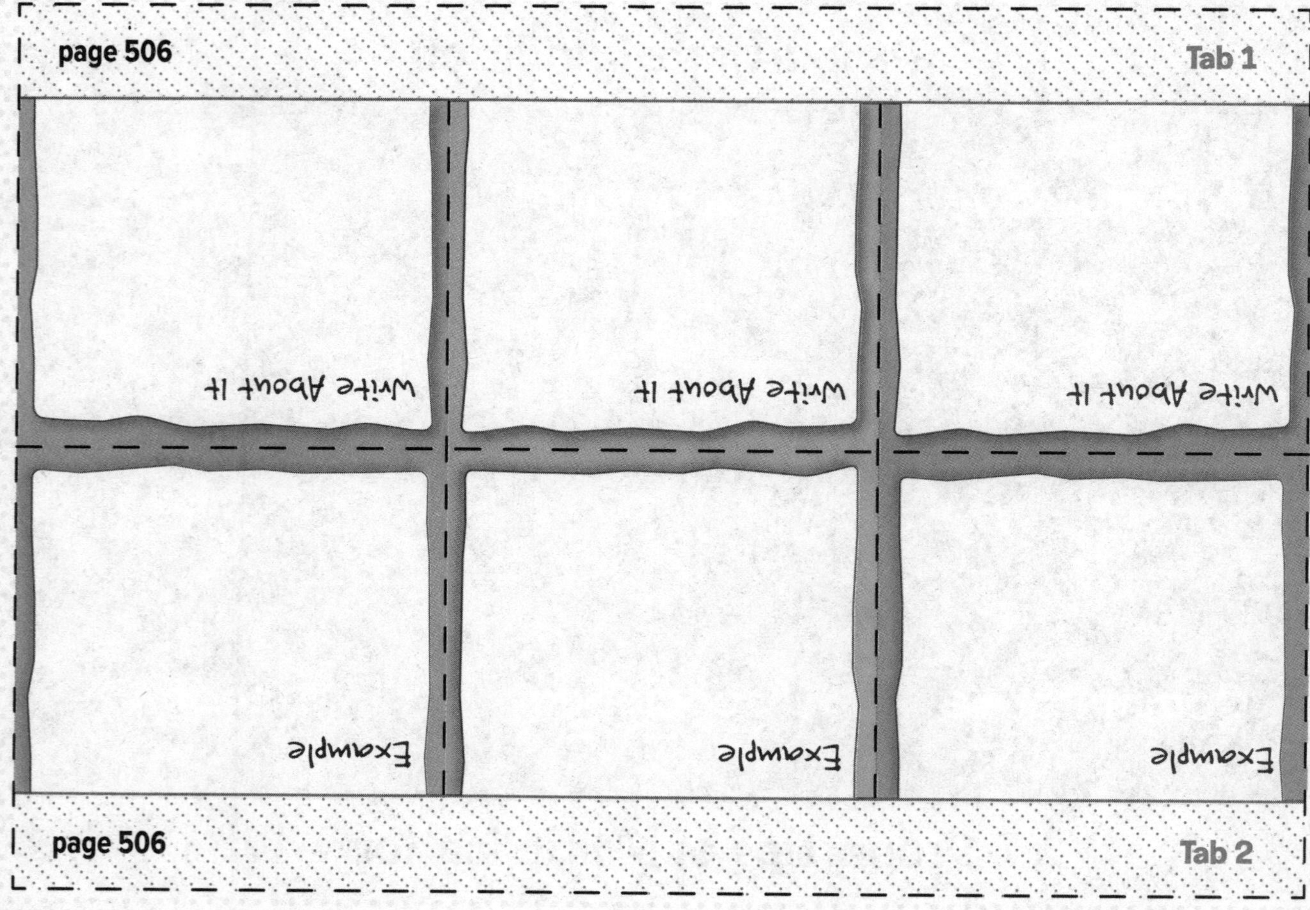

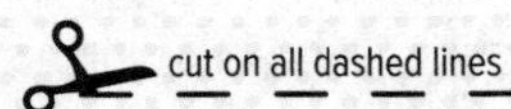
cut on all dashed lines

fold on all solid lines

tape to page 572

FOLDABLES®

equations

Models

Symbols

addition (+)

Models

Symbols

subtraction (−)

Models

Symbols

multiplication (×)

cut on all dashed lines | fold on all solid lines | tape to page 572

FOLDABLES

page 572 **Tab 4**

Write About It

page 572 **Tab 3**

Write About It

page 572 **Tab 2**

Write About It

page 572 **Tab 1**

Write About It

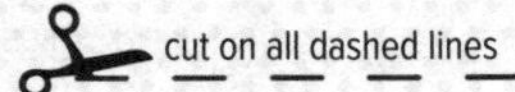

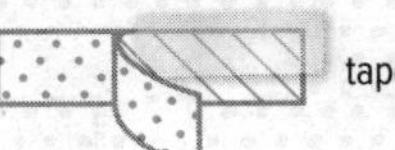

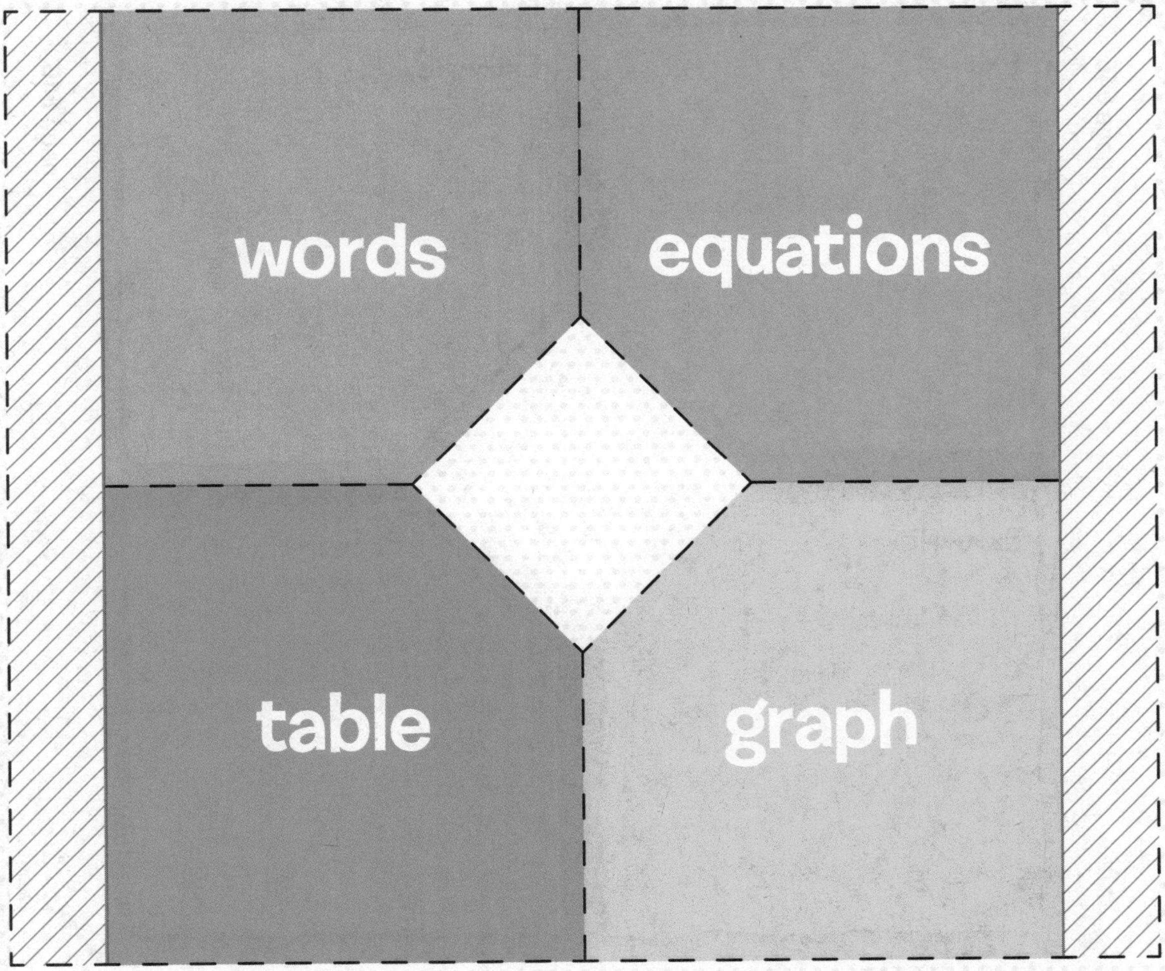

tape to page 646

words

equations

table

graph

cut on all dashed lines fold on all solid lines tape to page 646 FOLDABLES

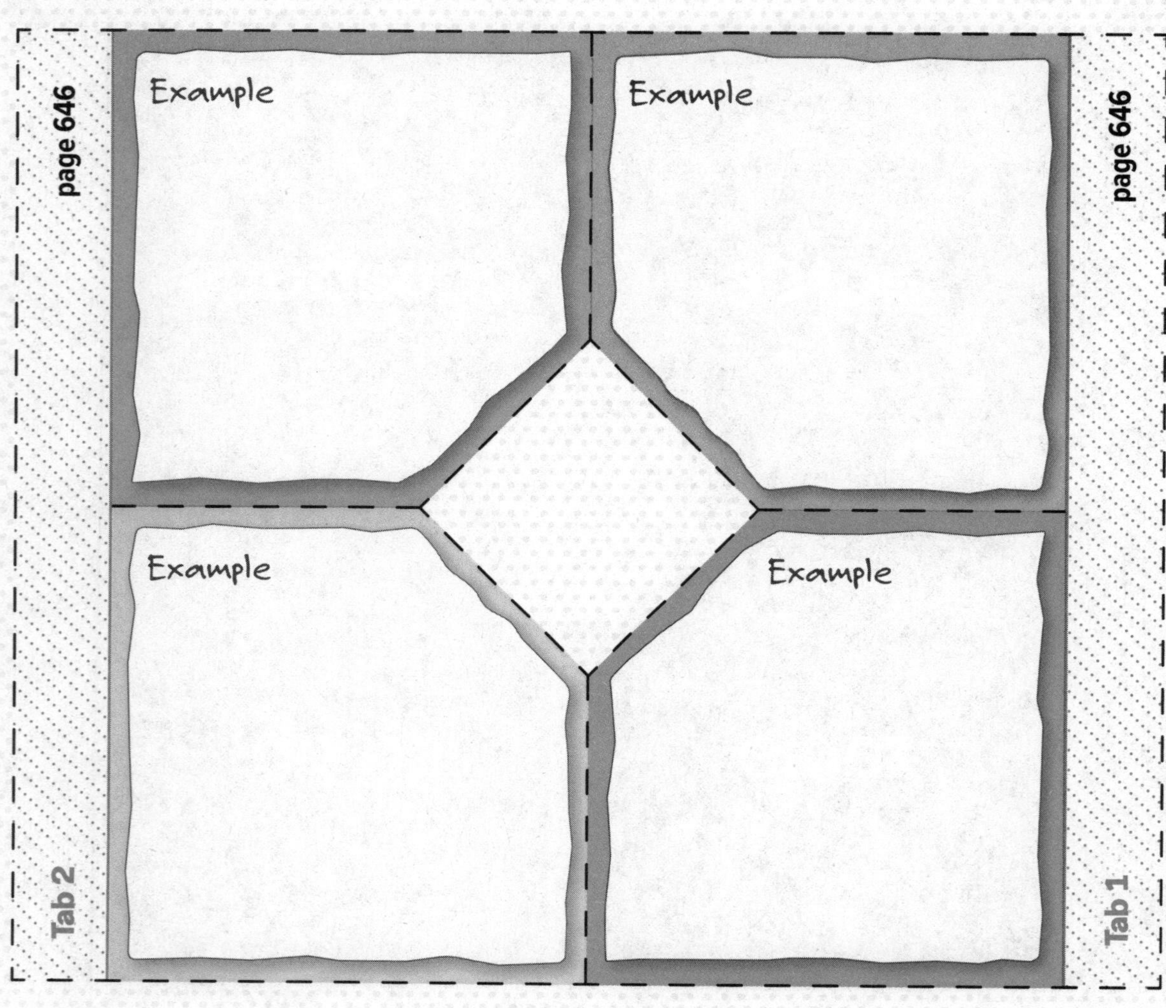

Area

parallelograms	triangles	trapezoids

cut on all dashed lines fold on all solid lines tape to page 728

FOLDABLES

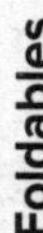

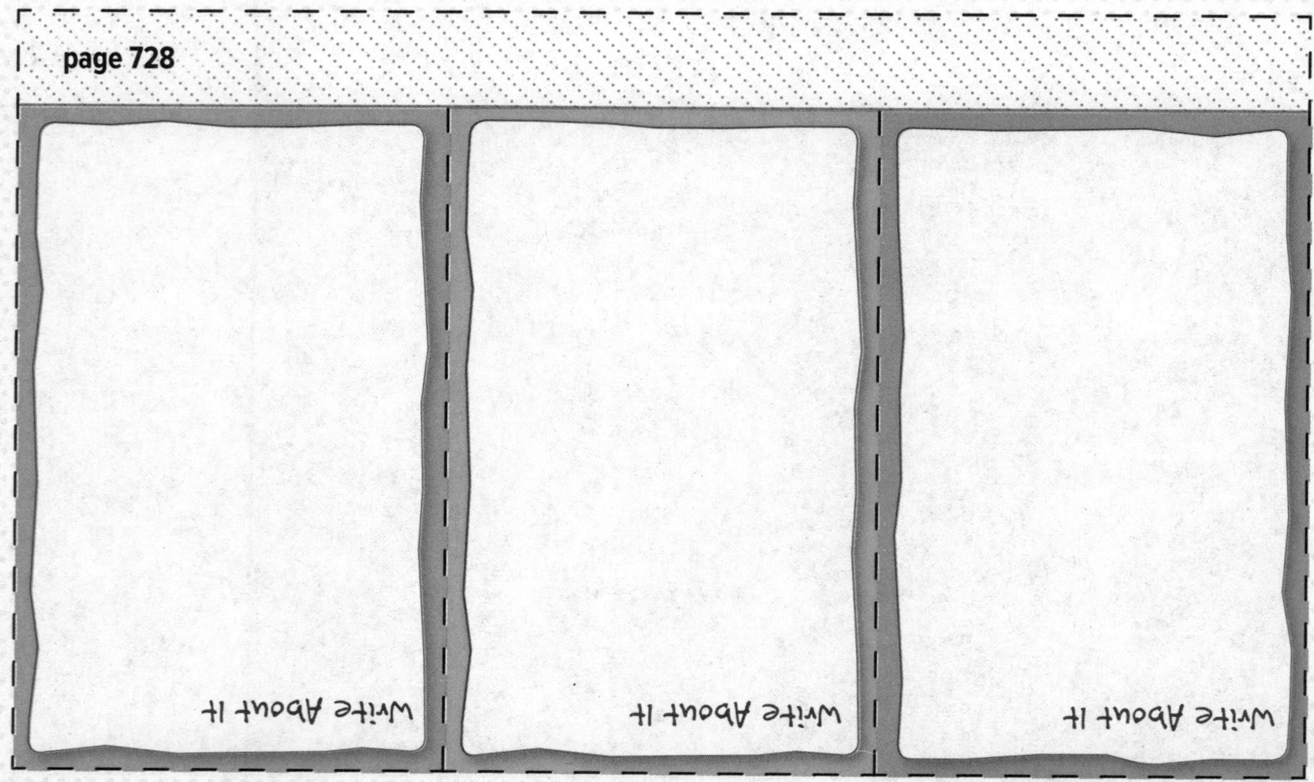

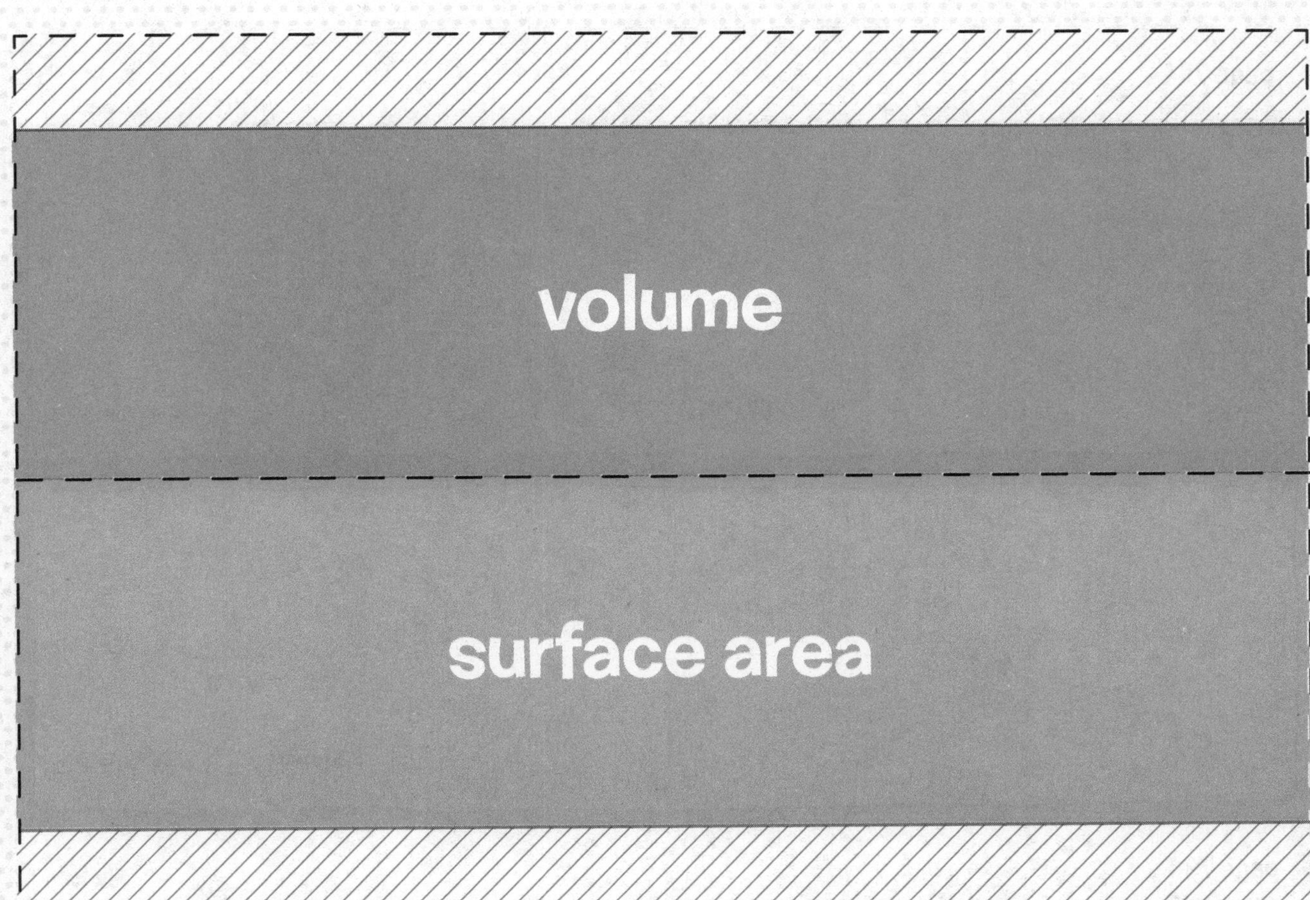
volume
surface area

cut on all dashed lines fold on all solid lines tape to page 794 FOLDABLES

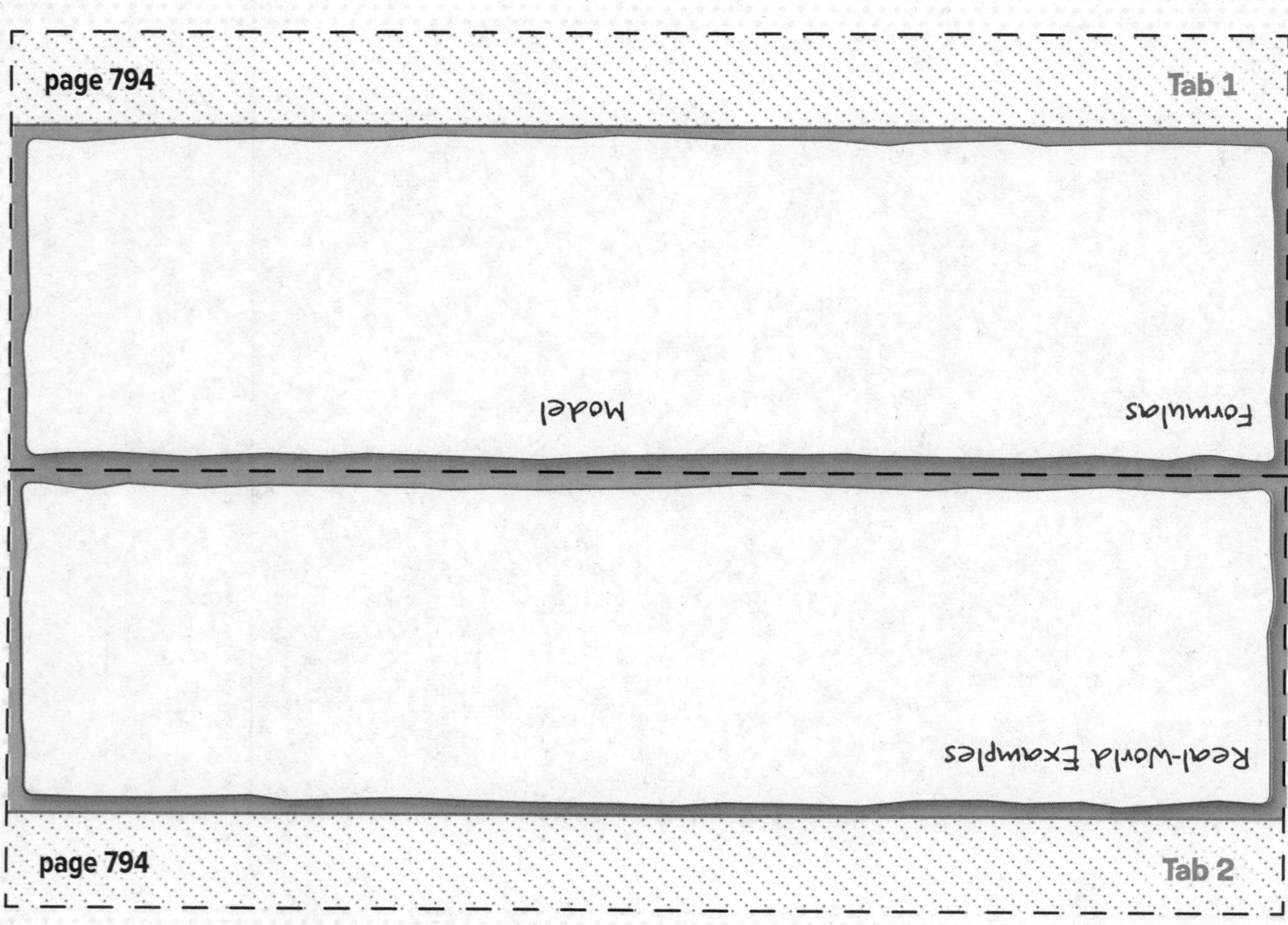

cut on all dashed lines

fold on all solid lines

tape to page 856

FOLDABLES®

Measures of Center

mean	range
median	quartiles
mode	mean absolute deviation

Measures of Variation

cut on all dashed lines

fold on all solid lines

tape to page 856

FOLDABLES®

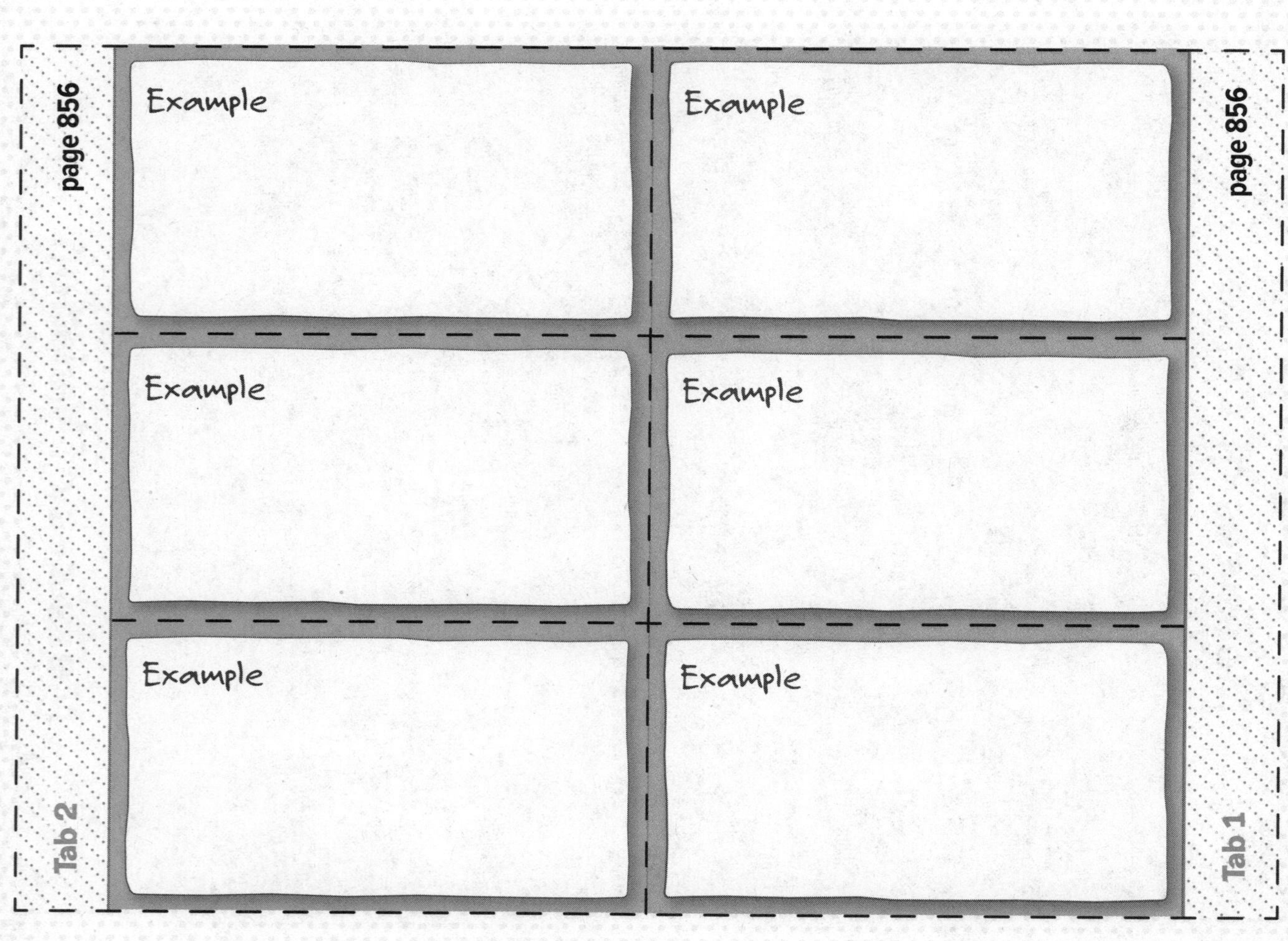

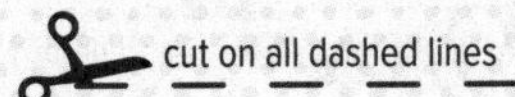
cut on all dashed lines

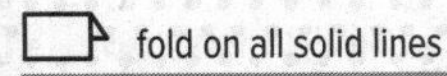
fold on all solid lines

tape to page 922

FOLDABLES®

Statistical Displays

line plot

histogram

box plot

line graph

cut on all dashed lines fold on all solid lines tape to page 922

FOLDABLES®

Foldables

Best used to...

Best used to...

Best used to...

Best used to...

page 922